High-Speed Rail in Poland

High-Speed Rail in Poland

Advances and Perspectives

Editor

Andrzej Zurkowski

Railway Research Institute (Instytut Kolejnictwa), Warsaw, Poland

CRC Press
Taylor & Francis Group
Boca Raton London New York

CRC Press is an imprint of the
Taylor & Francis Group, an **informa** business

A BALKEMA BOOK

Published by:
CRCPress/Balkema
P.O. Box 447, 2300 AK Leiden, The Netherlands
e-mail: Pub.NL@taylorandfrancis.com
www.crcpress.com – www.taylorandfrancis.com

First issued in paperback 2020

ISBN 13: 978-0-367-57128-3 (pbk)
ISBN 13: 978-1-138-54469-7 (hbk)

Visit the Taylor & Francis Web site at
http://www.taylorandfrancis.com

and the CRC Press Web site at
http://www.crcpress.com

Typeset by Apex CoVantage, LLC

Library of Congress Cataloging-in-Publication Data
Names: Zurkowski, Andrzej, editor.
Title: High-speed rail in Poland : advances and perspectives / editor,
    Andrzej Zurkowski, Railway Research Institute (Instytut Kolejnictwa),
    Warsaw, Poland.
Description: Leiden, The Netherlands ; Boca Raton : CRC Press/Balkema, [2018] |
    Includes bibliographical references and index.
Identifiers: LCCN 2018015843 (print) | LCCN 2018025140 (ebook) |
    ISBN 9781351003308 (ebook) | ISBN 9781138544697 (hardcover : alk. paper)
Subjects: LCSH: Railroads—Poland. | High speed trains—Poland. | High speed ground
    transportation—Poland.
Classification: LCC HE3060.5 (ebook) | LCC HE3060.5 .H54 2018 (print) | DDC 385.09438—dc23
LC record available at https://lccn.loc.gov/2018015843

Contents

List of authors

Henryk Bałuch, Prof. Ph.D. habil. Eng. In the early years of his professional career, Henryk was involved in construction and maintenance works on railway lines. Since 1963, he has been an employee of the Railway Research Institute. He obtained his doctorate in 1969, and in 1974 became a professor. For 8 years held the post of the Director of the Railway Research Institute and was a member of the Steering Committee of the European Rail Research Institute (ORE) in Utrecht. He was repeatedly elected as a member of the Committee on Civil Engineering and Hydroengineering and the Committee on Transport of the Polish Academy of Sciences. For 25 years, he was a lecturer at the Military University of Technology in Warsaw and is the author of more than 300 articles and lectures in Polish, English, and Russian and seven monographs and is the co-author of five other books.

His most important monographs are *Diagnostics of Railway Superstructure*, *Optimization of Track Geometric Layout*, and *Decision Support on Railway Lines*

Marek Bartosik, Prof. Ph.D. D.Sc. Eng. Professor at the Łódź University of Technology, The Vice – President of The Academy of Engineering in Poland. Marek is a specialist in electrical engineering, especially in the fields of theory and technology of limiting the short-circuit currents and switching overvoltages, rail transport, and global crisis of primary energy. He has made vast scientific, inventive, and implementational achievements in his fields. His scientific and engineering activity has been highly concentrated around the new methods of ultra-high-speed switching off high DC in a vacuum by means of countercurrent and with an active semiconductor element, on the synchronized switching off high AC, and on the limiting switching overvoltages of high energy. He works extensively with the electrotechnical industry, particularly in the new generation of DC and AC vacuum and hybrid circuit breakers for electric rail and urban traction, including single and multi-system vehicles for high-speed rails.

Juliusz Cieśla, Ph.D., Civil Engineer graduated with an M.Sc. in Civil Engineering from the Warsaw University of Technology, Poland, in 1959. He also obtained his Ph.D. from the Warsaw University of Technology, in 1975. Currently he works as Professor at the Road and Bridge Research Institute in Poland. He is an expert in railway engineering structures. He is the co-author of ten Polish and two European patents on structural health monitoring systems, and is the author of over 130 scientific publications. He is constantly involved in the development and practical implementation of procedures for railway structures management in Poland, including evaluation of real load capacity and durability of the structure based on different monitoring systems.

Selected Publication:

Cieśla J.: Experimental assessment of prestressing force in concrete bridges. *Bridge Maintenance, Safety, Management and Life-Cycle Optimization; Edited by Taylor & Francis Group*, London, UK 2010; ISBN 978-0-415-87786-2.

Andrzej Chudzikiewicz, Prof. Ph.D. habil. Eng. Andrzej is Head of Department of Bases of the Transport Devices Construction, Faculty of Transport at Warsaw University of Technology. His current research interests include dynamics of rail vehicle-track mechanical systems, diagnosis of rail vehicles and track, vibrations and stability of mechanical systems, and problems of contact and wear in rail-vehicle systems. His main teaching experience is in mechanics, dynamics of machines, vibrations, signals, and measurements.

Since 2002, he led projects financed by the Polish Ministry of Science and Polish industrial plants.

Effects of these projects:

- prototype of modernization tram type 105N
- prototype of four-part rail goods wagon
- prototype of electrical passenger train type ED74
- prototype of the trams (trams for Elbląg, Warsaw)
- Since 2006 to 2009, the main co-ordinator of European project INTERGAUGE, "Interoperability, Security and Safety of Goods Movement with 1435 and 1520 (1524) mm Track Gauge Railways: New Technology in Freight Transport Including Hazardous Products," Proposal/Contract no.: PL 516205, VI PR UE
- Since 2009, the main co-ordinator of project "Monitoring of Technical State of Construction and Evaluation of Its Lifespan" – MONIT, Financed by Polish Ministry of Sciences and Higher Education in action *Operational Programme Innovative Economy*

Visiting appointments:

- Fellowship – Delft University of Technology, The Netherlands, 12 months, 1987–88
- Visiting Professor – University of Hannover, Germany, 2 months, 1991
- Visiting Professor – ITESM San Luis Potosi, Mexico, 18 months, 1993–94

Marek Czarnecki, M.Sc. Eng. Graduate of Politechnika Warszawska (Warsaw University of Technology), Mechanical Design Faculty (1959). Since his graduation, Marek has been involved in matters of structural design, operation, and testing of rolling stock. In 1990–1996, he was on the post of Head of the Department of Traction Vehicles and then of the Rolling Stock of the CNTK. Since his retirement in 1996, he has been employed at the Railway Institute (earlier CNTK) on a part-time basis as Senior Research Technical Specialist. Among others, he headed research projects concerning crashworthiness of the rolling stock (ORE B165, SAFETRAIN). He took part in drafting and evaluating technical specifications as well as tender boards dealing with purchases of new traction vehicles for the PKP. He is the co-author of the chapter of the 2015 monograph *High Speed Railways in Poland – Investigation of Vehicle Dynamics* by the Railway Research Institute.

Tadeusz Dyr, Prof. Ph.D. habil. Eng. Tadeusz is a graduate of the Faculty of Transport of the Radom University of Technology. In 1995 he defended the doctoral degree at the Faculty of Economics of the Katowice University of Economics. At the same faculty in 2010 he also obtained a postdoctoral degree in economics. Since 1988, he has been a member of research and didactic staff at the Radom University of Technology (renamed in 2012 Kazimierz Pułaski University of Technology and Humanities), subsequently as an assistant, assistant professor, and since 1 November 2010 as an associate professor. Since 1 October 2014, he has been serving as the Head of the Department of Economics. Currently, he is Vice-dean for Science of the Faculty of Economics and Legal Sciences. Tadeusz has published results of his scientific work in more than 150 original publications. They include the authorship of four monographs and three academic textbooks, and authorship or co-authorship of chapters of monographs, articles reviewed in scientific journals, and conference papers. He is also the author of 25 didactic and popular science publications. He is a member of editorial teams in four journals. He is also a reviewer of several national and foreign scientific journals. His expert and consulting activities concern mainly problems of structural and ownership transformations, infrastructure and rolling stock investments financing, investments effectiveness, and development strategies of transport undertakings. In 2012–2015, he was a member of the Scientific Board of the Railway Research Institute in Warsaw.

Alina Giedryś, M.Sc. Eng. Alina is a graduate of Civil Engineering, Architecture, and Environmental Engineering Faculty of Technical University of Łódź and postgraduate management studies. She has wide experience in design engineering gained at a town planning office and during UNDP scholarship in Germany, as well as in management acquired while as President of the Board of the Motorway Investment Consortium Stock Company. Subsequently, she spent 10 years in the Łódź City Hall in executive functions within the Roads and Transport Management Board. She is a member of the Management Board in charge of investment in PKP Polish Railway Lines Stock Company, a member of the Management Board of the Program Council for monthly Urban and Regional Transport, and the author of publications in professional journals.

Michał Głowacz, M.Sc. Eng., M.Sc. in Electric Traction. Graduating in 2014 from the Warsaw University of Technology, Faculty of Electrical Engineering, Michał was employed for 4 years in the Railway Research Institute (Instytut Kolejnictwa). He holds basic experience in product certification, field and laboratory tests of pantographs, overhead contact line systems and OCL equipment. He has general knowledge about standards relevant for PKP and technical solutions used in OCL systems. He is a co-author of four publications in Polish and English, concerning pantograph-OCL interaction tests and system description. He participated in important projects, for example in the development of technical standards and detailed requirements for CMK line modernization plan, in pantograph-OCL interaction dynamic tests performed for Siemens Vectron locomotive (up to 220 km/h) and for ED250 Pendolino EMU (up to 300 km/h). He also participated in dynamic tests of pantographs 160EC, WBL 85, and DSA family performed on 3 kV DC system. Privately, he is a rail transport enthusiast.

Witold Groll, M.Sc. Eng. Witold is a graduate of Politechnika Warszawska (Warsaw University of Technology), Faculty of Automobiles and Construction Machinery. He is currently employed at the Railway Research Institute as Deputy Head of the Laboratory of

Rolling Stock Testing. Since his graduation, he has been involved in rolling stock testing, particularly issues of dynamic vehicle–track interaction as well as the influence of vibration on the human body. As a test manager, he took part in the majority of test runs of new rolling stock in connection with introduction of increased speeds on the PKP network. Witold is the author and co-author of 34 publications concerning rolling stock and influence of vibrations on human body.

He is the chief executive of, among others, the following national projects:

> KBN 9T12C08214 – "Investigation on rail vehicles dynamics at the speed range of 140–250 km/h in PKP conditions" – partly financed by the EU.
> CARGOVIBES – "Methods of evaluation of influence of vibrations caused by passing freight trains on inhabitants in the vicinity of railway lines. – VII EU Programme.

Marek Kaniewski, M.Sc. Eng., M.Sc. in Electric Traction. Marek graduated in 1974 from Warsaw University of Technology, Faculty of Electrical Engineering. He has been employed for 41 years in the Railway Research Institute, and has long-term experience in theoretical studies, product certification and field (dynamic) and laboratory tests of pantographs, overhead contact line (OCL) systems, and OCL equipment. He has comprehensive knowledge about standards relevant for the PKP network and about technical solutions used in OCL systems. Marek is an author or co-author of 75 Polish, English, and Russian publications concerning pantograph-OCL interaction tests, implementation process description and research studies and is a lecturer at the Warsaw University of Technology.

He participated in important projects, including

* development of technical standards and detailed technical requirements for CMK line modernization plan, for 200 km/h and 250 km/h cases
* dynamic tests of OCLs type 2C120–2C-3 performed at 200 km/h and 250 km/h
* dynamic tests of pantograph-OCL interaction performed for Siemens ES64U4 locomotive (up to 235 km/h) and for ED250 Pendolino EMU (up to 300 km/h)
* field (dynamic) tests of DSA 250, DSA 200, DSA 150, 160EC and WBL 85 pantographs performed on 3 kV DC system
* development of design guidelines for 2×25 kV 50 Hz electrification plan to be implemented in Poland.

Tomasz Komornicki, Prof. Ph.D. habil. Born in 1963, a graduate of the Warsaw University (1988), PhD in Polish Academy of Sciences (1998), head of the Department of Spatial Organization in the Institute of Geography and Spatial Organization Polish Academy of Sciences (PAS), professor at the Faculty of Earth Sciences and Spatial Management, Maria Curie-Sklodowska University in Lublin; member of the executive body of the Committee for Spatial Economy and Regional Planning, PAS, head and participant of many Polish and international research projects, including HORIZON, ESPON and INTERREG projects; member of the Consulting Board for preparation of the new Spatial Development Concept of Poland up to 2030 (in the Ministry of Regional Development); member of the international scientific team preparing the Territorial Agenda of the European Union 2020. Author of more than 400 scientific publications. His main area of interest are, geography of international interactions, transport geography and spatial planning.

Tadeusz Maciołek, Ph.D.habil. Eng. Born in 1952, Tadeusz is a Polish researcher and lecturer who graduated from Electrical Engineering Faculty, Warsaw University of Technology (WUT), in 1977; he received a Ph.D. and a D.Sc. degree in electric traction in 1986. Current position held: since 1977 at the Electric Traction Division at the Electrical Engineering Faculty at WUT, Head of postgraduate study (–2014), Vice Dean of the Electrical Engineering Faculty (2005–2012). He was a member of the Technical Certification Board at Railway Research Institute (IK) (2003–2016), and is a specialist in electrical engineering in transport, and the author of 23 applied patents and new technical solutions implemented in mass transport. Tadeusz is the author, co-author and coordinator over 50 research studies for electric rail transport, and has been widely published in periodicals and journals (over 50 papers and articles). He has been a co-organizer and member of the Organization Committee and Chairman of international conferences, including Modern Electric Transport-MET organized every other year from 1993 by the Warsaw University of Technology.

Andrzej Massel, Ph.D. Eng. Railway Research Institute's Deputy Director for Research Projects and Studies. Born in 1965, he graduated from TU of Gdansk in 1989 and received a Ph.D. in Civil Engineering from Gdansk TU in 1997. He is a specialist in railway track and railway operation. In 2005–2010 and since 2014, he is the Deputy Director of CNTK (currently Railway Research Institute) in Warsaw, responsible for research programs and studies. In 2005, Andrzej managed the pre-feasibility study for construction of the Wroclaw/Poznan – Łódź – Warsaw high-speed rail. In 2007–2008, he was responsible for the Master Plan for railway transport in Poland until 2030. In 2010–2013, he was the Undersecretary of State in the Ministry of Infrastructure responsible for railway transport.

Recent publications:

1 Measures for assessment of railway network condition – Polish example. Railway Engineering – 2015. Edinburgh, 30 June–1 July 2015
2 Gradual implementation of high-speed operation in Poland. UIC High Speed. 9th World Congress on High Speed Rail. Tokyo, 7–10 July 2015
3 *Koleje Dużych Prędkości w Polsce*. IK, Warszawa 2015

Marek Pawlik, Ph.D. Eng. Railway Research Institute's Deputy Director for Railway Interoperability. Since 1994, Marek has been involved in international railway development works conducted by the European Rail Research Institute ERRI, International Union of Railways UIC, European Association for Railway Interoperability AEIF, and European Railway Agency ERA. He is the author of four books, co-author of the *Railway Lexicon*, co-author of the European control command specifications, and author of over 30 scientific articles published in conference proceedings and technical magazines. He is a member of the signaling and telematics group, as well as member of the traction group of the Polish Academy of Sciences. Marek served from January 2002 to June 2008 as the Railway Scientific and Technical Centre CNTK (presently Railway Research Institute) deputy director for railway interoperability; and from June 2008 to June 2012 as a member of the PKP Polish Railway Lines Board responsible for strategy and development, especially in relation to full application of Railway Interoperability Directive and Railway Safety Directive both in technical and legal aspects. Since June 2012, he served as the Railway Research Institute's Deputy Director for Railway Interoperability, dealing especially with European and Polish certification of individual rail products, railway lines, and railway vehicles, as well as with risk evaluation and assessment. Since 2004 he was the chairman of the Polish Standardization Organization PKN, Traction Equipment Technical Committee KT61, and since 2007 was the chairman of the Polish Standardization Organization PKN, Railway Technical Committee KT138. Since 2005, he served as a member of the board of the National Association for Railway Interoperability and Development SIRTS, which represents Polish Railway Industry in UNIFE – the European Rail Industry Association.

Krzysztof Polak, M.Sc. He graduated from Environmental Protection Master's degree studies at Cardinal S. Wyszyński University in 2009. In 2010–2017, he worked as a senior specialist in the Office for Environmental Protection in PKP Polish Railway Lines and was responsible for investment projects relating to environmental protection, such as ERTMS projects as well as *Modernization of E65 Line South – Section Grodzisk Mazowiecki – Kraków/Katowice – Zwardoń/ Zebrzydowice – the border of the country*. The main theme of his specialization is the assessment of railway activities' impact on the environment. He is currently responsible for environmental matters working as an engineering and technical specialist at the Railway Research Institute.

Agata Pomykała, M.Sc. Eng. Agata works at the Railway Research Institute in Warsaw and is a graduate of Technical University in Lublin and postgraduate of French Polish School Organisation and Management/EDHEC Business School (Master en Management Europeen). In 2009–2011 as a project director in the High Speed Centre, PKP PLK S.A., she was in charge of strategy planning and communication. In 2011–2012 she was the project manager in PKP Intercity and was responsible for development and marketing of passenger service between Poland and the Czech Republic, Germany, and Austria. In 2013–2014, she was the General Secretary of the Polish Association of Transport Engineers & Technicians. She is the author of articles concerning high-speed rail.

Jan Raczyński, M.Sc. Eng. From 2014 in the Railway Research Institute in Warsaw, Jan has been engaged in strategic and research projects. In 2011–2014 in PKP Intercity, he was a Director responsible for strategy and development, including rolling stock; in 2009–2011 was Director of the High Speed Department in PKP PLK S.A.; and earlier was a project manager for the development of rolling stock and interoperability in PKP CARGO.

From 2011, Jan has been a member of the Intercity & High Speed Committee in UIC. Previously he was a member of UIC teams related to rolling stock. In 2005–2014, he was Speaker of Community of European Railways in the working group in the European Railway Agency responsible for preparation of TSI for rolling stock and the 1520 mm system.

He is the author of about 100 articles in the Polish railway press and international press.

Artur Rojek, Ph.D. Eng. Artur is a specialist in systems and devices of electric traction power supply, with 19 years of experience in conducting research and tests. Since 2000, he has been the head of the Electrical Power Division, and in 2011–2014 was Deputy Director for Research Projects and Studies. He is the author or co-author of over 90 publications and six monographs concerning rolling stock and electric power supply systems, and author or co-author of over 70 research and technical works and 14 patent applications and a chapter in a monograph on high-speed rails in Poland. He is also a Laureate of two awards from the Prime Minister for outstanding scientific and technical achievements and a member of the UIC Research and Innovation Coordination Group.

Mirosław Siergiejczyk, Prof. Ph.D. habil. Eng. Mirosław graduated from Warsaw University of Technology, Faculty of Transport, in 1977. In 1981 he obtained his doctorate, and in 2009 received a postdoctoral degree. Since 2010, he has been working as Professor of Warsaw University of Technology. In addition, he has been working in the Railway Research Institute as Scientific Secretary. Other posts he holds include Head of Department of Telecommunications in Transport and Faculty of Transport Head-manager of several research subjects. He has supervised over 100 graduate master's theses and engineering work in the telecommunication and teleinformatics fields. Mirosław is the co-author of five studies in the domain of telecommunication of the means of transport. He is a member of the Association of Polish ITS, and other professional appointments include Editorial Office Secretary of the scientific journal *Archives of Transport* (2007–2012), Subject Editor of *Archives of Transport*, Chairman of the organizing committees of several scientific conferences, and Chairman of the scientific committees of several international scientific conferences. In the past few years, he was the editor of two monographs and the author of over 50 scientific publications (articles, chapters in monographs), including in the field of telecommunications in rail transport.

Adam Szeląg, Prof. Ph.D. habil. Eng. Born in 1958, Adam is a Polish researcher and lecturer. He graduated from Electrical Engineering Faculty, Warsaw University of Technology (WUT), in 1982; received a Ph.D. and a D.Sc. degree in electric traction in 1990 and 2003 respectively, and was a research fellow at the University of Bath (UK) (1992–1993). Current position held: since 1982 at the Electric Traction Division of the Electrical Engineering Faculty of WUT, Head of the Electric Traction Division (2005–2008), Director of the Institute of Electric Machines (2008–2012, 2012–2016), full professor of technical sciences (2015). He is a member (2005–2011) and Chairman of the Electric Traction Section of the Electrotechnical Committee (EC) of the Polish Academy of Sciences – PAN (2012–2020), Vice-Chairman of EC PAN (since 2016), member of IET, and member of Scientific-Technical Council of Warsaw Underground, Institute of Electrotechnical Engineering, Railway Research Institute (IK), and others. He is a specialist in electrical engineering in transport, co-author of three applied patents and new research-technical solutions implemented in mass transport (over 100 studies, projects, expertises). He has been widely published in periodicals, journals (over 220 papers and articles), three student handbooks, and four research monographs, including a chapter on ground transportation systems in the 22-volume *Encyclopaedia of Electrical and Electronic Engineering* published by John Wiley & Sons in 1999. He was a reviewer of 13 as well as a supervisor of four completed Ph.D. theses. He has been a co-organizer and member of the Scientific Committee and a Chairman of international conferences, including Modern Electric Transport-MET, organized every other year from 1993 by the Warsaw University of Technology.

Przemysław Śleszyński, Ph.D. habil. Born in 1973, Przemysław is a geographer, Professor at the Institute of Geography and Spatial Organization of the Polish Academy of Sciences (PAS), and head of the Department of Urban and Population Studies. He deals with socioeconomic geography and spatial economy, in particular on the issues of urban development, spatial planning, population and settlement processes, economic location, spatial accessibility, electoral geography and cartography. He is the author of about 400 publications in this field, including more than 20 monographs. Przemysław is the manager and expert of many projects for international institutions, the Polish government, and local governments, including participant works on the ESPON Programme, National Spatial Development Concept 2030, and Responsible Development Strategy. He is a member of the Committee of Geographical Sciences PAS, Committee for Spatial Economy and Regional Planning PAS, the Committee of Demographic Sciences PAS, and Committee for Migration Studies PAS, as well as Vice-Chairman of the Polish Geographical Society and a member of the Society of Polish Town Planners.

Selected last publications:

Śleszyński P., Bański J., Degórski M., Komornicki T., 2017, Delimitation of problem areas in Poland, *Geographia Polonica*, 90, 2, s. 131–138.
Śleszyński P., 2015, Expected traffic speed in Poland using Corine land cover, SRTM-3 and detailed population places data, *Journal of Maps*, 11, 2, pp. 245–254.

Łukasz Topczewski, Ph.D., Civil Engineer Łukasz graduated with an M.Sc. in Civil Engineering from the Warsaw University of Technology, Poland, in 2003. He obtained his Ph.D. in civil engineering in the Department of Civil Engineering of the University of Minho, Guimaraes, Portugal. He is scientifically affiliated with the Road and Bridge Research Institute in Poland. Currently he works as the Business Development Manager at the Escort Ltd. in Poland. He is an expert in railway engineering structures. He is the co-author of several patents on structural health monitoring systems, co-author of the maintenance and construction standards for the existing and newly built engineering structures on all railway lines in Poland, and co-author of existing procedures for scour assessment near bridge supports. He is constantly involved in the development and practical implementation of procedures for life-cycle railway management in Poland, including implementation of degradation mechanisms, durability and structural health monitoring systems into assessment of real load resistance and life-cycle assessment.

Sławomir Walczak, M.Sc. Eng. Sławomir is a graduate of Politechnika Warszawska (Warsaw University of Technology), Faculty of Automobiles and Construction Machinery, and is currently employed at the Railway Research Institute as Head of the Department of Rail Vehicles and Head of Laboratory of Rolling Stock vehicles structural strength and running safety, as well as acceptance for operation processes. As head of the department and laboratory, he supervised testing and technical evaluations in certification processes of the majority of new types of vehicles being accepted for operation in Poland. He is the author and co-author of numerous publications concerning testing and certification of rolling stock.

Examples of publications:

1   Co-author of the chapter of the 2014 monograph *Experimental and Numerical Investigation of Oscillations in Brake System* by the Editorial Office of Warsaw University of Technology.
2   Co-author of the chapter in the 2015 monograph: *High Speed Railways in Poland – Investigation of Vehicle Dynamics* by the Railway Research Institute.

Wojciech Wawrzyński, Prof. Ph.D. habil. Eng. He graduated from the Faculty of Electronics at the Warsaw University of Technology. He obtained doctoral and postdoctoral degrees at the Faculty of Transport at his alma mater. Since 2006, he has held the title of full professor. His scientific interests include safety in transport, telematics in transport, and intelligent transport systems. In 2008, Wojciech was appointed to the post of the Dean of the Faculty of Transport of the Warsaw University of Technology. Since February 2016, he has been serving as President of the Transport Committee of the Polish Academy of Sciences. He is a member of commissions on didactic processes development and their quality on transport faculties at technical universities. He is an author of many scientific publications concerning subjects of transport, especially safety and implementation of new technologies in widely understood transport.

Professor Sławomir Wiak, DSc, Ph.D., MEng, MIEEE, MICS. Rector of Łódź university of technology, Former Vice-Rector and Former Dean of the Faculty of Electrical, Electronic, Computer and Control Engineering, Director of Institute of Mechatronics and Information Systems, Łódź University of Technology. **Doctor Honoris Causa of Great Nowgorod State University (Russia), Doctor Honoris Causa of University of Arras, France.** Specialization: Computer Science and Electrical Engineering. Honorary distinctions: Award of Scientific Secretary of the Polish Academy of Sciences, Individual Award of Scientific Secretary of IV Branch of Polish Academy of Sciences, Silver and Gold Cross of Merit, Silver Medal of Allesandro Volta of the University of Pavia, Italy, Honorary Medal of Łódź City Council. Visiting Professor to: University of Pavia, Italy, University of Arras, France. Membership of International Steering Committees and Boards (selected – 17 Committees). Referee Reports: Total number of review reports – 600 (Journals, Ph.D., DSc, books, professor titles, papers of the Conferences, grants). Published over 400 papers and 18 monographs. Contribution to over 21 grants (Polish and international). Member of the Committees: National Board for FORESIGHT POLAND 2020 (2012–2016), National Board for Scientific Projects (since-2016) – Nomination by Polish Ministry of Science and Higher Education Scientific, Technical Committee of Ministry of Economic Development, Committee of Tele and Radio Research Institute, Warsaw, Advisory Committee of Łódź province governor, Advisory Committee of Marshal of Łódź voivodship. Promoter of 11 Ph.D. projects.

Karolina Ziółkowska, Ph.D. Born in 1977, Karolina was a distinguished pupil and a year before the graduation from the secondary school she attended in the USA for the annual scholarship as a Rotary Club student, where in Dryden High School, New York, she obtained an International Baccalaureate. In 2005, she received her master's degree at the Faculty of Organization and Management at Politechnika Łódzka. In 2015, she received a Ph.D. in economics at the Faculty of Economics UTH in Radom. Since 2015 she has been Assistant Professor at Społeczna Akademia Nauk in Łódź (University of Social Sciences). Her research focuses on organization, management, economics, spatial development, and tourism, in particular managing public and economic organizations, local and regional development, entrepreneurship, and Polish and European transport issues. She is the author and co-author of 13 publications.

Andrzej Żurkowski, Ph.D. Eng. Managing Director of the Railway Research Institute since 2006. Andrzej is a graduate of the Warsaw University of Technology (1980), where he obtained a doctoral degree in Technical Sciences (2008), and is Lecturer at the Military University of Technology, Warsaw University of Technology, and the Warsaw School of Economics.

Specialist of railway traffic and organization of passenger services, he started his professional career in 1980 in COBiRTK (currently the Railway Research Institute). In 1989–2000 he headed a department in the General Directorate of PKP; and in 2001 created the PKP Intercity Ltd.; and became President of the Board in this company (2001–2004).

Andrzej represented the Ministry of Transport in working group EKTM at OECD and PKP and in many other working groups of the UIC. In 2003–2006, he held the position of Vice-Chairman of the Passenger Commission of UIC. Since 2006, he has been a member of the HS Steering Committee of UIC and a member of Scientific Committee of HS World Congresses. He was the President of Polish Association of Transport Engineers and Technicians in 2012–2014, and since 2009 is a member of the Transport Committee of the Polish Academy of Sciences and has been the Vice-President of the International Railway Research Board (UIC) since 2016.

He is the author of 130 technical and scientific articles and over 100 lectures at various conferences. He is also a co-author of the *Railway Traffic and Traffic Control* handbook and the author of a few chapters in the monographs on railways (including *HSR in Poland*).

Foreword

High-speed rail (HSR) has recently been a dynamically expanding sector of public transport, not only in the technically advanced countries of Western Europe, such as France, Germany, or Spain, but also in developing countries like Turkey, Morocco, China, and Argentina. They are a material alternative to private road transport, which in many cases has reached its limits, failing to satisfy needs resulting from expanding mobility of societies and limitations of transport infrastructure. In large countries, creating an efficient transport system without HSR is not possible. Average transit speed in a well-developed road network is 80–90 km/h, whereas in the existing HSR networks speeds exceed 150–200 km/h.

The Polish transport system is becoming less efficient. In addition to low-quality roads, including roads connecting major cities or metropolises, the issue of urban congestion and increasing number of accidents is ever more pronounced. The transport needs of the society are met in a less satisfactory manner that is also more costly and has greater environmental impact.

In August 2008, the Ministry of Infrastructure published the "Program for Establishment and Launching of HSR Transport System in Poland" (the Program), whose major purpose is to establish and launch high-speed rail operations in Poland, which is connected with the rail network of the European Union. In addition, implementation of the Program shall contribute to the establishment of a cohesive and efficient system of rail transport in Poland. Such an enterprise has to be implemented pursuant to detailed requirements as specified in European Community and Polish program documents. The Program shall be implemented simultaneously with an upgrade and revitalization of the conventional rail network, assumptions of which are specified in a Master Plan for rail transport by 2030. Actions included in the Program and Master Plan shall be mutually complementary, producing a modern system for rail transport of passengers and goods. Coordination of the conventional rail system upgrade and implementation of the HSR Program shall account for assumptions of governmental development plans for other modes of transport and complement such plans. This shall contribute to modernization and enhancing the effectiveness of the whole transport system in Poland pursuant to assumptions of EU transport policy. The goals of sustainable development of the transport sector in Poland shall be achieved.

Considering goals of European Community transport policy and findings of analyses, the program for the development of the rail system in Poland requires material reorientation, which is also indicated in other national expert documents regarding the development of Poland. The need to build high-speed lines results not only from the need to improve the market and economic situation of carriers but also to improve the cohesion of the country and the competitiveness of regions on the European market. Poland is too big a country to ensure fast connections solely by constructing motorways and expressways and upgrading railway connections. The current railway network is generally a thing of the 19th century, one that is unevenly distributed throughout the country and does not ensure direct rapid connections between major metropolitan areas.

HSR programs in other countries usually were integrated with a comprehensive development strategy and investments that stimulated an increase in GDP, as well as indirectly enhancing the economic attractiveness of regions and improving their mutual connections, which produced a synergy effect and augmented their international integration. Development plans for Polish railways should also account for the fact that member states took actions to improve rail freight within major transport corridors by establishing a priority network for this type of railway line (Regulation by European Parliament and the Council 913/2010). The following corridors were established in Poland, which partly overlap passenger transport corridors: line E20 Kunowice – Poznań – Warszawa – Terespol/Kaunas (E75) and line C-E65/E65 Gdynia – Katowice – Ostrava. In the light of such decisions, construction of HSR line Warszawa – Łódź – Poznań/Wrocław, concentrating rapid passenger transit and releasing the capacity for rail freight and regional transit between Warszawa and Poznań, gains more significance. Pursuant to regulations also establishing new transport corridors, it has to be assumed that first on the line is E59/C-E59 Szczecin – Wrocław – Międzylesie and corridor Warszawa – Łódź – Wrocław, as the extension of the line E75 Rail Baltica.

Major assumptions for development of HSR in Poland are as follows: (1) implementation of the governmental Program for establishment and launching of HSR transport system in Poland, regarding construction of the line Warszawa – Łódź – Poznań/Wrocław and adaptation of the E65 South (CMK) Warszawa – Kraków/Katowice – state border to HSR specification; (2) extension of national HSR lines to Germany, Slovakia, and Czech Republic; and (3) development of programs for the development of HSR in Poland for the years 2020–2050, covering other cities in Poland. PKP Polskie Linie Kolejowe S.A., working with the Railway Institute, prepared the concept for HSR development in Poland, offering to reduce the distance from EU railways by 2050 and meet the expectations of a modern economy and society.

Presently implemented upgrade projects do not sufficiently reduce transit times, with a very high cost of investment per minute of reduction in transit time, which is sometimes much higher than the costs of constructing a new HSR line. According to European standards, the capital city of Warsaw should be accessible from all major metropolitan areas in Poland within a 300 km radius in 90 minutes. This would be possible only after the construction of HSR lines, which should materially contribute to the integration and cooperation of regions and produce a synergy effect, enhancing mobility of the society and thus stimulating the labor and services markets, as well as dramatically improving conditions for business operations.

Construction of an HSR line is an enormous investment project, requiring mobilization of Poland's engineering, economic, and operational circles in many areas. Even though HSR lines have been engineered throughout the EU and the world and the majority of engineering and process solutions can be purchased, all HSR subsystems (infrastructure, rolling stock, engineering facilities, organization) are vast, and the application of such subsystems requires extensive knowledge, multiple specific decisions, and the adoption of a series of unique solutions. Despite the existence of standardization documentation, solutions proposed by individual suppliers differ significantly in engineering aspects, suitability for local conditions, and operating costs. It is therefore critical to learn the range of possible choices and to formulate decision-making criteria to obtain the most rational target solution for the HSR system.

To meet the above-described challenges, the Railway Institute came up with the initiative to develop a monograph which would approximate the aspects of organizational, engineering, socioeconomic, and economic demands for transport services and the formation of human resources for constructing and operating an HSR system in Poland.

The scope of such a monograph would include following problems:

- Polish and European dimension of HSR in the light of regulations;
- problems referring to the organization of HSR transport in Polish conditions;
- development of the HSR concept in Poland;
- analysis of previous projects (upgrade, construction of new sections) within the Polish rail network, for the purpose of the HSR project;
- railroad quality of HSR in Poland;
- rail traffic control and communication systems for high-speed rails;
- contact system in Poland for launching HSR operations;
- mating of the current collector and contact system as the element of an HSR power supply system;
- monitoring the technical conditions of HSR rolling stock;
- dynamics of vehicles in the context of increased operating speed in Poland;
- socioeconomic considerations of an HSR system in Poland;
- economic efficiency of an HSR system;
- international dimension of a Polish HSR project;
- demand, accessibility, and planning considerations of HSR system development in Poland;
- issues regarding formation of personnel for HSR operations;
- requirements concerning design and upgrade of engineering works on HSR lines;
- adaptation of the Łódź Metropolitan Railway exchange to the new role within the HSR system.

Due to capacity limitations, this monograph presents only selected problems related to an HSR system. Authors accounted for a variety of topics in individual sections, focusing on issues related to Polish conditions.

Having analyzed the existing condition of Polish railways, it is obvious that Poland needs overall restructuring of its rail network, patterned after the countries of Western Europe. This should be done by upgrading and restructuring current railway lines, but most of all by creating high-speed rail – agreeing with transport and environmental policies and plans imposed by EU bodies, but also having a positive impact on the economic development of Poland.

Chapter 1

Development of high-speed rail concept in Poland

Jan Raczyński

1.1. Introduction

The idea to build a high-speed rail system in Poland first emerged in the late 1980s and 90s. The initial effects of the construction of high-speed lines in Europe were already known, since this was when new lines were designed and construction began on them not only in France, but also in Germany, Belgium, Italy, and Spain.

A discussion was begun in Poland on the future of railways based on the idea of a new journey quality and a new infrastructure – with high-speed lines of up to 300 km/h.

The first Polish trains with a maximum speed of 160 km/h began their service on the Central Trunk Line, which opened in 1984. This line had geometric parameters that would allow a maximum speed of 300 km/h.

In Europe, concepts of trans-European transport corridors also emerged, with a few of them running through Poland. These corridors were to be used not only for freight but also for high-speed passenger transport.

The first concepts for a high-speed system in Poland were based on the experience of Western Europe. The objective of building new lines were similar – to maintain the share of rail transport market in the face of spontaneously growing automotive and fast-growing air transport. Observing market behavior in the transport sector in Europe, it was obvious that the mere modernization of the existing railway network, without building new lines, would not be able to stop the degradation process and the loss of market share. This prevision was already confirmed in the 1990s, and the degradation process continues to the present time, marginalizing railways in Poland on a scale unique in Europe, despite undertaking modernization projects.

1.1.1. First concepts of high-speed rail construction in Poland

In 1995 a document titled *The directional program for high-speed lines in Poland* [8] was published. This program was created in the early 90s, but the horizon of its implementation has been designated to 2030. This was the first concept of building a high-speed rail system in Poland. The document indicated the construction of two lines:

- Warsaw – Łódź – Poznan with a maximum speed of 300 km/h as the trans-European axis from Berlin to Moscow;
- Central Trunk Line (CMK) from Grodzisk on the west from Warsaw to Gdansk through Plock and to the south to the border with the Czech Republic, but excluding Krakow and Katowice.

The document proposed building a branch from the CMK line in Idzikowice to Piotrkow as a solution for the lack of good connection between Warsaw and southwestern Poland. This was to create, together with the line Piotrkow – Belchatow and further with new and existing sections, a connection between Warsaw and Wroclaw.

Choosing this last solution already at the stage of its planning was controversial. A disadvantage was that it indicated a peripheral location of such a connection created with sections of different technical parameters, which did not conform with the idea of high-speed rail because it did not ensure a reasonably short travel time between Warsaw and Wroclaw, despite large investments required.

Another weak point of the project was the location of the new line from Warsaw to Poznan on the north of Łódź, along the A2 motorway. In this concept, the Łódź railway hub would be included in this line primarily mainly by switchboards, and there would be no possibility of creating a large interchange, generating large streams of travelers for the new line. This project was putting under question the economic justification for the construction of the new line, as forecasts showed that passenger transport between Łódź and Warsaw can provide more than half of the new line passenger volume. To ensure such participation, it would be necessary, therefore, to install a new line through the center of Łódź.

The concept presented also adopted the principle of a large autonomy of new lines that would run autonomously from other lines regarding the existing rail network and stations.

Work on the new line location began under the lead of Kolprojekt and with provincial planning offices [30]. A necessary land reservation was done. For the Warsaw – Łódź – Poznan – Berlin line, a new line was drawn next to a planned A2 motorway. Until the construction of this new railway, it was assumed that passenger traffic will run on the existing E20 line, which was to be modernized in a large part to the speed of 160 km/h. Finally, the E20 line was to be dedicated to freight traffic (with regional passenger traffic), which was confirmed also in the latest Regulation No. 1315/2013 of the European Parliament and of the Council of 11 December 2013 on Union guidelines for the development of the Trans-European Transport Network [24].

In the years after the proposal was first accepted the construction of the Warsaw – Gdansk line was postponed, and priority was given to modernizing an existing line for a maximum speed of 160 km/h and partially to 200 km/h for trains with tilt-body. Modernization works on this line are at present being finished, and the maximum speed on particular sections from 100 km/h to 200 km/h will be adapted to local technical conditions. Finally, no tilt-body trains will be purchased, as they were judged too costly. The construction project of a Warsaw – east Polish border line was postponed due to low forecasted traffic.

The construction project of the Warsaw – Poznan – Łódź line was put in a document *State transport policy for 2001–2015* prepared by the Ministry of Transport and Maritime Economy in 2001 [16]. The new line from Warsaw to Łódź was planned to be built as the first.

The 1995 program was used a basis for analysis on the development of high-speed rail in Poland in the *Passenger Traffic Study 2020 Poland and Czech Republic* prepared for the International Railway Research Board (UIC) in 2003 [15]. In a summary of this study were included conclusions that justified for the construction of high-speed line Warsaw – Łódź – Poznan – Berlin and the extension of the CMK line.

Similar conclusions in the broader context of socioeconomic development were put forward in the next UIC study, *L'opportunité pour la Grande Vitesse dans l'espace PECO* [12]. In the study, similar high-speed rail systems in Western Europe were compared to Polish projects that demonstrated their high efficiency, especially regarding the Warsaw – Łódź line [38].

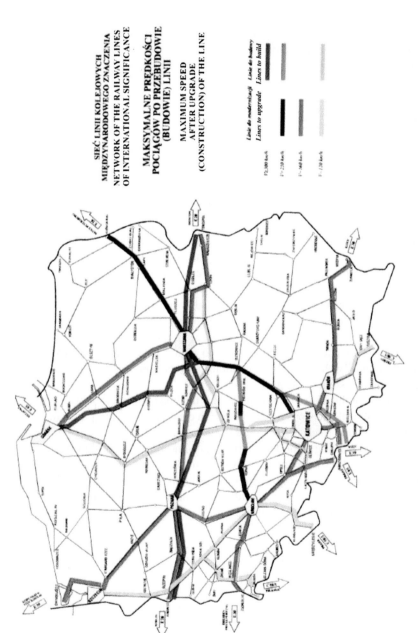

Figure 1.1 The concept of a high-speed rail system in Poland in view of main railway lines [18]

1.1.2. The concept of the Warsaw – Łódź – Poznan/Wroclaw line

Following the publication of the Program in 1995, a discussion began on the concept of another way of connecting Warsaw and Wroclaw – through Łódź. According to this proposal, the line between Warsaw and Łódź should coincide in part with a line to Wroclaw. The junction of branches to Poznan and Wroclaw would be located on the west from Łódź. As a result, the cost of investment would be reduced due to the common section for both the connection (over 220 km) and the travel time from Warsaw to Wroclaw would be much shorter.

This concept was publicly presented in a seminar titled *High-Speed Railway Łódź – Warsaw* in 2002 [20] and then published in the monthly railway magazine *Technika Transportu Szynowego* [19], [22].

The concept was partially related to 19th-century projects, which assumed connecting Warsaw with Kalisz through Łódź with the possibility of continuing the journey to Poznan and Wroclaw. That is the reason for choosing Kalisz as a junction for lines to Wroclaw and to Poznan. The choice of Kalisz as a hub was also related to its location – it is a large city located as far at east as possible without the necessity of extending the line. A well-designed connection system would make the line useful for the entire southern Wielkopolska region and for cities like Kalisz, Ostrow, Leszno, and Kepno. According to the concept, the new station Kalisz (KDP) would serve high-speed trains. Subsequent analysis indicated that the optimum location of the station would be between Kalisz and Ostrow Wielkopolski.

Another important part of the new network was supposed to be a rail link between a new line and the Central Trunk Line, which was to become a high-speed line. The shortest possible connection between these lines could be made with an Opoczno – Łódź section (about 80 km). The current line from Łódź to Opoczno consists of sections of different parameters. Adapting them to speed of 160 km/h is feasible at a relatively low cost. The advantage of this connection is the inclusion of the Opoczno region to the new network. The planned branch was to radically change travel conditions between southeastern Poland and Łódź and Wielkopolska and Western Pomerania.

The third part of the project was to modernize the Łódź node. Its aim was to correct faults resulting from political decisions made in the 19th century, which put Łódź outside the main Polish railway network. Such action would result in creating an intersection of the new east–west line with the existing important north–south line running through relatively densely populated regions of Katowice and Czestochowa to Bydgoszcz and further to Gdansk. This could generate greater passenger flows for the new high-speed line in the Łódź node. The key action in such modernization was to build a new underground line with a tunnel of length about 4 km under the city center and a construction of a new underground station at Łódź Fabryczna – replacing the old one.

The construction of a new line together with adapting the Central Trunk Line to high-speed rail parameters and modernization of other lines would create a well-developed high-speed network in Poland (Figure 1.2) [21].

The new Warsaw – Łódź – Wroclaw/Poznan line was to connect four main Polish agglomerations that together with Kalisz and Ostrow includes about five million inhabitants. The new line would create a vast railway network between east and west Poland.

Next to the new line construction was planned a modernization of other lines from Wroclaw to the west and south, as well as from Poznan to the west and north and from Warsaw to the east. This would create attractive interchanges from high-speed trains to existing conventional railways. The whole system based on a new Warsaw – Łódź – Wroclaw/Poznan line and a modernized Central Trunk Line would be accessible to about 15 million of inhabitants of Poland (Figure 1.3).

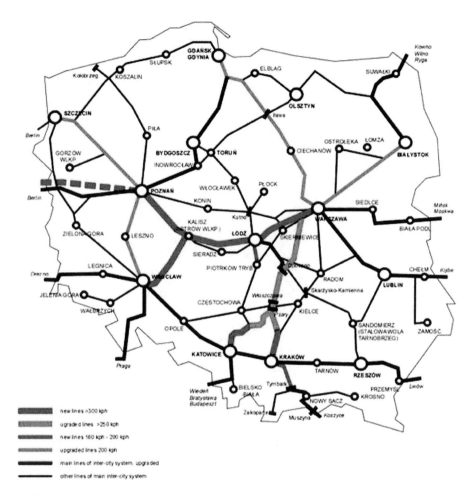

Figure 1.2 The concept of a high-speed rail system in Poland in view of main railway lines [21]

The new line construction and modernization of the Central Trunk Line were to lead to the development of international railway connections.

A Warsaw – Berlin relation was to become the first one to benefit from the construction of the new Warsaw – Łódź – Wroclaw/Poznan line. Presented in the project, a high-speed system would allow access from Warsaw to Berlin in approximately 3 hours. It would be also possible to travel by high-speed trains from Cracow to Berlin through Łódź and Poznan in about 4 hours and 20 minutes. Radical improvement of journey times from Warsaw to Wroclaw and modernization of the E30 from Wroclaw to the border would create good conditions to again run trains to Dresden and Leipzig.

Also considered was building a new line to Prague through Walbrzych and further through a tunnel under the Sudeten Mountains. This line was supposed to run further as a new high-speed line to Nuremberg.

Also, after constructing a Rail Baltica line, it would be possible to get convenient connections from Wroclaw and Poznan to Lithuania, at least. The Warsaw – Łódź – Wroclaw line is a natural extension of E75 (first pan-European corridor).

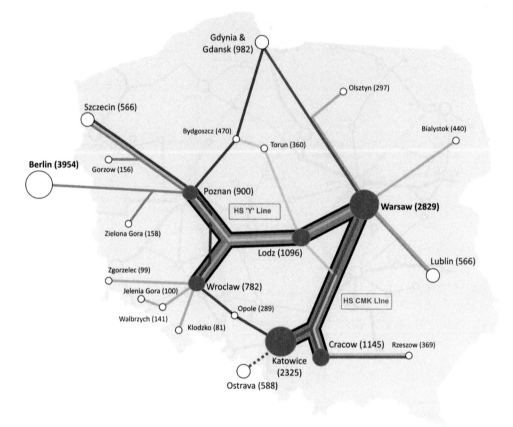

Figure 1.3 Possible train relations using the new Warsaw – Łódź – Wroclaw/Poznan line and a modernized Central Trunk Line Warsaw – Katowice/Krakow [21] (Population of city/agglomeration in parentheses)

1.2. Preliminary feasibility study for the high-speed line Warsaw – Łódź – Poznan/Wroclaw construction

In 2005, the Railway Technical and Research Centre prepared for PKP Polskie Linie Kolejowe S.A. a preliminary feasibility study for the construction of a high-speed Warsaw – Łódź – Poznan/Wroclaw line [36]. In this study, two groups of possible choices were defined in general to connect these four cities (Figure 1.4):

*   four options for the construction of high-speed lines for fast passenger traffic at a speed of 300 km/h or more (options 1, 2, 3, 4);
*   three options based on the modernization of existing railway lines (options 5, 6, 7) – in two options the construction of new sections were assumed (options 6, 7).

In each option, it was assumed the mixed traffic on railways included high-speed lines, regional passenger trains, and freight traffic. For modernized sections, a speed of 160–200

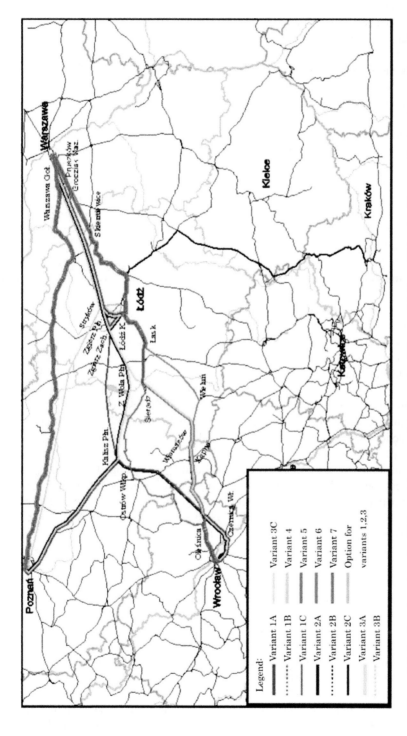

Figure 1.4 Analyzed options for high-speed connections between Warsaw, Łódź, and Wrocław/Poznan. Option 1 was indicated to be the best[36].

km/h was assumed, and 200–250 km/h was assumed for the new sections (Łask – Wieluń or Sieradz – Wieruszów).

Differences between each option concerned the location and use of an existing railway infrastructure level. It has to be emphasized that only in proposed solutions assuming the construction of new high-speed lines was possible to connect with new lines for all four agglomerations. Choosing modernization options means that the traffic between Poznan and Warsaw would still have to be put on a present line – E20 through Kutno.

A multi-criteria analysis was used to classify indicated options. They were analyzed in a study from the point of view of six criteria: financial, legal, social, environmental, technical, and transport. In order to obtain summary values for each detailed criterion, a weight was assigned that reflected its role and importance for the project and each subject. Preferences were then assessed of specified social groups regarding chosen scenarios.

The analysis showed the advantage of options 1, 2, and 3, which involved the construction of high-speed lines, and of those options 1 and 3 were identified as the best. For most criteria, option 1 was described as the best, but option 3 was better regarding the environmental impact. Option 1 assumed the construction of a new line from Warsaw through Łódź to Kalisz and then a branch to Poznan and Wroclaw. Option 3 was close to option 1, but the branch was to be next to Sieradz and a further line to the Wroclaw location was to partially use an existing line – so the line technical parameters would be considerably lower than in option 1.

As a main conclusion of this preliminary feasibility study was assumed to recommend options 1 and 3 to further detailed analysis.

It was pointed out that the project closely related to the new high-speed Warsaw – Łódź – Poznan/Wroclaw line ought to be connecting Łódź with the Central Trunk Line. Such a project would include three tasks:

• modernization of an existing Łódź – Tomaszów Mazowiecki line;
• modernization and electrification of a Tomaszów Mazowiecki – Opoczno section;
• construction of a Słomianka – Opoczno Południe section (about 3 km).

Modernization of the section between Łódź and Opoczno together with the construction of a Opoczno Południe – Central Trunk Line section should substantially improve the connection between Łódź, the Kujawy region, Wielkopolska, and Cracow. Modernization of a Łódź – Cracow connection together with the construction of new high-speed line between Łódź and Poznan was assumed to create an important new transport corridor: Cracow – Łódź/Bydgoszcz – Poznan – Szczecin.

After the publication of the preliminary feasibility study, further more detailed works were undertaken on a possible location of a high-speed line in Łódź. A study analyzing the line's possible course in Łódź was done for city transport authorities of Łódź [28], and subsequent studies were completed on the possibility of creating a national multimodal transport hub [29], construction of a new central railway station [1], and a tunnel under the city center [2]. The regional transport authority's Marshal Office ordered a feasibility study for the Łódź Agglomeration Railway [32]. Projects described in these studies are being realized at present.

The Kalisz and Ostrow Wielkopolski government also ordered a study to determine the optimal course of high-speed lines next to these cities. The study proposed a high-speed rail station between Kalisz and Ostrow Wielkopolski as a regional multimodal hub [11].

The concept of building a new Warsaw – Łódź – Wroclaw/Poznan line met with great public acceptance. It was included in an Infrastructure Ministry document titled *Transport Development Strategy for 2007–2013* (December 2005) [27]. It was also highly appreciated in March 2006 at the first meeting in Poland of the High-speed UIC Committee [14].

In August 2006 the city presidents of Łódź, Wroclaw, Poznan, and voivodeship Marshals of Łódźkie, Dolnośląskie and Wielkopolskie signed an agreement to cooperate on the construction of a high-speed Warsaw – Łódź – Wroclaw/Poznan line.

In 2007, the project was listed in the Infrastructure and Environment Programme as a task financed from EU funds for 2007–2013 (pos. 68) [26]. In December, this program was accepted by the European Commission. For the construction preparation, 80 million euros was reserved for 2008–2015.

At the same time, an analysis was done on the possibility of constructing the lines within a public-private partnership. At first, this would concern the Warsaw – Łódź section. The concept was first made public in June 2007 [6]. It was assumed that the private investor would take part in the construction phase of the new line, as well as during the subsequent operation and with the acquisition of high-speed rolling stock.

In order to accelerate this investment, a draft of a government regulation on the government plenipotentiary for the construction of a high-speed Warsaw – Łódź – Wroclaw/Poznan line was also prepared. Project implementation and supervision of the investment process would be done by a special purpose company. Due to the earlier parliamentary elections, the draft regulation was not discussed at the government meeting as planned in September 2007.

The new government, appointed after the parliamentary elections in 2008, accepted the project of the new lines construction as part of their program. On 19 December 2008, the Council of Ministers adopted Resolution 276/2008 on the adoption of supraregional strategy for the construction and operation of high-speed rail in Poland. It described tasks for the attainment of the purpose and set a schedule for the construction of a new Warsaw – Łódź – Poznan/Wroclaw line and traffic inauguration by 2020 [35].

The resolution together with the Master Plan for Polish railways until 2030 closed the discussion phase on social justification of the new line construction and high-speed rail development program. It also set tasks included in the adopted schedule. According to the program, by 2020 a high-speed rail network in Poland was to be composed of:

- a new Warsaw – Łódź – Poznan/Wroclaw line;
- a modernized to high-speed line E65 South (Central Trunk Line) from Warsaw to Katowice and Cracow with the possibility of its extension to the Czech and Slovak borders.

By 2015, studies should be prepared for these lines and for railway hubs – Warsaw, Łódź (including a tunnel under the city center), Poznan, Wroclaw, Katowice, and Krakow – in order to prepare them for high-speed services (and to address station infrastructure).

1.3. Feasibility study results for the construction of Warsaw – Łódź – Poznan/Wroclaw line

In 2008, work began on the preparation of tender documents for a feasibility study for the construction of the new line from Warsaw through Łódź to Wroclaw and Poznan. The results of public consultations and expert opinions were included for reference. Finally, the contractor had to analyze the two options of new line construction recommended in CNTK preliminary feasibility study and the following conclusions accepted as a result of public and expert consultations.

- On a section from Warszawa Zachodnia to Grodzisk Mazowiecki, the new line was also to be used for trains from Warsaw to Katowice and to Cracow through the Central Trunk Line (CMK). This would allow discharging line no. 1 in the Warsaw hub for the purpose of conventional trains and some regional trains with fewer stops. At the same time, it would reduce the time of departure of trains from Warsaw to Katowice and Wroclaw;
- On the west from Ostrow Wielkopolski, a rail link was planned that would make possible the use of new line sections from Wroclaw to Kalisz and from Poznan to Kalisz for high-speed passenger trains from Wroclaw to Poznan running in less than one hour;
- Before Wroclaw was planned a rail link to the line E30 which would make it possible to include Opole in the high-speed rail network with a travel time to Warsaw of 2 hours and to Poznan around 1 hour and 20 minutes;
- Regional stops on the new line were planned – Sieradz and Jarocin were indicated for further analysis.

For the purposes of the feasibility study, it was assumed that in the future the line would be extended from Poznan to Berlin and from Wroclaw to Praha.

The government project to construct and launch high-speed rail required creating in four cities central multimodal hubs that would integrate railway high-speed, regional, and agglomeration systems with city public transport. Detailed feasibility studies were needed for adapting these hubs for high-speed train operation with full integration to an existing conventional railway system. This was especially important in the case of Łódź, as according to forecasts more than 50% of the new line's passengers would be travelling between Łódź and Warsaw. Furthermore, the new line's success was supposed to depend on the effective implementation of the new line through Łódź. A feasibility study indicated that the line was to be put directly through the city center. In order to improve transport in this part of Poland, an investment package with regional and national funds was prepared. The most important railway projects were:

- the construction of an agglomeration railway system with regional and EU funding;
- modernization and revitalization of railways with regional and PKP PLK funding;
- the construction of a railway tunnel under the city center.

The new central station project as a regional multimodal hub joins those projects. The construction was planned as a second stage of modernization of an existing railway Łódź – Warsaw line.

These actions would allow the creation of an important high-speed rail system hub for west–east and north–south connections. See Figure 1.5 for a map of the final railway system.

At present, the Łódź Agglomeration Railway is has already been created and the construction of a new station is about to be finished. The conventional railway tunnel is planned to be opened in 2020.

Feasibility study results were presented to an interministerial group for high-speed rail. In March 2011, option 1 from Warsaw by Łódź in a tunnel, then to Kalisz with a branch to Wroclaw and Poznan, was chosen.

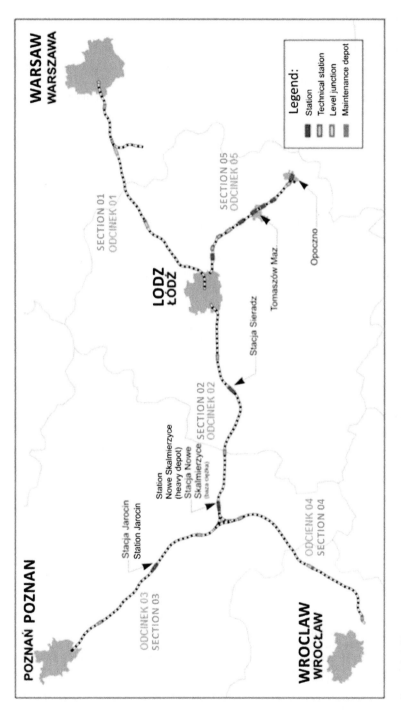

Figure 1.5 The high-speed Warsaw – Łódź – Poznan/Wroclaw line location [31]

The line location is as follows:

- in Warsaw node from Warszawa Zachodnia station through railway area Odolany and further to the surroundings of highways node Konotopa;
- along A2 highway, north from Grodzisk Mazowiecki where there will be a line junction to Katowice and Cracow (CMK line – E65 South) and then to A1 highway on the east of Łódź and along it to the Łódź Widzew station;
- in Łódź node, on the north side of Łódź Widzew station and then in a tunnel to a new railway station in the center of the city, then from a station in a tunnel to the west Łódź border;
- from Łódź to Kalisz/Ostrow Wielkopolski, on the south from Jeziorsko Lake with a regional station between Sieradz and Zduńska Wola;
- next to Kalisz/Ostrow Wielkopolski (regional station localization), where there will be a line junction to Poznan and Wroclaw;
- on the west from Ostrow was planned a rail link of 9 km, allowing to connect Wroclaw to Poznan using both high-speed line branches;
- from Kalisz/Ostrow to Wroclaw on the east of Oleśnica to the line Katowice – Wroclaw and entering the Wroclaw Główny station;
- building a rail link of 4 km to the Wroclaw – Katowice line in the direction of Opole;
- from Kalisz/Ostrow to Poznan with a regional Jarocin station to the Poznan Starołęka station and then to the Poznan Główny station.

The total Y line Warsaw – Łódź – Poznan/Wroclaw length together with rail links will be about 484 km. The length of each section and its construction cost are listed in Table 1.1.

The line was designed with very strict parameters regarding travel comfort so that in the future it would be possible to increase the speed to 400 km/h.

The total cost of the line construction was estimated at 21,213 million PLN. Costs of potential works in railway nodes are subject to separate studies and depend on chosen solutions. The highest cost will be of the Łódź node due to the cost of tunnel construction (estimated cost is 770 million PLN).

The results of analysis indicate that the project will generate a positive economic net present value (ENPV) amounting to 4,571,876 PLN and an economic rate of return higher than discount rate equal to 6.3%. The analysis showed that the project generates socioeconomic added value.

Table 1.1 Construction costs of each line section based on [31]

Section	Line/track length	Net cost [mln PLN]
Warszawa Zach. – Łódź	119 km	6491*
Łódź – Kalisz	102 km	4294
Kalisz – Wroclaw	102 km	4244
Kalisz – Poznan	104 km	4378
Rail link next to Kalisz for Wroclaw – Poznan	9 km	378
Rail link to Y line to Korytów station (CMK)	18 km	771
Node with C-E20 Skierniewice – Łowicz line	6.9 km	391

* In section costs, the rail link cost from the Y line to the Korytów station (CMK) is included.

Note: Construction costs can be changed as a result of a concurrence in tender offering. They are estimated using an indicator based on Spanish investments in more difficult topographical conditions than in Poland.

An important factor affecting the social, economic, and financial efficiency level is adopted in the analysis. The schedule was established for 11 years (2019–2029) and agreed with PKP PLK. The schedule also takes into account the possibility of its shortening by about 6–8 years, which will have a positive influence on economic and financial efficiency.

1.4. Adapting Central Trunk Line to high-speed line parameters

Central Trunk Line (CMK) was to join Upper Silesia with Gdansk, going next to Warsaw on its western side through Grodzisk Mazowiecki. It was supposed to facilitate travel in the north–south direction, mostly for freight travel. Tracks are located far from bigger cities. Its geometrical parameters were supposed to allow the 250 km/h speed but further analysis showed that even the speed 300 km/h is possible – of course after changing the power supply system to 25 kV AC. The first passenger trains were running on this line at 160 km/h in the late 1980s. The line was not entirely built – only the section from Zawiercie to Grodzisk Mazowiecki surroundings. The extension of the line to Gdansk on the north or to the south was not accomplished.

After 1990, it turned out that as a result of changes in the structure of the transport market in Poland, CMK became a line with the dominant passenger traffic from Warsaw to Katowice and Krakow and from Łódź to Krakow. This was why the idea emerged of adapting the line to the higher speed of 200–250 km/h or even 300 km/h.

In 2009, a contract was signed on the design project to adapt the Warsaw – Katowice/Krakow line to high-speed parameters. The contract was won by a consortium led by the company Halcrow [33]. Works were divided into four tasks:

- Task 1: section Grodzisk Mazowiecki – Zawiercie (CMK Line);
- Task 2: section from CMK to Katowice;
- Task 3: section from CMK to Cracow;
- Task 4: sections from Katowice to the Polish border (to Zebrzydowice and Zwardon).

At the beginning of 2011 ended the stage of recommending the best solutions after a cost-benefit analysis. The PKP PLK management board accepted following recommendations presented by the consortium as best:

- For task 1: section Grodzisk – Zawiercie (CMK line) – modernization to 300 km/h speed, including changing the power supply system to 25 kV AC;
- For tasks 2 and 3: section from CMK to Katowice and Cracow:
  - new line construction branching from the CMK line south of the village of Psary;
  - a branch next to Olkusz to Katowice and Cracow;
  - ultimately a construction of a rail link on the south of Olkusz joining both branches to create a high-speed Cracow – Katowice connection;
- For task 4: sections from Katowice to the Polish border (to Zebrzydowice and Zwardon):
  - modernization of existing lines to 160 km/h;
  - a project of a new line from Katowice to Ostrava – in the future was to be a subject of an international agreement between Poland and Czech Republic.

E65 South line location: Warsaw – Katowice/Cracow after choosing the best solutions analyzed in a feasibility study was recommended as follows (Fig. 1.6):

- CMK junction to the Y line Warszawa – Łódź – Poznan/Wroclaw will be on the north of Grodzisk Mazowiecki where a turnout will allow passing with 250 km/h to a new section to Korytów in a place of a planned CMK extension on the north to Gdansk;
- then on a CMK line to Nakło surroundings (on the South of Psary station);
- from Nakło on a new line which will branch out to Katowice and Cracow on the west of Olkusz;
- from a junction next to Olkusz to Katowice by Sosnowiec;
- from a junction next to Olkusz (with a regional station) to a Cracow Central Station;
- on the south of Olkusz is planned a rail link of 8 km which will allow to travel on a high-speed line between Katowice and Cracow using both branches.

It will be possible at present to use an existing section of a CMK Nakło – Zawiercie line for trains to Zawiercie, and in case of a Zawiercie – Pyrzowice line modernization for the Katowice – Pyrzowice airport service (Fig. 1.7).

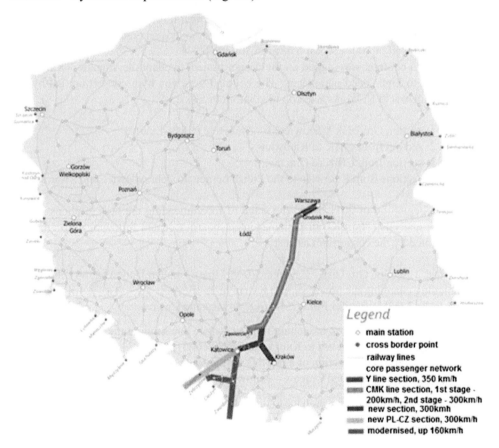

Figure 1.6 An E65 South (CMK) line location Warszawa – Katowice/Krakow based on [33]

Figure 1.7 An E65 South (CMK) line location Warszawa – Katowice/Krakow based on [33]

The total length of the E65 South: Warsaw – Katowice/Cracow, together with rail links will be about 373 km, including 149 km of new lines.

A project is coherent with a construction project of a high-speed Warsaw – Łódź – Poznan/Wroclaw line. Both lines have planned following connection points:

- common section Warsaw – Grodzisk Maz. surroundings, which will allow travel time from Warsaw and giving up travel on an overloaded section Warsaw – Grodzisk Wlkp. of a maximum speed only 160 km/h;
- joining together both lines with a modernized Opoczno – Łódź line (ultimately a high-speed line) in order to create a coherent system of high-speed rail for the whole country.

A deadline of CMK modernization according to a chosen option was not defined. The **completion** of modernization is estimated to be done by 2030, which will by that time have more than 30 years of exploitation.

Continued at present modernization stage includes as follows:

- increasing the maximum speed to 200 km/h;
- installing an ETCS system;
- adapting the power supply to operate the new electric multiple units purchased by PKP for speed above 200 km/h.

1.5. A high-speed system based on a Warsaw – Łódź – Poznan/Wroclaw line and a Central Trunk Line

In order to create a connection between high-speed Warszawa – Łódź – Poznan/Wroclaw and E65 South Warszawa – Katowice/Krakow lines, a solution for the best possible location was analyzed. Both new line construction and existing line modernization were considered. The concept of a conventional line 25 Łódź – Opoczno modernization (74 km long) [31] was chosen as best:

- from a new Łódź Fabryczna railway station to a Łódź Widzew station (modernized at present with a tunnel in construction) and then to Gałkówek – line already modernized to 140 km/h;
- Gałkówek – Mikołajów – Tomaszów Maz. section (31 km): modernization of an existing line to the speed of 160 km/h;
- Tomaszów Maz. – Opoczno section (26 km) and a rail link construction Słomianka – Opoczno Płd. (5 km):
    - line remains as a one-track line (sufficient capacity for the predicted traffic);
    - railroad modernization and electrification in order to increase speed to 160 km/h;
    - a construction of a one-track rail link Słomianka – Opoczno Płd.;
    - modernization of a Opoczno Południowe station on CMK (platform and pedestrian crossing construction).

Until 2030, it would be possible to create a high-speed rail system in Poland connecting major Polish agglomerations. In this system, travel times from Warsaw to Cracow, Katowice, Wroclaw, and Poznan will be about 1.5 hour and to Łódź about 35 minutes. The total length of high-speed lines will be 857 km (Fig. 1.8).

A high-speed lines system, similarly to foreign solutions could be a basis for creating a fast railway connections network in Poland. Such a possible network including modernized lines is presented in Figure 1.9. Expected travel times between main cities in Poland are listed in table 1.2.

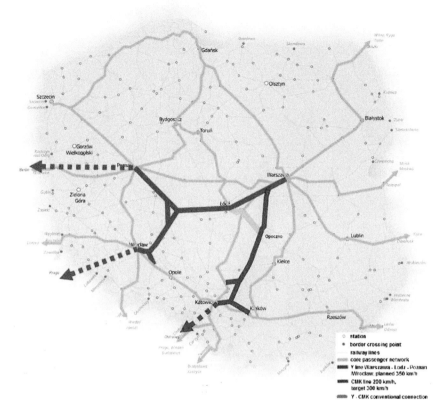

Figure 1.8 The best Warsaw – Łódź – Poznan/Wroclaw line location regarding the Warsaw – Katowice/Cracow line

Source: PKP PLK S.A.

Table 1.2 Line length and travel time for high-speed rail in Poland based on [31], [33]

Service	Distance	Travel time
Warsaw – Katowice (by Olkusz)	311	1 h 25 min
Warsaw – Katowice (by Zawiercie)	299	1 h 55 min
Warsaw – Cracow (by Olkusz)	297	1 h 25 min
Katowice – Łódź	272	1 h 25 min
Cracow – Łódź	267	1 h 25 min
Cracow – Katowice	79	0 h 35 min
Poznan – Cracow (by Łódź)*	450	2 h 30 min
Poznan – Katowice (by Łódź)*	464	2 h 30 min
Warsaw – Poznan	350**	1 h 35 min
Warsaw – Wroclaw	354**	1 h 40 min
Warsaw – Łódź	123	0 h 35 min
Poznan – Łódź	227	1 h 00 min
Wroclaw – Łódź	230	1 h 05 min
Poznan – Wroclaw	220	0 h 55 min
Poznan – Opole	282	1 h 30 min
Warsaw – Opole (by Łódź)*	414	2 h 00 min

Note: Travel times with a minimum margin of 2 minutes for 100 km with stops only in voivodship capitals.

* Relation fastest out of possible not excluding functioning of the existing ones
** Relation by Sieradz, Nowe Skalmierzyce

Source: PKP PLK S.A.

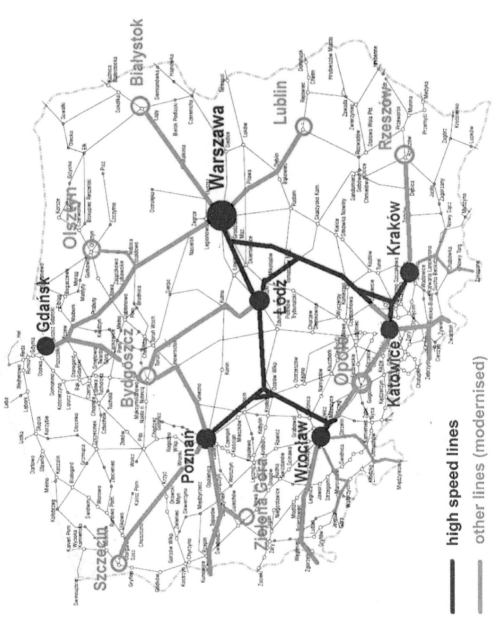

high speed lines

other lines (modernised)

Figure 1.9 A possible fast lines network in Poland based on high-speed and modernized railways [10]

1.6. Directional high-speed rail development program in Poland

In 2010, in reference to a current discussion about the future of a Polish transport system, PKP Polskie Linie Kolejowe S.A. ordered analysis aimed at identifying main directions of national high parameters railway network development allowing creating an attractive transport offer and preventing further loss of the share of rail in the transport market [7], [23]. The basis for these analyses was prepared in a Railway Institute program of high-speed rail development in Poland by 2040 [9].

Analyses were based on an existing Polish railway network diagnosis, which indicated its main faults as follows:

- the present railway network in Poland doesn't allow creating a competitive and sufficiently attractive offer in passenger transport, which dooms railway companies to a constant dependence on budgetary subsidies. Mature high-speed rail systems in the world are self-financing;
- the railway network comes in its basic shape from the period of the partitions and doesn't provide the required consistency of the state, particularly in the east–west direction; for this reason there is a clear tendency of proclivity of the western part of Poland to its western neighbors at the expense of the development of relations with eastern and central Poland;
- the density of the railway network in Poland is the lowest among the countries of central Europe and lines with high technical parameters are missing.

Presented in a document concept of high-speed rail system development in Poland, main transport corridors use modernized and revitalized conventional railways lines. Presented new high-speed lines proposals were supposed to become a ground for further studies in order to choose their best location and necessary consultations. It was also assumed that in accordance with latest European high-speed trends, there would be smaller regional stations for high-speed trains.

Studies were made in accordance with key Polish strategic documents guidelines created in that time:

1   Poland 2030. Development challenges (2009) [17]
    This document sets out directions and conditions of economic and social development. Regarding high-speed rail development plans, it confirms the necessity of the construction of the Warszawa – Łódź – Poznan/Wroclaw line and modernization of the E65 South (CMK).
2   Concept of spatial development of a state until 2030 (2011) [34]
    This document presents comprehensive social, economic, and spatial analysis and also confirms the necessity of the construction of the Warszawa – Łódź – Poznan/Wroclaw line and modernization of the E65 South (CMK). Additionally, the concept anticipates high-speed line construction until 2030 from Warsaw to Gdansk in accordance with a plan of CMK extension dating from the 1980s. In this document were defined international relations of Polish high-speed rail system according to the results of work on future Trans-European Transport Network (TEN-T) (Fig. 1.10).

Results of the analyses defined main transport corridors in Poland for fast passenger traffic. These corridors were to be created by new high-speed lines with more dense traffic and lines upgraded for high technical parameters for less congested sections. In a final document corridors were chosen which should be finalized in the perspective of 2040 [7].

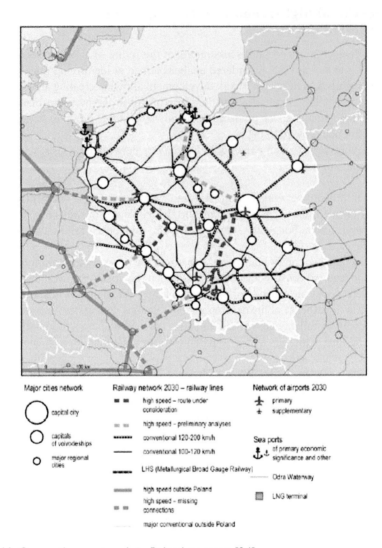

Major cities network

◯ capital city

◯ capitals of voivodeships

○ major regional cities

Railway network 2030 – railway lines

■ ■ high speed – route under consideration

≈ ≈ high speed - preliminary analyses

•••••• conventional 120-200 km/h

——— conventional 100-120 km/h

▬ ▬ LHS (Metallurgical Broad Gauge Railway)

════ high speed outside Poland

▥ ▥ high speed - missing connections

——— major conventional outside Poland

Network of airports 2030

✈ primary

✈ supplementary

Sea ports

⚓ of primary economic significance and other

——— Odra Waterway

▦ LNG terminal

Figure 1.10 Core railway network in Poland concept [34]

The implementation of this concept would result in a high-speed rail network in most of the large Polish cities. The travel time from the cities of western to eastern Poland would be significantly reduced to about 3 hours. Zones of transport exclusion would be reduced in a radical way. This would particularly concern eastern Poland, especially the regions of Lubelskie, Podkarpackie, and Podlaskie. All regions would see a significant improvement in travel times, which would allow travelling to and from the country's capital in one day.

An important result of that analysis was a creation of a new railway system which would be related to an existing railway network and lead to its better use (Fig. 1.11). The construction of new lines was supposed to create main corridors or to complete missing parts of the Polish railway network. A preliminary analysis regarding selecting sections for construction or modernization has been done (Figure 1.12).

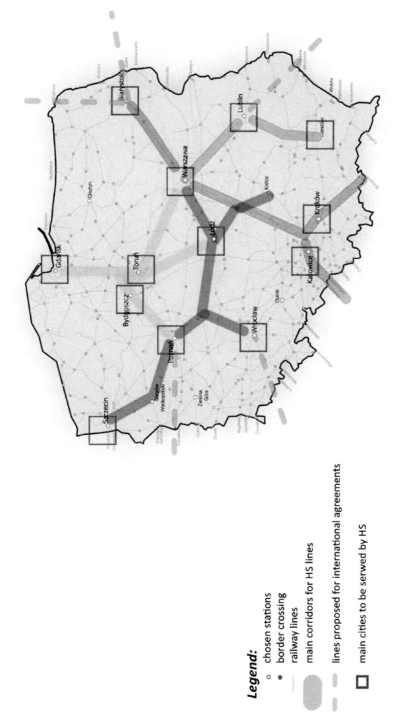

Figure 1.11   Main corridors for high-speed lines construction [27]

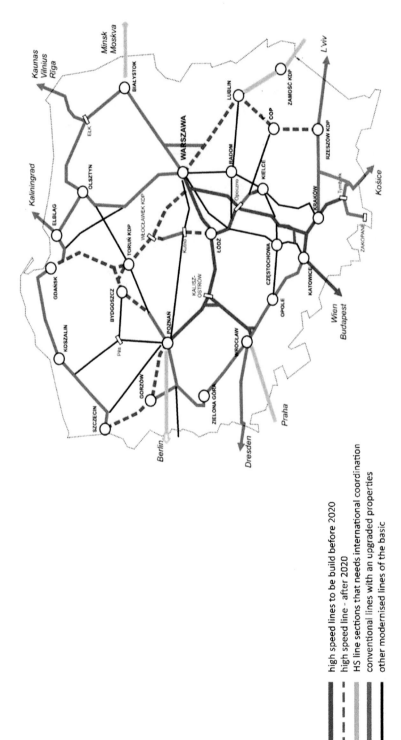

Figure 1.12 High-speed rail network in relation to a conventional one [27]

high speed lines to be build before 2020
high speed line - after 2020
HS line sections that needs international coordination
conventional lines with an upgraded properties
other modernised lines of the basic

It was assumed that it would be necessary to build 1870 km of new lines and to modernize 1870 km of lines. In Table 1.3, proposed investment is described. The line classification is in conformity with latest TSI Infrastructure specifications [7].

The high-speed rail network concept was put to a public consultation at the beginning of 2011 and was approved, and further analysis to refine the localization and planned line parameters were recommended [7]. In 2013, the local governments of the Lódzkie and Świętokrzyskie regions announced the plan of connecting Kielce to Opoczno – with a high-speed rail hub – with line no. 25 modernized on the Opoczno – Końskie section and further next to Kielce joined with line no. 8 by a new line of 30 km [18], [3].

Table 1.3 Polish high-speed rail length

Line	Section length [km]	Line category according to TSI	Notes
Warszawa – Łódź – Poznan/Wroclaw	484	I	Including a rail link for Poznan – Wroclaw connection west from Ostrow Wlkp. (about 10 km). Additionally, line Łódź – Opoczno, 90 km, category V-M ultimately new line of cat. I
Wroclaw – Polish border (toPraha)	About 110	I	Length depends on a localization option chosen according to Polish – Czech studies
Warszawa – Krakow/Katowice	373	I	Option with a branch next to Olkusz, including 149 km of new lines
Katowice – Polish border (to Ostrava)	60	I	
Gdansk – Poznan/ Łódź/Warszawa	673	I, II, V-M	New lines 492 km
Poznan – Szczecin	224	I	
Poznan – Berlin	About 150	I	Construction and localization depends on Germany decision (approximate length)
Warszawa – Lublin – Rzeszow	301	I	In a shortest option
Lublin – Zamosc – Polish border	140	I	Construction and localization depends on Ukrainian decision (approximate length)
Warszawa – Bialystok – Polish border	347	II, V-M	Only modernized railway. Project Bialystok – Polish border depends on international agreements
Kielce – Opoczno	68	II, V-M	Partial modernization and about 30 km of a new line
Krakow – Piekiełko	49	I	
Together – new lines	About 2190		
Other modernized lines	About 790		

1.7. International high-speed connection in a TEN-T network

Polish accession to the European Union in 2004 was associated with the inclusion of the Polish transport network to the Trans-European Transport Network. Selected national lines which became the TEN-T network have been identified in the Accession Treaty. These were conventional lines primarily indicated as parts of the freight transport corridors in the framework of agreements AGC (European Agreement on Main International Railway Lines) and AGTC (European Agreement on Important International Combined Transport Lines and Related Installations). For a TEN-T network amendment in 2010, the Polish government didn't report any new proposals. For the next network amendment, proposals for new lines to be included in a TEN-T network were already submitted. This comprehensive amendment concentrated largely on creating a European high-speed lines network according to White Paper published in 2011.

A regulation draft was prepared in 2010 as a result of an existing and planned railway network analysis. This network was supposed to create a coherent transport network in the entire European Union and neighboring countries. As Polish high-speed rail studies advanced, the possibility of including them in a new trans-European railway network was analyzed. Such inclusion was to be done by proposed by European Commission lines: Poznan – Berlin, Wroclaw – Praha, and Katowice – Ostrava. All high-speed lines are therefore a part of a comprehensive passenger railway network. The Warszawa – Łódź – Poznan/ Wroclaw and Warszawa – Katowice/Krakow lines, due to an advanced works stage, are included into a core network which should be completed in 2030. The Polish high-speed line project was generally viewed positively concerning both social and economic utility and technical parameters as well as lines location.

In December 2013, Regulation No. 1315/2013 of the European Parliament and of the Council of 11 December 2013 was published on Union guidelines for the development of the Trans-European Transport Network. Map were attached to this regulation that showed the high-speed rail network for both the Warszawa – Łódź – Poznan/Wroclaw and Warszawa – Katowice/Krakow lines (Figure 1.13) [24].

A second Regulation No. 1316/2013 defines funding principles for TEN-T investments. It makes possible to apply for funds from the newly created financial instrument CEF (Common European Facility). This fund can therefore be used for financing high-speed lines construction and existing lines modernization to high-speed parameters, including study and preparatory works important for cross border relations [25].

Construction of a new Warsaw – Łódź – Wroclaw/Poznan line and modernization of the existing CMK Warsaw – Katowice/Cracow line should lead to the development of international rail connections. It will be also possible to create convenient connections from Warsaw to Praha by Wroclaw. This line location, however, must be the subject of international agreements and joint Polish-Czech studies [5].

The idea of international high-speed connections with neighboring countries is a subject of the study *Directional program for the development of high-speed rail in Poland until 2040* [9].

Preliminary approvals were done with the Czech Republic Transport Ministry and railways. A letter of intent was signed between the Polish and the Czech Republic ministers (2010). Approvals regarding high-speed rail line locations were also a subject of the Visegrad Group agreement from 2010.

In 2011, a feasibility study was completed for the Czech Republic Transport Ministry that recommends options of localization of cross-border high-speed lines [4]. In 2015, a

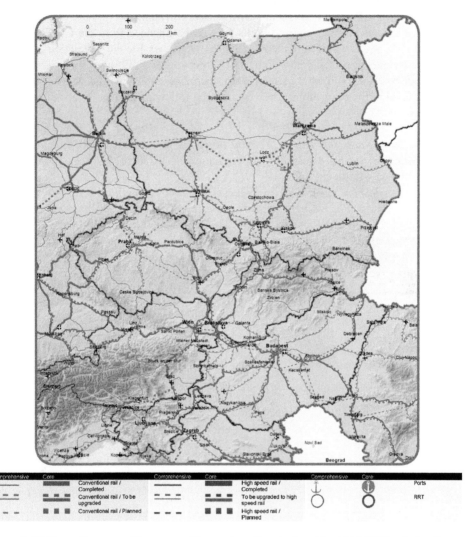

Figure 1.13 Trans-European Transport Network – Regulation No 1315/2013 of the European Parliament and of the Council of 11 December 2013 on Union guidelines for the development of the Trans-European Transport Network [25]

preliminary study was completed for the construction of a new high-speed line from Poznan to Berlin and from Wroclaw to Praha [37].

There are two possible line locations:

• Praha – Mlada Boleslav – Liberec – Jelenia Gora – Wroclaw;
• Praha – Hradec Kralove – Lubawska Pass – Wroclaw.

The first one would be using an already planned Praha – Liberec line and reach larger urban centers of highly important for tourism. The Praha – Wroclaw line would become a part of the Munich – Praha – Wroclaw – Łódź – Warsaw – Baltic countries corridor and the Praha – Wroclaw – Poznan – Szczecin – Swinoujscie corridor.

Further studies and preparatory works may all be financed from the CEF instrument, according to Annex 1 of the Regulation 1315/2013 and Annex 1 to the Regulation 1316/2013, because these lines are already part of the TEN-T network [24], [25].

Further works should be started in 2016, including extended studies and preparatory works required in EU regulations. They concern the following projects financed in the CEF framework:

- *Preparatory works for the construction of a high-speed Warszawa – Łódź – Poznan/ Wroclaw line.* Specified according to Regulation No. 1316/2013, Annex 1, part 1, point 2. North Sea – Baltic corridor.
- *Studies for the construction of a high-speed Wroclaw – Praha line as a high-speed Warszawa – Łódź – Poznan/Wroclaw line extension according to Regulation 1315/2013.* Specified according to Regulation No. 1316/2013, Annex 1, part 1, point 3. Other core network sections. In view of the above, an application should urgently be prepared for funds for this purpose (Fig. 1.14).

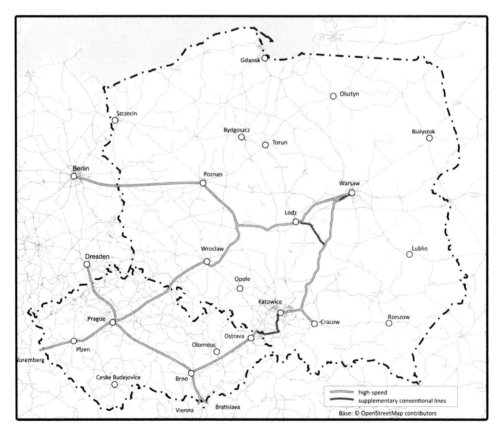

Figure 1.14 Connections between Polish and Czech high-speed network in TEN-T

Drawn by T. Bużałek

1.8. Possibilities of using high-speed lines for regional and agglomeration transport

The possibility of using high-speed lines for regional transport was also analyzed. It is possible that new line capacity will allow trains with lower speeds to enter on a line. Assuming that the maximum speed of high-speed trains will be 300 to 350 km/h, the maximum speed of regional trains should be at least 160–200 km/h.

The possibility of regional train operations using a high-speed line section that comes into big agglomerations for regional traffic purposes was considered. After travelling 50–100 km, such trains would be leaving high-speed lines and further would benefit from the existing conventional lines. In this way, services to cities located far (about 60 minutes or more) from agglomerations would be significantly improved. For this purpose, the Warszawa – Łódź – Poznan/Wroclaw, a modernized CMK Warszawa – Katowice/Krakow, and Wroclaw – Praha lines were analyzed.

1.8.1. Warszawa

A high-speed Warszawa – Łódź line can be used for fast regional trains from Warszawa to Łowicz, Kutno, and even Pock. After leaving a new line, trains would use line 11 and then line 3 from Łowicz to Kutno. This would shorten significantly travel time from Warszawa to Łowicz and Kutno as well as villages located between them (Fig. 1.15).

1.8.2. Łódź

For Łódź railway node few such relations were analyzed. The highest possibility is to connect Brzeziny and Łódź. Brzeziny is a city located 30 km from Łódź city center that is deprived of access to railway lines. A high-speed line will run at a distance of about 5 km west of this city. One concept considered involves the construction of a single-track rail link from the high-speed line to the center of Brzeziny. In this case, the question is not to shorten the travel time but whether to include this city in the railway system. Journey time from Brzeziny to the center of Łódź would decrease to about 15 minutes, compared to the current 40 minutes for the bus trip. A second equally interesting connection would be from Łódź to Łowicz by Nieborów. Regional trains which now use a conventional line to Łowicz would be completed by a fast connection with the city of Łowicz with a travel time of approximately 25 minutes. A similar connection could be created for the cities of Zduńska Wola, Sieradz, and Szadek. Regional trains would benefit from a high-speed line on a section from Łódź to a junction with line 131, and using this line would be getting to the currently used line 14. This high-speed line section could be used for trains from Łódź to Poddębice currently not connected with Łódź (Fig. 1.15).

1.8.3. Poznan

A high-speed line could be used for fast regional trains in relations: Poznan – Kalisz, Kepno by Ostrow Wielkopolski, Pleszew and Jarocin to Poznan, and Krotoszyn by Pleszew to Poznan. Particularly important is the possibility to connect Kalisz and Poznan. This connection, due to the route length, is not competitive with road transport. Using

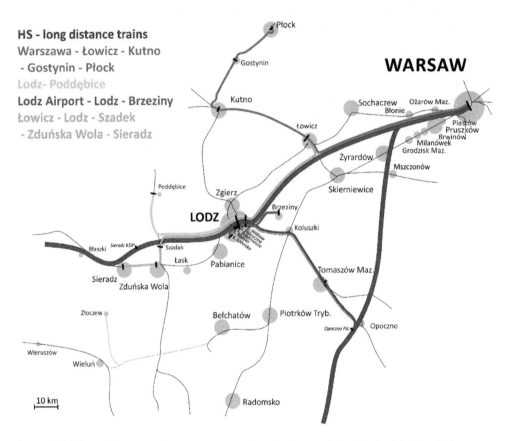

Figure 1.15 Possible regional connections using a high-speed line in the Łódź and Warszawa regions

a high-speed line would enable traveling at a higher speed and shortening the distance travelled (Fig. 1.16).

1.8.4. Wroclaw

Using a new high-speed line from Warszawa to Wroclaw would enable running fast regional trains from Kepno and Kluczbork to Wroclaw.

Construction of a high-speed line from Wroclaw to Praha would be of great importance for railway services development in the Lower Silesia region and would significantly improve travel from cities located next to the southern Polish border. Use of a high-speed line could shorten travel times from Walbrzych, Jelenia Gora, Świdnica, Bielawa, Dzierżoniów, Lubań Śląski, and Kamienna Gora to Wroclaw (Fig. 1.17).

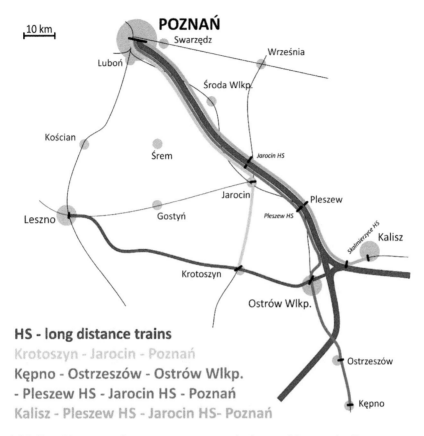

POZNAŃ

10 km

Swarzędz

Września

Luboń

Środa Wlkp.

Kościan

Śrem

Jarocin HS

Leszno

Gostyń

Jarocin

Pleszew

Pleszew HS

Skalmierzyce HS

Kalisz

Krotoszyn

Ostrów Wlkp.

HS - long distance trains

Krotoszyn - Jarocin - Poznań

Ostrzeszów

Kępno - Ostrzeszów - Ostrów Wlkp.

- Pleszew HS - Jarocin HS - Poznań

Kępno

Kalisz - Pleszew HS - Jarocin HS- Poznań

Figure 1.16 Possible regional connections using a high-speed line in the Poznan region

1.8.5. Katowice and Krakow

A CMK extension to Katowice and Krakow will allow opening new fast regional connections from Katowice to Krakow by Olkusz. It would also be possible to run trains with shorter travel times from Czestochowa to Krakow using the current line no. 1 and then sections of an extended CMK line. It would be also possible to run fast trains from Kielce to Katowice partially using a new line.

Construction of a new line from Katowice to Ostrava running through densely populated areas would enable creating a new fast train network connecting cities located in the southern part of the Silesia region. This would allow railway connections to Jastrzębie Zdrój to be restored and to significantly shorten travel times to cities like Rybnik, Raciborz, and Żory. It would be possible also to run fast trains from Katowice to Cieszyn. Some relations can be extended to Krakow (Fig. 1.18).

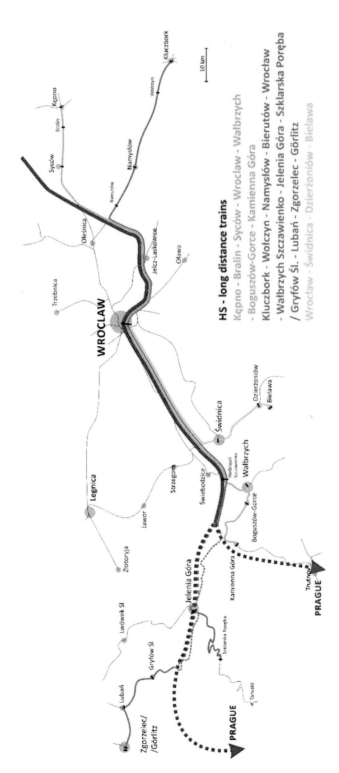

HS - long distance trains

Kępno - Bralin - Syców - Wrocław - Wałbrzych - Boguszów-Gorce - Kamienna Góra
Kluczbork - Wołczyn - Namysłów - Bierutów - Wrocław - Wałbrzych Szczawienko - Jelenia Góra - Szklarska Poręba / Gryfów Śl. - Lubań - Zgorzelec - Görlitz
Wrocław - Świdnica - Dzierżoniów - Bielawa

Figure 1.17 Possible regional connections using a high-speed line in Wrocław region, including a Wroclaw – Praha line

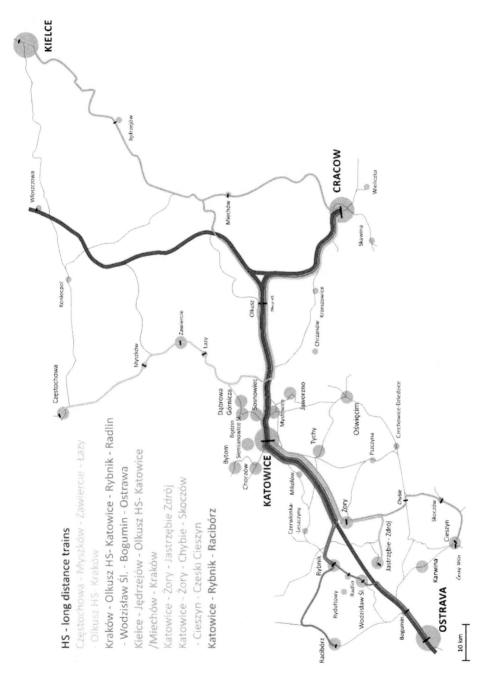

HS - long distance trains

Częstochowa - Myszków - Zawiercie - Łazy

- Olkusz HS- Kraków

Kraków - Olkusz HS- Katowice - Rybnik - Radlin

- Wodzisław Śl. - Bogumin - Ostrava

Kielce - Jędrzejów - Olkusz HS- Katowice

/Miechów - Kraków

Katowice - Żory - Jastrzębie Zdrój

Katowice - Żory - Chybie - Skoczów

- Cieszyn - Czeski Cieszyn

Katowice - Rybnik - Racibórz

Figure 1.18 Possible regional connections using a CMK extension and a new Katowice – Ostrava line in the Małopolska and Upper Silesia regions

1.9. Summary

The concept of building a high-speed rail system in Poland in an evolutionary way changed its shape in the direction of the model, which was aimed to join together as many conurbations in Poland as possible. The analyses were based on the best of the European experience. Foreign experts rated very positively the idea of connecting an existing CMK line that could be upgraded to the parameters of a high-speed line to a new Warsaw – Łódź – Poznan/Wroclaw line, creating the so-called double Y. The construction of high-speed lines in Poland should be considered in the context of the restructuring of the entire railway network and its international links. Including Polish projects into a Trans-European Transport Network would undoubtedly be greatly successful, giving a possibility of receiving European funds for their implementation. Recent years have shown a significant disparity in infrastructure investments between road and rail transport in Poland. In a situation where road system investments are already bringing tangible results, and the road and highway network became more developed and consistent, railway interregional traffic began to rapidly lose market share. Western countries' experience shows that it is not possible to maintain a high railway market share without the construction of high-speed lines. The forecast presented in a government's *Transport development strategy until 2020* indicate that a decline of the role of railways in Poland is fully realistic, and it is expected that this share will decrease to approximately 4% and will be one of the lowest in Europe.

Studies for the construction of high-speed lines in Poland listed in the government program in 2008 were completed in 2018. In this way, all the conditions to start the investment process aimed at opening a high-speed Warsaw – Łódź – Poznan/Wroclaw line in the next decade will be fulfilled according to Polish government commitments regarding EU Regulation 1315/2013 on the implementation of the Trans-European Transport Network. At the same time, further studies should be undertaken on the development of a high-speed rail network in Poland covering other urban areas, especially in western and eastern Poland.

Bibliography

[1] *Analiza funkcjonalna nowego dworca centralnego w Łodzi* [Functional analysis of the new central station in Łódź], SITK RP Oddział w Łodzi ordered by Zarządu Dróg i Transportu w Łodzi. Łódź 2007.

[2] *Analiza warunków budowy kolejowego tunelu średnicowego pod centrum Łodzi.* [Analysis of the conditions for the construction of railway diametral tunnel under the center of Łódź], SITK RP Oddział w Łodzi ordered by Zarządu Dróg i Transportu w Łodzi. Łódź 2008.

[3] Bużałek, T., Raczyński, J.: Połączenie Łodzi i Kielc jako element krajowej sieci kolejowej – stan obecny i perspektywy rozwoju [Connection Łódź and Kielce as an element of national railway network – today's state and perspectives of development]. *Technika Transportu Szynowego*, no. 9/2013.

[4] Dopracování studie: *"Praha – H.Králové/Liberec, rychlostní spojení"* [Praha – H. Kralowe high-speed rail connection]. *Prověření propojení České republiky a Polska tratěmi vyšších rychlostí. Česká republika. Ministerstvo dopravy.* Zhotovitel: IKP Consulting Engineers, s.r.o., Praha 2011.

[5] Gugalka, E., and Pomykala, A.: *Koleje dużych prędkości w Polsce – element systemu transeuropejskiego* [High speed rail in Poland – an element of the Trans-European system]. Diploma thesis, Szkoła Główna Handlowa w Warszawie, Warszawa, 2010.

[6] Jarosiewicz, W., Kozłowski, M., and Meller, M.: *Szybka kolej Warszawa – Łódź w systemie BOOT (Build, Own Operate, Transfer)* [High-speed rail Warszawa – Łódź in system BOOT]. Seminar

Budowa kolei dużych prędkości w Polsce [Building high-speed rail in Poland], Warszawa, 21 June, 2007.

[7] *Kierunki rozwoju kolei dużych prędkości w Polsce* [Direction of development oh high-speed rail in Poland]. PKP PLK S.A., Warszawa, 2011.

[8] *Kierunkowy program linii dużych prędkości w Polsce* [Directional program high-speed rail in Poland]. PKP, Warszawa 1995.

[9] *Kierunkowy program rozwoju kolei dużych prędkości w Polsce do roku 2040* [Directional program for the development of high-speed rail in Poland until 2040]. Instytut Kolejnictwa, Warszawa, 2010.

[10] *Koleje dużych prędkości – szansa dla Polski i Europy Centralnej* [High-speed rail – a chance of Poland and Central Europe]. UIC Seminar. Warszawa, November, 2011.

[11] *Koncepcja przebiegu trasy linii KDP wraz z lokalizacją węzła intermodalnego w obrębie aglomeracji kalisko-ostrowskiej* [The concept of route of the high-speed line along the location of multimodal node within the Kalisz/Ostrów Wlkp. Agglomeration]. Zakład Usług Projektowych, Inwestycyjnych i Eksploatacyjnych, Szczecin, 2009.

[12] *L'opportunite pour la Grande Vitesse dans l'espace PECO* [The possibilities of developing of high-speed rail in the countries of Central and Eastern Europe]. CENIT Barcelona ordered by UIC, 2004.

[13] Massel, A., and Raczyński, J.: Czy kolejowe przewozy regionalne w Polsce mają przyszłość? [Does long-distance railway service have a future?]. *Technika Transportu Szynowego* no. 10/2003.

[14] *Obwieszczenie Ministra Rozwoju Regionalnego w sprawie listy projektów indywidualnych dla Programu Operacyjnego Infrastruktura i Środowisko na lata 2007–2013 z dnia 29 sierpnia 2007 r.* [The notice of the Minister for Regional Development concerning a list of individual Project for Operational Program Infrastructure and Environment for 2007–2013 of 29 August 2007]. *Monitor Polski* no. 69, item. 757.

[15] *Passenger Traffic Study 2020 Poland and Czech Republic.* Commissioned by UIC conducted by Intraplan Consult GmbH, IMT Trans and INRETS, 2003.

[16] *Polityka transportowa państwa na lata 2001–2015* [Transport policy of the state for the years 2001–2015]. Ministry of Transport and Maritime Economy, 2001.

[17] *Polska 2030.* Wyzwania rozwojowe [Poland 2030. Development challenges], Ministry of Development, Warszawa, 2009.

[18] *Połączenie kolejowe Łódź – Kielce* [Railway connection Łódź – Kielce]. Seminar, Kielce, March, 2013.

[19] *Projekt PEGAZ* [Project PEGAZ]. Concordia. Warszawa, 2003.

[20] Raczyński, J.: *Łódź – Warszawa w sieci szybkich połączeń kolejowych Polski na tle rozwiązań europejskich* [Łódź – Warszawa in the network of fast Polish connection against background European solution]. Seminar *Szybkie połączenie kolejowe Łódź – Warszawa w sieci połączeń kolejowych Polski.* Organized by Zarząd Oddziału SITK w Łodzi, Urząd Miasta Łodzi and Biuro Projektów Kolejowych i Usług Inwestycyjnych Sp. z o.o. in Łódź. 2 October, 2002.

[21] Raczyński, J.: Rządowy program budowy linii dużych prędkości w Polsce [Government Program for the construction of high-speed lines in Poland]. *Technika Transportu Szynowego,* no. 9/2008.

[22] Raczyński, J., Boryczka, J., and Szafrański, Z.: Łódź – Warszawa w sieci szybkich połączeń kolejowych Polski [Łódź – Warszawa in the network of fast Polish railway connection]. *Technika Transportu Szynowego,* nos. 3–4/2002.

[23] Raczyński, J., Wróbel, I., and Pomykała, A.: Kierunki rozwoju kolei dużych prędkości w Polsce [Directions of the development of high-speed rail in Poland]. *Technika Transportu Szynowego* nos 11–12/2010.

[24] REGULATION (EU) No 1315/2013 OF THE EUROPEAN PARLIAMENT AND OF THE COUNCIL of 11 December 2013 on Union guidelines for the development of the trans-European transport network and repealing Decision No 661/2010/EU.

[25] REGULATION (EU) No 1316/2013 OF THE EUROPEAN PARLIAMENT AND OF THE COUNCIL of 11 December 2013 establishing the Connecting Europe Facility, amending Regulation (EU) No 913/2010 and repealing Regulations (EC) No 680/2007 and (EC) No 67/2010.

[26] *Seminarium Koleje dużych prędkości w Europie i projekty budowy nowych linii w Polsce* [Seminar: High-speed rail in Europe line construction in Europe]. *Technika Transportu Szynowego*, no.4/2006.

[27] *Strategia Rozwoju Transportu na lata 2007–2013* [Transport Development Strategy for 2007–2013]. Ministry of Infrastructure, Warszawa, 2004.

[28] *Studium przebiegu przez Łódź kolei dużych prędkości V-300* [Study of the course of high-speed line V300]. Teren Sp. z o.o. Przedsiębiorstwo Zagospodarowania Miast i Osiedli w Łodzi commissioned by Zarząd Dróg i Transportu w Łodzi. Łódź, 2006.

[29] *Studium rozwoju funkcjonalnego łódzkiego węzła kolejowego w aspekcie budowy linii dużych prędkości* [Functional development study of the Łodź railway node in the aspect of building high-speed line]. SITK RP Oddział w Łodzi commissioned by Zarząd Dróg i Transportu w Łodzi. Łódź, 2007.

[30] *Studium trasowania linii kolejowych dla V>300 km/h (Berlin –) Kunowice – Warszawa – Terespol – (Mińsk – Moskwa) na terytorium RP.* [Study of routing railways lines V300 (Berlin –) Kunowice – Warszawa – Terespol – (Mińsk – Moskwa]. Railway project commissioned by Dyrekcja Generalna PKP, Warszawa 1993.

[31] *Studium wykonalności dla budowy linii kolejowej dużych prędkości "Warszawa – Łódź – Poznań/ Wrocław"* [Feasibility study for the construction high-speed line Warszawa – Łódź – Poznań/ Wrocław"]. IDOM, Warszawa, 2013.

[32] *Studium wykonalności dla budowy systemu Łódzkiej Kolei Aglomeracyjnej, Etap I.* [Feasibility study for the construction of Łódź Agglomeration Railway. Stage 1]. SITK RP Oddział w Łodzi commissioned by Urząd Marszałkowski w Łodzi. Łódź, 2009.

[33] *Studium wykonalności – dokumentacja przedprojektowa dla modernizacji linii kolejowej E65-Południe odcinek Grodzisk Mazowiecki – Kraków/Katowice – Zwardoń/Zebrzydowice – granica państwa* [Feasibility study – Pre-design documentation for modernisation of the railway line E65-South section Grodzisk Mazowiecki – Kraków/Katowice – Zwardoń/Zebrzydowice – border PL/ CZ]. Hallcrow, Warszawa 2011.

[34] *Uchwała nr 239/2011 Rady Ministrów z 13 dnia grudnia 2011 r. w sprawie przyjęcia Koncepcji Przestrzennego Zagospodarowania Kraju 2030* [Resolution 239/2011 of the adoption of the concept of spatial development of Poland 2030]. Ministry for Regional Development, Warszawa, 2011.

[35] *Uchwała Nr 276/2008 Rady Ministrów z dnia 19 grudnia 2008 r. w sprawie przyjęcia strategii ponadregionalnej "Programu budowy i uruchomienia przewozów kolejami dużych prędkości w Polsce"* [Resolution 276/2008 of the adoption of supra-regional strategy "Program of the construction and launch of the high-speed rail in Poland"]. Ministry of Infrastructure, Warszawa, 2008.

[36] *Wstępne studium wykonalności budowy linii dużych prędkości Wrocław/Poznań – Łódź – Warszawa* [Pre-feasibility study for the construction of high-speed line Wrocław/Poznań – Łódź – Warszawa]. Centrum Naukowo-Techniczne Kolejnictwa commissioned by PKP PLK S.A. Warszawa 2005.

[37] *Wstępne studium wykonalności dla przedłużenia linii dużych prędkości Warszawa Łódź – Poznań/Wrocław do granicy z Niemcami w kierunku Berlina oraz do granicy z Republiką Czeską w kierunku Pragi* [Pre-feasibility study for the extension of high-speed line Warszawa Łódź – Poznań/Wrocław to the border with Germany and Czech Republic"]. Performer by IDOM in cooperation with Railway Research Institute. Warszawa, 2015.

[38] Żurkowski, A.: Duże szybkości. UIC, Polska [High-speed rail, UIC, Poland]. *Technika Transportu Szynowego*, nos. 5–6/2005.

Chapter 2

Development of the high-speed rail infrastructure – Polish experience

Andrzej Massel

2.1. Introduction

Infrastructure is the key factor enabling implementation of high-speed rail services. This is also the most expensive and long-standing element of the railway system. According to annex I of the Directive 2008/57/EC, the high-speed lines comprise [9]:

- specially built high-speed lines equipped for speeds generally equal to or greater than 250 km/h;
- specially upgraded high-speed lines equipped for speeds on the order of 200 km/h;
- specially upgraded high-speed lines which have special features as a result of topographical, relief, or town planning constraints, on which the speed must be adapted to each case.

A rather flexible approach adopted in the 2008/57/EC Directive shows that there is no uniform template on how to design, develop, and expand the European high-speed rail network. Potential options include [12]:

- new very high-speed rail lines (VHS) with design speeds at or above 300 km/h;
- new medium high-speed rail lines (MHS) with design speeds at 250–280 km/h; and
- upgrades of conventional lines (CUP) typically at 200–220 km/h.

It seems necessary to stress the importance of the 200 km/h speed resulting from the fact that it is sometimes considered as a kind of "sound barrier"; in other words, a threshold for the technology based on [3]:

- existing (conventional) railway lines;
- traditional, locomotive-hauled trains.

It is also worth mentioning that this "threshold" is to a certain extent artificial, as there are conventional lines in Europe upgraded for speeds greater than 200 km/h. In Germany, on some sections of the Berlin – Hamburg line, built in the 19th century, the maximum speed is 230 km/h. Also, in France, there are sections where the TGV trains may run at 220 km/h; this is particularly the case on the Le Mans – Nantes and Tours – Bordeaux lines. More details on this aspect are given in later sections of this chapter. As far as the rolling stock is concerned, the sets of Austrian coaches Railjet, hauled by class 1116 "Taurus" locomotives, are an exception, with a maximum speed of 230 km/h.

2.2. Upgraded high-speed lines – international experience

The scope of upgrading a conventional railway line depends on numerous factors. Typically it covers:

- adjusting geometrical layout (modifying alignment);
- civil works, including strengthening of bridges or their replacement with new structures;
- elimination of level crossings;
- renewal of existing rails;
- equipping or renewal of concrete sleepers;
- adjusting or replacing catenary;
- installation of new signaling and telecommunication systems.

Significant examples of the European best practice related to upgrading conventional railway lines to the speeds of 200 km/h or more are presented below.

In France, the 200 km/h speed in day-to-day operation was introduced on 28 May 1967 on the section between les Aubrais (close to Orleans) and Vierzon. The length of the section cleared for 200 km/h (from km 118.9 to km 195.5) is 76.6 km. The line has been specially adapted to that speed. It was equipped with a cab signaling system, indicating the permissible speed on three consecutive block sections. An innovative solution was the installation of a transverse force measuring system, which was mounted on the wheelset in the first bogie of the locomotive (BB 9200 series) [28].

The speed of 200 km/h was achieved by the "Le Capitole" express train from Paris to Toulouse, covering the entire route (713 km) in 6 hours with stops at Limoges, Brive-la-Gaillarde, Cahors, and Montauban. Relatively low commercial speed (118 km/h, later increased to 120 km/h) resulted from geometry constraints on the southern part of the line, especially south of Limoges, where the typical maximum speed is just 110 km/h. Average start-to-stop speed at the Paris-Austerlitz – Limoges-Benedictins section (400 km), travelled non-stop, was significantly higher and equaled 138 km/h (141 km/h in the year 1971).

In May 1971, the TEE "Aquitaine" train was introduced on the route from Paris to Bordeaux and operated at a maximum speed of 200 km/h. The train covered the 581 km distance without any intermediate stops in 4 hours, which resulted in a commercial speed of 145.3 km/h.

It has to be mentioned that in both cases, Paris – Toulouse and Paris – Bordeaux, the lines are electrified with 1.5 kV direct current. Very dense location of substations is required at this system. For example, in the case of the Paris – Bordeaux main line, the electric energy is supplied through 52 substations, and the average distance between substations is 11 km. Since 1971, the TEE trains "Le Capitole," "Aquitaine," and "Étendard" were hauled by CC 6500 series electric locomotives with continuous power output of 5900 kW [14].

German railways have collected significant experience concerning increasing train speed. It is noteworthy that a few years before World War II (since 1933), the operation of scheduled express trains at a speed of 160 km/h was started. These trains were served with diesel motor units. Similar speeds were also achieved after World War II on the railways of the Federal Republic of Germany. Before introduction of the maximum speed of 200 km/h on the Deutsche Bundesbahn network, comprehensive research was performed in the years 1963–1964 [5]. The tests supervised by F. Birmann were made on the Forcheim – Bamberg test section. The total length of the section was 24 km, of which 20 km were cleared for 200 km/h running. Four intermediate stations with various types of switches and crossings were located on the test section. Track-side tests included stress measurements in five locations in tracks and turnouts. The test train was composed of the adapted E10 electric locomotive, the measuring car, and the passenger car. The tests confirmed the suitability of track superstructure with S54 rail, being introduced on German railways at that time. Decreasing the sleeper spacing to a value of 0.58 m and setting stricter values for limits, however, were proposed. Moreover, decreasing the nominal value of track gauge to 1432 mm was recommended.

For the first time, a speed of 200 km/h for trains carrying passengers was introduced in the year 1965 on the Munich – Augsburg line. From 26 June until 3 October, the International Transport Exhibition (IVA) took place in Munich [11]. During the exhibition, special express trains from Munich to Augsburg were operated. Running 200 km/h was allowed on two sections 10.5 km and 34.3 km long, respectively. They were separated with the section with the speed of 120 km/h [8]. The special trains were hauled with an E03 electric locomotive, coming from the first production batch (four locomotives). In the following years, a speed of 200 km/h was allowed for selected express trains in the framework of testing of the new train control system (Linienzugbeeinflussung).

In normal operation, a speed of 200 km/h was implemented on the Munich – Augsburg line according to a special regulation of the Minister of Transport from 25 September 1977. According to the winter timetable for the year 1977/1978, IC and TEE trains covered that section (61.9 km) in 30 minutes, giving the start-to-stop value of 123.8 km/h. A much better start-to-stop average (136.3 km/h) was recorded for IC trains stopping at the Munich-Pasing station. Journey time for these trains on the Munich-Pasing – Augsburg section (54.5 km) was 24 minutes.

Around the year 1980, train speeds were increased on further sections of German railways, including the Hamm – Bielefeld section (66.9 km) forming part of the trunk route from the Ruhr industrial area to Hannover. It should be noted that the section is equipped with four tracks specialized according to type of traffic, which creates particularly favorable conditions for faster running. A speed of 200 km/h has been implemented since 1980 on fast tracks at the total length of 58 km in each direction. The start-to-stop average speed achieved by IC trains on this section in 1981 was 167.3 km/h.

One of the most interesting examples of upgrading a conventional railway to high-speed standards is the modernization of the Berlin – Hamburg main line (286 km). It was opened to traffic in December 1846 as one of the first connections of the principal cities in Germany. The modernization of the line was performed in two stages. In 1991, just after the reunification of Germany, the project for renewal and upgrading the line to a speed of 160 km/h was started. It was done in the framework of German Unity (*Deutsche Einheit*) transportation projects. In May 1997, the entire line started to be operated at a speed of 160 km/h using electric traction. The shortest journey time between Berlin and Hamburg was then reduced to 2 hours 14 minutes.

In 2000, the federal government decided to go with the second phase of the Berlin – Hamburg project, involving modernization of the line for a speed of 230 km/h. The cost of the additional works was around 65 million EUR. The total length of the modernized section was 263 km. The planning and construction works included [29]:

- rebuilding of 21 platforms at 14 stations to enhance passenger safety (platform barriers, markings, train-controlled warning messages);
- installation of 27 new turnouts and enhancement of 135 existing turnouts;

- substitution of 56 level crossings with rail underpasses or overpasses;
- installation of 560 km of cable loops for LZB (Linienzugbeeinflussung);
- rebuilding or construction of 517 km of overhead catenary lines.

The Berlin – Hamburg line was the first case of upgrading a conventional railway to a speed up to 230 km/h. On 12 December 2004, the upgraded line went to regular service. The shortest journey time for a non-stop train in the 2004/2005 timetable was only 1 hour 30 minutes (for the train without any stops on intermediate stations) and resulted in the extraordinary value of commercial speed: 189 km/h. In the following years, however, the journey time was gradually extended to 1 hour 42 minutes in the 2017/2018 timetable, however, with one stop on route (see Figure 2.1).

In Great Britain, implementation of regular train services at the speed of 125 mph (201 km/h) was related with the introduction of new HST (high-speed train) diesel trainsets, branded as Intercity 125. On 4 October 1976, the Intercity 125 trains started to operate at the Great Western Main Line from Paddington station in London to the west of England and South Wales. Before that time, maximum speeds on the British railway network did not exceed 100 mph (160 km/h). At the beginning, London – Bristol and London – Cardiff services were served with the new rolling stock [29]. It is noteworthy that the Great Western Main Line is characterized by extremely favorable alignment; moreover, specialization of tracks exists (fast lines and slow lines). Therefore, this line is particularly suitable for high-speed running. Despite their age of more than 40 years, the HST trains are still operated on the line at a speed of 201 km/h. They are also operated on other main lines in Britain, for example on the East Coast Main Line from London to Scotland, as well as on the Midland Main Line from London to Nottingham and Sheffield.

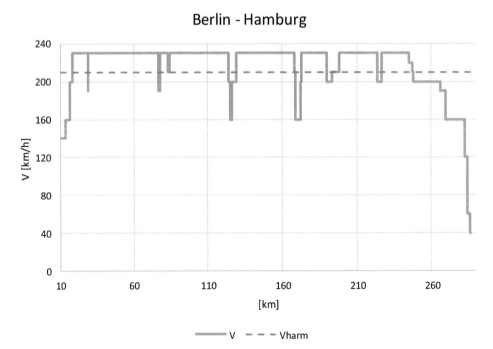

Figure 2.1 The speed profile on the Berlin – Hamburg railway line

In Italy, the first railway line with the operating speed of 200 km/h was the Direttissima high-speed line from Rome to Florence. The line was constructed from 1970 until 1992. Its first section from Sette Bagni (in the outskirts of Rome) to Città della Pieve (close to Chiusi), 121.9 km long, was opened to traffic on 24 February 1977 [15]. Initially the trains were operated at a maximum speed of 180 km/h. The speed was increased to 200 km/h in 1985 and to 250 km/h in 1988, i.e. after putting the ETR450 (first generation of Pendolino) trains into service.

A remarkable example of adaptation of the railway line constructed in the 19th century to high-speed operation is the modernization of the Sankt Petersburg – Moscow railway. This line was opened to traffic in 1851 and its length is around 650 km. A characteristic feature of the line is its favorable alignment: the total length of curved sections is 47.6 km, only 7.5% of the overall length of the line. The minimum curve radius is 1600 m for open sections and 1065 m for stations. The ruling gradient is 6‰.

Traditionally the train speeds on the Sankt Petersburg – Moscow line were higher than other railways in Russia. In June 1963, just after completion of electrification of the line, the express train "Aurora" was put into service. Initially its journey time was 5 hours 27 minutes, but two years later, in the 1965 timetable, it was shortened to 4 hours 59 minutes. The average commercial speed of the train was 130 km/h (at a maximum speed 160 km/h).

In the early 1970s, the line became the test section for high-speed traffic. At that time the first electric motor unit for high speed was constructed in Riga works. It was designated as ER200. Starting on 1 March 1984, this EMU was put into scheduled service on the Sankt Petersburg – Moscow route. This trainset achieved 200 km/h on (rather short) selected sections; however, only one train per week in each direction was operated.

The design for complex reconstruction of the Sankt Petersburg – Moscow line and for its adaptation for high speed was elaborated in "Giprotransput" Institute. Several studies have been made. Their scope covered [24]:

- improvement of continuously welded rail (CWR) track structure, with particular attention to bridges;
- designing new types of turnouts with swing nose crossings, laid on concrete sleepers;
- structure of ballast bed and upper part of subgrade;
- designing transition zones between bridge and embankment;
- elaboration of solutions for drainage of subgrade;
- elaboration of solutions assuring stability of high embankments;
- elaboration of modern technologies for works at subgrade, structures, and renewal of track.

Requirements for track components have been set as follows:

- first grade R65 rails with very tight tolerances for straightness of rail head;
- concrete sleepers with increased frost resistance;
- ballast with high durability and cleanliness;
- overhead catenary lines with cross-section area in the range of 120–150 mm^2 with cadmium and silver supplement.

In order to strengthen track structure, heat-treated and continuously welded rails without insulated joints have been adopted along the entire line. Swing nose crossings have been installed in main lines at stations. Ballast cleaning (or complete ballast replacement, where necessary) have been done, together with installation of geotextiles and a layer made of mixture of gravel and sand. The works were generally performed in 4-hour possessions of one of the main tracks.

Modernization of track layouts was restricted to necessary changes in alignment and in level of tracks. The scope of changes included:

- increasing curve radiuses to 2000 m;
- elimination of multi-radius curves;
- increasing the distance between axes of main line and any remaining tracks to 5.3 m;
- increasing minimum length of track section with constant gradient to 300 m;
- decreasing the maximum difference of gradients to 2‰;
- increasing radiuses of vertical curves to 20,000 m.

Since 17 December 2009, high-speed train services have been operated on the Sankt Petersburg – Moscow railway, branded as Sapsan. Velaro RUS EMUs delivered by Siemens are used at this service. It is noteworthy that almost 600 km of the line are covered at a speed of 200 km/h or more (some sections are cleared for 220 km/h and even 250 km/h). The shortest journey time in the 2017/2018 timetable is just 3 hours 30 minutes, giving the remarkable commercial speed of 185 km/h.

2.3. Upgraded high-speed line – the Polish case

December 2014 saw the commencement of the commercial operation of the first high-speed trains on the Polish railways: electric multiple units class ED250 Pendolino, built for PKP Intercity by Alstom. To make it possible, several important infrastructural projects, or their important parts, needed completion [21]. The ETCS level 1 automatic train control system was put in operation on the Central Trunk Line (CMK), for the first time on the Polish network. That enabled operation of the passenger trains over the CMK line at maximum speed 200 km/h.

Achieving 200 km/h on the Central Trunk Line makes this line the first high-speed line in Poland. This chapter presents the conditions for introducing trains running at 200 km/h in commercial operation on the Polish rail system. In particular, preparation of the track infrastructure will be discussed here.

CMK is a specific line from the point of view of its technical characteristics, as well as the way it is operated. When the first concept of the line was being developed in 1970, operation of both high-speed passenger services and heavy freight trains was envisaged. Therefore, the following basic assumptions were adopted [4]:

- maximum speed of passenger trains v = 200–250 km/h;
- freight trains weight – up to 5000 tons;
- length of freight trains – 150 axles (equivalent to 750 m).

The geometrical layout of the line corresponded with the above operation requirements:

- minimum curve radius R_{min} = 4000 m;
- maximum cant h = 100 mm;

- length of transition curves $l = 12Vh$;
- minimum vertical curve radius $R_v = 15,000$ m;
- maximum gradient on the open line 6.0‰;
- maximum gradient in the station 1.0‰;
- minimum straight intersection 100 m;
- width of track formation 10.9 m;
- distance between the main track center lines in the stations and on the open line 4.5 m.

Derogation from these parameters, related to alignment, were only accepted on the sections connecting the CMK with Zawiercie station ($R_{min} = 1900$ m), and with the Warsaw Railway Node near Grodzisk Mazowiecki ($R_{min} = 2600$ m). It should be emphasized that with adopting in 1970 such large radii for horizontal curves, the newest (at that time) trends and the best practice in the field of the railway design have been applied. The minimum curve radius on the CMK line (4000 m) is the same as those on the first French high-speed line from Paris to Lyon, being built in the 1970s, and on the San-Yo line from Osaka to Fukuoka Hakata in Japan. Even if the cant was limited to 100 mm (due to operation of slower trains), the non-compensated acceleration at 250 km/h is 0.55 m/s^2. The distance between the track centers on the open line and in the stations (4.50 m) is bigger than the one applied on the Direttissima line Rome – Florence (4.00 m), Paris – Lyon (4.20 m), or Osaka – Okayama section of the San-Yo line. Therefore, the alignment of the CMK is as good as the layout of numerous high-speed lines in Europe and in Japan, and regarding some parameters (gradient), even more favorable, compared with other lines, at which the maximum speed is now 250 km/h (Rome – Florence), or 300 km/h (Paris – Lyon, Osaka – Okayama). Unlike on the above lines abroad, where concrete sleepers of mono-block (Italy, Japan) or bi-block (France) type were used, the sleepers (1733 sleepers per 1 km of track) made of hard wood were installed initially on the CMK line. At the beginning the S49-type rails were laid down; they were replaced a few years later with the S60 (UIC60) rails, considered as a target solution.

Two types of turnouts have been built in the main through tracks of the stations [4]:

- S60-type turnouts with 1:18.5 crossing angle and 1200 m curve radius, suitable for 100 km/h running on a diverging track;
- S60-type turnouts with 1:12 crossing angle and 500 m curve radius, suitable for 60 km/h operation on a diverging track.

Only in Idzikowice station, due to its complex layout, the standard S60 1:9, 300 m turnouts were used.

Formally the investment on the CMK line started with the decision of the Minister of Transport on 7 August 1970. The line was intended to be part of the future connection of Upper Silesia with Gdansk and was built in the two stages:

- stage I – Zawiercie – Idzikowice (Radzice), 143 km section built in 1971–1974 (electrification completed in 1975–1976);
- stage II – Idzikowice – Grodzisk Mazowiecki, 80 km section including connecting towards Warsaw, built in 1974–1977.

The final section from Szeligi station to Grodzisk Mazowiecki was opened to traffic on 28 December 1977. It should be noted that special attention was paid to the quality of works, appropriate to the required operational conditions. In 1972, the Main Railway Technology Research and Development Centre (today's Railway Institute) had developed the terms for commissioning the superstructure work, and in 1973 the recommendations for commissioning the turnouts on the CMK line [4]. Three-stage commissioning was foreseen: initial (ODB-1), intermediate (ODB-2) after passage of 2 million tons, and final (ODB-3) after having 5 million tons passed over the line.

Despite very favorable alignment of the CMK, initially only freight trains operated over the line. It was not until 3 June 1984 that the new 1984/1985 timetable entered into force and the first two pairs of express trains commenced operation over the CMK line. These trains, for the first time in Poland, run at a maximum speed of 140 km/h [19]. In the following years, further preparations have been carried out for increasing the speed of passenger trains to 160 km/h. The necessary adaptation of the signaling systems was essential in this regard. In 1985–1986, the entire line was equipped with a 4-aspect automatic line block system of Eac type. The first two pairs of trains commenced commercial operation at 160 km/h on 29 May 1988 [20].

In 1990, preparations began for introducing a speed of 200–250 km/h on the CMK line. At first, track construction technology was tested on the experimental section, located on track no. 2 in the area of Biała Rawska post (km 39.6–44.6). In 1992, the track superstructure was changed on this section [13]. UIC60-type rails were laid on concrete sleepers equipped with SB3 fastenings. There was only one horizontal curve on the section with a radius of 4984 meters and two crossovers UIC60–1200–1: 18.5. The grinding of rails was performed in 1993. It was proved that deviations of geometric track parameters were within the limit values established for a speed of 200 km/h. Synthetic track quality index J, calculated for 1 kilometer sections was 1.1–1.4 for the open line, and 1.8 for a kilometer with turnouts [22]. On 11 May 1994, an ETR 460 Pendolino electric multiple unit (hired from Italian railways) reached a speed of 250.1 km/h on the experimental section [6].

In 1993, comprehensive track renewal aimed at preparing tracks for speeds of 200–250 km/h began at CMK. Modernization of the CMK line for high speed was initially planned for the years 1995–2005 and was divided into two phases. In general, the scope of the first stage of works was to rehabilitate the entire line for speed $V = 160$ km/h with the assumption that modernized infrastructure elements will be adjusted to the speed $V = 200$ km/h for the loco-hauled trains and $V = 250$ km/h for the electric motor units and simultaneously they would meet interoperability requirements [23]. The scope of the first stage was changed several times, mainly because of the financial constraints and to take the EU directives on interoperability into account. By the end of 2007, a following works were completed:

- replacement of tracks at length of 332.8 km;
- modernization of traffic posts Psary, Góra Włodowska, Knapówka, and Korytów;
- modernization of 19 engineering structures, including seven structures for the speed $V = 300$ km/h;
- modernization of the overhead catenary lines to $V = 200$–250 km/h at the length of 126.6 km (in the case of both tracks on the Góra Włodowska – Zawiercie section with the possibility of future transition to the alternate current supply).

Track renewal at the entire CMK length was performed until the year 2000. The concrete sleepers with SB-type fastenings have been laid in place of wooden sleepers. Due to the change of type of sleepers, the thickness of the ballast layer has been increased from 0.30 m to 0.35 m.

The scope of works on Psary, Góra Włodowska, and Korytów stations and on Knapówka post covered changes in track layouts. Fifty-nine new turnouts were laid, including 28 turnouts with a swinging crossing nose. It is noteworthy that the first rebuilt station, Psary, was treated as a test site for high-speed turnouts. In 2002, the Scientific and Technical Railway Centre (now the Railway Institute) carried out the dynamic tests at the speeds exceeding 200 km/h. The measurements were made either on the specially equipped train (in particular vertical and horizontal forces) or on the track side (for example accelerations on the crossing and on the turnout sleepers) (Figure 2.2). The UIC60–1200–1: 18.5 and UIC60–500–1 12 turnouts, produced by three different manufacturers, were tested [18].

EBILOCK-type STC interlockings have been installed on the upgraded stations and the buildings of the signal boxes have been modernized. Moreover, overhead catenary lines have been replaced on these stations.

In April 2008, the parameters of modernization were changed once again as it was decided to modernize the line to the speed V = 300 km/h and axle load of 245 kN. The scope of work planned for the period 2008–2010 included, among others [23]:

- rebuilding viaducts and bridges for a speed V = 300 km/h on the section from Włoszczowa Północ (km 155) until km 212.3;
- upgrade of culverts on the Włoszczowa Północna – Zawiercie section;
- reconstruction of catenary with poles piled at a distance of 3.2 m from the track and with the possibility of transition to the AC power system 2×252×25 kV, 50Hz;
- installation of ERTMS/ETCS level 1.

It was assumed that as a result of the above scope of work, it would be possible to increase the speed up to 200 km/h on the Włoszczowa Północna – Zawiercie section (to km 212.3).

Figure 2.2 Dynamic tests at the Psary station. The electric locomotive series 124 from VUZ Praha was used for tests.

Figure 2.3 The construction of the road viaduct in order to eliminate a rail-level crossing at CMK line (June 2013)

The planned works have been fulfilled; however, the installation of ETCS took more time than originally planned. In addition, reconstruction of viaducts and bridges on the section, carried out in 2008–2009, led to a significant disruption to train traffic (due to severe speed restrictions to 30 km/h).

Another part of the Central Trunk Line modernized for higher speed was the Olszamowice – Włoszczowa Północ section. On this section, the complete centenary replacement was made in 2011. Also in 2011 began works on the modernization of engineering structures: bridges, viaducts, and culverts. An important innovation consisted of use for works of the new type temporary spans, allowing passage of trains at a speed of 100 km/h. A characteristic feature of the Olszamowice – Włoszczowa Północ section was the location of a large number of level crossings. All but two crossings have been eliminated and replaced with newly built road viaducts (Figure 2.3).

Simultaneously with upgrading the southern section of the Central Trunk Line, the works on the northern section of the line between Grodzisk Mazowiecki and Idzikowice were undertaken. Three technical stations – Korytów, Szeligi, and Strzałki – and a diversion post, Biała Rawska, are located on this section.

Korytów station was modernized in 2007. Worth mentioning is that this station, like Psary station, served for testing new types of turnouts. 60E1–1200–1:18.5 and 60E1–500–1:12 turnouts with fixed crossing noses designed for 200 km/h and manufactured by VAE, Cogifer, and KolTram have been installed in this case. They were intended for use on other modernized railway lines, especially on the E65 Warsaw – Gdańsk line. Instytut Kolejnictwa (Railway Research Institute) carried out tests including the measurements in the track and on the rolling stock by the specially instrumented train, running at speeds up to 220 km/h.

In 2012–2014, reconstruction of the two remaining stations (Szeligi and Strzałki) on the Grodzisk Mazowiecki – Idzikowice section was undertaken. At these stations, similarly to the southern section of the Central Trunk Line (CMK), the 60E1–1200–1: 18.5 and 60E1–500–1 12 turnouts with swing crossing noses were used. The modernization of Biała Rawska post (two crossovers composed of four turnouts with 1200 m radius) took place in the year 2016.

Modernization of the catenary on the northern section of the Central Trunk Line, with the exception of stations, was carried out in 2012–2013. On both tracks the catenary has been replaced. The 2C120–2C-3 catenary, adapted to the speed of 250 km/h, was implemented. Reconstruction of the catenary on particular stations (Korytów, Szeligi, Strzałki) has been performed in the framework of modernization works on these stations.

The most difficult part of the modernization of the CMK, limiting its capacity and imposing reduction of the speed of trains in the course of works, is the reconstruction of engineering structures. On the northern section between Grodzisk Mazowiecki and Idzikowice are 86 structures, including 7 bridges, 28 overpasses, and 51 culverts. Modernization of these facilities was completed until 2017. The largest structures are:

- railway viaduct with arch structure over the S8 expressway at km 26.571. Works on this viaduct were completed in 2014 in track no. 1 and in 2015 in track no. 2;
- railway bridge over the Pilica river at km 63.728. Construction of the new bridge was completed in 2015.

On the northern section of the Central Trunk Line – Grodzisk Mazowiecki – Idzikowice, six level crossings were located. By 2016, all these crossings had been replaced by multilevel intersections.

According to the timetable for the years 2015/2016, a speed of 200 km/h was applied to both tracks between Olszamowice and Zawiercie, from km 125.200 to km 214.800, with the exception of the following locations:

- Włoszczowa Północ station at km 151.900–155.430,
- Two category A level crossings on the Olszamowice – Włoszczowa Północ open section, with a speed limit of 160 km/h at km 142.850 and km 149.500.

In conclusion, it can be stated that the introduction of a speed of 200 km/h on the Olszamowice – Zawiercie section required significant investment activities (see Table 2.1 and Figure 2.4). This included:

- reconstruction of Góra Włodowska and Psary stations and Knapówka diversion post with installation of turnouts adapted to a speed of 250 km/h (with swing crossing nose);
- modernization of structures (railway bridges, viaducts, and culverts);
- replacement of catenary;
- installation of the train control system ERTMS/ETCS level 1 on the whole length of the Central Trunk Line by the consortium formed by Thales Rail Signalling Solutions Ltd. and Thales Rail Signalling Solutions GmbH Austria (project co-financed by the EU under the TEN-T program);
- replacement of category A or B level crossings at km 127.190, km 129.357, km 132.853, and km 147.548 with road viaducts (two level crossings remained);
- modernization of the power supply system (modernization of existing substations and construction of new ones).

Table 2.1 Completion of modernization works on CMK line – December 2017

Section	km	Track	Turnouts	Structures	Level crossings	Catenary	Signaling	Remarks
Grodzisk Maz. – Korytów		Done		Done	Done	Done	Done	
Korytów	14.574	Done	Done			Done	Done	
Korytów – Szeligi		Done		Done	Done	Done	Done	
Szeligi	23.542	Done	Done			Done	Done	
Szeligi – Biała Rawska		Done		Done	Done	Done	Done	
Biała Rawska	40.122	Done	Done			Done	Done	
Biała Rawska – Strzałki		Done		Done	Done	Done	Done	
Strzałki	57.381	Done	Done			Done	Done	
Strzałki – Idzikowice		Done		Done	Done	Done	Done	
Idzikowice	80.608	Done	t.b.d.	t.b.d.	t.b.d.	t.b.d.	t.b.d.	LC in the station
Idzikowice – Opoczno Płd.		Done		t.b.d.	Done	Done	Done	
Opoczno Płd.	92.142	Done	t.b.d.	t.b.d.	t.b.d.	t.b.d.	t.b.d.	
Opoczno Płd. – Olszamowice		Done		Done	Done	Done	Done	LC in km 95.136, 98.550
Olszamowice	124.808	Done	Done			Done	Done	
Olszamowice – Włoszczowa Płn.		Done		Done	Under way	Done	Done	LC in km 142.850, 149.500
Włoszczowa Płn.	154.390	Done	Done			Done	Done	
Włoszczowa Płn. – Knapówka		Done		Done	Done	Done	Done	
Knapówka	160.588	Done	Done			Done	Done	
Knapówka – Psary		Done		Done	Done	Done	Done	
Psary	170.479	Done	Done			Done	Done	
Psary – Góra Włodowska		Done		Done	Done	Done	Done	
Góra Włodowska	206.688	Done	Done			Done	Done	
Góra Włodowska – Zawiercie		Done		Done	Done	Done	Done	LC in km 215.500 (t.2)

Source: Author's own elaboration

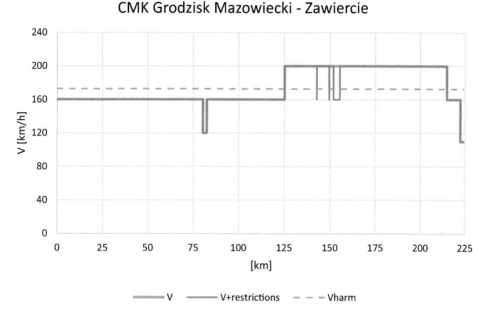

Figure 2.4 The speed profile on CMK line (as of December 2015)

The confirmation of the correct adoption of solutions in the field of track structure, engineering structures, and catenary on the modernized section of the Central Trunk Line were the test runs of the high-speed electric motor units of ED250 series. They were made by Instytut Kolejnictwa (Railway Research Institute) in the framework of tests necessary to obtain authorization to operate vehicles in Poland. During the tests, on 24 November 2013, the ED250 trainset reached the speed of 293 km/h.

Also very important were the tests of the ERTMS/ETCS system at the first section in Poland. The tests covered in particular the cooperation of ETCS track-side and on-board equipment and were conducted by the Railway Research Institute. According to the certificate issued by the Institute acting as notified body, the subsystem stays in accordance with the TSI and meets the safety requirements. On 21 November 2013, the president of the Rail Transport Office (Urząd Transportu Kolejowego) issued the approval for the structural subsystem: control – track-side ERTMS/ETCS level 1 equipment, installed on the CMK line, on the section from Grodzisk Mazowiecki to Zawiercie.

One of the conditions to increase the speed on the Central Trunk Line up to 200 km/h was also alignment of national legislation. Changes in the legal framework related to the implementation of ETCS included mainly an amendment to the regulation on general conditions of traffic and signaling signed on 11 April 2014 [25]:

- Chapter 2a of the regulation defines the rules for preparing of trains to run using ERTMS/ETCS system;
- Chapter 3a lays down the rules of train traffic and shunting with the use of ERTMS/ETCS system;
- Chapter 15a adjusts the principles of signaling, defining signal aspects and indicators necessary to operate the trains with the use of ERTMS/ETCS.

The new rules set by national legislation have been also adopted by internal regulations of PKP Polish Railway Lines. The Ir-1a instruction/manual has been developed, and was adopted by Ordinance of PKP PLK from 16 September 2014 [16].

In terms of railway track aspects, the regulation on technical conditions to be met by railway structures and their location [26], [27] has been revised. This regulation assured the consistency of national legislation with Community legislation on railway interoperability, especially with the new TSI INF [7].

2.4. Scheduled operation of high-speed trains in Poland – the first year of experience

The operation of train services operated with ED250 EMUs at a maximum speed of 200 km/h on the CMK line started from the first day of the annual timetable for the years 2014/2015, i.e. from 14 December 2014. The trains, branded as Express Intercity Premium (EIP), connect Warsaw with Cracow, Katowice, and Wroclaw. The following journey times were offered at that time:

- Warsaw – Cracow 2 hours 25 minutes;
- Warsaw – Katowice 2 hours 28 minutes;
- Warsaw – Wroclaw 3 hours 41 minutes.

In the 2014/2015 timetable, the highest value for commercial speed was achieved on the Warsaw Central – Cracow route, for which speed in both directions reached 121.4 km/h [21]. Despite continuation of modernization works, involving long-term track possessions on two sections of CMK (Szeligi – Biała Rawska and Strzałki – Idzikowice), the punctuality of EIP trains has proven very good. In the first year of operation, it reached 96%. In the 2015/2016 timetable, valid from 13 December 2015, the journey times of EIP trains were significantly reduced:

- Warsaw – Cracow route is covered in 2 hours 15 minutes (130.4 km/h),
- Warsaw – Katowice route is covered in 2 hours 19 minutes (128.5 km/h),
- Warsaw – Wroclaw route is covered in 3 hours 28 minutes (121.8 km/h).

It should be noted that the shortening of the journey time of trains using CMK was to some extent the result of a speed increase from 120 km/h to 160 km/h on the stretch incoming from Warsaw, between Warszawa Włochy and Grodzisk Mazowiecki (over a length of about 23 km) since December 2015. Its effect on travel time can be estimated as 3 minutes.

It should also be emphasized that the Central Trunk Line offers opportunities for further shortening of transit times; however, the conditions of railway traffic, and consequently also journey times achieved on this line in the next few years, will be influenced by investment processes being currently implemented or planned.

Taking into account the works already accomplished, the next section approved for the speed of 200 km/h has been the part of CMK between Grodzisk Mazowiecki and Idziko-wice. On this stretch, Korytów, Szeligi, and Strzałki stations have been rebuilt. On the whole length, the catenary has also been replaced.

The introduction of a speed of 200 km/h on the Grodzisk Mazowiecki – Idzikowice sec-tion took place in December 2017, which shortened the train journey time about 6 minutes (initially without shortening of overall journey times to Cracow, Katowice, and Wroclaw due to ongoing works at the Warsaw Railway Node). In coming years, a speed of 200 km/h should take effect on the whole of the CMK line (after the modernization works on the Idzikowice – Opoczno Południe – Olszamowice section), which will lead to a further shortening of travel time by about 4 minutes.

2.5. Conclusions

December 2014 saw the commencement of the commercial operation of passenger trains at a speed of 200 km/h on the Central Trunk Line. The period of preparation to increase the speed was disproportionately long and lasted for more than 20 years. One of the significant reasons was budgetary constraints in financing investment works on the CMK. However, several changes to the concept of modernization contributed as well to the delays.

In particular, debatable seems to be the decision from April 2008, according to which the line was to be modernized to a speed of V = 300 km/h and an axle load of 245 kN. It appears that such an assumption has led to the extension of the scope of the engineering works and, consequently, to the increase of their costs and greater difficulties in day-to-day traffic during the modernization. Very late, in 2009, the installation of ETCS began. Moreover, until 2014 there were no vehicles equipped with ETCS on-board systems, certified in Poland.

The introduction of a speed of 200 km/h on the southern section of the CMK can be regarded as the beginning of the creation of high-speed rail system in Poland. Foreign experience shows that such a system should evolve towards development in three directions:

*   Increasing the line speed to 200 km/h on subsequent parts on the Central Trunk Line (ultimately on the whole line) and on the upgraded Warsaw – Gdańsk railway;
*   increasing the speed on the CMK to more than 200 km/h, within the limits allowed by the modernized catenary power supply voltage 3 kV system (most likely to the speed of 230 km/h);
*   construction of the first section of the new high-speed line "Y" linking Warsaw with Poznań and Wrocław.

In this context, one can speak of an evolutionary approach to the development of an HSR system. This approach can be justified by the possibility of collecting experience in the field of the railway infrastructure maintenance in the train operation at ever higher speeds. In this respect, it is particularly important to provide good track quality along its whole lifecycle [2], [10], [17]. In addition, gradually increasing speed in the rail network will give all involved parties time necessary for the education and training of specialists.

Bibliography

[1] Bałuch, H., and Bałuch, M.: *Determinanty prędkości pociągów – układ geometryczny i wady toru* (Determinants of the train speed – track geometry and faults). Warszawa: Instytut Kolejnictwa, 2010.
[2] Bałuch, M., and Bałuch, H.: *Kształtowanie niezawodności nawierzchni w toku modernizacji linii kolejowych* (Shaping tarck reliability in the process of modernisation of railway lines). Problemy Kolejnictwa, 2014, nr 162, str. 7–20.
[3] Barron, I.: *High Speed Rail: Development Around the World.* Łódź, 29 stycznia 2008.
[4] Basiewicz, T., Łyżwam, J., and Modras, K.: *Centralna Magistrala Kolejowa Śląsk – Warszawa* (Central Trunk Line Silesia – Warsaw). Warszawa: WKiŁ, 1977.

[5] Birmann, F.: *Gleisbeanspruchung und Fahrzeuglauf bei Schnellfahrten.* Eisenbahntechnische Rundschau, 1965, nr 8, str. 335–351.

[6] Cejmer, J., and Massel, A.: Próby eksploatacyjne pociągu ETR 460 Pendolino na PKP – zagadnienia drogowe (Operational tests of ETR460 Pendolino trainset on PKP – track aspects). *Zeszyty Naukowe Politechniki Poznańskiej*, 1995, nr 41, str. 103–110.

[7] Commission Regulation (EU) No 1299/2014 of 18 November 2014 on the technical specifications for interoperability relating to the "Infrastructure" subsystem of the rail system in the European Union.

[8] Deutsche Bundesbahn: *Bundesbahndirektion.* Augsburg: Buchfahrplan. Heft 2A München – Augsburg – Treuchtlingen.

[9] Directive 2008/57/EC of the European Parliament and of the Council of 17 June 2008 on the interoperability of the rail system within the Community. OJ L 191, 18.7.2008, pp. 1–45.

[10] EN 13848–5. Edition: 2010–10–01. Railway applications — Track — Track geometry quality. Part 5: Geometric quality levels — Plain line.

[11] Europas schnellster Zug – die Sensation der IVA. Die Bundesbahn 1965, nr 15, str. 534–536.

[12] Further Development of the European High Speed Rail Network. System Economic Evaluation of Development Options. Civity. Paris, Hamburg, December, 2013.

[13] Gacka, J., and Schaefer, P.: *Centralna Magistrala Kolejowa – ważny element VI europejskiego korytarza transportowego* (Central Trunk Line – important element of VI Paneuropean Transport Corridor). Linie dużych prędkości na PKP – Centralna Magistrala Kolejowa. Materiały konferencyjne, str. 9–18. Kielce, 2004.

[14] Gielnik, J.P, and Rusak, R.: *Elektrowozy serii CC 6500 i CC 21000 kolei SNCF (2)* (Electric locomotives CC65 and CC21000 of SNCF). Świat kolei, 2014, nr 10, str. 30–35.

[15] Hardmeier, W., and Schneider, A.: *Direttissima Italien. Orell Füssli.* Zurich, Wiesbaden, 1989, str. 59.

[16] *Instrukcja o prowadzeniu ruchu pociągów z wykorzystaniem systemu ERTMS/ETCS poziomu 1 Ir-1a* (Instruction on railway traffic using ERTMS/ETCS level 1). PKP Polskie Linie Kolejowe, Warszawa, 2014.

[17] Kędra, Z.: *Ocena poziomu jakości geometrii toru kolejowego* (Assesment of quality of track geometry). Technika Transportu Szynowego, 2012, nr 2–3, str. 48–51.

[18] Korab, D.: *Rozjazdy kolejowe do dużych prędkości. Wybrane wymagania dla interoperacyjności oraz przegląd zastosowanych niektórych rozwiązań technicznych* (Railway turnouts for high speed. Selected requirements for interoperability and review of adopted technical solutions). Linie dużych prędkości na PKP – Centralna Magistrala Kolejowa. Materiały konferencyjne, str. 123–133. Kielce, 2004.

[19] Massel, A.: *Centralna Magistrala Kolejowa – 30 lat eksploatacji* (Central Trunk Line – 30 years of operation). Technika Transportu Szynowego, 2004, nr 10, str. 20–24.

[20] Massel, A.: *Modernizacja Centralnej Magistrali Kolejowej – zagadnienia ruchowe* (Modernisation of Central Trunk Line – operational aspects). Modernizacja południowej części międzynarodowego korytarza transportowego E65. Materiały konferencyjne str. 21–28. Podlesice 15–17 października, 2008.

[21] Massel, A.: *Przyspieszenie ruchu pasażerskiego w Polsce* (Acceleration of train traffic in Poland). Technika Transportu Szynowego, 2015, nr 1–2, str. 25–32.

[22] Oczykowski, A.: *Badania i rozwój przytwierdzenia sprężystego SB* (Tests and development of SB elastic fastening). Problemy Kolejnictwa, 2010. Zeszyt 150, str. 121–156.

[23] Pawlik, M., and Wojsław, Z.: *E65 – Południe – stan aktualny, założenia modernizacji i budowy nowych odcinków linii* (E65 southern section – present state, principles of modernisation and construction of new sections). Konferencja naukowo-techniczna. Modernizacja południowej części międzynarodowego korytarza transportowego E65, Materiały konferencyjne str. 7–20. Podlesice 15–17 października, 2008.

[24] Prokudin, I.W., Graczew, I.A., and Kołos, A.F.: *Organizacija pierieustrojstwa żeleznych dorog pod skorostnoje dwiżenie pojezdow* (Organisation of reconstruction of railways for higher speed train traffic). Marszrut. Moskwa, 2005.

[25] Rozporządzenie Ministra Infrastruktury i Rozwoju z dnia 11 kwietnia 2014 roku zmieniające rozporządzenie w sprawie ogólnych warunków prowadzenia ruchu i sygnalizacji (Regulation of Minister of Infrastructure and Development changing the regulation on general conditions of railway traffic and signalling). Dz.U. 2014, poz. 517.

[26] Rozporządzenie Ministra Infrastruktury i Rozwoju z dnia 5 czerwca 2014 roku zmieniające rozporządzenie w sprawie warunków technicznych, jakim powinny odpowiadać budowle kolejowe i ich usytuowanie (Regulation of Minister of Infrastructure and Development changing the regulation on technical conditions for railway structures and their location). Dz.U. 2014, poz. 867.

[27] Rozporządzenie Ministra Transportu i Gospodarki Morskiej w sprawie warunków technicznych, jakim powinny odpowiadać budowle kolejowe i ich usytuowanie (Regulation of Minister of Transport and Maritime Economy on technical conditions for railway strucures and their location). Dz.U. 1998, nr 151, poz. 987.

[28] Vuillet, G.: Comparative Speeds. *The Railway Gazette*. 1 września 1967, str. 648.

[29] 1976: New Train Speeds into Service. *BBC News Online*, 4 October 1976.

Chapter 3

The view on socioeconomic aspects of high-speed system

Agata Pomykała

3.1. Introduction

The year 2014 was a jubilee year for high-speed rails. Fifty years earlier in Japan, the first high-speed (HS) trains in the world had been launched. Following that example, European railways accelerated works in order to construct similar systems. The last years of the 20th century saw intensive growth of these systems throughout the world. Simultaneously, research involving the estimation of influence of this means of transport on the environment, national market, and regional development was carried out. Decades of high-speed rail systems' development in the world allowed conclusions to be drawn concerning the impact of the systems on the quality of social life and economic development and specifying their advantages and disadvantages. The experience of countries with the longest history of the systems' functioning contributed to their increasing popularity and, as a result, led to many countries, despite an economic crisis, deciding to realize high-speed projects.

Investments in rolling stock and infrastructure appliances led to higher and higher efficiency and environmental sustainability for this means of transport. These phenomena are of crucial importance in the context of global challenges connected with opposing climate changes and seeking oil-independent sources of energy. In 2012, high-speed passenger transport constituted 26% of passenger rail transport in the European Union, with almost 60% in France, 50% in Spain, and over 27% in Germany [5]. According to the UIC (International Railway Union, *fr. Union Internationale des Chemins de fer*) data presented at the 9th World Congress on High Speed Rail in Tokyo, the HS network is currently seeing extraordinary development. There are 29,792 km of high-speed lines in exploitation (as of 1st April 2015) and, moreover, the lines' length has increased by 12,245 km in the last three years. [28]

3.2. Beginning of high-speed rail

Already at the turn of 1950s and 1960s in many countries with a rapidly increasing motorization rate, railway stopped serving the transport needs of the society, due to, among other factors, social changes, economic development, and the increasing value of time. Temporary activities connected with improving the value of already existing infrastructure, a slight shortening of the journey time, and increasing the frequency of trains did not solve intensified problems, which, in turn, led to a less and less attractive rail offer and a loss of shares in the market. If in the 1950s and 1960s problems were mainly linked with insufficient capacity ability of the lines, in the 1970s and 1980s it was clearly visible that transportation, especially passenger transportation at medium and long distances, was not competitive in relation to motor or air transport. Rail companies' deteriorating situation, the increasing loss of market, and unpromising prognoses without explicit changes in the offer (travel time!) were the reason for building high-speed lines (HSL).

The most frequent arguments for making the decision to build high-speed lines were:

- improving the capacity of the existing lines and aiming to prevent transport "bottle-necks" through transferring onto new high-speed passenger lines;
- striving to restore competitiveness of railway transport, thanks to significant shortening of the travelling time;
- stimulating regional development;
- aiming to gain independence from oil imports and oil's increasing prices;
- limiting the negative impact of transport on the natural environment.

In the following years, the willingness to establish a trans-European network became significant as it would contribute to fulfilling the assumptions of creating a common economic area, and, especially, a free movement of passengers and foods; it would also support the intensification of cultural, economic, and political cooperation between the linked countries. The first high-speed international project was the construction of HSL North (LGV Nord, *fr. ligne à grande vitesse*) line from Paris to Lille and further to London and Brussels.

The establishment and development of high-speed rail systems not only in Europe but also throughout the world let us notice the convergence of conditions and expectations. The UIC, as a unit federating rail enterprises and supporting the sector with its activities, has been analyzing the operations in high-speed rail sector for years. The analyses and research enable better understanding of the working mechanisms, and the long-standing observations of the market have made it possible to draw conclusions. One of the latest studies presenting the history and conditions of the first decisions on constructing high-speed rails is the study titled *UIC Study on Origin and Financing of First High-Speed Lines in the World* by professor Andrés López Pita [22], whose text was used as a basis for presenting the following data.

3.2.1. Japan

The dramatic situation of post-war Japan led to a number of initiatives undertaken by the Japanese government in the 1950s and 1960s, which were characterized by big social projects aimed at improving the situation of inhabitants and striving for rebuilding the national economy. The main objectives of the projects were economic growth and fighting against unemployment. Building infrastructure became not only a necessity resulting from post-war devastations but also a way to create a firm economy and permanent development that would improve Japan's standard of life. Japan's phenomenal pace of growth is proof that the action undertaken was absolutely right. One of the elements of the rebuilding program was the realization of the Shinkansen rail network project. The necessity of building a new rail infrastructure resulted from Japan's growing population, its fast urbanization, and the concentration of economic activity in the Tokyo – Osaka corridor, which had been one of the main transport corridors of the country since the 17th century. Plans to organize the Olympics in Tokyo in 1964 accelerated the building of the first Shinkansen line.

In the 1950s, the line linking Tokyo and Osaka was the main rail line, and the rails were 1067 mm wide and 550 km long. In the middle of 1950, 22% of the whole passenger flow was realized through it, 34% of the population of Japan lived in the area along its line, and 60% of industry production was concentrated there. Every day, 130–190 trains ran through there, and the line capacity was nearly exhausted. There also existed a commercial problem with the availability of tickets, which were purchased three months in advance. A few-year long analyses and considerations involving the possible ways of solving the problem came to an end in December 1958 when the decision to build lines with high-speed parameters was reached. Works began in 1959, and the line was put in use on 1 October 1964. Adopting such a decision enabled the development of innovative branches of economy and became a stimulus to the growth of Japanese industry. The investment was not only a commercial success but also a financial one (in the third year of exploitation, income already exceeded incurred costs). The construction of consecutive lines was introduced in 1967 and 1970. Constructing and exploiting high-speed lines in the 1960s coincided with an annual 10% economic growth.

3.2.2. France

The 1970s in France was a decade of dramatic changes. After significant growth of passenger numbers in 1961–1974/75 (46% altogether in 1st and 2nd class), the second half of the 1970s saw a noticeable decrease of passenger numbers in 1st class, due to air travel which equaled 24% in 1975–1982.

On the Paris – Lyon line in 1964–1967, within 4 years 17% of market share was lost and the estimates for 1974 showed a further loss of 23% (despite increasing the speed and shortening the journey time by 15 minutes). In 1963–1976, railways decreased market shares 2.6 times, while the number of air journeys multiplied by five.

Simultaneously, the capacity of Paris – Lyon line was on the verge of exhaustion; on the St. Florentin section, approximately 250 trains ran daily.

The analyses conducted to compare the concerned variants of improving the situation (Table 3.1) showed significant predominance of the variant with constructing a new line. It was the best option in the context of benefits from shortening the journey time and refunds from investments, not only for SNCF (French National Railway Company, fr. Société nationale des chemins de fer français) but also for the national budget, at the same time taking into account the questions connected with energy consumption. Considerable domination of rail transport in this scope was of crucial importance for making final decisions.

As the new line embraced almost 40% of inhabitants in France within its scope, assuring the demand on this section did not constitute a problem.

Table 3.1 Comparison of investment variants on the Paris – Lyon line

Indicator	Duplication of St. Florentin – Dijon section (109 km)	New TGV line (409 km)
Necessary investment (MF1975)	1200	2600
– Infrastructure	1000	1550
– Material	–	340
Journey time	3 h 30 min	2 h
Return to SNCF	reduced	18%
Return to the community	very low	33%

Source: [22]

3.2.3. Germany

The post-war years in Germany were the time of rebuilding destroyed industry and infrastructure.

In 1952–1974, intensive development of rail transport took place. Growth by 49% was observed between Cologne and Frankfurt (length 219 km, journey time 2 h 15 min), 589 trains ran daily (in 1952–395) with an average speed of 97 km/h. Similarly, between Fulda and Flieden, in 1950–1974 the number of trains increased from 150 to 300. Heavy traffic meant the decline of punctuality and decrease of trade speeds; thus, it led to the deterioration of service quality, of both passenger and goods trains.

The same problems were visible on the Hannover – Würzburg and Mannheim – Stuttgart lines. It was clear that the only solution was the construction of new rail lines on the routes. The plans of building the line assumed a novel connection of satisfying the demand for passenger journeys (mainly during a day) with freight transport (mainly at night).

The unification of Germany in 1990 made Berlin an important destination for people travelling from both eastern and western parts of Germany. Constructing new rail lines with high technical parameters was one of the key elements of the program of uniting Germany. Considerable financial resources were assigned for this objective. Even now, works in the Berlin – Leipzig – Nuremberg – Munich corridor are being finalized. Presently, a high-speed rail network consisting of both new and modernized lines connects the majority of urban centers in Germany.

3.2.4. Italy

Regardless of the modernization of infrastructure undertaken in the post-war period, the market situation of Italian railways in the 1950s was deteriorating visibly. Already in 1950–1955, road transport doubled. In 1955–1970, the share of railways in the transport market at medium and long distances decreased from 38% to 18%, and at the same time the share of road transport increased from 61% to 80%.

The main transport corridor of Milan – Naples (5.3% of rail network) concentrated in the 1970s over 30% of rail transport, passenger and freight altogether. Landscape features made the Rome – Florence section extend the distance between the two cities in a straight line by 35%, and the speed at almost 60% of the route could only reach 105 km/h. Daily transport needs exceeded 96,000 tons of products, which meant launching over 200 trains (freight and passenger) a day. Improving this section of the Milan – Naples corridor in the first place became unquestionable. The accepted solution, consisting of constructing sections of the new line with high-speed parameters (250 km/h) between the points of interfaces with the old line, allowed taking advantage of both systems. Although the construction was begun in 1970, due to technical and economic difficulties, launching the line for use was postponed and divided into phases; the line has been fully exploited since 26 May 1992. As a result of the new offer of travelling between Rome and Florence within 82 minutes, the rate in rail market increased considerably (59%) in relation to road transport (34%).

3.2.5. Spain

The Spanish railway network, even in the 1970s, was poorly developed and consisted mainly of non-electrified lines with low technical parameters. Even if it satisfied the needs of the poorly developed country that Spain used to be before entering the European Union, it was clear that, with regard to fast growth, estimated due to accession, reorganization of the railway system and including it in the European transport system would be essential. Such a change required solving the problem of Spanish gauge (different from the European). Additionally, a better transportation connection of distant regions of Spain was a condition necessary for preserving the country's cohesion and economic development. One of the most important circumstances that accelerated the construction of the first high-speed Madrid – Seville line was a planned organization of the World Exhibition EXPO in 1992 in Seville. Entering the European Union in 1986 gave Spain an impulse and means for development.

It was especially visible with regard to Andalusia. Rail passenger transport in the Madrid – Seville connection had 20% and Madrid – Barcelona only 11% of market share (in the railway–air transport division). Within 1988 and 1991, the share in the Madrid – Seville connection decreased by 12.5%. The decision to build a high-speed line which would shorten the distance between Madrid and Barcelona to 621 km enabled the reduction of the journey time from 6 h to 2 h 50 min on a new Madrid – Seville line (471 km), and the market reacted with an increased demand, causing the growth of rail share from 22.5% to 51.6% already in the first month of exploitation, and after a year it equaled 81.7% in the train–plane line (121,000 passengers in 1993).

Taking the other means of transport into consideration, it can be assumed that the transport rate also significantly improved in relation to 1991, which is presented in Figure 3.1.

In the following years, further high-speed lines were launched and they connected Madrid with other regions, Valencia (2010) and Catalonia (2009). Currently, projects creating

Table 3.2 Modal split, comparison 1991 and 1993

Means of transport	1991	1993
Airplane	18%	7%
Bus & coach	11%	9%
Car	51%	39%
Train	20%	45%

Source: self-study, based on [22]

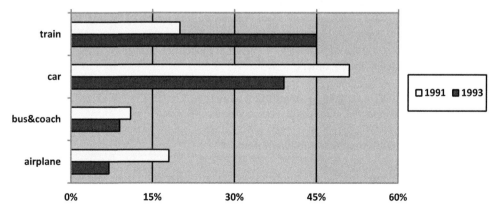

Figure 3.1 Division of market for the Madrid – Seville journey, comparing 1991 and 1993
Source: self-study, based on [22]

high-speed connections with the Basque Country and towards Portugal are being realized. In 2009, the first international service of Figueras – Perpignan was introduced, which linked Spain with France.

After the completion of the accomplished projects, in accordance with the approved assumptions, 80% of the inhabitants of the country will have access to the modern rail network, with a distance to the nearest railway station within 60 minutes.

3.3. High-speed rail as a means of local development

Transport, as an important factor determining efficient flows of resources and foods, has a significant influence on the national economy and the quality of social life. In developing countries, high-speed rail (HSR) systems are taken into consideration not only in terms of transport as a way of improving passenger carriage and assuring services between regions but also as a stimulator of economic growth, both at national and regional levels. More and more frequently, the ways of influence quantification/interaction of transport investment are sought [21].

High-speed rail generates social benefits, which come from time savings, an increase in reliability comfort and safety, and a reduction of congestion and accidents in alternative modes. Releasing capacity in the conventional network, which can be used for freight transport, is an additional benefit of the investment in the construction of new lines. Obviously, building new lines is expensive, as is their exploitation; thus, in making investment decisions one should in each case answer the question of whether the benefits outweigh the necessary expenses [7].

Taking into account in analyses the wider influence of high-speed rail on the economy (which results from, for example, the existence of agglomeration effects, i.e. benefits connected with the concentration of resources, professional skills, talents, and their easy accessibility, as well as business support, growth of effectiveness, changes in special arrangement of business activity, etc.) increases the scale of interaction of this means of transport on the socioeconomic environment by 12–25% [11].

Development of regions, determined by different market processes, is both an element conditioning the achievement of long-term aims of strategic plans of countries' development and a result of intended political operations aiming at its stimulation.

The years of functioning of high-speed rail systems allow to point out, on examples of the countries where they were created, the areas of their interaction and to differentiate effects.

Thus, investing the socioeconomic background, the effects can be divided into direct, indirect, and widely treated/induced ones [11].

During the preparation of lines and soon after launching carriages, direct effects are noticeable and include, among others:

- employment due to conducting investment works;
- time savings connected with a short journey time;
- improvement of safety;
- increase of infallibility and journey comfort.

Indirect effects are those which can be observed only after some time and whose importance is growing over time. To the group belong:

- changes in population mobilities;
- increasing the capacity for other means of transport and improvement of safety resulting from it.

The third group consists of induced effects connected with the widest approach to the results that influence the environment. The effects are:

* influence on location of enterprises;
* generating urban changes;
* changes of property values situated in the vicinity of HSR;
* development of additional infrastructure for HSR;
* extension of a network of other branches of transport;
* influence on natural environment;
* influence on touristic attractiveness;
* influence on economic growth of the country and regions.

The analyses carried out within the UIC works show benefits for cities and regions resulting from their inclusion into the HSR system. The comparison assessment of cities influenced by the high-speed system and those deprived of its impact, conducted by Deutsche Bahn (DB)' and commissioned by the UIC, clearly indicates a positive influence of HSR on the development of not only agglomerations but also smaller towns. This can be especially observed in the field of economy and tourism, in the development of economic structure, and the improvement of the brand connected with creating the image of a modern, innovative place that is tourist and entrepreneur friendly. The changes are the most visible in areas surrounding railway stations, which, thanks to investments, gain in value and improvement of their usage.

A shorter journey time and lower costs – emerging from the inclusion of the region in the high-speed rail network – are the starting point for additional development of towns and regions [10].

The observation of the Spanish market indicates mainly time savings and cost reduction of running business as social benefits. It is also noticeable that high-speed rail influences positively the budgets of the communities located in the vicinity of HSR, while the highest impact on local income, resulting from business activity growth, is visible in communities situated within 5 km from the station [9].

Launching the first high-speed service, Spanish AVE (*Sp. Alta Velocidad Española*) had a significant influence on smaller towns, located along the line, in the vicinity of the station. In Leida – a town situated between Barcelona and Madrid, since 2006 the number of tourists has risen by 15%, and moreover offices of information and communications technology (ICT) companies have been created. Ciudad Real lying on the Madrid – Seville line significantly profited from better transport access of the labor market in Madrid, for which the town serves as an accommodation base. Furthermore, a university center expanded in the town, and it was treated as a supply base for arising technological companies. A similar situation occurred in Castilla – La Mancha, where high access to qualified scientific staff and students facilitated fast development of the university [11].

The French city Lille situated on the crossroads of the line connecting Paris and Brussels with London constitutes another positive example. The Lille-Europe station is a center of economic activity offering office and retail spaces, apartments, and touristic and entertainment attractions.

In Lyon, within 10 years of launching the TGV line (1981) around the newly built station, a whole economic center developed: the quantity of office spaces increased by 43% and 60% of investments in the town were located in its vicinity.

Also in other towns in which high-speed rail stations appeared in the following years (Le Mans, Nantes, Vendome), attractiveness of the land surrounding the stations increased and the average renting prices of premises were approximately 20% higher than in other parts of the towns. The process of transferring locations of companies from Paris to towns remote from Paris but well connected thanks to TGV lines is also a noticeable phenomenon.

The impact on tourism development is clearly visible as well – the growth of accommodation facilities in Nantes by approximately 43% [11].

In Dutch towns, a significant increase in attractiveness and value of areas surrounding the stations has been observed. In Hague appeared about 80,000 m² of office spaces, 15,000 m² of retail spaces, 565 apartments, and 1,300 parking spots [11].

Furthermore, the example of German Montabaur (12,500 inhabitants) and Limburg (34,000), two small towns located on the high-speed Cologne – Frankfurt line, in which emerged stations servicing high-speed trains, indicates a positive effect of this investment on regional development. In areas adhering to the stations, both GDP and employment levels within 4 years since its launch have increased by approximately 2.7%. What is more, every 1% of growth of market availability caused 0.25% economic growth. Undoubtedly, the cost reduction in this case results only from the passenger flow through improving business relations, contacts with clients, and employment conditions, as the line is not used for freight transport [1].

Changes that took place in China resulting from intensive development of a high-speed rail system in the first decade of the 21st century show that the system influenced regions in a significant way. Research involving the impact of transport investment on regional economy pointed out positive relations between the level of public investment and the pace of accumulation of private capital, employment, and production growth. High-speed rail visibly stimulates the regional economy even in a short time, and its influence led to a 38% growth of GDP in towns within 2 years. The fears about "the tunnel effect" in relation to the influence of HSR did not find confirmation – just the opposite – in the example of the Wuhan – Guangzhou corridor, in which its dispersive impact was observed. In almost all regional towns, economic growth affected the GDP [16].

Other countries' experiences confirm the effects on the regional economy of China with regard to economic growth, industrial structure, and optimization of spatial structure of cities, as well as improvement of regional cooperation. Opinions about the dynamic influence of HSR on regions situated within the scope of a rail line, on the acceleration of industrialism and efficient flow of capital and technology between regions have been acknowledged [19].

Furthermore, analyses have indicated that HSR development influences positively the growth of industry. In the case of China, an impulse for research development and initiation of new technologies was of key importance. Innovations and technologies used during the creation of the high-speed rail system have been used in other branches of the economy as well. Together with the growth of the HSR system, its influence on industry in the region is increasing. Simultaneously, a phenomenon of improving economic connections between regions can be observed and, therefore, the stimulation effects for industry in one region are able to transfer quickly to other regions. This is illustrated in Figure 3.2 [19].

It should be noticed that when making the decision to launch the high-speed rail program, China did not dispose of relevant knowledge in the sphere of indispensable technologies. Therefore, taking advantage of other countries' experience was assumed in the first phase. In the next phase, the Chinese industry was supposed to gain essential competences through acquiring new technologies. Currently, it is already an exporter of appliances and rolling stock for high-speed rail. Other countries followed a similar path, especially Spain,

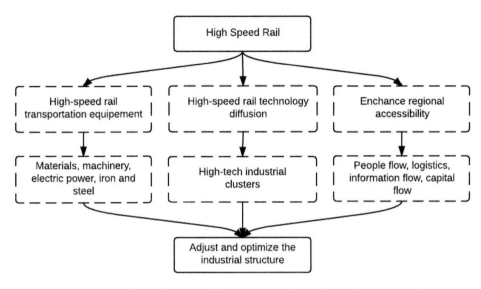

Figure 3.2 Mechanism of the influence of high-speed rail on industrial structure development
Source: [19]

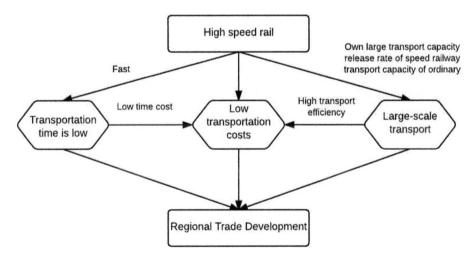

Figure 3.3 The mechanism of high-speed rail influence on regional exchange
Source: [19]

which already in the 1990s, constructing its first HSR, almost entirely made use of French and German experiences and presently is a leader in this field.

HSR development as a means of transport which enables fast movement of people activates the strengthening of interregional connections. The more regions participate in formulating and implementing new technologies and their exploitation, the faster the possibility of using the so-called scale effect rises. This mechanism is presented in the Figure 3.3.

3.4. Socioeconomic features

Taking into consideration the special role of transport as a factor of socioeconomic development and its role in creating the competitiveness of both individual regions and in the whole European Union in the middle of the 1990s, action was taken in order to elaborate a coherent concept of the TEN-T network, which was supposed to ensure efficient functioning of interior market, economic, social, and territorial coherence and better availability in the entire EU [4]

The works on the integration of transport systems and the liquidation of barriers constituting hindrances that followed European historic, political, and geographical conditions became the result of long-standing works on fulfilling the assumptions of the main EU freedoms: freedom of flow of services, goods, people, and capital within the federated countries. The amendment of Trans-European Transport Networks (TEN-T) in 2013 is one of the most recent indications of the European Community's interest in the issue [12]

The assumptions concerning the creation of a common transport area of the European Union made the UIC, while observing the development of HSR systems in Western Europe, pay attention to deficiencies of the HSR network in the countries of Central Europe and commission analyses regarding this market. The study from 2006 presenting the possibilities of high-speed rail development on the PECO (fr. Pays d'Europe Centrale et Orientale) area showed the potential and the needs of Poland and defined it as the most likely of the countries of East and Central Europe to build and launch the system quickly [17]. The current socioeconomic data confirm the potential and needs in this aspect. Specific data concerning the areas of NUTS2 and NUTS3, which are directly influenced by the planned high-speed line, are presented in section 3.4.2, and their basic demographic and economic rates can be seen in Tables 3.5 and 3.6.

Due to the Central Trunk Line (CMK), connecting the south of Poland with Warsaw, partly adjusted to high-speed line parameters and in view of the exploitation of the first high-speed trains [18] on the line since December 2014, the areas affected by the lines should be included in the analyses concerning HSR in Poland. Thus, not only voivodeships, cities, and regions connected with Y line (Warsaw – Łódź – Poznan/Wroclaw) will be included in the direct impact of HSR system, but also these connected with CMK line (Figure 3.4).

3.4.1. Agglomerations

Big cities together with adjacent areas play a significant role in the economic and social life. The biggest cities are regarded as the basic links of the global economy and are defined as locomotive engines of national economies. An over-proportionally large number of economic activities is concentrated in metropolitan areas, whereas the intensity of economic problems connected with, for example, unemployment is considerably smaller [40]. According to The National Strategy of Regional Development 2010–2020:

> in current development conditions the biggest urban centres and their functional areas are the motors of development of a country and particular regions. The centres which constitute junctions of contemporary socio-economic processes are of the highest importance – they can produce and attract the best human resources, investments in sectors ensuring the biggest productivity and they are also able to create innovations and join in the network of cooperation with other centres in international and national arrangements in order to increase their complementarity and specialisation and, at the same time, they can make use of agglomeration benefits more fully [15].

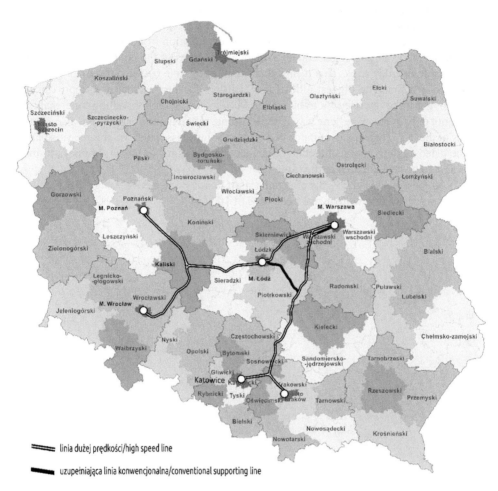

Figure 3.4 NUTS2 and NUTS3 with underlined main cities in service of HSR

Elaborated by Tomasz Bużałek based on [8]

The role of agglomerations and their development are based on mutual economic and social connections. In Polish conditions we face a very profitable polycentric arrangement of big cities. They consist of cities of medium size on the European scale (Kraków, Wrocław, Poznań, Łódź), a big city but not a dominant one – Warsaw, the Tri-City agglomeration (Gdańsk, Gdynia, Sopot), and the Silesia conurbation (a group of 14 cities in the south of Poland). Due to the fact that in the time of economic development and the creation of economic centers Poland did not have nationhood and its area was divided into three separate political units, sufficient transport connections have not been established between them. In order to increase the cohesion and development of the country, this deficiency should be eliminated as soon as possible. Further economic growth of Poland requires using the effect of synergy between the main Polish urban centers (See: Fig.3.5). The distances between them (Table 3.8) indicate that the operations of transportation connections can be successfully realized by land transport. Nowadays it is believed that rail transport is the most efficient means of transport in passenger carriage operations, provided it can reach average trade speeds of 200–250 km/h. It is possible to accomplish the speed through the exploitation of a high-speed rail system. Efficient operation of passenger connections between cities in the contemporary economy should enable one-day trips. This means that the journey time between destinations should not exceed 3 hours. The biggest cities for which the distance oscillates at approximately 300 km ought to be operated by trains within 90 minutes. Such a travelling time enables a journey not only in the cycle "one way in the morning, return in the evening" but also in other periods of the day. This creates suitable conditions for business travel. It should be also noted that the listed agglomerations function in connection with smaller towns. Journeys to these towns should be standardized in the same way.

It is not possible to achieve the effect of a short journey time between big cities by means of the motorway system with an average speed of 100 km/h.

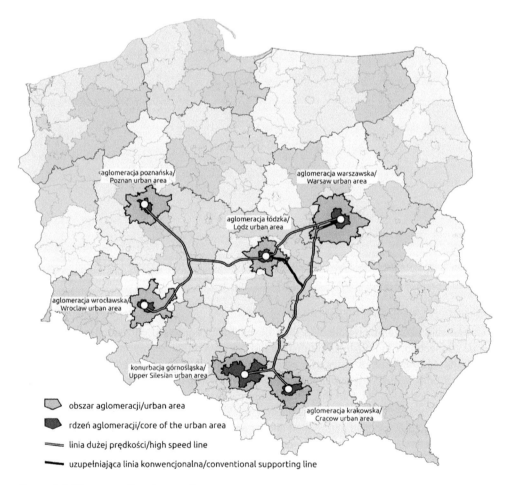

Figure 3.5 The area of agglomerations

Elaborated by Tomasz Bużałek based on [8]

Although the first Polish proposed high-speed line will connect only four agglomerations, Kraków and Katowice should also be included in the analyses concerning the Polish HSR system, because the Central Trunk Line has been adjusted to high-speed line parameters [18] and both lines have been connected by the Łódź – Opoczno line.

Thus, in the first part of the elaboration, agglomerations: Warsaw, Wrocław, Łódź, Kraków, and Silesia conurbation have been taken into account and indicated according to methodology based on functional connections of the main urban center and its surrounding areas [41].

Adjusting CMK to HSL parameters and the established line route allotted in the *Preliminary Feasibility Study for High-speed Line Warsaw – Łódź – Poznań/Wrocław* leads to the fact that the created HSL system will use the potential of six out of seven biggest Polish agglomerations embracing 24.1% of the population and 5.6% of the area of Poland.

Table 3.3 The population within areas around Polish agglomerations, 2013

Agglomeration	Numbers of gminas [communes]			Area [km²]			Number of registered inhabitants [thous.]		
	total	core	external zone	total	core	external zone	total	core	external zone
Warsaw	50	1	49	3817	517	3300	2827	1724	1103
Łódź	19	1	18	1828	293	1535	1027	711	316
Poznań	21	1	20	2500	262	2237	922	548	374
Wrocław	15	1	14	2239	293	1946	875	632	243
Kraków	23	1	22	1992	327	1664	1184	759	425
Katowice*	46	14	32	2979	1218	1761	2451	1905	546

* Silesia conurbation/Slaska conurbation (including towns: Bytom, Chorzów, D browa Górnicza, Jaworzno, Gliwice, Katowice, Mysłowice, Piekary Śląskie, Ruda Śląska, Siemianowice Śląskie, Sosnowiec, Świętochłowice, Tychy, Zabrze)

Source: self-study based on [41 Table 3.2]

Table 3.4 Chosen socioeconomic indicators for the areas of agglomerations, 31.12.2013

Agglomeration	Registered unemployment * [%]	Income of communes' budget per capita [EUR]	Flats commissioned in newly built buildings [%]	Accommodation in hotels [%]	Length of cycle tracks per 10,000 km² [km]	Children under nursery care [%]	Children under preschool care [%]
Poland	8.8	903.45	100	100	247.1	4.8	74.1
Warsaw	6.1	890.19	14.4	10.4	1477.9	21.3	82.9
Poznań	3.6	833.44	4.5	3.4	1086.8	23.0	79.3
Wrocław	5.6	815.96	7.2	3.2	1098.2	13.4	70.6
Łódź	8.8	759.16	1.9	2.8	748.9	10.6	67.8
Kraków	6.1	737.16	6.4	7.6	766.6	9.2	72.5
Katowice*	6.3	868.45	3.1	3.9	937.2	5.3	76.5

* rate of the registered unemployed in the population number in their productive age: women aged 15–59, men aged 15–64 1€= 4.1472 PLN according to NBP (National Polish Bank) from 31.12.2013

Source: self-study based on CSO (Central Statistical Office) [8]

The agglomerations which have been included into high-speed rail are characterized by a relatively high level of economic development and the wealth of inhabitants. Additionally, their infrastructure for visitors coming for business and touristic reasons is satisfying. In terms of accommodation, the agglomerations constitute 1/3 of the potential of Poland. Over 37% of newly built flats are located in their area. Unemployment in agglomerations is generally significantly lower than the national average, but (for example) in Poznań agglomeration it is lower by almost 60%. The presented indicators show that the agglomerations have a big market potential for high-speed carriages.

Good transportation of agglomerations by high-speed rail network is of crucial importance for economic development and transport service of regions. Well-shaped transportation junctions are one of the characteristic features of agglomerations. The distribution of not only interregional but also intraregional journeys to other important towns of the region, as well as local ones that ensure everyday access, is realized there. In Polish conditions, transportation junctions of some agglomerations have not been adjusted to their new role yet. Efforts undertaken by local authorities led to the fact that more and more agglomeration centers ensure intermodality in transport and fulfill the expectations of passengers, considering the accessibility of different means of transport.

The quantity of multi-agglomerative labor market constitutes additional development potential of public transport and is connected with the level of transport accessibility. In Poland, due to long journey times between main cities, its quantity is still insufficient. Nevertheless, this situation may change as a result of meaningful shortening of travelling time, as the shorter the journey time, the higher the attractiveness of rail transport.

The upper limit of attractiveness is a 1.5-hour journey in one direction, which currently occurs only between Łódź and Warsaw. After constructing the high-speed line, the travelling time will be shortened by 35 minutes and between the remaining agglomerations up to under 1.5 h, except for some relations in which the travelling time will be shortened only up to 2.5 hours (Table 3.8).

According to the research on commuting conducted in 2011 by the National Census [3], in 2011 in Poland 3,130,600 employed people commuted to work, which amounted to 32.5% of their total number. The quantity and intensity of commuting indicated meaningful spatial differentiation. Voivodeship cities play a significant role in forming circular migration connected with employment. Below, information involving the main directions of commuting in the analyzed area is presented.

Of people living outside Warsaw, the capital of Poland, 276,366 came to the city to work in 2011. Employers outside the Mazowieckie Voivodeship (130,458) constituted 47.2% of the total number of people coming to work in Warsaw. Inhabitants of the Śląskie Voivodeship were the most numerous (18,200), then the Łódzkie Voivodeship (16,700) and the Małopolskie Voivodeship (14,100). From the capital, 26,299 people went to work in other territorial units. Employed people living in Warsaw and going to work outside the Mazowieckie Voivodeship constituted 32.3% of the total number of people leaving the city. Workplace of the most numerous group of people was in Kraków – 0.20% of the number of employed people living in Warsaw, and then respectively in Wrocław (0.13%), Gdańsk (0.12%), and Poznań (0.09%).

To Łódź, the capital of the Łódzkie Voivodeship, in 2011 came 49,206 inhabitants of other communes. Employed workers living in the area of other voivodeships (9396) constituted 19.1% of the total number of people coming to work to Łódź. Most of them came from the Mazowieckie Voivodeship (1600 people), the Wielkopolskie Voivodeship (1300 people), and the Małopolskie Voivodeship (1000).

Of employed people living in Łódź, 19,407 left the capital of the voivodeship in 2011. Most of them went to Warsaw (2.38%), Poznań (0.22%), Kraków (0.21%), and Wrocław (0.17%).

Of employed people living in other territorial units, 90,755 came to work in Poznań in 2011.

Workers living outside the Wielkopolskie Voivodeship (17,244 people) constituted 19.0% of the total number of people coming to work to Poznań. Most often came inhabitants of the Kujawsko-Pomorskie Voivodeship (2600), Lubuskie Voivodeship (2400) and the Dolnośląskie Voivodeship (1900). Most workers coming to work in Poznań lived in the Wielkopolskie Voivodeship (81.0%). The highest percentage lived in the communes of Poznań district: Czerwonak (rural commune), Luboń (urban commune), Swarzędz (urban commune), Kostrzyn (urban part), Suchy Las (rural commune), Kleszczew (rural commune), and Rokietnica (rural commune). Over 30% of employed people living in communes came to work in Poznań.

The most numerous group of employed people living in Poznań went to Warsaw (2.01%). The following cities took the next places: Wrocław (0.26%), Szczecin (0.17%), and Kraków (0.16%).

In 2011, 65,219 people from other communes of the country came to work in Wrocław – the capital of the Dolnośląskie Voivodeship, mainly from the Opolskie Voivodeship (3100 people), the Śląskie Voivodeship (3000) and the Wielkopolskie Voivodeship (2700). Distinctly, most workers coming to work to Wrocław lived in the Dolnośląskie Voivodeship (70.2%). Over 30% of employed people living there came from the neighboring communes: Siechnice (rural area), Długołęka (rural commune), Wisznia Mała (rural commune), Miekinia (rural commune), and Czernica (rural commune).

Of Dolnośląskie Voivodeship inhabitants, 14,278 went to work outside the voivodeship. They went mainly to Warsaw (1.74%) and less frequently to Kraków (0.22%), Poznań (0.15%), and Opole and Łódź (both 0.09%).

Of employed people whose place of residence was in another commune in the country, 122,315 came to work to Katowice, the capital of the Śląskie Voivodeship, in 2011.

The inhabitants of voivodeships other than the Śląskie Voivodeship (9052 people) constituted 7.4% of total number of people coming to work in Katowice.

Most people came to work in Katowice from the Małopolskie Voivodeship (3900) and the Wielkopolskie Voivodeship (800).

In 2011, 17,819 people left Katowice and went to other communes in Poland – most of them went to Warsaw, and their share in the number of employed people living in Katowice equaled 2.17%. The rate of employed people who went to work in Kraków or Warsaw was much lower and it equaled respectively 0.35% and 0.19%.

Of people living in other territorial units of the country, 96,540 came to work to Kraków in 2011, mainly from the Śląskie Voivodeship (5100), the Podkarpackie Voivodeship (3800), and the Świętokrzyskie Voivodeship (2600).

Communes, from which came to Kraków the highest percentage of workers in 2011, are concentrated around the borders of the city. In 47 of the communes, the workers coming to Kraków to work constituted over 20% of all employed people living in their territory, and in four communes (Iwanowice, Kocmyrzów-Luborzyca, Wielka Wieś, and Liszki) their share exceeded 40%.

Of people who went to work outside the administrative borders of Kraków in 2011, 30,823 went mainly to Warsaw, Dąbrowa Górnicza, Wrocław, Poznań, Katowice, and Łódź (See: Fig.3.6).

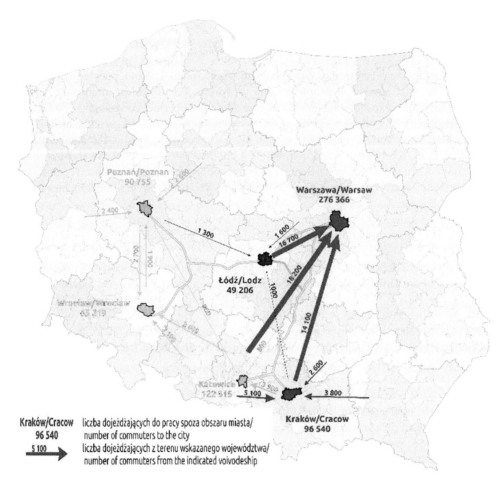

Figure 3.6 Main directions of commuting

Elaborated by: Tomasz Bużałek based on [8]

3.4.2. NUTS2 and NUTS3 – characteristics

The classification of NUTS (*fr. Nomenclature des Unités Territoriales Statistiques*) was formally introduced in Poland on 26 November 2005 at the moment of the entry into force of the regulation (WE) no. 1888/2005 [31]; nonetheless, the classification had been used since 1 May 2004 when Poland joined the European Union. Six non-administrative units of NUTS1 (regions), 16 units of administrative level NUTS2 (voivodeships), and 45 units of non-administrative level NUTS3 (subregions) were introduced. During further revision of NUTS, the division of Poland into NUTS1 and NUTS2 did not change in contrast to the division into subregions NUTS3. In the period between 1 January 2008 and 31 December 2014, the number of NUTS3 equaled 66, whereas since 01 January 2015 it has amounted to 72 [32, 33].

This chapter includes a description and characteristics of the area (voivodeships – NUTS2 and subregions – NUTS3) which will be directly influenced by the HSL system in Poland. The route of the newly designed line has been chosen to take into consideration technical, environmental, and financial abilities as well as the demand assumptions that result from the potential and needs of the regions. Different variants of the line route have been analyzed, and the recommended variant includes the benefits to a wide group of stakeholders.

The route of Y line (Warsaw – Łódź – Poznan/Wroclaw) assigned in the feasibility study by IDOM and the adjustment of CMK to high-speed line parameters have led to the fact that the area directly influenced by the HSL system is located in the voivodeships (NUTS2): the Mazowieckie, Łódzkie, Dolnośląskie, Wielkopolskie, Małopolskie, and Śląskie Voivodeships as well as in subregions (NUTS3), which are characterized in Tables 3.5 and 3.6.

3.4.2.1. Voivodeships (NUTS2) and Subregions (NUTS3)

3.4.2.1.1. MAZOWIECKIE VOIVODESHIP

The Mazowieckie Voivodeship is the biggest in the country and comprises 11.4% of the country's area. It is also the most populous voivodeship (13.8% of the overall population of Poland). In 2013, 14.7% of the urban population of the country lived in towns within the voivodeship. It is the area of active economic development and strongly proceeding urbanization – 64.2% of population are the inhabitants of towns and cities. The area can be also distinguished by a high rate of highly educated people (23.6%), a high level of wealth of the citizens, and the lowest level of unemployment in the country (11.1%). At the same time, it is the voivodeship with the highest spatial disproportions of socioeconomic development in Poland. It consists of two remarkably different parts – Warsaw agglomeration (subregion Warsaw city, eastern Warsaw subregion, and western Warsaw subregion) and the remaining part (composed of Ciechanow-Plock, Ostroleka-Siedlce, and Radomsko subregions). These two parts differ significantly in terms of demography, employment structure, economic and cultural activity, and touristic attractiveness. In western Warsaw subregion, the feminization rate equals 108% and the demographic load rate equals 59.9%, whereas in Warsaw city subregion these two rates are higher: for every 100 men there are 117.9 women, and for 100 people in their productive age there 62.4 people in the unproductive age.

In 2013, compared with 2012, the population increased as a result of positive natural growth and positive domestic migration rate, at negative foreign migration rate. In the voivodeship, 2.2 million people work, which, despite the fall in employment rate in comparison with the year 2012, makes the voivodeship one of the biggest labor markets in Poland (16.34% of working people in Poland). The region created over 1/5 of the national GDP in 2013, and the value of GDP per capita was higher by 60.5% than the average in the country. Almost 73% of GDP in Mazowieckie Voivodeship, according to the gross added value structure, is created by the service sector, 17.4% by industry, 6.9% by building, and a little over 3.0% by agriculture. The main sectors of the voivodeship's economy are trade, telecommunications, financial services, insurance, IT, the car industry, and the petrochemicals industry, with the largest petroleum refinery in Poland located in Plock.

The voivodeship, mainly thanks to the dynamically developing capital, possesses a relatively modern structure of economy. Modern and innovative branches of services and industry are developing quickly. Investment expenditure in the voivodeship amounts to more than the national average (over 142%). The Mazowieckie Voivodeship is a pioneer in the field of innovation, concentrating (mainly Warsaw) approximately 1/3 of national potential of research and development units and people working there. The expenses incurred in the research and development (R&D) sphere are increasing and constitute almost 40% percent of national expenses (2013). Their share per capita equals over 258 EUR – two and a half times higher than the average in the country. These means derive mainly from the budget, which is a substantial cause of a relatively weak cooperation of R&D sector and the economy of the voivodeship. One hundred seven academies and universities are located in the area of the voivodeship, where in 2013 there were 29,413 students and a teaching staff of 17,960.

In 2013, there were 283,196 unemployed people registered in Masovian (Mazowieckie) unemployment agencies, which is by 4.14% more than in 2012.

In 2013, the average monthly gross salary equaled 1150 EUR which increased by 2.93% in relation to 2012. The amount of income translates into the quality of life and the opportunity of fulfilling people's aspirations. The Mazowieckie Voivodeship can be distinguished in the country on one hand by the highest level of salary (the average monthly salary exceeds the national average by 23%) and on the other hand by the highest spatial disproportions in favor of the metropolitan area and the former voivodeship cities (Ciechanów, Siedlce, Ostrołęka, Radom, and Płock).

The biggest city of the region, Warsaw, had 1724.4 thousand inhabitants in 2013, which is 32.4% of the voivodeship's population. The average monthly gross salary amounted to 1260.43 EUR in December 2013 and was 4.9% higher in relation to November 2013 and 2.5% higher than in an analogical period in 2012.

3.4.2.1.2. ŁÓDZKIE VOIVODESHIP

The Łódzkie Voivodeship has a disadvantageous situation among the analyzed regions; it has a negative migration balance which equaled 5794 people in 2013. Within 10 years, there was a population decline of inhabitants in their productive age; compared to 2004, the population of this group decreased by 6%. The number of inhabitants with a higher level of education is relatively low – a little over 16%, with secondary education over 30% and 20% inhabitants with only a middle school education or lower. The voivodeship has a high unemployment rate – 14.1%, which is better by 0.1% compared to 2012, but worse by 1.2% compared to 2011. In the voivodeship, 22.73% are employed in industry and 19.37% in agriculture and forestry. The voivodeship encompasses 5.8% of the country's area, and the population density exceeds the national average. Similarly, the urbanization rate exceeds the national average, for which the agglomeration of Łódź is mainly responsible. The capital of the region, Łódź, with a population of 714.2 thousand inhabitants (the third highest in the country), is going through a period of transformation connected with the economic reorganization in the 1990s. The average population density in Łódź equals 2426 people/km^2. Apart from Łódź, bigger cities of the regions are Piotrków Trybunalski (75,903 inhabitants), Pabianice (67,688), Tomaszów Mazowiecki (64,783), Belchatów (59,565), and Zgierz (57,503). In terms of development level, measured by the amount of GDP per capita, the Łódzkie Voivodeship has an unexceptional sixth position in the country. The structure of gross added value tells about the economic structure: almost 60% of Łódź GDP is created by the service sector, 30% by industry, 6.4% by building industry, and 3.8% by agriculture. Investment expenditure per inhabitant is extraordinary and constitutes 116.2% of average national expenditure, which put the voivodeship in second place in the country and indicates high dynamics of development. Investment on research and development activity is lower than the national average – eighth place in Poland, despite the significant role of the Łódź academic center – 22 universities with 83,471 students in 2013. Łódź, as an important academic and cultural center, paradoxically belongs at the same time to cities with a low rate of highly educated people, who, as a result of poor local economy, often search for work outside the region. In 2013, the average monthly salary in Łódź was 5.5% higher than in the previous year and amounted to 859.32 EUR (in the voivodeship it equaled 846.40 EUR).

The voivodeship has seen an increase in economic activity and the growth of the number of enterprises in recent years: 228,537 in 2011, 234,079 in 2012, and 237,915 in 2013. The voivodeship has a strong research center which consists of 29 universities, 7 of which are state universities, and in 2013 there were 92,153 students in all of higher education. The biggest academies in the voivodeship are the University of Łódź, The Technical University of Łódź, and the Medical University. These universities are also among the biggest academies in the country.

In the investment area, one of the most important projects in Łódź is the rebuilding of the railway station Łódź Fabryczna. On a surface area of 90 ha, in the neighborhood of the railway station, the New Center of Łódź is going to be located (an investment comprising the building of the railway station with road and parking infrastructure). The area has been divided into two spheres – in the first one, with a surface area of 60 ha, a housing and business estate is supposed to be built; the second one will be dominated by culture.

The Łódź Agglomeration Railway (ŁKA) is a very ambitious and already advanced communication project and will serve the integration of the region, but, at the same time, it is an element of preparing the regional transport system for supplying the future high-speed line [39].

The structure of the economy is moving in a good direction. The appearance of significant investors in the region is proof of this process (mainly in Łódź and its neighborhood and on the terrains of Łódź Special Economic Zone).

The Łódzkie Voivodeship promotes the region's capital as a place for creating opportunities for young people, with a focus on students. The city authorities undertake many initiatives that encourage young people to study, live, and work in Łódź.

3.4.2.1.3. DOLNOŚLĄSKIE VOIVODESHIP

The feature that distinguishes Lower Silesia (the Dolnośląskie Voivodeship) is a modern, dynamically developing economy, combining traditional industry and the newest technologies. Economic development is based on highly qualified and educated human resources, rich natural heritage, and investors supported by convenient conditions of industrial zones and technological parks. This is one of the most industrialized regions of Poland. As far as the share in creating GDP of the national economy is concerned, the voivodeship is placed in fourth position in the country, and in terms of GDP per capita it is in second position, giving way only to Mazowieckie Voivodeship.

According to the gross value added structure, 56.8% of Lower Silesian GDP is produced by the service sector, 34.7% by industry, 7.1% by the building industry, and 1.5% by agriculture. The distinguishing feature of the region is high touristic attractiveness. Subregions situated in the Sudetes (the Wałbrzych subregion and the Jelenia Góra subregion) are attractive for both winter and summer tourism. Health resorts are also located there. The attractiveness of the subregions is measured by the number of accommodations, which is significantly higher than in the remaining subregions. Also, the number of visitors in museums can be a proof of development of touristic infrastructure. It is much higher in mountainous subregions than in the remaining ones (apart from Wrocław). Unfortunately, poor transport accessibility for tourists in this attractive part of the voivodeship leads to underestimating the potential of the Sudetes. It is crucial for tourism development to better connect these regions not only with Wrocław but also with the rest of the country. A planned fast railway connection with Prague will also open the region for tourists from the southern part of Europe.

Wrocław is perceived as a historic city with a great development potential, especially for business. It is the third largest and most influential academic center in Poland (almost 30 universities showing a significant potential in the field of training engineering and management staff). It belongs to the leaders of Polish computer services market, and it concentrates important research and development centers and service centers. The population density of Wrocław amounts to 2159 people/km². The Dolnośląskie Voivodeship is characterized by a high level of urbanization. The urban network is well expanded, especially in comparison with other voivodeships. Apart from Wrocław in the region, towns with significant potential include, among others, Wałbrzych (117,926 inhabitants), Legnica (101,992), Jelenia Góra (81,985), Lubin (74,053), Głogów (68,997), and Świdnica (59,182).

In 2013, 24.56% (250,110) of all the employed people worked in industry. The unemployment rate in 2013 was below the national average and was lower by 0.4% than in the previous year, and by 0.7% than in 2011. Investment expenditure per inhabitant constitutes 109% of the national average and is in third position in the country. Despite the significance of Wrocław on the academic map of the country, the region cannot boast high expenditure in research and development (219.1 million EUR, sixth position in the country). Per capita they constitute 83.34% of the national average, which gives the voivodeship fifth position in the country.

3.4.2.1.4. WIELKOPOLSKIE VOIVODESHIP

The Wielkopolskie Voivodeship, situated in the western part of the country, takes advantage of the availability of important east–west and north–south transportation routes. It belongs to the most highly developed regions of Poland. The capital of the voivodeship, Poznań (548,028 inhabitants), is one of the key development centers in the country. The voivodeship is the second in terms of size in Poland, and the third in terms of the number of inhabitants. The population density is lower than the average. More important cities of the region, apart from Poznan, are Kalisz (103,997 inhabitants), Konin (77,224), Piła (74,609), Ostrów Wielkopolski (72,890), Gniezno (69,883), and Leszno (64,589). In 2010–2013, the population increased (which is an opposite tendency to the country as a whole) by 20,271 people, that is slightly over 0.6%. The level of feminization in the subregion of Poznań city is above average – the number of women exceeds the number of men by 15%.

In 2013, the average monthly salary in the voivodeship was 3.5% higher than in the previous year and amounted to 847.63 EUR, which constitutes 90.7% of the average national pay.

The average employment in 2013 equaled 1,367,192 people. At the end of 2013, 144,832 unemployed people were registered in the Wielkopolskie Voivodeship. This means a 2% improvement of the situation in the labor market in relation to 2012. The unemployment rate in the voivodeship at the end of 2013, similar to the previous years, was the lowest in the country. A characteristic feature of Greater Poland's (the Wielkopolskie Voivodeship) industry is a considerable predominance of small and medium enterprises whose main advantage is mobility and flexibility in adjusting to the rules of the market. In 2013, the voivodeship built over 9.5% of the national GDP (third position in the country), and the value of GDP per capita was higher by 7.2% than the average in the country. Almost 60% of Greater Poland's GDP is created by the service sector, 28.5% by industry, 7.4% by the building industry, and 4.4% by agriculture (according to the gross value added structure). The industry structure in Greater Poland is differentiated. High agriculture encourages food processing.

In the sphere of education and development, Poznań has at its disposal a great academic and scientific potential, and as such it promotes itself as an ideal place for young and ambitious people. Concerning culture and entertainment, authorities promote Poznań as an attractive touristic city with rich cultural backgrounds. Poznań also wants to be a dynamically changing, modern, and elegant place offering a high standard of living, possessing cultural diversity, and being open to new ideas and mainstreams [36].

The voivodeship is characterized by investment expenditure below the average (fifth position in the country). In terms of expenditure on research and development (240.3 million EUR), the region's position is slightly better (fourth place).

3.4.2.1.5. MAŁOPOLSKIE VOIVODESHIP

The voivodeship constitutes 4.86% of the country's area. In 2013, in terms of the number of inhabitants it comes fourth (8.73% of the population in Poland). The population density is almost twice as high as the national average. In 2010–2013, the number of inhabitants in the region increased (which is an opposite tendency to the country as a whole) by 23,889 people, i.e. over 0.73%.

The voivodeship is characterized by a strongly developed city network – urban areas constitute 10.9% of the voivodeship's area (the third place in Poland) where 48.45% of inhabitants live. Kraków has the country's second highest population (after Warsaw) – 758,992 people. The biggest cities in the region are Tarnów (112,120 inhabitants), Nowy Sącz (83,943), Oświęcim (39,664), and Olkusz (36,724).

In 2013, an average salary in the voivodeship was 3.42% higher than in the previous year and equaled 861.84 EUR, which is 92.2% of the national average.

The voivodeship in terms of creating the national GDP is in fifth position in the country. According to the structure of gross value added, 66.7% of GDP in the Małopolskie Voivodeship is created by the service sector, 22.2% by industry, 9.6% by the building industry, and 1.5% by agriculture.

Great educational and development potential has been connected for years with the created image of Kraków as a dynamically developing and friendly center on the map of Europe, where history and tradition meet modernity.

In the voivodeship in 2013, 32 universities educated 189,609 students (the highest number in the country) and employed 12,563 academic staff. The share of expenditure paid in the voivodeship on research and development is still increasing (compared to 2010 it has increased by 50%) and in 2013 it amounted to 400.3 million EUR (second position in terms of national expenditure). Their share per capita equals almost 119.3 EUR and exceeds the average in the country.

The educational level is relatively high – 20.9% people with a degree, over 33% secondary school graduates.

The unemployment rate in Kraków in 2012 and 2013 maintained a level of 5.9%, with, respectively, 23,900 and 24,700 registered unemployed people. An average salary in the enterprises sector equaled 944.93 EUR in 2013, increasing by 3.82% compared to the previous year.

In the voivodeship, for every 106 women there are 100 men, and the rate of people in their non-productive years equals 58%. In the Kraków subregion, the level of feminization is 104 women per 100 men and the rate of demographic burden is 58%, whereas in the subregion of Kraków city there are 115 women per 100 men, and the rate for people in non-productive age is the same for the whole voivodeship.

The adjustment of CMK to high-speed line parameters and the exploitation of a new rolling stock since 2014 have led to the shortening of journey time between main cities and Warsaw. The completion of the established works will make it possible to improve significantly the availability of Krakow itself, as well as the popular tourist destinations situated in the south.

3.4.2.1.6. ŚLĄSKIE VOIVODESHIP

The voivodeship, although one of the smallest in area, has the second largest population in the country. Its population density is the highest in Poland, exceeding three times the national average. Urban areas are inhabited by 77% of the region's population. The Slask conurbation is decisive in the voivodeship and, according to the delimitation conducted by the Institute of Geography and Land Management of the Polish Scientific Academy in 2012, it consists of 14 neighboring cities. The conurbation occupies 24% of the voivodeship's surface (2979 km²) and it is inhabited by 1,904,611 people, which constitutes 41.4% of the voivodeship's population. The population density is remarkably higher than the national average and amounts to 1564 people/km².

Bigger cities of the voivodeship are Katowice (304,362 inhabitants), Sosnowiec (211,275), Gliwice (185,450), Zabrze (178,357), Częstochowa (232,318), Bielsko-Biała (173,699), Bytom (173,439), Ruda Śląska (141,521), Rybnik (140,173), and Tychy (128,812).

An average salary in the sector of enterprises in the voivodeship's capital, Katowice, equaled 1,308 EUR in 2013 and was the highest of all voivodeships' capitals. Simultaneously, the cities were characterized by a 5.4% unemployment rate (0.2% more than in the previous year), with 11,300 registered unemployed people.

The Śląskie Voivodeship is the most urbanized in Poland – the area of urban communities is 3241 km² and the area of cities constitutes 30.7% of the voivodeship's surface (almost 4.5 times more than the national average). The voivodeship is regarded as the most attractive from the point of view of industrial activities. Three out of four subregions of the voivodeship are among the subregions with the highest investment attractiveness for industrial activity. The voivodeship is characterized by relatively high expenditure per capita – sixth position in the country, 97.10% of the national average. However, the region's position in terms of expenditure on research and development is much higher (the third place in Poland), 306 million EUR in 2013.

The region in 2013 was in second place in the country in respect of created GDP, and the value of GDP per capita was higher than the average in the country. In compliance with the structure of gross value added, 57.2% of Silesian GDP is created by the service sector, 34.2% by industry, 7.8% by the building industry, and less than 1% by agriculture.

In 2013, the average monthly salary in the voivodeship was 4.35% (40.40 EUR) higher than in the previous year and equaled 970 EUR, exceeding by 3.7% the national average and the second highest in the country.

Table 3.5 Basic demographic rates of NUTS2 and NUTS3 directly influenced by HSR system, 31 December 2013

Territorial unit	Population [people]	Population density [person/km²]	Population before productive age [people]	Population in productive age [people]	Population after productive age [people]	Higher education rate*[%]	Total net migration per 1000 people
Poland	38,495,659	123	6,995,362	24,422,146	7,078,151	17.0	-0.5
Mazowieckie V.	**5,316,840**	**150**	**984,077**	**3,319,937**	**1,012,826**	**23.6**	**2.5**
Warsaw city	1,724,404	3334	275,960	1,061,842	386,602	37.8	4.8
Western Warsaw	789,645	184	156,509	493,867	139,269	21.7	7.5
Łódź V.	**2,513,093**	**138**	**426,305**	**1,568,769**	**518,019**	**16.2**	**-1.1**
Łódź	385,035	175	66,467	241,474	77,094	15.1	4.1
Łódź city	711,332	2426	100,567	438,012	172,753	22.6	-2.2
Sieradz	45,056	80	83,243	284,166	85,647	12.0	-1.5
Skierniewice	368,912	90	65,267	231,020	72,625	13.3	-2.2
Dolnośląskie V.	**2,909,997**	**146**	**490,561**	**1,867,85**	**552,151**	**16.8**	**0.0**
Jelenia Góra	576,219	103	96,904	371,553	107,762	12.2	-3.0
Wałbrzych	673,860	161	107,871	430,715	135,274	12.0	-3.3
Wrocław	574,545	89	109,902	371,490	93,153	14.1	6.5
Wrocław city	632,067	2159	94,232	401,817	136,018	30.0	2.2
Wielkopolskie V.	**3,467,016**	**116**	**669,381**	**2,206,381**	**591,254**	**16.4**	**0.0**
Kalisz	672,530	116	129,330	425,094	118,106	13.0	-1.2
Poznań	618,754	126	131,162	398,362	89,230	17.8	10.1
Poznań city	548,028	2092	84,968	346,120	116,940	30.4	-4.70
Małopolskie V.	**3,360,581**	**221**	**644,733**	**212,0442**	**595,406**	**17.2**	**0.8**
Kraków	707,788	175	143,957	448,087	115,744	14.2	5.4
Kraków city	758,992	2322	118,177	238,100	159,825	31.6	1.3
Śląskie V.	**4,599,447**	**373**	**776,393**	**2,934,496**	**888,558**	**15.9**	**-2.0**
Katowice	752,441	1980	119,741	477,535	155,165	17.9	-3.7
Sosnowiec	701,830	390	107,523	448,946	145,361	17.4	-2.2

Bold – NUTS2

*in the group of over 13 year olds, data based on the National Personal Register in 2011.

Source: self-study based on GUS data [8]

Table 3.6 Basic economic rates of NUT2 and NUTS3 directly influenced by HSR system, 31.12.2013

Territorial unit	GDP (current price) [million EUR]	GDP per capita (current price) [EUR]	GDP [%]	Entities of national economy in the REGON register – public sector	Entities of national economy in the REGON register – private sector	Registered unemployment rate [%]	Average monthly wages and salaries [EUR]	Investment expenditure per capita [EUR]	B&R investment per capita [EUR]
Poland	399,387.8	10,373.3	100	122,759	3,947,500	13.4	93.0	1447.7	90.3
Mazowieckie V.	**88,356.2**	**16,648.1**	**22.1**	**12,942**	**712,055**	**11.1**	**1151.0**	**2148.0**	**258.5**
Warsaw city	52,648.1	30,640.9	13.3	4413	367,063	4.8	1260.1	Bd	Bd
Western Warsaw	10,295.4	13,085.2	2.6	1610	107,716	9.0	985.6	Bd	Bd
Łódzkie V.	**24,361.5**	**9675.7**	**6.1**	**6629**	**231,286**	**14.1**	**846.4**	**1610.7**	**64.8**
Łódź	3287.8	8542.9	0.8	940	37,418	16.6	750.0	Bd	Bd
Łódź city	9159.7	12,804.1	2.3	1861	88,908	12.3	894.8	Bd	Bd
Sieradz	3174.4	7008.1	0.8	1317	34,845	14.4	730.1	Bd	Bd
Skierniewice	2804.3	7591.6	0.7	1071	27,236	13.3	764.9	Bd	Bd
Dolno ląskie V.	**33,792.0**	**11,608.1**	**8.5**	**15,548**	**332,013**	**13.1**	**932.9**	**1710.1**	**75.3**
Jelenia Góra	5026.5	8701.8	1.3	3492	60,819	17.8	783.7	Bd	Bd
Wałbrzych	5065.8	7491.8	1.3	5514	67,927	20.4	815.8	Bd	Bd
Wrocław	5903.3	10,310.3	1.5	1758	55,808	12.9	827.3	Bd	Bd
Wrocław city	10,527.1	16,676	2.6	3082	105,267	5.5	995.7	Bd	Bd
Wielkopolskie V.	**38,528.2**	**11,124.4**	**9.6**	**9734**	**388,121**	**9.6**	**847.6**	**1314.1**	**69.4**
Kalisz	5893.4	8763.5	1.5	1926	60,727	9.8	710.4	Bd	Bd
Poznań	7596.9	12,355.8	1.9	1292	78,015	7.2	842.2	Bd	Bd
Poznań city	11,440.7	20,836.0	2.9	1538	103,545	4.2	1026.4	Bd	Bd
Małopolskie V.	**30,884.7**	**9,200.7**	**7.7**	**7961**	**343,113**	**11.5**	**861.8**	**1261.3**	**119.3**
Kraków	5259.2	7454.9	1.3	1679	66,126	11.6	828.5	Bd	Bd
Kraków city	12,630.7	16,642.6	3.2	421	123,080	5.8	964.0	Bd	Bd
Śląskie V.	**49,716.0**	**10,792.8**	**12.4**	**16,179**	**444,171**	**11.3**	**970.0**	**1447.2**	**66.4**
Katowice	10,817.7	14,323.2	2.7	2990	80,606	8.0	1156.7	Bd	Bd
Sosnowiec	6692.7	9505.7	1.7	1928	69,321	14.7	926.9	Bd	Bd

Bold – **NUTS2** 1€= 4.1472 PLN according to NBP from 31 December 2013

Source: self-study based on GUS data [8]

The rate of professional activity in the Śląskie Voivodeship in 2013 was slightly below the national average (55.9%) and amounted to 53.5% – in relation to 2012, it did not change.

A total of 1,638,657 people were employed in the voivodeship, 481,311 in industry, which equals 29.4% of the employed and exceeds the average national level by 8.8%.

In 2013, in the voivodeship 41 higher schools functioned, with 144,545 students and 9063 academic teachers, and 19.7% of inhabitants have a higher education, and 35.1% a secondary education. The number of people with vocational schooling is high and constitutes 28.4%, which exceeds the national average. In spite of a growing tendency (since 2009), the voivodeship is still among the regions with the lowest (the third place in Poland) unemployment rate.

3.4.3. Development trends in the areas of HSR impact

The analyzed areas are characterized by high wealth and the fastest level of development. Agglomerations under direct influence of Y and CMK lines concentrate a significant percentage of Poland's potential. According to the available prognoses, the level of GDP per capita in Poland within the next 30–40 years should reach the average level noted in the countries of the Eurozone, among others, in Germany. As a result, essential changes in consumer behavior of the Polish people in the direction of characteristic features of highly developed countries should be estimated (e.g. bigger expenses on goods of a higher level). The greatest positive changes of the level of GDP per capita in regard to the national average in 2009–2050 are predicted in the Mazowieckie Voivodeship and the Łódzkie Voivodeship (increase respectively by 19% and 14%) [2].

The results of analyses conducted by IDOM for the needs of Feasibility Study, in the Łódzkie, Wielkopolskie, Mazowieckie, and Dolnośląskie Voivodeships, indicate the occurrence of concentrating socioeconomic activities in agglomerations – Łódź, Poznań, Warsaw, and Wrocław. It should be expected that the phenomenon will continue. The common factor of the analyzed areas is the increase of education level of inhabitants and the decrease in the number of people employed in agriculture. The tendency is estimated to continue for 10–30 years. In all voivodeships, a decline in people employed in agriculture and forestry and an increase of employment in services are predicted. Permanent change of age structure is a characteristic factor – an increase in the rate of people in their post-productive age and a decrease in the rate of people in their under-productive age. This phenomenon will continue in the next 10–30 years and will even increase.

The main cities of the analyzed regions are academic centers. It can be expected that universities with long tradition and rich history will maintain their dominant position despite the forthcoming demographic changes. Agglomerations have a clearly higher population density than the remaining subregions, higher education level, and richer economic and cultural activities. It can be forecast that the prevalence will persist for the next 10–30 years [38].

Below, other characteristic elements for particular voivodeships are specified. For the Mazowieckie, Łódzkie, Dolnośląskie, and Wielkopolskie Voivodeships, data included in the Feasibility Study of Y line have been used, whereas for the Małopolskie Voivodeship, the information has been prepared for the needs of strategic programs of the voivodeship development until 2020 and for the Śląskie Voivodeship on the basis of the Concept of National Land Management and Planning 2030 [14].

3.4.3.1. Mazowieckie Voivodeship

- An increase of population in the voivodeship, resulting from both population growth and a positive migration rate.
- The attractiveness of living and working in the capital of Poland and in the nearest neighborhood (agglomeration) will still be significant in the next 10–20 years. Migration of people in their productive age will lead to the growth of the number of inhabitants in their under-productive age.
- Warsaw will remain the center of culture. Modern multimedia museums (Warsaw Uprising, Copernicus Science Centre, and Chopin's Museum) and research centers will attract people from all over the country [38].

3.4.3.2. Łódzkie Voivodeship

- The maintenance of the existing disadvantageous development trends will lead to an increase in the number of people searching for jobs outside Łódź. In the Łódź – Warsaw rail service, there are approximately 13,000 passengers daily (in 2004). Some of these journeys are for commuting reasons (daily and every few days). According to the Blue Book Railways [43], the value of time in commuting is almost twice as low as that for business journeys. Such passengers will make use of HSR, provided the existing price of the connection continues. Each increase in the ticket price will cause a drastic reduction in demand for this category of passengers, and this factor ought to be taken in consideration in the conducted analyses and research.
- The reversal of disadvantageous development trends will cause a fall in passenger flow connected with work – inhabitants will not look for a job outside Łódź. On the other hand, the number of journeys connected with business will rise, and passengers travelling for this reason are less sensitive to ticket costs. Undoubtedly, they will use HSR willingly.
- There is a possibility of changing the disadvantageous trends by means of the concept of Łódź development as a film center, trade and exhibition center of the clothing industry, or an outsourcing center. These changes (if they appear) will see effects in a few years' time [38].

3.4.3.3. Dolnośląskie Voivodeship

- High touristic attractiveness of Jelenia Góra and Wałbrzych subregions will persist, which will lead to the growth of the number of accommodation places. It is a permanent phenomenon and its continuation should be anticipated in the next 10–20 years. This will influence the growth of passenger movement for leisure purposes, both on a national scale (movement from other agglomerations) and a foreign scale (international tourism, especially from Germany) [38].

3.4.3.4. Wielkopolskie Voivodeship

- The increase of the population in the voivodeship will continue as a result of birth rate and positive migration rate.
- Poznan agglomeration will maintain the position of leader as a business, scientific, and cultural center of the voivodeship.
- Attractiveness of Poznań as an exhibition and trade fair center will be maintained [38].

3.4.3.5. Małopolskie Voivodeship

• The growth of population will occur [2].
• In the perspective of the year 2020, the Małopolskie Voivodeship will reach 60.1% of the average level of GDP per capita for the EU-27. In that respect, the region will be in eighth place in the country [35].
• The average annual growth rate of real GDP per capita for the Małopolskie Voivodeship in 2010–2020 will amount to 3.1% and will account for approximately 0.1 percentage points higher than the average predicted for the country [35].
• Lesser Poland will be among the voivodeships in which the growth of gross added value will be at least equal to the average for the country [35].
• The highest growth rate of employed people (5.5%) is reported in the region. Likewise, advantageous changes occurring in the structure of working according to sectors will be a positive tendency and will confirm the increase of the meaning of sector III (services, trade, transport) with simultaneous decline in the share of working in agriculture [35].

3.4.3.6. Śląskie Voivodeship

• The biggest decline of population in 2010–2050 (in absolute terms) – by 945,000 people [2].
• Śląsk agglomeration will create a functionally and spatially integrated area which will possess a rich development potential [14].
• Rich cooperation with the metropolitan area of Kraków and Ostrava (in the Czech Republic), which will form together a transnational area of intensive growth of European significance, will be an impulse for the development of Śląsk agglomeration (in the perspective of the year 2030) [14].

The analysis of six voivodeships – Mazowieckie, Łódzkie, Dolnośląskie, Wielkopolskie, Małopolskie, and Śląskie – indicates that they all belong to a group of the richest and the most successfully developing regions. According to the CSO (Central Statistical Office) data from 2013[8]:

• population density, apart from the Wielkopolskie Voivodeship, exceeds the national average, the number of inhabitants in their productive age in all the voivodeships is close to the national average and in three voivodeships exceeds it: in the Dolnośląskie Voivodeship it amounts to 64.2%, in the Śląskie Voivodeship – 63.8%, in the Wielkopolskie Voivodeship – 63.6%;
• they all exceed 90% of average monthly gross pay, and two of them exceed the average (the Mazowieckie Voivodeship by 23.1%, the Śląskie Voivodeship by 3.7%);
• the registered unemployment rate is below the average (apart from the Łódzkie Voivodeship);
• altogether they have created 66.4% of GDP, taking consecutive highest positions in the country;
• with regard to GDP per capita (higher or slightly lower than the national average), they occupy the highest positions in the country;
• with regard to investment and expenditure on R&D they are the frontrunners;
• altogether they possess 64.6% of universities;
• the total number of all academic teachers in these regions exceeds 65% in the scale of the country;

- the rate of participants of doctorate courses exceeds 75%;
- the percentage of students at technical and scientific universities exceeds or is insignificantly lower than the national average – 27% (Mazowiekie – 23.4%, Łódzkie – 25.3%, Dolnośląskie – 33.4%, Wielkopolskie – 22.8%, Małopolskie – 33.1%, Śląskie – 29.4%) and the tendency has been rising for years;
- they possess over 62% of national theatres, almost 60% of museums, and almost 58% of cinemas;
- they all have significant natural benefits: landscape parks, preserved landscape centers, and areas of Nature 2000 (*Natura 2000*), among others.

On the basis of the above analyses and development prognoses, high interest in transport services realized by high-speed trains can be assumed.

The HSR route has been assigned on the basis of foreign experience, according to a classical approach used in choosing the routes of high-speed lines, i.e. in choosing in the first place routes with potentially high effectiveness, connecting big agglomerations. The connection of regions with highest economic rates into one cohesive transport system is justified by economic reasons. It should ensure the demand for the offered service of high-speed rails and provide a fast return on incurred investment costs.

Simultaneously, as the experience of other countries shows, the improvement of transport possibilities will create better conditions for the development of cities and regions, and, what is more, can be an additional development impulse.

3.4.4. Demand for service

Demand for high-speed rail transport was the main subject of research and analyses in both The Feasibility Study for Y line [38] and in Egis elaboration [2].

Within the IDOM study, an analysis of passenger flow with demand estimation was performed. The analysis was based on investigating passengers' preferences and the existing carriage offer. The prognoses, conducted for a 30-year time horizon, proved that the potential passenger flow in the first year of the line exploitation (year 2020) will equal 8.1 million passengers, which will constitute 19% in the department division in the analyzed corridor. With the growth of the society's prosperity, the demand for services on the Y line will increase. In 2050, the estimated demand on the line will exceed 12 million passengers annually.

As a result of changing the assumptions concerning the deadline for launching HSR in the second part of the study, the date of beginning the line exploitation was changed for 2030 and the growth rate of GDP was reduced due to the worldwide crisis. Both factors influenced the outcome of the updated demand analyses. The obtained results indicated that the passenger flow on the line within the first year of its exploitation (year 2030 due to a 10-year postponement of the investment) will reach 9.8 million passengers, which will correspond to 18% in the department division in HSR corridor. In 2060 the estimated demand on the line will equal approximately 13.5 million passengers annually.

In foreign experts' opinion [20], the estimated induction of passenger flow seems too low. The flow prognoses indicate minor influence of the investment on total traffic on the Y line. A very low level of induction leads to an overall reduction of the level of flow and profits.

A realistic scenario predicts that the total flow in 2020 will reach 42.3 million journeys in the non-investment variant and 43 million in the investment variant. According to the

Table 3.7 Rates of economic effectiveness

Parameter	Unit	Value
Social discount rate	%	5.0
Cash flow netto	thousand zlotys	48,801,075
ENPV	thousand zlotys	4,571,878
ERR	%	6.3
B/C	X	1.2

Source: [38, final document, p. 263]

reviewers, small difference of traffic between the referenced situation and the estimated one does not correspond to the situation of any other HSR projects. The elicited demand constitutes 8.85 of the HSR market. Therefore, the reviewers recommend revising the traffic prognoses [20].

The reiterated analyses, conducted in the second stage of the study, did not take into consideration the UIC experts' comments and recommendations concerning the increase of the estimated level of inducted demand.

In *The Analyses of Social Conditions of High-Speed Rail Development in Poland in the Perspective of Year 2050* [2], one of the established objectives was the specification of economic efficiency and socioeconomic results of building and exploiting a high-speed network in Poland. The elaboration included, among others, the comparison of the variants of improving passenger transport between four big Polish agglomerations. Approving the assumption about the marginal participation of bus transport as a means of long distance transport, due to its uncompetitiveness in relation to rail transport, and ignoring it in analyses is questionable. The observation of both the foreign market and the Polish one points out a completely opposite phenomenon, and the dynamic development of Polski Bus, competing successfully with PKP Intercity, can set an example here. Omitting this segment of the market hindered the adequate comparison of the variants, only one of which considered building the HSR system.

The study of independent consultants from December 2011 [20] considers the bus as one of the obvious means of transport competing with the rail. It is even proposed by them to use the methods of survey analyses in order to specify the preferences of journeys, if one could choose between high-speed trains, planes, coaches, and cars.

Nevertheless, the analyses conducted by Egis proved that the project realization of building the Warsaw – Łódź – Poznań/Wrocław line is justified in terms of socioeconomic efficiency.

Regardless of the above, IDOM in its study indicated that the realization of the project provides socioeconomic added value; the rates of economic effectiveness are presented in Table 3.7.

3.5. Effects for regions and the country

The analysis conducted in the Feasibility Study for Y line enabled the distinction of assignments and objectives (Figure 3.7) that will be realized by means of the creation and exploitation of a high-speed system. However, according to the Author, the effects should be expanded, taking into consideration the areas of Central Trunk Line influence (modernized to HSL parameters) as the element of high-speed rail system.

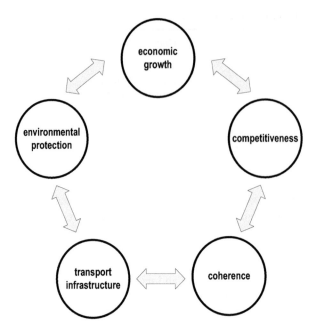

Figure 3.7 Objectives and assignments within the analysis of strategic and program documents
Source: self-study based on [38 Report 3, Figure 48, p. 351]

Building and launching a high-speed rail transport support particular objectives established in strategic documents:

a.   increase of competitiveness of Polish regions, respecting the rules of balanced development through:

- reinforcing the development potential of regions;
- improving the image of Poland and Polish regions;
- increasing the international importance of cities and regions;
- raising the standard of living of the inhabitants;
- openness to Europe;

b.   raising the level of social, economic, and territorial cohesion through:

- ensuring transport connections in international relations and between the biggest cities in Poland;
- ensuring availability of journey destinations which would enable inhabitants to realize all forms of activities;
- stimulating metropolitan functions of cities through strengthening the connections with regional, national, and international settings;

c.   stimulating economic growth and employment through:

- facilitating passenger flow, maintaining high level of comfort and safety;
- supporting the modernization process of the labor market;

d. development of transport infrastructure, mainly:

- improvement of safety in transport;
- raising the quality of transport and passenger service;
- increasing the possible speeds of carriages;

e. limitation of negative impact of transport on the environment through:

- increasing the participation of environmentally friendly branches of transport in general transportation of people and foods;
- supporting means of transport which are an alternative to road transport [38].

The effects of a significant shortening of the journey time, which is a basic condition of preparing a much better offer and fulfilling passengers' expectations connected with the increasing value of time, are presented in Table 3.8 [38]. The modification of connections determines the upgrading of cooperation between regions and contributes to the improvement of the country's cohesion.

In accordance with the experience of other countries, from the point of view of regions, it can be assumed that high-speed rail will have a positive impact on:

- improvement or creation of access infrastructure to stations;
- improvement of touristic attractiveness and availability;
- improvement of the usage and growth of value of the grounds surrounding a station;
- improvement of regional transport;

Table 3.8 Journey time between cities after launching high-speed lines and finishing the modernization of conventional lines

Relation	Line length [km]	Journey time [h]
Warsaw – Poznań	350**	1.35
Warsaw – Wrocław	354**	1.40
Warsaw – Łódź	123	0.35
Poznań – Łódź	227	1.00
Wrocław – Łódź	231	1.05
Poznań – Wrocław	220	0.55
Poznań – Opole	282	1.30
Poznań – Kraków (through Łódź)*	465	2.45
Poznań – Katowice (through Łódź)*	479	2.45
Warsaw – Opole (through Łódź)	414	2.00
Warsaw – Szczecin	563	3.25
Warsaw – Zielona Góra	485	2.55
Poznań – Lublin	529	3.10
Szczecin – Lublin	742	4.55
Białystok – Lublin	304	2.10
Białystok – Szczecin	744	5.00
Wrocław – Białystok	535	3.45
Wrocław – Lublin	533	3.15

Source: [38 Final report p. 73.]

Attention: Journey times with minimal 2-minute reserve for 100 km with stops only in voivodeship cities.
* The fastest relation which does not eliminate the functioning of others.
** Variant through Sieradz, Nowe Skalmierzyce

- intensification of local enterprises;
- increased incomings to the budgets of communities located in the vicinity of HSR stations;
- optimization of the structure of industry;
- the image of a place as modern, innovative, and easily available.

Both strategic and planning documents of voivodeships and cities include expectations of regions with regard to passenger transport and especially fast railway connections. They concern the improvement of linkage with neighboring regions and national and international surroundings. The connections are supposed to support the development of metropolitan functions of cities and raising the quality of life of the inhabitants. Some of them require directly the creation of high-speed lines [23–27, 34–35, 37, 42].

According to *The Concept of Land Management of the Country 2030* [14], the transport system should assure time availability between main cities within less than 2 hours and the backbone of the railway part should be the Y line, which constitutes the central element of the system and guarantees significant shortening of time availability. In this document, it was specified that the impact of the line will be meaningful on the national level, through increasing the availability of regions and their better integration.

The simulation of the effect of improving availability as a result of building the Y line conducted by PAN Institute of Geography and Land Management experts indicates that the majority of communities in the country are the beneficiaries of the new line. Definitely the Dolnośląskie Voivodeship and Wrocław itself will benefit most, but generally speaking the whole western part will take advantage of the investment [13].

Taking into account the fact that, similar to foreign considerations of HSR systems, high-speed trains will be able to take advantage of conventional networks as well, in order to lengthen their relations, the scope of impact of the HSR project will be wider than the six voivodeships considered in analyses. In an indirect way these voivodeships can be included in an HSR network: especially the Zachodnio-Pomorskie and the Lubuskie Voivodeships, connected with Poznań by relatively good conventional lines, but also Opole, Świętokrzyskie, Lublin, Podkarpackie, Pomorskie, Kujawsko-Pomorskie, Warmińsko-Mazurskie, and Podlaskie Voivodeships. This is presented in Figure 3.8.

3.6. Summary

The improvement of accessibility to transport networks and especially to high-speed rails is a subject of interest to the European Commission. Expectations connected with creating a cohesive, modern transport TEN-T network result not only from its ecological values but also from perceiving transport as a stimulus of economic development. Research conducted in the countries where high-speed rail systems have been functioning for years, with regard to the influence of transport investment on regional economies, has indicated positive relations between the level of public investment and the speed of accumulating private capital, employment, and production growth. Their impact on improving the attractiveness of serviced regions is noticeable.

The creation of a high-speed rail system in Poland is a necessity, as it was in the examples of other countries which searched solutions for the improvement of connections between regions to activate economic growth. Moreover, it is also an opportunity for the dynamic development of innovative technologies and science, as well as for the training of qualified

Figure 3.8 The map of areas influence by high-speed rails.

Elaborated by Tomasz Bużałek

staff. Economic prognoses until the end of the first half of 20th century point at the increasing wealth of the Polish society and at reaching the European standards as far as GDP per capita is concerned, which lets us assume that the following changes in the society's movement will generate additional passenger flows, and the increasing value of time will prompt the choice of the fastest means of transport.

The creation of high-speed rail in Poland is connected with, as in other countries, shortening the journey time, providing security and travel comfort, decreasing harmful effects to the environment, and raising the competitiveness of Polish regions. Undoubtedly, HSR will be an impulse for innovative branches of industry and will contribute to the development of regional connections. From the point of view of the economy, this project, unprecedented in its field, is of key importance for the development of not only the regions under direct influence of the line but also the whole country. Big cities and urban agglomerations, with centers of regional development, are not only administrative centers but more importantly are the engine of growth – they stimulate social and economic development of the region and the country. As centers of development, innovation, and technology, they are the headquarters of the most significant companies, public administration offices, and academic and cultural centers. The role of big cities as regional development centers is crucial here, as well as their significance as generators of new worksites and enterprise development. Poland, in order to take the opportunity to develop, should aim at territorial cohesion through fuller usage of endogenous potential of the biggest urban centers and at strengthening the relations between cities and urban areas and smaller towns and villages surrounding them. The establishment of the HSR system in Poland will enable efficient and economically effective transportation between the crucial national agglomerations. This is of key importance for their further development and, taking into consideration the fact that in the contemporary world economy big cities are the factor that stimulates economic growth, it will translate into the development of the whole country. The HSR system will facilitate the improvement of transport accessibility not only of cities connected directly by Y and CMK high-speed lines. The cities of eastern Poland can be included in the network of high-speed rail connections through the network of modernized rail lines. Ultimately, as the example of other countries shows, building further high-speed lines should be considered [29, 30].

Taking into consideration the changes within the meaning of increasing value of time, it should be expected that time will be the key factor in planning a journey, as it is playing a more and more significant role in everyday life. Transport systems are built in a long-standing perspective and the effects of their influence on economy are permanent. High-speed rail in Poland contributes to the improvement of the cohesion of the country and to the elimination of transport barriers – after-partition remnants.

Thanks to the effective connection of six voivodeships, high-speed rail will directly affect the area of almost half of the country and indirectly will have an impact on the whole country.

The construction of high-speed rail system is cohesive with the objectives and directions presented in strategic and policy documents of the European Union and Polish documents of transport policy on national and regional levels. Looking forward, the considered connections via the high-speed line to Prague and Berlin are already the subject of studies and analyses following the expectations of the European Commission.

Bibliography

[1] Ahlfeldt, G.M., and Feddersen, A.: *From Periphery to Core: Economic Adjustments to High-Speed Rail*, LSE, University of Hamburg, 2010, pp. 9, 48–50.

[2] *Analizy uwarunkowań społecznych rozwoju kolei dużych prędkości w Polsce w perspektywie roku 2050* [The analyses of social conditions of high-speed rail development in Poland in the perspective of year 2050]. Consorcium: Egis Poland Sp. z o.o., Ernst & Young Corporate Finance Sp. z o.o. i DHV POLSKA Sp. z o.o., 2012, Etap I Raport metodologiczny p. 30, Etap III Raport końcowy, pp. 32–72, Podsumowanie raportu etapu III, pp. 6–15 [Stage I Methodological Report p. 30, Stage III Final Report, pp. 32–72, Summary of the III stage report, pp. 6–15].

[3] *Dojazdy do pracy: Narodowy Spis Powszechny Ludności i Mieszkań 2011* [Commuting: National Population and Housing Census 2011]. GUS, Warszawa, 2014, www.stat.gov.pl/.

[4] Dyr, T.: *Strategia rozwoju transeuropejskiej sieci transportowej* [Strategy for the Development of the Trans-European Transport Network]. Technika Transportu Szynowego (TTS Technology-Transport-Systems), 1–2/2012.

[5] *EU Transport in Figures: Statistical Pocketbook*, European Commission, 2014, p. 52.
[6] Giedryś, A.: *Łódź as a Future HSR TEN-T Node*. Technika Transportu Szynowego (TTS Technology-Transport-Systems), 6/2015.
[7] Ginés de RUS (ed.), *Economic Analysis of HS Rail in Europe*. Economía y Sociedad, Informes, 2009, p. 119.
[8] GUS, Bank Danych Lokalnych www.stat.gov.pl/.
[9] Hernández, A., and Jiménez, J.L.: Does High-Speed Rail Generate Spillovers on Local Budgets? *Transport Policy*, Vol. 35, September 2014, p. 20. www.sciencedirect.com/science/article/pii/S0967070X14001243 [access: 12.06.2015]
[10] *High-Speed Rail as a Tool for Regional Development*, UIC Paris, 2011.
[11] *High-Speed Rail Wider Economic Benefits Study*. Final Report 16 October 2009, Halcrow Group, pp. 11–19.
[12] Ilik, J., and Pomykala, A.: *The Concept of High-Speed Connection Between Poland and Czech Republic*. Technika Transportu Szynowego (TTS Technology-Transport-Systems), 6/2015.
[13] *Koleje dużych prędkości w Polsce* [High speed rail in Poland] (ed.): Instytut Kolejnictwa, 2015, Komornicki, T. and Sleszynski, P., chapter 14, pp. 319–330.
[14] *Koncepcja Przestrzennego Zagospodarowania Kraju 2030* [The concept of land management of the country 2030]. Uchwała Nr 239 Rady Ministrów, 13 December 2011, Monitor Polski 27.IV.2012 poz. 252, p. 49.
[15] *Krajowa Strategia Rozwoju Regionalnego 2010–2020. Regiony, Miasta, Obszary wiejskie* [National Strategy for Regional Development 2010–2020. regions, cities, rural areas], pp. 92–94. Uchwała Rady Ministrów, 13 July 2010. Monitor Polski no 36, poz. 423 p. 1374.
[16] Kuang, Wenbo, Huapu, L.U.: *The Different Impacts of High-Speed Trains and Expressway on Chinese Population and Economic Spatial Distribution: An Empirical Study of Wuhan-Guangzhou Corridor*, Beijing, 2013, pp. 7–8.
[17] *L'opportunité pour la Grande Vitesse dans l'espace PECO* [Opportunities for high-speed rail in the Central-East Europe]. CENIT Barcelona for UIC, 2006.
[18] Massel, A.: *Development of the High-Speed Railway Infrastructure in Poland*. Technika Transportu Szynowego (TTS Technology-Transport-Systems), 6/2015.
[19] Ming, Zhang, Wu, Qing, Wu, Dianting, Zhao, Lin, and Liu, Xi: Analysis of the Influence on Regional Economic Development of High-Speed Railway. *Journal of Chemical and Pharmaceutical Research*, 2014, pp. 246–248.
[20] *PEER REVIEW Warszawa-Łódź-Poznan/Wroclaw: High-Speed Rail Feasibility Study. Preferred Route*. Paris: Transport Research Institute, XII, 2011, pp. 8–9, 16–24, 62–63.
[21] *Perry C. Interview*: Technika Transportu Szynowego (TTS Technology-Transport-Systems), 12/2015.
[22] Pita, A.L.: *UIC Study on Origin and Financing of First High-Speed Lines in the World, 2014*, pp. 7–110.
[23] *Program Rozwoju Infrastruktury Transportowej i Komunikacji dla Województwa Dolnośląskiego* [Program for the Development of Transport Infrastructure and Service for the Lower Silesian Voivodship], UM, Wroclaw 2015.
[24] *Plan Rozwoju Lokalnego Miasta Łodzi na lata 2007–2013* [City Development Plan for Łódź for perspective 2007–2013], UM, Łódź 2008.
[25] *Plan Zagospodarowania Przestrzennego Województwa Łódzkiego* [Spatial Development Plan of the Łódź Voivodship], UM, Łódź 2010.
[26] *Plan Zagospodarowania Przestrzennego Województwa Mazowieckiego* [Spatial Development Plan of the Mazowieckie Voivodship], UM, Warsaw 2013.
[27] *Plan Zagospodarowania Przestrzennego Województwa Wielkopolskiego* [Spatial Development Plan for the Wielkopolskie Voivodship], UM, Poznan 2014.
[28] Pomykała, A.: *9. Światowy Kongres Kolei Dużych Prędkości w Tokyo* [9. World Congress of high speed rail in Tokyo]. Technika Transportu Szynowego (TTS Technology-Transport-Systems) 1–2/2016.

[29] Pomykała, A., Raczyński, J., and Wróbel, I.: *Kierunki rozwoju kolei dużych prędkości w Polsce* [Directions of high-speed rail development in Poland]. Technika Transportu Szynowego (TTS Technology-Transport-Systems), 11/2010.

[30] Raczyński, J.: *Projekt kolei dużych prędkości w Polsce w kontekście trendów rozwojowych kolei w Europie* [High-speed rail project in Poland in the context of railway development trends in Europe]. Technika Transportu Szynowego (TTS Technology-Transport-Systems, 4/2015.

[31] Regulation (EC) No 1888/2005 of the European Parliament and of the Council of 26 October 2005 amending Regulation (EC) No 1059/2003 on the establishment of a common classification of territorial units for statistics (NUTS) by reason of the accession of the Czech Republic, Estonia, Cyprus, Latvia, Lithuania, Hungary, Malta, Poland, Slovenia and Slovakia to the European Union. https://publications.europa.eu/pl/publication-detail/-/publication/e9c64dac-e66a-4227-ae98-9155df09bc4a/language-en

[32] Commission Regulation (EC) No 105/2007 of 1 February 2007 amending the annexes to Regulation (EC) No 1059/2003 of the European Parliament and of the Council on the establishment of a common classification of territorial units for statistics (NUTS). https://eur-lex.europa.eu/legal-content/EN/TXT/?uri=CELEX:32007R0105

[33] Commission Regulation (EU) No 1319/2013 of 9 December 2013 amending annexes to Regulation (EC) No 1059/2003 of the European Parliament and of the Council on the establishment of a common classification of territorial units for statistics (NUTS) https://eur-lex.europa.eu/legal-content/EN/TXT/?uri=CELEX:32013R1319.

[34] *Strategia Rozwoju Województwa Łódzkiego na lata 2007–2020* [Development strategy for the Lodzkie Voivodship for the years 2007–2020], UM, Łódź 2006.

[35] Strategia rozwoju województwa małopolskiego 2011–2020 [Development strategy for the Małopolskie Voivodship 2011–2020], UM, Kraków 2011, pp. 18–25.

[36] *Strategia komunikacji i promocji projektu "Przygotowanie budowy linii dużych prędkości"* [Communication and promotion strategy for the project "Preparing the construction of high-speed lines"]. Martis CONSULTING Sp. z o.o., Warsaw 2011, pp. 13–27.

[37] *Studium rozwoju i modernizacji technologicznej transportu szynowego na Mazowszu w kontekście polityki transportowej Województwa Mazowieckiego* [Study on the development and technological modernization of rail transport in Mazovia in the context of the transport policy of the Mazowieckie Voivodeship]. UM, Warsaw 2009.

[38] *Studium Wykonalności dla budowy linii kolejowej dużych prędkości "Warszawa – Łódź – Poznań/Wrocław"* [Feasibility study for high-speed line Warsaw – Łódź – Poznań/Wrocław]. Ingenieria IDOM Internacional S.A., commissioned by PKP PLK S.A., Report III, February, 2011, pp. 24–355, Final document, July, 2013, pp. 73, 338–349.

[39] *Studium wykonalności dla budowy system Łódzkiej Kolei Aglomeracyjnej, I etap* [Feasibility study for the construction of the Łódź Agglomeration railway system, I stage], Łódź, 2009.

[40] Swianiewicz, P., and Klimska, U.: *Społeczne i polityczne zróżnicowanie aglomeracji w Polsce – waniliowe centrum, mozaika przedmieść* [Social and political diversification of agglomerations in Poland – Vanilla Center, Mosaic Suburbs], Warsaw, 2005.

[41] Sleszynski, P.: *Delimitacja miejskich obszarów funkcjonalnych stolic województw* [Delimitation of urban functional areas of Voivodeship CAPITALS]. Przegląd Geograficzny [Geografic Review], 2013, 85, 2, pp. 173–197.

[42] *Zrównoważony Plan Rozwoju Transportu Publicznego Poznania na lata 2007–2015* [Sustainable Plan for the Development of Public Transport in Poznań for the perspective 2007–2015]. UM, Poznan 2006.

[43] *Blue Book Railways*, Jaspers, December 2008, https://www.pois.2007-2013.gov.pl/WstepDoFunduszyEuropejskich/Documents/Blue_Book_Railways_1512_TC_without.pdf [access: 21.03.2011]

Chapter 4

Operational effectiveness of high-speed rails

Tadeusz Dyr and Karolina Ziółkowska

4.1. Introduction

The development of a high-speed rail system is one of significant establishments of the European transport policy for the second decade of the 21st century [2] [15]. High-speed rails are regarded as the factor of the rail transport revitalization and improvement competitiveness on the European passenger transport market. System developments justify the current market effects, in particular a dynamic growth of the number of transported passengers and the transport performance [6].

Among the strategic aims of creating a competitive and resource-efficient transport system in the European Union, a threefold length of the existing high-speed rail network by 2030 was established, while keeping a dense railway network in all member states [2]. New lines equipped for high-speed rails in Poland should be an important component of the European high-speed rails network. So far, their structure hasn't extended beyond studies, and most of arguments of supporters and opponents of these railways are political in character, supported by selective information about the functioning of high-speed rails in other countries. They concern in particular high investment expenditures on building new lines and purchasing means of transport. Investment decisions should be backed up by an assessment of the effectiveness of planned projects, in which the amount of expenditure is only one element. Investment decisions are compared, within the cost-effectiveness calculation, with forecasted and expressed in monetary units benefits related to the investment and future operating costs [4].

Credibility of the cost-effectiveness calculation – considering forecasted and expressed in monetary units costs and benefits – depends on being diligent in estimating the cash flows used in calculation formulas. In the case of the investment projects associated with the structure of the high-speed rail system in Poland, there is a lack of reliable data, which should be the base for forecasting costs and benefits. This problem concerns, however, largely forecasting the operational and maintenance costs of modernized lines. Historic costs and average costs estimated by JASPERS experts based on international comparisons are the base of the forecast [13], [19]. This approach is applicable for projects in which both the revitalization of the railway line and its modernization allowing for movement of trains with different speeds are assumed. Such an approach, although acceptable in projects co-financed from European Union funds, cannot be regarded as correct for evaluating the effectiveness of the investment in high-speed rails. Using this approach, the unit costs of the operation and maintenance of the railway line do not depend on train speed (the difference in the total cost in considered investment variants results from forecasting load on the line of passenger and freight transport, expressed in gross km-tons). Meanwhile, experiences of infrastructure managers exploiting high-speed rails lines show that there is a strong relationship between the high-speed trains and the unit costs of the operation and maintenance of the railway lines [10], [11], as well as for operating the rolling stock.

Considering restrictions in the access to credible data concerning operating costs of the infrastructure and means of transport, based on accessible reports, results of costs and benefits studies for different train speeds are presented in this chapter. It allows for conducting the simplified effectiveness analysis and indicates which investment variant is the most profitable.

4.2. Operating costs in the functioning of train speed

4.2.1. Methodological guidelines of estimating operating costs

The function of operating cost for the high-speed rail system was built on the basis of the *Relationship Between Rail Service Operating Direct Costs and Speed* report [10]. Only direct costs were included in it. In the report, however, indirect costs which are independent of train speed were omitted, although they can be associated with the standard of services carried out, which are reflected in the ticket price (e.g. meals on board). Also worth mentioning is that costs may differ inside explored countries depending on specific lines. In analysis, infrastructure costs not covered by charges, a revenue tax, and costs of running empty trains and related to shunting maneuvers were omitted. In spite of these restrictions, identified costs reflect their dependence on train speed, number of stations, trainset length, or route configuration.

"Capital costs" included in operating costs are understood to be the resulting cost needed to remunerate any economic resources used in the purchase of trains. This is done to make capital costs in the amount involved independent of the financial structure. It also makes it possible to put trains owned on the same level as rented trains, used upon lease agreements or other agreements of a similar nature (e.g. leasing).

The unit of measurement to be used for operating cost is the "Euro cent per seat-kilometre" (€-ct/seat-km). It allows for direct comparison of costs associated with the supply of services. Establishing the course of trains (number of kilometers), only a distance covered in providing the service was taken into account. All monetary values were given in net prices.

In building the model of cost identification, various configurations of rolling stock were assumed. Representative 100, 200, and 300 m-long trains intended for regional and long-distance transports were used in the study. With respect to comfort, two types were used: a single-class train with comfort for regional services (denoted by the "r" subscript) and another with first and second class and a cafeteria for long-distance services (denoted by the subscript "l"). Basic parameters of analyzed trains are described in Table 4.1.

The total number of seats for regional trains was calculated from the formula:

$$s_r = 3 \cdot (L - 7) \tag{4.1}$$

and for long-distance trains:

$$s_l = 1,7 \cdot (L - 7) \tag{4.2}$$

where:
L – train length [m].

In order to estimate unit costs in the quoted report, the annual distance covered by a train in terms of route length, turnaround time at final stations, and number of daily hours of service were estimated. In the case study it was accepted that average daily use would be 7 hours. This parameter is independent from the train speed. It takes account of the seasonality of demand, frequency, the peak time service policy and reservation (maximum requirements of transport), and maintenance policy. Based on conducted calculations, the annual distance covered by a train in the function of the commercial speed was established. Results of these calculations are presented in Table 4.2. They point at the improvement of train use along with the increase in the average commercial speed and the distance length. At low

Table 4.1 Relevant parameters of train used in operating costs identification [10]

Symbol	Parameters	Train type					
		T_{100l}	T_{100r}	T_{200l}	T_{200r}	T_{300l}	T_{300r}
L	Train length [m]	100	100	200	200	300	300
M	Tare [t]	160	160	320	320	480	480
s	Number of seats	158	279	328	579	498	879
CC	Comfort coefficient	1.5	1.0	1.5	1.0	1.5	1.0
C	Number of cabs	2	2	2	2	2	2

Table 4.2 Annual distances covered by a train in terms of commercial speed and average turnaround time [10]

Commercial speed [km/h]	Long distance trains (D = 500 km)			Regional trains (D = 200 km)		
	R = 90 [min]	R = 60 [min]	R = 30 [min]	R = 45 [min]	R = 30 [min]	R = 15 [min]
75	156 429	166 630	178 256	149 561	161 368	175 200
100	196 538	212 917	232 273	185 818	204 400	227 111
125	232 273	255 500	283 889	217 447	243 333	276 216
150	264 310	294 808	333 261	245 280	278 727	322 737
175	293 197	331 204	380 532	269 962	311 043	366 872
200	319 375	365 000	425 833	292 000	340 667	408 800
225	349 209	396 466	469 286	311 797	367 920	448 683
250	365 000	425 833	511 000	329 677	393 077	486 667

R – turnaround times
D – distances covered by a train [km]

speeds, differences of distance covered by a train are relatively small. They increase along with the average train speed.

4.2.2. Costs of train purchase

The maximum train speed is a crucial factor determining investments for the purchase of trains. Their value is possible to calculate from the formula:

$$TP = 150000 \cdot C + 26500 \cdot M + 1000 \cdot P + 25000 \cdot NM + 6000 \cdot s \cdot CC \ [euro] \qquad (4.3)$$

where:
C – number of cabs,
M – empty mass [t],
P – continuous power [kW],
NM – number of motors
s – number of seats,
CC – comfort coefficient.

Out of the factors listed, only the continuous power has an essential relation with the train speed. Based on carried theoretical distances, it was established that there is an essential relation between the power and the train speed (Table 4.3). Simultaneously, the relation between the speed and train mass is small (Table 4.4). Admittedly, the number of motors influences the volume of mass – as the power function, as well as the speed – but their mass compared

Table 4.3 Relation between power and speed [10]

Train type	Train length [m]	Power [kW] at speed [km/h]			
		160	200	250	300
T100$_l$	100	1 462	1 959	2 624	3 332
T100$_r$	100	1 462	1 959	2 624	3 332
T200$_l$	200	3 183	4 264	5 711	7 252
T200$_r$	200	3 183	4 264	5 711	7 252
T300$_l$	300	4 903	6 569	8 799	11 173
T300$_r$	300	4 903	6 569	8 799	11 173

Table 4.4 Relation between mass of the empty train and speed [10]

Train type	Train length [m]	Mass [t] at speed [km/h]			
		160	200	250	300
T100$_l$	100	194	199	206	213
T100$_r$	100	194	199	206	213
T200$_l$	200	391	402	417	432
T200$_r$	200	391	402	417	432
T300$_l$	300	589	605	627	651
T300$_r$	300	194	199	206	213

Table 4.5 Influence of the speed on the price of purchasing trains for passenger transports [10]

Train type	Train length [m]	Purchase price [thous. euro] at speed [km/h]			
		160	200	250	300
T100$_l$	100	7701.3	8291.6	9082.2	9923.8
T100$_r$	100	7617.6	8207.9	8998.5	9840.1
T200$_l$	200	15,516.1	16,800.6	18,521.2	20,352.6
T200$_r$	200	15,342.4	16,626.9	18,347.5	20,178.9
T300$_l$	300	23,330.9	25,309.6	27,960.2	30,781.4
T300$_r$	300	23,067.2	25,045.9	27,696.5	30,517.7

with the mass of cars without motors is relatively low. In consequence, in the quoted report it was assumed that mass depended mainly on the train length. Whereas provided the power installed in trains moving with a speed of 300 km/h is over twice that of trains with a speed of 160 km/h, the difference of mass does not exceed 10%.

Considering the assumptions, the cost of train purchase depending on the permissible speed and the train type and length was estimated. The results of calculations are presented in Table 4.5. They show that the price has a relatively small dependence on the speed. The difference in price of the trains adapted to speeds of 300 km/h and 160 km/h is only about 30%.

4.2.3. Costs arising from train possession

The price of purchasing trains provides the basis to estimate annual costs resulting from their possession. They are dependent on the predicted operating period, the cost of capital, and the insurance rate. For calculations, the following values were assumed:

operating period – 25 years,
residual value – 0,
insurance rate – 2%,
interest rate – 6%.

At accepted assumptions, annual costs resulting from the possession amount to 9% of the purchase price. The influence of the maximum speed on annual costs is shown in Table 4.6.

The values described in Table 4.6 form the basis for estimating annual unit costs. They were calculated from the formula:

$$OC = \frac{TC}{s \cdot RA \cdot 100} \; [eurocent/seat - km] \tag{4.4}$$

where:
TC – total annual cost arising from train possession,
s – number of seats,
RA – annual distance covered by the train (km/year).

Table 4.6 Influence of the speed on annual costs resulting from the possession of trains for passenger transport [10]

Train type	Train length [m]	Costs [thous. euro] at speed [km/h]			
		160	200	250	300
T100$_l$	100	693.1	746.2	817.4	893.1
T100$_r$	100	685.6	738.7	809.9	885.6
T200$_l$	200	1396.4	1512.1	1666.9	1831.7
T200$_r$	200	1380.8	1496.4	1651.3	1816.1
T300$_l$	300	2099.8	2277.9	2516.4	2770.3
T300$_r$	300	2076.0	2254.1	2492.7	2746.6

Table 4.7 The influence of speed on annual unit costs of train possession for passenger transports [10]

Train type	Train length [m]	Unit costs [€-ct/seat-km] at speed [km/h]			
		160	200	250	300
T100$_l$	100	1190	1081	1001	961
T100$_r$	100	650	591	545	596
T200$_l$	200	1155	1055	984	950
T200$_r$	200	644	577	535	515
T300$_l$	300	1151	1047	978	946
T300$_r$	300	631	572	532	513

Considering presented assumptions, unit costs of train possession by speed function were calculated (Table 4.7). Results of the calculations show that unit costs decrease with an increase in speed. The decrease in these costs reflects the higher cost-effectiveness of the use of high-speed trains. It results from the fact that, even with identical ticket prices, a profit margin will be higher as the speed increases.

4.2.4. Rolling stock maintenance and cleaning costs

Maintenance costs for vehicles include conservational and repair operations, interior and exterior cleaning and repair, the purchase of materials, and removing damages caused by vandals. These actions require the use of workshops, warehouses, pits, examination facilities, train washes and staff, spare parts, and other materials (water, pens, paper, etc.). Total maintenance costs of trains are a sum of fixed and variable costs.

The study results presented in the quoted report show that annual fixed costs of train maintenance (CF) are on the level of 1.175 euro/1 m of rolling stock. Considering this value, individual fixed costs of train maintenance per seat-km may be calculated by the formula:

$$cf = \frac{CF*L}{s*RA}*100\left[(€ - \frac{ct}{seat} - km)\right] \tag{4.5}$$

where:
CF – fixed maintenance cost per linear train meter [euro/1 m of train],
L – length of the train [m],
s – number of train seats (seats),
RA – annual distance covered by the train (km/year).

Variable costs refer mainly to consumption of materials and labor requirements corresponding (in order of importance) to worn wheels, shoes, pantographs, disc brakes, and axles. Greater radius of the bend, lower level of pneumatic brakes use, and low number of traction energy supply system, along with an increase in speed, limits the costs of consuming these materials. A significant decrease was observed, particularly in the case of pantographs, after the change from conventional lines to high-speed lines. The study results presented in the quoted report show that annual variable costs of train maintenance (*CV*) are on the level of 0.98 €-ct per linear meter and per kilometer covered. Considering this value, individual variable costs of train maintenance per seat-km may be calculated from the formula:

$$cv = \frac{CV \cdot L}{s} \quad \left[\text{€} - ct \, / \, seat - km \right]$$
(4.6)

where:
CV – variable costs of train maintenance per linear meter and per kilometer covered (€ – ct/m-km),
L – train length [m],
s – number of seats.

Maintenance costs also include variable costs of a workshop. The variable costs of workshop maintenance include electricity, water, and other charges for workshop management. The study results presented in the quoted report show that variable costs of workshop maintenance (*CT*) are on the level 17.5 98 €-ct per linear train meter and per kilometer covered between inspections (*RR*) on the level of 6 thousand km. Considering these values, the unit fixed cost of the workshop per seat-km may be calculated from the formula:

$$ct = \frac{CT \cdot L \cdot RR}{s} \quad \left[(\text{€} - ct / seat - km \right]$$
(4.7)

where:
CT – workshop maintenance cost per linear train meter and per kilometer covered between inspections (*RR*) (€/m-km),
RR – kilometers covered between inspections [km],
L – train length [m],
s – number of seats.

Cleaning costs cover the costs of cleaning work and interior and exterior preparation of the train for each journey made. They depend on the gross area of the train, which is directly related to its length and the route length. The study results presented in the quoted report show that unit cleaning cost (*cc*) are on the level of 1.33 euro per train meter. Considering this value, unit cleaning costs per seat-km may be calculated from the formula:

$$CC = \frac{cc \cdot L}{D \cdot s} \cdot 100 \; \left[\text{€} - ct / seat - km \right] \tag{4.8}$$

where:
cc – unit cleaning costs [euro/train meter],
L – length of the train [m],
D – route distance [km],
s – number of train seats.

Considering presented assumptions, unit variable costs of train maintenance and unit cleaning costs in the function of train speed and type of transport were calculated. Results of calculations are presented in Table 4.8.

4.2.5. Energy costs

Energy costs analysis conducted in the report [10] comprised:

charges for energy consumed,
benefits (revenue) received from energy returned at the substation,
distribution charges.

Charges for the traction energy consumption are strongly correlated with the mass of the train and its speed. They depend also on energy loss in the network and the unit price of electricity. Analogous factors affect benefits from the sale of returned energy.

Table 4.8 Unit variable costs of train maintenance and unit cleaning costs in the function of train speed and type of transport covered by 200 m-long trains [€-ct/seat−km]

Commercial speed [km/h]	Long-distance trains (D = 500 km);			Regional trains (D = 200 km)		
	R = 90 [min]	R = 60 [min]	R = 30 [min]	R = 45 [min]	R = 30 [min]	R = 15 [min]
75	1.28	1.25	1.23	0.88	0.86	0.84
100	1.19	1.16	1.13	0.82	0.80	0.78
125	1.13	1.10	1.08	0.79	0.77	0.75
150	1.09	1.07	1.04	0.77	0.75	0.73
175	1.07	1.04	1.01	0.75	0.73	0.72
200	1.05	1.02	0.99	0.74	0.72	0.70
225	1.03	1.00	0.98	0.73	0.71	0.69
250	1.02	0.99	0.96	0.73	0.71	0.69

R – turnaround time

D – travel distance [km]

Source: Own study based on [10]

The cost of energy described in the report [10] was calculated with the formula:

$$EC_s = E_{SM} \cdot EP_{SM} = E_{PM} \cdot \pi \cdot EP_{SM} \left[\frac{eurocent}{train} - km \right] \qquad (4.9)$$

where:

E_{SM} – energy imported on entering the substation [kWh/train-km],
EP_{SM} – energy price per km [€-ct/kWh],
E_{PM} – pantograph imported energy [kWh/train-km],
π – Ratio of energy imported on entering the substation to pantograph imported energy:

$$\pi = \frac{E_{SM}}{E_{PM}} \qquad (4.10)$$

The cost of energy per seat-km therefore being:

$$ec = \frac{E_{SM}}{S} = \frac{E_{PM} \cdot \pi \cdot EP_{SM}}{S} \left[\frac{eurocent}{train} - km \right] \qquad (4.11)$$

Energy consumption (in the pantograph) for a train with a mass of M tons running at an average speed of S_{av} km/h can be estimated as:

$$E_{PM} = \left(5,867 \cdot 10^{-7} \cdot S_{av}^2 - 2,878 \cdot 10^{-5} \cdot S_{av} + 2,163 \cdot 10^{-2} \right) \cdot M \left[\frac{kWh}{train} - km \right] \qquad (4.12)$$

If the value of $\pi = 1.03$ adopted for AC current and $\pi = 1.11$ for DC current and unit energy price is 9 eurocent, cost of energy per seat-km is:

$$ec = \frac{1,03 \cdot 9 \cdot \left(5,867 \cdot 10^{-7} \cdot S_{av}^2 - 2,878 \cdot 10^{-5} \cdot S_{av} + 2,163 \cdot 10^{-2} \right) \cdot M}{S} \qquad (4.13)$$

Cost of energy returned (€-ct/seat-km) described in the report was calculated with the formula:

$$rec = \frac{E_{Px} \cdot \dfrac{1}{\pi} \cdot EP_{SM} \cdot \beta}{S} \qquad (4.14)$$

where:

E_{Px} – amount of energy returned [kWh/train-km],
π – loss factor between substation and pantograph,
EP_{SM} – price of imported energy [€-ct/kWh],
β – coefficient of the price of energy exported/price of imported energy,
s – number of train seats.

Energy exported for a train with a mass of M tons can be estimated as:

$$E_{Px} = \left(-1,76 \cdot 10^{-8} \cdot S_{av}^2 - 1,76 \cdot 10^{-8} \cdot S_{av}^2 + 0,004 \right) \cdot M \left[\frac{kWh}{train} - km \right] \qquad (4.15)$$

where:

S_{av} – average train speed [km/h].

So for the train in the above case and assuming that $\beta = 1$ and $\pi = 1.03$, cost of energy returned per seat-km was calculated with the formula:

$$rec = \frac{9 \cdot \left(-1,76 \cdot 10^{-8} \cdot S_{av}^2 - 1,76 \cdot 10^{-8} \cdot S_{av}^2 + 0,004\right) \cdot M}{1,03 \cdot s} \quad \left[\frac{eurocent}{train} - km\right] \quad (4.16)$$

The cost of electric energy distribution is assumed proportional to the cost of imported energy:

$$ced = \alpha \cdot ec \quad \left[\frac{eurocent}{train} - km\right] \quad (4.17)$$

where:
ec – cost of energy per seat-km [€–ct/seat–km], calculated according to above rules,
α – energy distribution coefficient (its typical value is 0.015).

The total energy cost will be obtained by adding the cost of traction energy and auxiliary power forms consumed per seat-km, subtracting the cost of energy returned to the network and adding the distribution cost:

$$tec = ec - rec + ced = ec - rec + \alpha \cdot ec = (1+\alpha) \cdot ec - rec \quad \left[\frac{eurocent}{train} - km\right] \quad (4.18)$$

Considering presented methodical assumptions, the unit cost of traction energy in the function of train speed and type of transport were calculated. Results of the calculations are presented in Table 4.9.

Table 4.9 Unit cost of traction energy in the function of train speed and type of transport covered by 200 m-long trains [€–ct/seat–km]

Commercial speed [km/h]	Long distance trains	Regional trains
Energy charges		
75	0,207	0,117
100	0,223	0,127
125	0,247	0,140
150	0,277	0,157
175	0,314	0,178
200	0,357	0,202
225	0,408	0,231
250	0,464	0,263
Benefits from returned energy		
75	0,03213	0,01820
100	0,03110	0,01762
125	0,02988	0,01693
150	0,02847	0,01613
175	0,02688	0,01523
200	0,02510	0,01422
225	0,02313	0,01310
250	0,02097	0,01188

Commercial speed [km/h]	Long distance trains	Regional trains
Cost of energy distribution		
75	0,00310	0,00176
100	0,00335	0,00190
125	0,00371	0,00210
150	0,00416	0,00236
175	0,00471	0,00267
200	0,00536	0,00304
225	0,00611	0,00346
250	0,00696	0,00394
Total energy cost		
75	0,17753	0,10057
100	0,19573	0,11088
125	0,22085	0,12511
150	0,25288	0,14326
175	0,29184	0,16533
200	0,33772	0,19131
225	0,39051	0,22122
250	0,45023	0,25505

Source: Own study based on [10]

4.2.6. Train operation personnel costs

Personnel providing services in the operation can be classified into two main groups:

1. personnel providing ground services: this may in turn be operating personnel (e.g. personnel for reception and welcoming at stations, information and ticket sales, after-sales service) or indirect personnel (dedicated to administration or commercial activities).
2. personnel providing services on board the train: this is the case of personnel for train operation, ticket inspection, and those serving passengers on board.

Study results described in the report [10] show that in the first group there is no relationship between costs and the speed of trains, so we should recognize these costs as independent of the speed.

With personnel working on board the trains, there is a clear link between their costs and the average speed of the trains. This personnel does in fact have a fixed working timetable (typically 40 hours a week or 1,880 hours a year). If the trains move more quickly, employees who travel on the trains will cover more kilometers in the same time, thereby reducing the number of people required to offer the same service and, therefore, the labor cost per unit of supply.

As proposed in the report [10], the method of setting the unit cost of the on-board personnel casts serious doubts. Therefore, using assumptions made in this report, an author's method was suggested.

Based on the report [10], the following assumptions, resulting from empirical studies conducted by authors, were adopted:

annual cost to the company of a driver – LCd = 40 000 €/person,
annual cost to the company of accompanying personnel – LC_a = 30 000 €/person,
average number of days a year the driver works J_d = 210 days/year,
average number of days a year each accompanying person works on the train – J_a = 210 days/year,
average number of days use of trains at the draught work – J_t = 365 days/year,
average hours a day the driver actually operates the train – H_d = 4 hours/day,
average hours a day each accompanying person works on the train – H_a – 4.5 hours/day,
average hours a day use of trains at the draught work – H_t – 7 hours/day.

Based on made assumptions, it is possible to calculate annual working hours of the staff and train use:
train driver working time:

$$Y_d = J_d \cdot H_d \quad \left[\frac{hours}{year} \right] \tag{4.19}$$

on-board personnel for passenger services working time:

$$Y_a = J_a \cdot H_a \quad \left[\frac{hours}{year} \right] \tag{4.20}$$

use of trains at the draught work working time:

$$Y_t = J_t \cdot H_t \quad \left[\frac{hours}{year} \right] \tag{4.21}$$

Placing accepted average values on the basis of the report [10], we receive:
train driver working time – 840 hours/year,
on-board personnel for passenger services working time – 950 hours/year,
use of trains at the draught work working time – 2,555 hours/year.
Total number of employees in train operation is:
train drivers:

$$N_d = n_d \cdot \frac{Y_t}{Y_d} \tag{4.22}$$

on-board personnel for passengers services:

$$N_a = n_a \cdot \frac{Y_t}{Y_a} \tag{4.23}$$

where:
n_d – number of train drivers actually driving the train,
n_a – number of on-board personnel for passengers services in the train.

Using made assumptions and markings, unit personnel cost is possible to calculate from the formula:

$$sc = \left(\frac{LC_d \cdot N_d + LC_a \cdot N_a}{s \cdot RA} \right) \cdot 100 \left[\epsilon - \frac{ct}{seat} - km \right] \tag{4.24}$$

Table 4.10 Unit cost of salaries of on-board personnel in the function of train speed and type of transport covered by 200 m-long trains [€-t/seat-km]

Commercial speed [km/h]	Long distance trains (D = 500 km);			Regional trains (D = 200 km)		
	R = 90 [min]	R = 60 [min]	R = 30 [min]	R = 45 [min]	R = 30 [min]	R = 15 [min]
75	0,550	0,516	0,482	0,326	0,302	0,278
100	0,437	0,404	0,370	0,262	0,238	0,214
125	0,370	0,336	0,303	0,224	0,200	0,176
150	0,464	0,416	0,368	0,283	0,249	0,215
175	0,418	0,370	0,322	0,257	0,223	0,189
200	0,384	0,336	0,288	0,238	0,204	0,170
225	0,351	0,309	0,261	0,223	0,189	0,155
250	0,336	0,288	0,240	0,211	0,177	0,143

R – turnaround time
D – travel distance [km]

Source: Own study based on [10]

Considering presented methodical assumptions, the unit cost of salaries of on-board personnel in the function of train speed and type of transport covered by 200 m-long trains were calculated. Results of calculations are presented in Table 4.10. In calculations, it was assumed that the train staff are two train drivers and two on-board personnel for passenger services (e.g. the train manager and the conductor). Conducted calculations confirm the thesis explicitly, that unit on-board personnel costs decrease along with the increase of the speed.

4.2.7. Infrastructure use charges

Rules of determining rail infrastructure use charges were formulated mainly from Directive 2001/14/WE [7]. The main rule requires basing the charges on costs arising directly from the infrastructure manager as a result of making available to the operator of the train route at determined line and due to a train movement. These costs were not specified closely in the directive, but they are often identified with so-called marginal costs. Charges based on marginal costs should include the minimum service package (minimum access package and track access to service facilities), specified by the infrastructure manager [17].

Direct impact on the level of unit charges for the access to the rail infrastructure has state policy in financing the maintenance and the development of rail infrastructure from public means. Because these charges are an instrument influencing a branch distribution of transports – they can stimulate growth or create a decline in the participation of the rail transport in transport systems of individual countries and in the entire Union.

The latest regulations of the European Union don't change the rules of manager costs identification, being the base to form charges for access to the train infrastructure. According to Directive 2012/34/UE [8], charges for minimum access package and track access to service facilities are set at the cost that is directly incurred as a result of operating the train service. Neither the scope of costs identification, considered as direct costs, nor common principles of the allocation of costs were implemented [18].

Considering EU regulations presented in the report [10] as the base for calculating charges for the access to the rail infrastructure in the speed function, a concept of

Table 4.11 Unit for access to the rail infrastructure in the function of train speed and type of transport covered by trains with length of 200 m and mass of 320 t [€-t/seat-km]

Commercial speed [km/h]	Long distance trains	Regional trains
75	0,230	0,227
100	0,229	0,227
125	0,229	0,227
150	0,232	0,228
175	0,235	0,230
200	0,240	0,233
225	0,247	0,237
250	0,256	0,242

Source: Own study based on [10]

marginal costs was adopted. It was recognized that other factors (e.g. transport policy, funding strategy of the train infrastructure, operators' ability to cover the costs) have nothing to do with the speed of the train. It was emphasized that high-speed trains have a higher ability to cover charges for access to rail infrastructure and therefore are often charged with the payment exceeding marginal cost. However, this doesn't have to be the rule.

According to empiric studies, unit charges for access to the rail infrastructure per seat-km were established. They can be calculated from the formula:

$$ci = \frac{M}{1000} \cdot \left(\frac{64}{s} \cdot (0,004 \cdot S_{av} - 0,83) \cdot 0,005 \cdot S_{av} + 0,7 \right) + \frac{6}{s} \left[€ - \frac{ct}{seat} - km \right] \qquad (4.25)$$

where:
M – empty mass [t],
s – number of seats,
S_{av} – average train speed [km/h].

Considering the presented methodical assumptions, the unit for access to the rail infrastructure in the function of train speed and type of transport covered by trains with length of 200 m and mass of 320 t were calculated. Results of calculations are presented in Table 4.11.

4.3. Cost-effectiveness of high-speed rails

Described in section 4.2, operating costs of the rail transport completion are the base for determining direct unit operating costs in the function of train speed. They are the amount of all analyzed costs (Table 4.12). Calculation results are presented for eight levels of the commercial speed and trains 200 m long. Considering the typical relationship between the maximum and commercial speed, operating costs were presented for the speed from 120 km/h up to 350 km/h.

Table 4.12 Direct unit operating costs the function of train speed and type of transport covered by trains with length of 200 m [€-t/seat-km]

Commercial speed [km/h]	Long distance trains (D = 500 km);		
	R = 90 [min]	R = 60 [min]	R = 30 [min]
75	3,524	3,463	3,401
100	3,242	3,181	3,119
125	3,108	2,946	2,813
150	3,098	3,022	2,946
175	3,048	2,972	2,896
200	2,994	2,918	2,842
225	2,968	2,846	1,875
250	2,956	2,880	2,804

R – turnaround time
D – travel distance [km]

Source: Own study based on [10]

Presented methodological assumptions of estimating particular elements of unit direct costs in the function of train speed show that unit costs per seat-km decrease along with the increase in commercial speed. Presented data contradict arguments often formulated about very high costs of implementation of high-speed trains transports.

Detailed analysis of the individual indicators of direct costs associated with the completion of passenger transports show that along with the increase in the speed, there's a growth in cost of traction energy consumption and charges for the access to infrastructure. However, unit costs of capital (associated with purchasing trains), cost of train maintenance, and cost of on-board personnel work decrease. These relationships – based on calculations conducted for the train: 200 m long, covering long distances in passenger transports, on the line of 500 km, with turnaround time of 60 min – are presented in Figure 4.1. They show that in the analyzed case, total unit costs decrease – with an increase in commercial speed from 70 km/h up to 250 km/h (corresponding to the maximum speed of 120–350 km/h) –about 17%. The biggest decrease of unit costs is noticeable up to about 160 km/h. At an increase in the maximum speed around 160 km/h to 350 km/h, unit costs decrease about 5.5%. Similar evolution of costs confirm study results led by other teams (cf. for example [9]).

Assessing the cost-effectiveness of high-speed rails, a current value of the direct costs stream for the passenger services on the hypothetical railway line of 350 km long was calculated.

It was assumed that to meet demand it would be necessary to use trains offering 820 million seat-km annually. 200 m-long trains with 328 seats will be used for the service.

Considering estimated unit direct costs, annual total costs of services implementation in the function of train speed (Table 4.13) were calculated from the formula:

$$C_{t,v} = 100 \cdot PE_t \cdot \sum_{k=1}^{n} c_{k,v} \ [euro] \tag{4.26}$$

where:
PE_t – exploitation work in the year t [train-km/year],
$c_{k,v}$ – unit direct costs of completion of long-distance transports k-category in the function of train speed [€-t/seat-km].

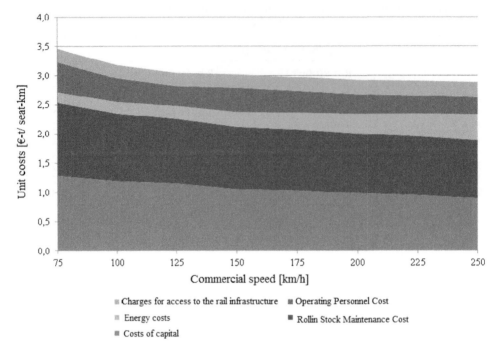

Figure 4.1 Structure of unit direct costs in long-distance transport in the function of train speed and type of transport [€-t/seat-km]

Source: Own study based on [10]

Table 4.13 Total direct costs of completion of transports on the hypothetical railway in the function of train speed

Commercial speed [km/h]	Total direct costs [euro/year]
75	28,393.2
100	26,081.4
125	24,976.8
150	24,779.9
175	24,779.9
200	24,369.7
225	23,927.5
250	23,791.1

Source: Own study based on [10]

Estimated total costs constitute the basis for determining the current net value of direct costs in the calculation period. This value is calculated from the formula:

$$NPV_{C,v} = \sum_{t=1}^{n} \frac{C_{t,v}}{(1+r)^t} [euro]$$ (4.27)

where:

$C_{t,v}$ – total direct costs when traveling long distances in the function of train speed in a year t [euro/year],

r – discount rate [%],
t – following year of calculation period,
n – number of years of calculation period.

Taking a 25-year calculation period and the 5% discount rate, a current value of direct operating cost was calculated. The calculation results are presented in Figure 4.2. They confirm the thesis about the higher cost-effectiveness of high-speed trains. Along with the speed costs increase, a net present value of the direct costs associated with rail transport decreases.

In order to set real differences of costs of transport implementation in the function of speed, an incremental (differencing) method was applied. Its application makes calculations independent of costs unrelated to train speed. These costs, as stated in the first part of the chapter, were omitted in the analysis, as their amount is identical for every speed of train.

According to the idea of the incremental method, cost increase was set in each year of the calculation period towards the reference variant:

$$CF_{C,t} = CF_{C,t}^{a} - CF_{C,t}^{r} \; [euro] \tag{4.28}$$

where:
$CF_{C,t}$ – incremental costs stream in t year [euro],
$CF_{C,t}^{a}$ – costs of the analyzed variant in t year [euro],
$CF_{C,t}^{r}$ – costs of reference variant in t year [euro].

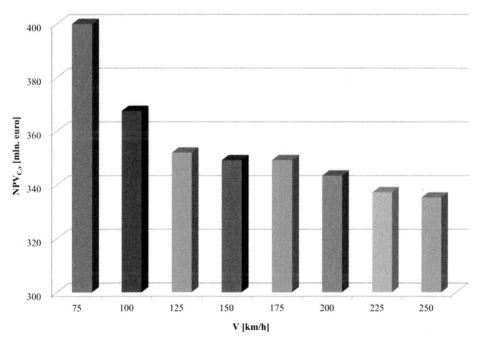

Figure 4.2 Net present value of the direct costs associated with long-distance rail transport in the function of speed and type of train [euro]

Source: Own study

In the evaluation of the cost-effectiveness with differencing method as the reference variant operating the traffic of trains with the commercial speed of 125 km/h was accepted (with a maximum speed of about 160 km/h).

Based on incremental cash flows, a net current value of the cost savings was calculated:

$$NPV_{CF,v} = -\sum_{t=1}^{n} \frac{CF_{C,t}}{(1+r)^t [euro]} \tag{4.29}$$

where:

$CF_{C,t}$ – incremental costs stream in t year [euro],
r – discount rate [%],
t – following year of calculation period,
n – number of years of calculation period.

Considering formulated assumptions, a current value of the incremental costs of transport implementation for 25 years of the calculation period and the discount rate $r = 5\%$ was calculated. Results of calculations are presented in Figure 4.3. They show that along with the speed increase, a current value of the cost saving increases. Positive values of this indicator for the commercial speed above 125 km/h, i.e. the maximum speed of about 160 km/h, show that these variants are more cost-efficient than providing the transport services with the maximum speed of 160 km/h. Results of this analysis confirm the thesis about the higher cost-effectiveness of the high-speed trains.

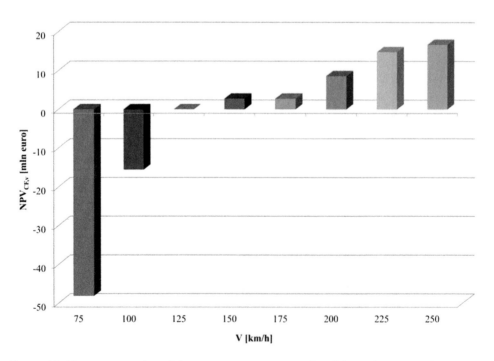

Figure 4.3 Net present value of the cost savings associated with long-distance rail transport in the function of speed and type of train [euro]

Source: Own study

4.4. Direct benefits on the passenger transport implementation with high-speed trains

So far, presented dissertations have concentrated on comparing costs of providing transport services with trains of different speeds. Aspects of the cost, although essential, cannot constitute a sufficient basis for making investment decisions. Providing transport services should also be included. They include mostly revenue of operators from the sale of services, generally called ticket sales revenues. Their value depends on the number of rail transport passengers and the price paid by the passenger for the ride (ticket price).

The number of passengers, being a measure of the demand for the railway service, is determined by many factors that are both dependent on the rail operator (e.g. charge, offer quality) and independent (e.g. willingness to pay, communications behaviors, needs and preferences of consumers). The price of railway services and travel time are the most crucial factors determining the volume of demand for rail transport services.

An increase in the speed of trains – shortening the travel time – results in an increase in the demand for services. To measure the relationship between the demand and the travel time, a temporal demand elasticity is used. It designates a relative demand variation as the reaction to a reaction to relative variation of travel time by 1%:

$$e_t = \frac{\Delta Q}{Q_0} : \frac{\Delta T}{T_0} \tag{4.30}$$

where:
ΔQ – demand change (number of transported passengers),
ΔT – travel time change,
Q_0 – demand by T_0 travel time.

According to the law of demand, along with a price rise, a demand for goods decreases (in further deliberations, atypical cases of demand forming will be omitted). Power of the relationship between these market elements reflects a price elasticity of demand calculated, like in case of travel time elasticity:

$$e_p = \frac{\Delta Q}{Q_0} : \frac{\Delta P}{P_0} \tag{4.31}$$

where:
ΔQ – demand change (number of transported passengers),
ΔP – price change,
Q_0 – demand by P_0 price.

A general practice in the train passenger transport market is a price rise along with an increase in trains' commercial speed. The price and the travel time have an opposite influence on the demand volume of transport services. Ultimate development of demand depends on the level of price and travel time elasticity.

Analysis of proposals in the train passenger transport market shows that the travel time elasticity is higher than the absolute value of price elasticity of demand (the first shows a positive value in typical cases; the second one is negative). Consequently, in spite of the price rise in transport services, the demand increases after increasing the speed of trains. The graphical presentation of these relationships are presented in Figure 4.4.

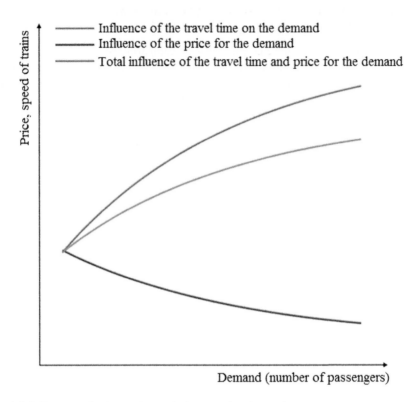

Figure 4.4 Influence of price and travel time on the demand

Source: Own study

In analysis of the cost-effectiveness of the high-speed rail system, a steady demand for services and steady services supply are measured by the number of seat-km. In light of the above deliberations on the elasticity of demand, maintaining a steady demand at the changing speed of trains requires implementing appropriate changes in prices. It allows a formulation of the thesis that revenues for the rail operator are higher for the shorter travel time. In consequence, it leads to a higher effectiveness of high-speed rails.

In order to assess the benefits from shortening the travel time based on the Civity Management Consultants studies [9], unit revenues from passengers' transport formation was accepted:

11 €-t/1 pas-km for maximum speed up to 200–220 km/h,
12,1 €-t/1 pas-km for maximum speed up to 250–280 km/h,
13,2 €-t/1 pas-km for maximum speed up to 300–350 km/h [9] [25].

However, the described rise in prices doesn't lead to hampering the demand. Shortening the travel time – resulting from the increase in train speed – shows that in spite of operators increasing the price, the demand increases. Adopting analogous trends of changes in prices and the demand, as in the quoted study [25] and assuming that with the maximum speed of

300–350 km/h and an average seat occupancy on the train of 60%, a volume of demand and revenue from the passenger transport for the hypothetical railway line was estimated. The amount of revenue was calculated as the transit work and the unit revenue:

$$R_{t,v} = Q_{t,v} \cdot r_{t,v} \cdot 100 \left[euro\right] \tag{4.32}$$

where:
$R_{t,v}$ – revenue from the passenger transport [euro] in *t-year* with *v-speed* [km/h],
$r_{t,v}$ – unit revenue [eurocent/pas-km] in *t-year* with *v-speed* [km/h],
$Q_{t,v}$ – revenue volume of passenger transport (transit work) [pas-km/year] in *t-year* with *v-speed* [km/h].

Considering presented methodological assumptions, a volume of demand and revenue from passenger transports on the hypothetical railway line was estimated (motor and transit parameters were the same, as in the analysis of the cost-effectiveness). Results of calculations were compared in Table 4.14. They confirm the thesis about the essential relationship between shortening the travel time and the sales of services revenue.

The growth in revenue, as the effect of shortening the travel time, isn't the only benefit from building the high-speed rail system. An extremely important effect is the improvement of railways competitiveness on the transport market. High-quality services support stimulating the growth in the economy.

Competitiveness is the ability of the enterprise to successfully and effectively achieve market purposes [21]. This ability lets the enterprise to avoid effects of presenting a more favorable offer by other company, influencing the decision of entering into a transaction [3]. From such a perspective, competitiveness is a factor of the company's market expansion, affecting its market development. Creating competitiveness can concern both the individual company and the entire branches of industry. In case of passenger rail transport, competitiveness is connected with the ability of companies from the transport branch to oppose the expansion of individual transport and other transport branches – particularly air transport. Thanks to the ability to present an effective transit experience, rail carriers can influence transport users' behavior, contributing to their resignation from using their own cars and the services of other operators [6]. Statistical data confirms that implementing a high-speed rail system contributes not only to the growth of the number of passengers transported by fast trains, but also to the growth of other segments of the train passenger transport market.

Table 4.14 Revenue from passenger transports on the hypothetical railway line in the function of the trains' speed

Maximum speed [km/h]	Demand [mln paskm/year]	Unit revenues [eurocent/paskm]	Revenues [thous. euro/year]
160	328.0	11.0	36,080
200–220	426.4	11.0	46,904
250–280	470.1	12.1	56,882
300–350	492.0	13.2	64,944

Source: Own study based on [25]

4.5. Financial effectiveness of high-speed rails

Estimated amounts of the revenues and costs base the assessment of the impact of the trains' speed to the effectiveness of high-speed rails. Net present value rate can be a measure of evaluating effectiveness. It's calculated from the formula:

$$NPV = \sum_{t=1}^{n} \frac{B_t - C_t}{(1+r)^t} \ [euro]$$

(4.33)

where:
B_t – benefits in t-year,
C_t – spending in t-year,
r – discount rate [%],
t – following year of calculation period,
n – number of years of calculation period.

To evaluate the effectiveness of the high-speed rail system, an incremental method was applied. Putting estimated revenues and costs together in one formula, under the above assumptions, would be incorrect. The estimated revenues are the most important part of the operator's direct benefits. Calculations of expenses were limited only to the direct costs associated with the speed of trains. It means that all indirect costs and those regarding direct costs of services completion, which remain steady along with the increase in the speed of trains, were omitted. Conducted calculations show that at every analyzed maximum speed above 160 km/h, direct costs are covered by the sales of services revenue (Figure 4.5). Along with the increase in the speed,

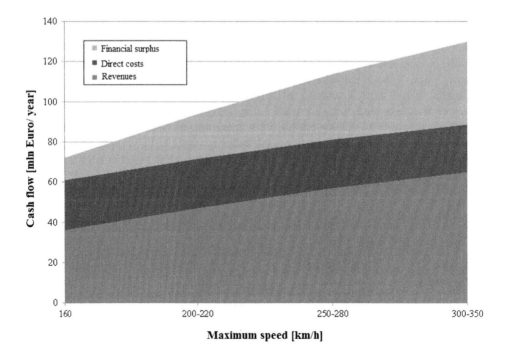

Figure 4.5 Direct revenues and costs in the function of train speed

Source: Own study

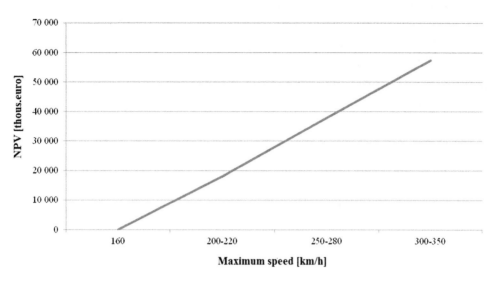

Figure 4.6 Net present value incremental (NPV) in the function of train speed

Source: Own study

cash flow increases considerably. Its high value for the maximum speed above 300 km/h confirms the ability of high-speed rails to generate a net profit. A relatively small surplus of revenue against direct costs at a speed of 160 km/h indicates that after including indirect costs and other direct costs (unrelated to train speed), reaching a net profit can constitute the relevant problem (this doesn't mean, however, that there will be losses generated in every case).

Using the incremental method, an increase in revenue and direct costs were estimated in analyzed variants towards the referential variant, i.e. the variant of operating train traffic with a commercial speed of 125 km/h (maximum speed of about 160 km/h). Such an approach allows omitting the revenues and costs that aren't dependent on train speed. In each variant, they have the same value, which means that the increase equals 0. According to incremental cash flows, the net present value for analyzed speed variants was calculated. Results of calculations are presented in Figure 4.6.

Calculated NPV rates indicate that with the increase in train speed (shortening the travel time), the current value of the net benefit increases. It is a result of the increase in the revenues from services implementation and the reduction of unit direct costs. At the permanent transport offer, decreasing unit costs results in the decrease of total costs.

Smaller demand for transport services at lower train speeds usually limits the transport offer, aimed at reducing costs of services. However, it usually triggers a further fall in the demand for services, and consequently a decline of the railways competitiveness in the transport market.

4.6. Social and economic benefits of high-speed rails

Investments in the transport infrastructure, apart from the direct benefits gained by the investor, generate other benefits, seen in widely understood environment, i.e. in the natural environment, transport users, region or transport operators. They are determined as external effects [5], [16].

External effects are one of the factors of market failure. They appear when one company's functioning affects other operators' situations. This influence may lead to a company's need to incur extra costs for which they don't obtain compensation from their authors of a violation [22].

External effects cause the occurrence of indirect costs and benefits determined also as outside, social, or economic. They are imposed mainly by the environment, rather than being caused by the company itself [24]. They generate real cash flows which cannot always be estimated precisely. A noise caused by vehicles (planes, cars, trains, trams) can be an example. It can cause hearing loss and nervous system illnesses. Medical costs constitute real expenses for institutions responsible for providing health care (e.g. National Health Fund). Simultaneously, it isn't possible to identify people who already have health problems as a result of exposure to excessive communication noise and other factors (e.g. listening to overly loud music with headphones).

External benefits, arising from transport infrastructure investments, are, in particular:

- cost saving of travel time in the passenger transport:

  for existing train passengers;
  for passengers taken over from other means of transport;
  for generated passengers;

- savings in operating costs of individual vehicles for users who have used other means of transport so far;

  - cost saving of the air pollution;
  - cost saving of the climate change;
  - decreasing expenses of road traffic accidents (as the effect of passengers' displacement from roads to rail).

The structure of the high-speed rail system can shorten travel time. Consequently, time saving is generated for passengers who already use rail transport and for those taken over from other means of transport. The graphical presentation of shortening the travel time of passengers taken over from cars, buses, and conventional trains are described in Figure 4.7.

Shortening the travel time generates savings in travel time, which causes change in the branch structures of transports. This change results from securing some car and bus passengers, and in some cases air passengers. In consequence, road traffic is reduced. Since rail transport is ranked as the safest means of transport and with the lowest negative influence to the natural environment, the takeover reduces costs from air pollution, climate change, and road traffic accidents. Detailed rules of forecasting these economic categories are presented in [4] and [20].

The volume of economic benefits of the high-speed rail structure project depends on transit and motor forecasts. These, in turn, depend not only on shortening travel time, but also on other factors affecting demand, including service prices, demographic and economic potential, and social and cultural factors. From studies conducted as part of the feasibility study of the structure of the new high-speed rail Warsaw – Łódź – Wrocław/Poznań line [23], results show the annual average value of economic benefits will amount 627.4 million euro. Such a high value of economic benefits will cause an economic rate of return (ERR) of 9.3% of this operation, but an economic net present value amounts to 2622.5 million euro.

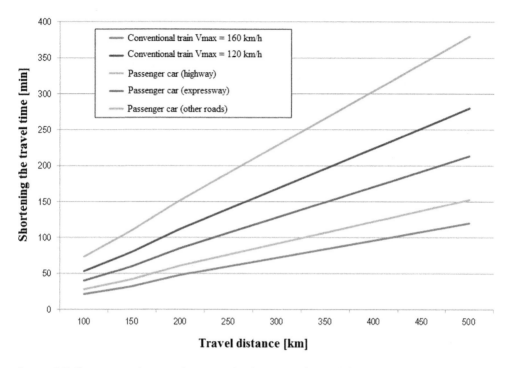

Figure 4.7 Shortening the travel time in the function of travel distance

Source: Own study

4.7. Summary

In searching sources for a competitive edge, we should focus on crucial factors affecting consumers' decisions. In the passenger transport market, time and travel costs can be included, particularly those of long-distance transport. Market successes of high-speed rails in many different countries all around the world confirms that. Growing market share, resulting from securing passengers from air transport and cars, shows that high-speed rails are an innovative means of transport that are able to meet the communication needs of economically developed societies of the 21st century.

High-speed rails guarantee benefits not only for passengers, but also for the states and regions they serve, because an effective transport system is crucial for regional development. The lack of efficient transport causes regions' marginalization and long-lasting exclusion from developmental processes (more [14], p. 18). Direct benefits from the functioning of high-speed rails are also a result of their high safety level and relatively low negative influence on the natural environment. And so they are an important factor for sustainable development.

Among the significant advantages associated with the influence on the natural environment are:

- relatively low level of occupying the area (on average 3.2 ha/1 km of the line, compared to 9.3 ha/1 km of highway);

- high energy efficiency (about 3.4 times higher than in passenger cars and 8.5 times higher than in air transport);
- low level of CO_2 emission;
- high level of safety;
- low external costs (around nine times lower than that generated by passenger cars and five times lower than air transport)[1].

The last factors – low external costs – at present are important mainly in public relations campaigns. Operators often pay attention to the low harmful effects of the rail transport for the environment. Soon, however, this factor could affect real cash flows. The planned internalization of external costs will cause, in the area of the European Union, the need to incur payments reflecting the level of generated costs. Consequently, implementation costs of services and car travel will rise, which will lower their competitiveness and support an increase in demand for fast train services.

Generating great social benefits is predicted for starting the high-speed rail system (300–350 km/h) in Poland, confirmed by the results of current feasibility studies (cf. e.g. [23]). Relatively high values of economic net present value ratio, including indirect costs and benefits (environmental, travel time savings, improvement in safety, savings of cars' operating costs), confirm the benefits of implementing a high-speed rail system in Poland. Its structure should be an important component of strategic investments for Europe, being a current priority of the European Commission, as an important instrument for increasing the competitiveness of the European Union.

Bibliography

[1] Barrón de Angoiti, I.: *The Sustainability of High-Speed*. Presentation of the 7th Training on High-speed Systems, UIC, Paris, 28 June 2010, www.uic.org/ spip.php?article2092 (accessed on 08.02.2011).

[2] The White Paper 'Roadmap to a single European transport area —Towards a competitive and resource efficient transport system' *COM*, 2011, p. 144.

[3] Bogdanowicz, S.: Warunki uczciwej konkurencji w transporcie [Conditions for fair competition in transport]. *Problemy Ekonomiki Transportu*, 1996, no. 2.

[4] Dyr, T., and Kozubek, P.R.: *Ocena transportowych inwestycji infrastrukturalnych współfinansowanych z funduszy Unii Europejskiej* [Assessment of infastructural transport investments co-financed by European Union's funds]. Spatium, Radom, 2013.

[5] Dyr, T., Kotowska-Jelonek, M., Zagożdżon, B., and Kozubek, P.: Problemy oceny efektywności inwestycji infrastrukturalnych w transporcie kolejowym współfinansowanych z funduszy Unii Europejskiej [Problems of assessing the effectiveness of infrastructure investments in railway transport co-financed by European Union Funds]. *Przegląd Komunikacyjny*, 2008, nos. 1–2.

[6] Dyr, T.: *Koleje dużych prędkości jako czynnik poprawy konkurencyjności kolei na rynku transportowym* [High speed-rail as a factor improving railway competitiveness on the transport market]. *Technika Transportu Szynowego*, 2010, nos. 11–12.

[7] Directive 2001/14/EU of the European Parliament and the Council of 26 February on the allocation of railway infrastructure capacity and the levying of charges for the use of railway infrastructure and safety certification. L75 of 15.3.200, pp. 29–46.

[8] Directive 2012/34/EU of the European Parliament and the Council of 21 November 2012 establishing a single European railway area. *Official Journal of the European Union*, 343 of 14.12.2012, pp. 32–77.

[9] *Further Development of the European High-Speed Rail Network: System Economic Evaluation of Development Options*. Summary Report. Study commissioned by Alstom and SNCF.

Civity Management Consultants, Paris–Hamburg, December 2013, www.civity.de/sites/civity/files/assets /downloadscivity_dev_eu_hsr_network_012014.pdf (dostęp 24.03.2015).

[10] Garcia, A.: *Relationship Between Rail Service Operating Direct Costs and Speed*. Fundación Ferrocarriles Españoles, UIC, 2010.

[11] Goossens, H.: *Maintenance of High-Speed Lines*. E-RAILCONSULT, UIC, 2010.

[12] HEATCO: *Developing Harmonised European Approaches for Transport Costing and Project Assessment*. Project funded under the 6th Framework Programme, coordinated by the University of Stuttgart, European Commission, 2006.

[13] Kazimierczak, S.: Ramowe zasady określania kosztów utrzymania w analizach kosztów i korzyści projektów kolejowych PKP PLK S.A. realizowanych w ramach Programu Operacyjnego Infrastruktura i Środowisko 2007–2013 [Framework rules for determining maintenance costs in the cost-benefit analyses of PKP PLK S.A. railway projects implemented under the operational program: Infrastructure and environment 2007–2013]. Warszawa, July 2012. The document was approved by JASPERS on 24 August 2012 and Centrum Unijnych Projektów Transportowych on 5 October 2012 .

[14] Klasik, A.: *Strategie regionalne. Formułowanie i wprowadzanie w życie* [Regional strategies: Formulation and implementation]. Katowice: Wydawnictwo Akademii Ekonomicznej, 2002.

[15] Komunikat Komisji: Zrównoważona przyszłość transportu: w kierunku zintegrowanego, zaawansowanego technologicznie i przyjaznego użytkownikowi systemu [Communication from the commission: A sustainable future transport: Towards an integrated technology-led and user friendly system]. *COM*, 2009, p. 279.

[16] Kotowska-Jelonek, M., and Dyr, T.: Problemy oceny efektywności inwestycji infrastrukturalnych w transporcie kolejowym [Problems of assessing the effectiveness of infrastructure investments in railway transport], in: S. Wrzosek (ed.): Praktyczne aspekty pomiaru efektywności [Practical aspects of effectiveness measurement]. Wrocław: Wydawnictwo Akademii Ekonomicznej, 2005.

[17] Kotowska-Jelonek, M.: Opłaty za dostęp do infrastruktury kolejowej w Europie [Fees for access to railway infrastructure in Europe]. *Technika Transportu Kolejowego*, 2006, nos. 7–8.

[18] Kotowska-Jelonek, M.: Rachunek kosztów a system kalkulacji opłat za dostęp do infrastruktury kolejowej: aspekty metodyczne [Cost accounting and the system of calculating fees for access to railway infrastructure: Methodological aspects] Zeszyty Naukowe. *Problemy Transportu i Logistyki*, 2014, no. 25.

[19] Niebieska Księga: *Sektor kolejowy: Infrastruktura i tabor* [Blue Paper: Railway sector: Infrastructure and rolling stock]. Warszawa: Jaspers, December, 2008.

[20] Niebieska Księga: Sektor kolejowy: Infrastruktura kolejowa [Blue Paper: Railway sector: Railway infrastructure]. Warszawa: Jaspers, September, 2015.

[21] Stankiewicz, M.J.: *Konkurencyjność przedsiębiorstwa* [Company competitiveness]. Toruń: TNOiK, 2002.

[22] Stiglitz, E.: *Ekonomia sektora publicznego* [Public sector economics]. Warszawa: Wydawnictwo Naukowe PWN, 2004.

[23] Studium Wykonalności dla budowy linii kolejowej dużych prędkości "Warszawa – Łódź – Poznań/Wrocław" [Feasibility study for building high-speed line Warszawa – Łódź – Poznań/Wrocław] commissioned by PKP Polskie Linie Kolejowe S.A. drawn up by Consortium Ingenieria IDOM Internacional S.A. and Biuro Projektów Komunikacyjnych in Poznan Sp. z.o.o., Warszawa, 2013.

[24] Zagożdżon, B.: Koszty i korzyści zewnętrzne w analizie efektywności ekonomicznej inwestycji infrastrukturalnych w transporcie kolejowym [Costs and external advantages in economic effectiveness analysis of infrastructural investments in railway transport], in: M. Kotowska-Jelonek (ed.): *Unia Europejska. Wpływ na rozwój Polski* [European Union. Impact on Poland's Development]. Kielce: Wydawnictwo Politechniki Świętokrzyskiej, 2008.

[25] Zschoche, F., Bente, H., Schilling, M., and Wittmeier, K.: *The Relevance of High-Speed Rail: System Economic Evaluation of Development Options. Civity Management Consultants*. Presented at "The Relevance of High-Speed Rail" Lunch Debate, Brussels, January 21, 2014.

Chapter 5

HSR in Poland

Demand, spatial accessibility, and local spatial planning conditions

Przemysław Śleszyński and Tomasz Komornicki

5.1. Introduction

This section presents the results of research conducted at the Institute of Geography and Spatial Organisation of the Polish Academy of Sciences (IGSO PAS) over the last few years on:

- demographic and business demand and hinterland gravitation as factors in the planning of the transport system;
- transport accessibility (temporal and potential);
- the status quo of local spatial planning, i.e. local development plans coverage along transport corridors, including the HSR corridor.

Demand-side research has focused on hinterland gravitation analysis, assuming that the greatest demand and synergy effects are generated by an interaction of masses (of population or enterprises) located near to each other. This method is particularly useful in identifying potential interactions between large urban centers and has been applied in an analysis performed for the purposes of Poland's National Spatial Development Concept 2030 [14], the main planning document at the national level of spatial policy.

Spatial accessibility is one of the most versatile indicators identifying the efficiency of spatial linkages [6], [13]. Research on accessibility presented in this section forms part of a group of larger projects, which have either been published or at least are available on the websites of the organizations commissioning these studies [4], [5], [7].

The progress of work on planning, the progress of conversion of land for development, and the degree to which land is reserved for infrastructure projects provide indications about potential problems and conflicts involved in their execution. For this purpose, this study has utilized data on local development plans in each Polish municipality (*gmina*) from sources at the Ministry of Infrastructure and Development and the Central Statistical Office (CSO). These data are also included in IGSO PAS annual reports, e.g. [16–18]. For the purpose of this study, IGSO PAS performed an additional assessment of progress on local plans that focused on municipalities intended to host the basic variant of the new "Y" HSR route (Warsaw – Łódź – Nowe Skalmierzyce – Poznań/Wrocław). In total, 44 municipalities were selected for the study, which number excludes large cities, i.e. the national and regional capital cities, where multiple other factors that influence development processes would have severely limited the usefulness of this kind of analysis.

Essential methodological aspects are covered in subsections devoted to each particular aspect.

5.2. Demand and gravitation

It is commonly agreed that the demand for transport services between two centers is proportional to their size and inversely proportional to their distance apart. This premise is used in building a gravitational model, which assumes that the likelihood of socioeconomic relationships and links is associated with physical proximity and size [1], [8]. Such a model can utilize a range of variables of size, also known as mass (typically the absolute value of the population, but alternatives include the number of enterprises, visitor overnight stays, purchasing power, etc.) and distance (physical, economic, or time). Physical distance can be measured using the mileage of the shortest route, an economic cost expressed in money, time, etc. The full gravitational model includes not just the two centers, but typically involves a matrix of gravitational attractions between all the centers within a territory (here: 18 capital cities of Polish voivodeships, or provinces, produce $18 \times 18 - 18 = 306$ links).

This study employs a compound indicator of mass M consisting of the population total and enterprises at 50:50% parity following a formula borrowed from another study [12]:

$$M = 0.5\frac{L_i}{L} + 0.5\frac{P_i}{P} \tag{5.1}$$

where:
L_i – population of a given unit of settlement (municipality),
L – the total population of all units in Poland,
P_i – the number of economic entities registered in the unit (municipality),
P – the number of economic entities registered in all units in Poland.

The latest data for 2013 were adopted. The calculation was performed on a matrix of all municipalities in Poland that was modified by consolidating certain units:

- The towns and cities forming the two largest conurbations, i.e. Tricity (Gdańsk, Gdynia, Sopot) and Katowice (14 cities and towns);
- 158 torus-shaped ("bagel") municipalities, i.e. rural municipalities surrounding an urban municipality with which they share both the seat and the name (e.g. Mińsk Mazowiecki and Tarnów); they were merged with their urban cores. This was due to the fact that in terms of land surveying (nesting), the "mass" of these rural municipalities is normally located within the core of a city anyway.

This exercise produced a set of 2298 municipal units and 5,278,506 ($2298 \times 2298 - 2298$) gravitational attractions. The map in Figure 5.1 features a matrix of the strongest gravitational attractions (links). Superimposed on this backdrop is the current proposed route of the Polish HSR, known as the "Y route" connecting Warsaw, Łódź, Kalisz/Ostrów Wielkopolski (Nowe Skalmierzyce), Poznań, and Wrocław.

This comparison of the route and the network of gravitational links reveals a mixed picture. Some of the route optimally matches the network, especially the segment connecting Łódź and Warsaw, while the split to Poznań and Wrocław is less efficient when compared to the gravitational network elsewhere.

This course of analysis can be reversed and a gravitational analysis can be used to identify the strongest potential links in Poland. The results of this approach are shown in Table 5.1, which is designed so that the highest value of gravitational pull (Katowice conurbation and

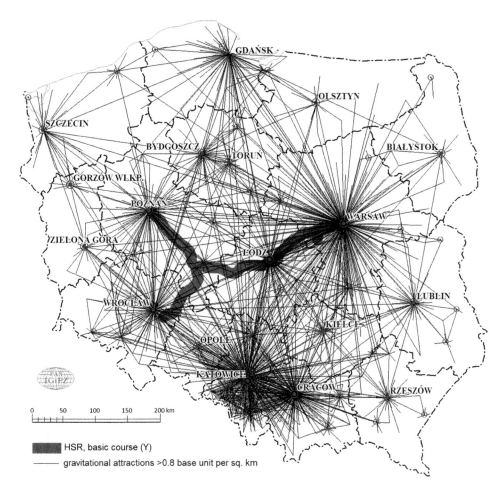

Figure 5.1 The HSR route and a network of major gravitational attraction in Poland in 2013
Source: Own study

Rybnik urban agglomeration) was adopted as 100 to enable all the other links to be compared as a percentage of the strongest one. The top five, except for the one mentioned, are Katowice (conurbation) – Cracow (61.4), Warsaw – Łódź (28.1), Katowice (conurbation) – Częstochowa (24.8), and Katowice (conurbation) – Warsaw (15.8) (Table 5.1).

It is worth noting the sums of gravitation of all 21 cities indicate the "nodality" of a given center, as a changeover station. If the planned railway were to serve all of these cities, the greatest benefits would be reported in Katowice, Rybnik, Warsaw, and Cracow. Also, Częstochowa, Łódź, Bydgoszcz, and Radom would stand to benefit when compared to their simple demographic or business potential.

By changing the exponent of the distance between cities from square (2) into "straight" (1) mileage in the gravitational model, one arrives at the efficiency of each segment in terms of its utility, which in this case means the shortest distance at maximized use of the existing gravitation (Table 5.2). The top 10 most efficient pairs are Katowice (conurbation) – Cracow (100.0), Katowice (conurbation) – Cracow (100.0), Katowice (conurbation) – Warsaw (79.6), Rybnik (urban

Table 5.1 Matrix of gravitational attractions between 21 major Polish cities (power exponent of distance = 2)

Cities	Białystok	Bydgoszcz	Częstochowa	Gorzów Wielkopolski	Kielce	Katowice (conurbation)	Cracow	Lublin	Łódź	Olsztyn	Opole	Poznań	Radom	Rybnik (agglomeration)	Rzeszów	Szczecin	Toruń	Tricity	Warsaw	Wrocław	Zielona Góra	Total
Białystok		0.2	0.1	0.0	0.1	0.6	0.3	0.5	0.6	0.3	0.0	0.2	0.2	0.1	0.1	0.1	0.1	0.5	5.2	0.2	0.0	9.7
Bydgoszcz	0.2		0.3	0.3	0.2	1.4	0.5	0.2	1.8	0.5	0.2	4.7	0.2	0.7	0.1	0.7	10.6	3.2	3.8	1.1	0.3	30.5
Częstochowa	0.1	0.3		0.1	1.0	24.8	4.1	0.3	3.1	0.1	1.0	0.7	0.4	2.1	0.2	0.1	0.2	0.3	2.9	1.7	0.1	43.5
Gorzów Wielkopolski	0.0	0.3	0.1		0.0	0.4	0.1	0.0	0.2	0.0	0.1	1.6	0.0	0.1	0.0	2.0	0.1	0.3	0.5	0.5	0.6	7.0
Kielce	0.1	0.2	1.0	0.0		5.1	4.5	0.9	2.2	0.1	0.2	0.4	2.0	0.6	0.5	0.1	0.1	0.2	5.0	0.6	0.1	23.8
Katowice (conurbation)	0.6	1.4	24.8	0.4	5.1		61.4	1.9	10.2	0.5	7.9	3.7	2.2	100.0	1.6	0.9	0.9	1.6	14.4	10.9	0.7	250.9
Cracow	0.3	0.5	4.1	0.1	4.5	61.4		1.4	3.9	0.2	1.1	1.3	1.4	6.2	1.9	0.4	0.4	0.7	7.7	2.6	0.2	100.5
Lublin	0.5	0.2	0.3	0.0	0.9	1.9	1.4		1.2	0.1	0.1	0.3	1.6	0.3	0.8	0.1	0.1	0.3	8.1	0.4	0.0	18.8
Łódź	0.6	1.8	3.1	0.2	2.2	10.2	3.9	1.2		0.6	0.9	3.3	2.3	1.4	0.6	0.6	1.5	1.8	28.1	3.2	0.3	67.8
Olsztyn	0.3	0.5	0.1	0.0	0.1	0.5	0.2	0.1	0.6		0.0	0.4	0.1	0.1	0.0	0.1	0.4	1.8	3.2	0.2	0.0	8.8
Opole	0.0	0.2	1.0	0.1	0.2	7.9	1.1	0.1	0.9	0.0		0.5	0.1	1.9	0.1	0.1	0.1	0.2	1.0	3.6	0.1	19.3
Poznań	0.2	4.7	0.7	1.6	0.4	3.7	1.3	0.3	3.3	0.4	0.5		0.4	0.7	0.2	2.2	1.9	2.2	4.9	6.2	2.1	37.8
Radom	0.2	0.2	0.4	0.0	2.0	2.2	1.4	1.6	2.3	0.1	0.1	0.4		0.4	0.3	0.1	0.1	0.3	14.0	0.4	0.0	26.6
Rybnik (agglomeration)	0.1	0.7	2.1	0.1	0.6	100.0	6.2	0.3	1.4	0.1	1.9	0.7	0.3		0.2	0.2	0.2	0.3	2.1	3.7	0.3	119.2
Rzeszów	0.1	0.1	0.2	0.0	0.5	1.6	1.9	0.8	0.6	0.0	0.1	0.2	0.3	0.2		0.1	0.1	0.1	1.6	0.4	0.1	8.6
Szczecin	0.1	0.7	0.1	2.0	0.1	0.9	0.4	0.1	0.6	0.1	0.1	2.2	0.1	0.2	0.1		0.3	1.2	1.3	0.9	0.5	12.0
Toruń	0.1	10.6	0.2	0.1	0.1	0.9	0.4	0.1	1.5	0.4	0.1	1.9	0.1	0.2	0.1	0.3		1.7	3.2	0.6	0.1	22.7
Tricity	0.5	3.2	0.3	0.3	0.2	1.6	0.7	0.3	1.8	1.8	0.2	2.2	0.3	0.3	0.1	1.2	1.7		5.5	1.0	0.3	23.3
Warsaw	5.2	3.8	2.9	0.5	5.0	14.4	7.7	8.1	28.1	3.2	1.0	4.9	14.0	2.1	1.6	1.3	3.2	5.5		4.5	0.6	117.7
Wrocław	0.2	1.1	1.7	0.5	0.6	10.9	2.6	0.4	3.2	0.2	3.6	6.2	0.4	3.7	0.4	0.9	0.6	1.0	4.5		1.4	43.3
Zielona Góra	0.0	0.3	0.1	0.6	0.1	0.7	0.2	0.0	0.3	0.0	0.1	2.1	0.0	0.3	0.1	0.5	0.1	0.3	0.6	1.4		7.7
Total	9.7	30.5	43.5	7.0	23.8	250.9	100.5	18.8	67.8	8.8	19.3	37.8	26.6	119.2	8.6	12.0	22.7	23.3	117.7	43.3	7.7	995.1

NB: Values converted to bring the largest value (Katowice-Rybnik) to 100.

Source: Own study

Table 5.2 Matrix of gravitational attractions between 21 major Polish cities (power exponent of distance = 1)

City	Białystok	Bydgoszcz	Częstochowa	Gorzów Wielkopolski	Kielce	Katowice (conurb.)	Cracow	Lublin	Łódź	Olsztyn	Opole	Poznań	Radom	Rybnik (aggl.)	Rzeszow	Szczecin	Torun	Tricity	Warsaw	Wrocław	Zielona Góra	Total
Białystok		1.4	0.8	0.3	1.0	5.6	3.0	2.2	3.4	1.3	0.4	2.3	1.2	1.0	0.7	1.2	0.9	3.5	19.1	2.2	0.4	51.9
Bydgoszcz	1.4		1.4	1.4	1.2	9.4	4.2	1.6	7.0	1.8	0.9	11.0	1.3	1.7	0.7	3.6	8.9	10.0	18.4	5.7	1.2	92.7
Częstochowa	0.8	1.4		0.4	2.3	31.8	9.4	1.5	7.2	0.6	1.8	3.3	1.5	4.0	0.9	1.2	0.9	2.3	12.7	5.5	0.6	90.1
Gorzów Wlkp.	0.3	1.4	0.4		0.3	3.1	3.0	0.4	1.6	0.6	0.6	3.9	0.3	0.6	0.2	3.7	0.6	1.9	4.0	2.3	1.1	28.1
Kielce	1.0	1.2	2.3	0.3		14.2	9.5	2.6	6.0	0.6	0.6	2.4	3.1	2.1	1.5	1.0	0.8	2.2	16.4	3.2	0.4	71.5
Katowice (con.)	5.6	9.4	31.8	3.1	14.2		100.0	10.9	36.6	3.8	14.4	22.0	9.1	77.0	7.5	9.0	5.7	15.9	79.6	38.9	4.3	498.8
Cracow	3.0	4.2	9.4	3.0	9.5	100.0		6.8	16.1	1.9	3.8	9.1	3.6	13.6	5.8	4.0	2.6	7.5	41.4	13.6	1.7	260.6
Lublin	2.2	1.6	1.5	0.4	2.6	10.9	6.8		6.8	1.0	0.7	2.9	3.5	1.8	2.3	1.4	1.0	3.2	26.5	3.3	0.5	79.8
Łódź	3.4	7.0	7.2	1.6	6.0	36.6	16.1	6.8		2.7	3.1	13.2	6.0	6.0	2.5	4.5	4.8	10.0	71.2	12.0	2.0	223.8
Olsztyn	1.3	1.8	0.6	0.6	0.6	3.8	1.9	1.0	2.7		0.3	2.2	0.7	0.7	0.4	1.1	1.2	5.3	12.0	1.8	0.3	39.9
Opole	0.4	0.9	1.8	0.6	0.6	14.4	3.8	0.7	3.1	0.3		2.4	0.6	3.3	0.4	0.9	0.5	1.4	6.1	6.4	0.5	49.1
Poznań	2.3	11.0	3.3	3.9	2.4	22.0	9.1	2.9	13.2	2.2	2.4		2.4	4.0	1.1	8.8	5.3	11.8	29.6	18.6	4.9	161.7
Radom	1.2	1.3	1.5	0.3	3.1	9.1	3.6	3.5	6.0	0.7	0.6	2.4		2.3	1.5	1.0	1.0	2.3	26.9	2.7	0.4	71.4
Rybnik (aggl.)	1.0	1.7	4.0	0.6	2.1	77.0	13.6	1.8	6.0	0.7	3.3	4.0	2.3		1.4	1.0	0.8	2.8	13.0	7.7	0.8	144.7
Rzeszow	0.7	0.7	0.9	0.2	1.5	7.5	5.8	2.3	2.5	0.4	0.4	1.1	1.5	1.4		0.5	0.5	1.4	8.6	1.9	0.3	40.1
Szczecin	1.2	3.6	1.2	3.7	1.0	9.0	4.0	1.4	4.5	1.1	0.9	8.8	1.0	1.0	0.5		1.8	7.1	12.6	5.9	2.0	73.0
Torun	0.9	8.9	0.9	0.6	0.8	5.7	2.6	1.0	4.8	1.2	0.5	5.3	1.0	0.8	0.5	1.8		5.5	12.9	3.2	0.6	59.6
Tricity	3.5	10.0	2.3	1.9	2.2	15.9	7.5	3.2	10.0	5.3	1.4	11.8	2.3	2.8	1.4	7.1	5.5		34.3	8.3	1.8	138.5
Warsaw	19.1	18.4	12.7	4.0	16.4	79.6	41.4	26.5	71.2	12.0	6.1	29.6	26.9	13.0	8.6	12.6	12.9	34.3		29.4	4.7	479.5
Wrocław	2.2	5.7	5.5	2.3	3.2	38.9	13.6	3.3	12.0	1.8	6.4	18.6	2.7	7.7	1.9	5.9	3.2	8.3	29.4		4.7	179.2
Zielona Góra	0.4	1.2	0.6	1.1	0.4	4.3	1.7	0.5	2.0	0.3	0.5	4.9	0.4	0.8	0.3	2.0	0.6	1.8	4.7	4.7		32.8
Total	51.9	92.7	90.1	28.1	71.5	498.8	260.6	79.8	223.8	39.9	49.1	161.7	71.4	144.7	40.1	73.0	59.6	138.5	479.5	179.2	32.8	2866.9

NB: Values converted to bring the largest value (Katowice–Cracow) to 100.

Source: Own study

agglomeration) – Katowice (conurbation) (77.0), Warsaw – Łódź (71.2), Warsaw – Cracow (41.4), Katowice (conurbation) – Wrocław (38.9), Katowice (conurbation) – Łódź (36.6), Tricity – Warsaw (34.3), Częstochowa – Katowice (conurbation) (31.8), and Poznań – Warsaw (29.6).

This approach can be pursued further for a more detailed settlement model that also takes into account cities of subregional significance (capitals of *poviat*/county and of pre-1999 voivodeships). To maximize the efficiency of a transport network vis-à-vis the network of gravitational attractions, some of the gravitational links from the full matrix should be included in links of a higher order (e.g. Opole – Wrocław and Opole – Katowice should be included in the gravitation attraction between Wrocław and Katowice). Theoretically, there are a great many potential combinations in terms of the number of cities that are taken into account, but the main goal would be to find a transport network that would provide the shortest links between bipolar, tripolar, etc. constellations of cities with the strongest gravitational pull between them.

Gravitation analysis thus allows an assessment in which variants of high-level transport route planning, such as HSR, are effective in terms of synergy effects, potential relationships, etc. In this case, high efficiency levels were obtained for the Warsaw-Łódź connection, which scored several times higher (6–7x) than the connections between Łódź and Wrocław/Poznań. Far from denying the desirability of providing HSR services along these latter segments, this suggests the priorities and hierarchy of projects along the entire planned line. Ultimately, it would be optimal if an HSR network connected all the major urban centers (urban agglomerations) in Poland, as is proposed in detailed studies [3]. A gravitational analysis helps establish the specific number of these urban centers to be served and which connections would be the most efficient. With HSR construction costs running into tens of billion zlotys [2], these considerations should be among the most important in designing the routes.

5.3. Spatial accessibility

Future plans for capital projects for land-based transport infrastructure can be evaluated by analyzing their projected impact on potential accessibility. This section uses a simulation of the effects of a "Y" HSR variant performed [4] for the purposes of the 2020 Transport Development Strategy document (*Strategia Rozwoju Transportu 2020*) [10]. The baseline for this assessment was the situation in mid-2010.

The potential accessibility was estimated using our own IGSO PAS software to compute what is known as the Multi-Modal Transport Accessibility Indicator, or WMDT [6], built on a gravitational-pull model. This compound indicator identifies the time accessibility between any nodes (municipalities) of a transport network weighted by physical distance. It takes into consideration four transport networks, i.e. road, rail, air, and inland waterways, and their respective share of the total transport influences the indicator values. It differentiates the attractiveness of the destination depending on time distance from the source. WMDT follows the formula:

$$A_i = \sum_{1 \leq j \leq n, j \neq i} \frac{1}{t_{ij}} \frac{m_j}{M} \tag{5.2}$$

where:
A_i – potential accessibility of a node (poviat or a group of poviats in an urban agglomerations) i,
t_{ij} – travel time between nodes i and j,

m_j – mass of node j,
M – total mass of all nodes.

The travelling times were calculated separately for each of the transport networks, i.e. road, rail, air, and water. The road travel times were derived using a model of traffic speed developed by IGSO PAS [6], [9], [13], while the Polish railways' own standard times were used for the railway network.

Figure 5.2 illustrates a simulation of a controlled effect of the Y HSR on accessibility. This effect varies from place to place depending on the gain in travel time. Importantly,

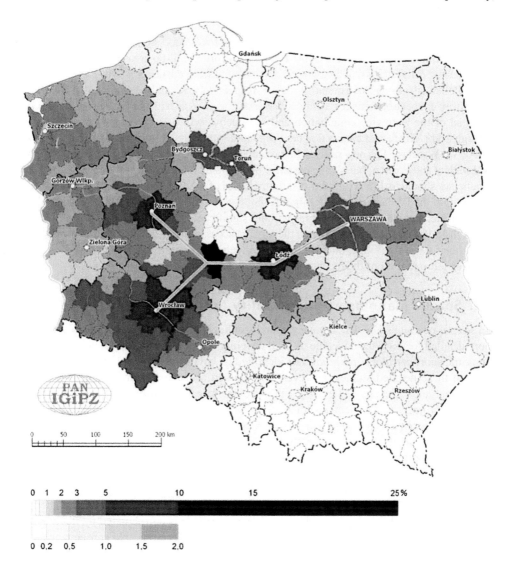

Figure 5.2 Simulated accessibility (WMDT) improvement as a result of the Y HSR line

Source: Komornicki, Rosik, and Stępniak 2010 [4]

the measure of this gain is WMDT, an indicator including all modes of transport. This means that the effect of the Y connection on railway accessibility must be proportionally stronger. Despite these reservations, the new connection can benefit a clear majority of poviats in Poland. This is an illustration of an improvement in accessibility offered by this new project. The largest gains would be observed in the Dolnośląskie Voivodeship and its capital city of Wrocław. The effect is also clearly visible, if not as spatially extensive, around other large cities connected by the HSR line (Poznań, Łódź, and Warsaw). In general, the whole of the western half of the country stands to gain from the project. Indeed, WMDT would increase by more than 2% in nearly all poviats of the Dolnośląskie, Lubuskie, Wielkopolskie, and Zachodniopomorskie voivodeships, and in many of them it would grow by 5–10%.

What is characteristic of this project is that it improves accessibility in places seemingly far away from the new line, such as the Bydgoszcz-Toruń built-up areas and in the Opolskie Voivodeship. This means that the two areas will benefit from the new line (in services to other major Polish cities) via Poznań and Wrocław, respectively. It can be assumed that the shortest railway connection between Toruń and Warsaw, as well as with the entire socioeconomic potential of the Katowice conurbation and of Cracow, will pass through Poznań and then through Łódź. The fastest connection between Opole and Warsaw, and on to northern Poland, will be via Wrocław. The new railway line would draw traffic from poviats located to the west of Wrocław and Poznań, but also in eastern Poland (fanned out to the east of Warsaw). Clear beneficiaries include, but are not limited to, the cities of Siedlce and Lublin. A positive effect reaches into the Podlaskie Voivodeship, leaving just the southern parts of the Podkarpackie, Małopolskie, and Śląskie voivodeships without significant benefits.

Kaliski Poviat would stand to gain the most in accessibility. This is explained by its location at the fork of the Y line, but also by its relatively low overall accessibility, as the Kalisz has been left out of all of the recent motorway and expressway projects.

5.4. Spatial planning at municipality (gmina) level

On several occasions [16–18], annual research studies into spatial planning have included a more detailed focus on the coverage of local development plans and the pressure on property along transport corridors and in gminas identified as having developed transport functions. Most of these detailed studies, however, only looked at the road infrastructure due to its more direct impact on spatial development. Also, the research mostly focused on the impact of capital projects on the demand for land. Nevertheless, the studies suggested that there might be a reversed dependency whereby inadequacies in the spatial planning systems and dispersed development (especially in built-up areas not covered by a local development plan) could constitute a barrier to major capital projects.

In the most recent of these studies, covering 2013 [17] gminas with a transport function (i.e. hosting motorway or expressway nodes, or nodes of an intermodal nature), came out with a relatively low coverage of local development plans, but also a relatively high degree of coverage of local development plans in elaboration. This is corroborated by fluctuations in the change of status from farmland to development land as a possible consequence of a modification to earlier local development plans. Their ratio of farmland earmarked for such conversion to population is among the highest among the various functional types of gminas. This demonstrates that despite their often-peripheral location in less populated areas, gminas with a transport function experience a pressure for land development. Also the low overall numbers of administrative decisions granting conditions for land development (an obligatory document for development in areas not covered by a local development plan) issued in such gminas, when converted into a ratio per 1000 inhabitants, turns out to be quite considerable and close to those of suburban gminas.

Looking at the linkages between road-building projects and the coverage by local development plans [17], it was concluded that it was a feedback loop that resulted in a sequential order of processes. The process would begin with heavy traffic that generates local pressure for infrastructure projects. Launching of such a major project often requires adjustments to the local development plan, including the adoption of new documents and the cancelling of some old ones. Upon completion, the new road improves accessibility at a supra-local level, thus attracting more traffic and generating new pressures for infrastructure. After a while, although with considerable delay, the planning work is expedited and the new documents produced tend to feature a large scale of conversion from farmland to development land, especially in densely populated areas. Gradually, all aspects of making and using local development plans are aligned, including a diminishing number of construction projects approved on the grounds of an administrative decision granting conditions of land development.

These observations could equally as well be made in relation to railway projects, such as HSR. In this case the first phase of the sequence, the traffic-generated pressure on infrastructure projects, does not apply. Once a decision is made, however, interest in land, including speculative interest, can be induced and may indirectly expedite the planning work. After commissioning, new pressure on property may occur, mainly near HSR station towns. Assuming that in major cities this sort of pressure already exists, an increased drive to buy property ought to be expected near the towns of Kalisz and Nowe Skalmierzyce.

Table 5.3 summarizes the basic planning-related characteristics of gminas located along the Y line. Their coverage with local development plans is high, much above the national average, along the Kalisz – Wrocław and Łódź – Warsaw segments. The status is much worse along the other two segments, especially Łódź – Kalisz. Not only that, but virtually no planning work has even been conducted along this segment (less than 1% of land is covered by draft plans). The data on the proportion of the gminas covered by local development plans reflect highly varied situations in each of them (Figure 5.3). A relatively better situation is typically found near each of the four large cities connected by the new line. For example, the gmina of Ożarów Mazowiecki, near Warsaw, had nearly 90% of its territory covered by a local development plan, while the gmina of Stryków near Łódź is fully covered.

Table 5.3 Spatial planning status in gminas along the HSR Y line

HSR segment	Local development plan coverage (%)		Proportion of farmland earmarked for conversion to development land	Decisions of land development conditions per 1000 population
	Adopted	Draft		
Warsaw – Łódź	41.8	6.6	13.7	4.7
Łódź – Kalisz	13.9	0.8	1.3	7.4
Kalisz – Poznań	23.7	7.1	5.3	7.5
Kalisz – Wrocław	45.4	12.9	1.4	6.1
Total HSR	31.1	7.0	5.5	6.5

Source: Own study based on data [17]

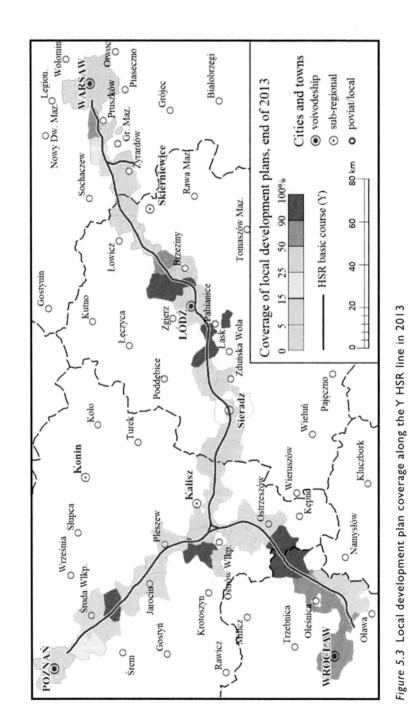

Figure 5.3 Local development plan coverage along the Y HSR line in 2013

Source: Own study based on data [17]

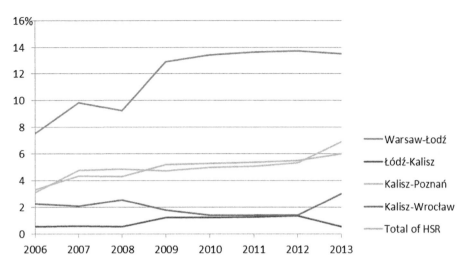

Figure 5.4 Change in local development plan coverage (%) along the Y HSR line in 2004–2013 period

Source: Own study based on data [17]

The average rate of change in local development plan coverage of the units included in the study (Figure 5.4) suggests a nearly linear increase from approximately 15% in 2004 to 30% in 2013, which is similar to the national average. A closer look reveals several differences within the area. Initially, development was fastest along the Warsaw – Łódź segment, but it has gradually plateaued. Currently, a relatively fast growth in coverage is observed between Kalisz and Wrocław, while between Łódź and Kalisz the coverage remains very low.

The scale of current pressure on property is illustrated by the proportion of farmland earmarked for conversion into development land in the local development plans. Along the HSR Warsaw – Łódź segment, this indicator is much higher than the national average, which is explained by suburbanization, especially near the national capital. For example, nearly 50% of farmland has been reclassified in the gmina Ożarów Mazowiecki. Similar land conversion levels are found around the midway point between these cities in gminas with high levels of residential and leisure attractiveness, especially Jaktorów and Nieborów (20% each). All other areas along the HSR have significantly lower land conversion levels.

Another measure of the actual risk to the project is provided by the number of administrative decisions granting conditions of land development issued to developers. In principle, this kind of decision is only issued in areas not covered by a local development plan. A comparison of this value with the figures for the local population totals shows that the scale of development permits issued in areas not covered by plans is considerable along all segments of the planned railway line. Its relatively lowest point was observed along the Warsaw – Łódź segment and is explained by the generally higher plan coverage. A relatively high level along the Kalisz – Wrocław segment is rather surprising, as the local development plan coverage there is also quite high. This may be suggesting certain irregularities in the spatial planning system and, consequently, a prospect of difficulties in the construction of the HSR. The number of administrative decisions granting conditions for land development is generally high in gminas located within metropolitan areas, especially Warsaw, but also in residentially attractive areas that are also not covered by local development plans (e.g. the gmina Bolimów between Warszawa and Łódź).

In summary, it seems that the new HSR line has not yet become a determining factor in local planning. In the future, this may prove damaging for the efficiency and cost of the project when it finally materializes. This may be equally true along segments with low planning coverage (e.g. Łódź – Kalisz), where the project may trigger uncontrolled property speculation, as well as in areas within major cities or in the vicinity of attractive natural complexes (e.g. Bolimowska Forest), where the large scale of conversion from farmland to development land and numerous administrative decisions granting conditions of land development may lead to an increase in the buyout cost of land for the project compared to planned figures.

It should be noted that the HSR feasibility study of 2013 [11] made no direct reference to the planning status at gmina level. It silently assumed that the project would be performed on the basis of a special-purpose law. The issue of spatial planning and its impact on buyout costs is also missing from the final SWOT analysis.

The feasibility study [11] does bring up a risk of social protests encapsulated in three typical syndromes well known from the literature:

- NIMBY (Not in My Back Yard), where a project is acceptable, but outside the direct vicinity of one's own place of residence;
- LULU (Locally Unacceptable Land Use), where a community is generally reluctant to host a certain kind of development;
- BANANA (Build Absolutely Nothing, Anywhere, Near Anything), where any large project will be opposed.

According to the feasibility study quoted above [11], the HSR line will cross 117 built-up areas where more than 700 structures will be condemned and expropriated. It also notes that the buyout of land in agricultural areas may face not just the opposition of the individual owners, but also protests from local authorities, especially where good and very good soils are involved. What it fails to notice, however, is a connection between local planning and a potential for a reduction in the scale of protests and, eventually, also to costs. Several other studies looking at the costs and benefits of high-speed rails in Poland also fail to give this area due credit (e.g. [2], [3]).

It seems that the authors of the feasibility study and of the other studies failed to fully appreciate the importance of local planning and its direct link to potentially high future costs stemming from increased property prices in areas developed under procedures other than those involved in local development plans and where existing plans formally convert farmland into development land.

This situation may be addressed by often-advocated measures, including expediting the rolling-out of local development plans in uncovered areas and, alternatively, changing the legal status of local studies of conditions and directions of spatial development. These measures should achieve:

- A reduction of buyout costs, especially where capital projects are planned along the new line;
- A reduction in the number and scale of social conflicts;
- Better conditions for executing long-term transport policies.

Well carried out public participation could also help the situation, as confirmed by evaluation surveys of previous road-building projects [5]. In practice, consultations have often been scheduled too late in the process.

The future HSR line follows an entirely new course without reusing existing lines. This makes it a candidate for a paradox known from new motorway projects. A greenfield project like this is more likely to clear the hurdle of the public participation process as it has a lower risk of falling victim to the NIMBY syndrome, but its execution phase is more difficult to complete due to obstacles encountered in areas not covered by local development plans.

5.5. Conclusions

The main conclusions of the study are as follows:

- Gravitation analysis can be useful in assessing both existing and planned transport routes, including high-speed rails. It has been harnessed for the design of optimal and most efficient intercity networks in terms of potential synergy effects. Such analysis could look at the currently planned HSR network, as well as its expanded and supplementary versions;
- The Y HSR line would produce a wide-ranging improvement in spatial accessibility. This would cover nearly all of Poland's territory, especially in the vicinity of the connected metropolises and across western Poland where the Dolnośląskie Voivodeship and Kalisz stand to gain the most;
- An expansion of the network effect to other Polish regions would require good integration with the traditional railway system, especially in the south and southeast, including the integration of the central trunk line (CMK; *Centralna Magistrala Kolejowa*);
- The scale of hidden costs from an inadequate spatial planning system is considerable and requires extended research to provide more accurate estimates. Large-scale planning for the purpose of major linear projects is "held hostage" by inadequate local spatial planning and flawed spatial policy.

Bibliography

[1] Chojnicki, Z.: *Zastosowanie modeli grawitacji i potencjału w badaniach przestrzenno-ekonomicznych* [Application of gravity and potential models in spatial and economic research]. Studia Komitetu Przestrzennego Zagospodarowania Kraju PAN, vol. 14, Warsaw: Państwowe Wydawnictwo Naukowe, 1966, p. 126.

[2] Gorlewski, B.: *Kolej Dużych Prędkości: Uwarunkowania ekonomiczne* [High Velocity railway: Economic conditions]. Warsaw: Oficyna Wydawnicza SGH, 2012, p. 192.

[3] *Kierunki rozwoju kolei dużych prędkości w Polsce. PKP PLK S.A.* [Directions of High-Speed Rail development in Poland. PKP PLK S.A.]. Centrum Kolei Dużych Prędkości. Warsaw: Instytut Kolejnictwa, 2011, p. 59.

[4] Komornicki, T., Rosik, P., and Stępniak, M.: *Analiza dostępności transportowej w poszczególnych gałęziach transportu* [Analysis of transport accessibility in particular branches of transport]. Ekspertyza wykonana dla Ministerstwa Infrastruktury w ramach prac nad Strategią Rozwoju Transportu do roku 2030. Instytut Geografii i Przestrzennego Zagospodarowania PAN, Warsaw, 2010, maszynopis, 41.

[5] Komornicki, T., Rosik, P., Śleszyński, P., Solon, J., Wiśniewski, R., Stępniak, M., Czapiewski, K., and Goliszek, S.: *Wpływ budowy autostrad i dróg ekspresowych na rozwój społeczno-gospodarczy i terytorialny Polski.* [The impact of the construction of motorways and expressways

on the socio-economic and territorial development of Poland]. Warsaw: Ministerstwo Rozwoju Regionalnego, 2013, p. 215.

[6] Komornicki, T., Śleszyński, P., Rosik, P., Pomianowski, W., przy współpracy (in cooperation) Stępniaka, M., and Siłki, P., *Dostępność przestrzenna jako przesłanka kształtowania polskiej polityki transportowej* [Spatial accessibility as a background for Polish transport policy]. Biuletyn KPZK PAN, 2010, 241, p. 163.

[7] Komornicki, T., Śleszyński, P., Siłka, P., and Stępniak, M.: *Wariantowa analiza dostępności w transporcie lądowym* [A variant analysis of availability in land transport], in: K. Saganowski, M. Zagrzejewska-Fiedorowicz, and P. Żuber (eds.), Ekspertyzy do Koncepcji Przestrzennego Zagospodarowania Kraju 2008–2033. (Expertises for the concept of spatial development of the country 2008–2033) Tom II, Warsaw: Ministerstwo Rozwoju Regionalnego, 2008, pp. 133–334.

[8] Ratajczak, W.: *Modelowanie sieci transportowych* [Modelling transportation networks]. Poznań: Wydawnictwo Naukowe UAM, 1999, p. 274.

[9] Rosik, P., and Śleszyński, P.: *Wpływ zaludnienia w otoczeniu drogi, ukształtowania powierzchni terenu oraz natężenia ruchu na średnią prędkość jazdy samochodem osobowym* [Impact of population in the road surrounding, area surface configuration and traffic intensity on average speed of private car], Transport Miejski i Regionalny, 2009, 10, pp. 26–31.

[10] *Strategia Rozwoju Transportu do roku 2020 (z perspektywą do 2030 roku).* [Transport development strategy until 2020 (with a prospect until 2030)], Ministerstwo Transportu, Warsaw: Budownictwa i Gospodarki Morskiej, 2013.

[11] *Studium wykonalności dla budowy linii kolejowej dużych prędkości "Warsaw – Łódź – Poznań/ Wrocław"* [Feasibility study for the construction of the High-Speed Railway Line "Warsaw – Łódź – Poznań/Wrocław"], Raport nr 16, Rewizja 2.1., 2013.

[12] Śleszyński, P.: *W sprawie optymalnego podziału terytorialnego Polski: zastosowanie analizy grawitacyjnej* [Regarding the optimal territorial division of Poland: application of gravitational analysis]. Przegląd Geograficzny, 2015, 87, 2, pp. 343–259.

[13] Śleszyński, P.: *Dostępność czasowa i jej zastosowania* [Temporal accessibility and its applications]. Przegląd Geograficzny, 2014, 86, 2, pp. 171–215.

[14] Śleszyński, P.: *Ocena powiązań gospodarczych i kapitałowych między miastami,* [Assessment of economic and capital links between cities], in K. Saganowski, M. Zagrzejewska-Fiedorowicz, and P. Żuber (eds.): *Ekspertyzy do Koncepcji Przestrzennego Zagospodarowania Kraju 2008–2033* [Expertises for the concept of spatial development of the country 2008–2033], Tom I, Warsaw: Ministerstwo Rozwoju Regionalnego, 2008, pp. 335–391.

[15] Śleszyński, P.: *Expected Traffic Speed in Poland Using Corine Land Cover, SRTM-3 and Detailed population places data.* Journal of Maps, 2015, 11, 2, pp. 26–32.

[16] Śleszyński, P., Bański, J., Degórski, M., Komornicki, T., and Więckowski, M.: *Stan zaawansowania planowania przestrzennego w gminach* [Progress of spatial planning in gminas (municipalities)]. Prace Geograficzne, 211. Instytut Geografii i Przestrzennego Zagospodarowania PAN, Warsaw, 2007, p.284.

[17] Śleszyński, P., Komornicki, T., Deręgowska, A., and Zielińska, B.: *Analiza stanu i uwarunkowań prac planistycznych w gminach w 2013 roku* [Analysis of the condition and determinants of planning works in gminas (municipalities) in 2013]. Instytut Geografii i Przestrzennego Zagospodarowania PAN na zlecenie Ministerstwa Infrastruktury i Rozwoju, Warsaw, 2015, p. 124.

[18] Śleszyński, P., Komornicki, T., Solon, J., and Więckowski, M.: *Planowanie przestrzenne w gminach* [Spatial planning in municipalities (gminas)]. Wydawnictwo Sedno. Instytut Geografii i Przestrzennego Zagospodarowania PAN, Warsaw, 2012, p. 239.

Chapter 6

Adaptation of the Łódź Agglomeration Railway node to a new role in the high-speed rail system

Alina Giedryś and Jan Raczyński

6.1. Introduction

Łódź, a post-industrial city, is the third largest agglomeration in Poland. Currently, the city runs a wide range of vital investments aimed at reversing the negative trends connected with unemployment, depopulation, and downtown degradation. Along with neighboring cities and municipalities, Łódź at present has a population of over a million inhabitants and is also a strong metropolitan, academic, and cultural center with considerable economic potential. It is a center for business process outsourcing (BPO) industries, logistics, and household appliances.

Łódź is currently one of the four undisputed strategic nodes on the planned line of high-speed rails (HSR), the so-called Y line in Poland. The concept of routing of a new railway line to the east–west axle, however, required substantial effort on the local level in order to break from the historically conditioned isolation of the city from the basic railway network.

In the last two decades, Poland has taken up a number of studies, projects, and infrastructural investments in the field of transport, most of which favor a modern road infrastructure; thus, rail transport has been confronted with the spontaneously growing automotive industry, which intensified the current imbalance in the transport market share. The low technical level of the railway infrastructure and a further decrease in subsequent years on the efficiency of the existing lines, which have only been partially renovated and modernized, has led to the disappearance of tram traffic on some lines of the Łódź railway node (e.g. on the Zgierz – Łowicz line) and to the development of commercial transport of bus services. During this period, passenger traffic on the Łódź railway node fell to its lowest point in the post-war period.

Analyses carried out on how to fulfill the pro-development policies for sustainable transport development of the Łódź region proved that it is high time for a radical restructuring of the system, the construction of new sections of the railway infrastructure, and its integration with other public transportation. It is planned that the construction of the underground central station Łódź Fabryczna and diametrical tunnel for conventional rail will allow the introduction of routes through Łódź for interregional trains qualified for east–west and north–south directions. It will also enable efficient exchange of their passengers with the agglomeration railway. The main investment for the region that will open the region to the world will be the high-speed Warsaw – Poznan – Łódź/Wrocław line with the new central train station Łódź Fabryczna.

6.2. Location of Łódź within the transport network of Poland and Europe

Łódź is situated in the center of Poland, at the intersection of historical, contemporarily rebuilt transportation routes. It is also a central point on the map of the region of Central and Eastern Europe. The distance from Łódź to Warsaw is 120 km, to Berlin – 460 km, to Vienna – 540 km, to Hamburg – 740 km, to Paris – 1110 km, to Moscow – 1430 km, and to Rome – 1690 km.

Currently, across the Łódź province run main transport corridors TEN-T – both road and rail corridors. In December 2013, the European Parliament and the European Council published Regulation 1315/2013 on guidelines for the development of a Trans-European Transport Network, according to which six lines belonging to the TEN-T network, including five lines of the core network, have been set out through the Łódź region. Łódź is also specified as one of eight nodes of the TEN-T network in Poland (Fig. 6.1).

Figure 6.1 Core transport corridors of the Trans-European Transport Network (TEN-T) – in the territory of Poland

Source: BPPWŁ on the basis of the European Parliament and the European Council Regulation no. 1316/2013

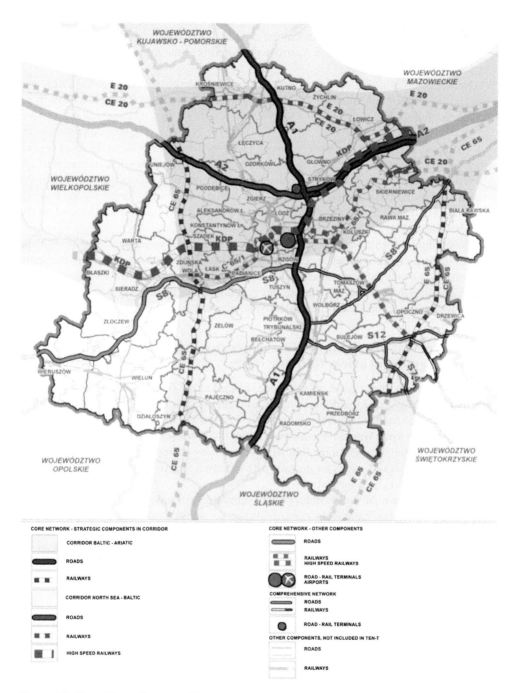

Figure 6.2 Core Trans-European Transport Network corridors – view on the Łódź Voivod-ship orientation

Source: BPPWŁ, Transport Plan of the Łódź Voivodship

Two main corridors of the European network of the TEN-T are:

1   Baltic – Adriatic corridor: connecting the ports of the Baltic Sea through Poland, the Czech Republic, Slovakia, and Austria with the ports of the Adriatic Sea – Trieste, Venice, Ravenna, and Koper.
2   North Sea – Baltic Sea corridor: connecting ports of the eastern coast of the Baltic Sea (Helsinki and Tallinn) via Estonia, Latvia, Lithuania, Poland, and Germany with ports on the North Sea coast.

Therefore, two transport corridors intersect with the freight network in the Łódź region, which seems a huge advantage in terms of logistics. Łódź becomes, therefore, the point which will ultimately have access to priority transport connections of goods in the direction of the southern coast of the Baltic Sea and the Eastern Adriatic coast (and further by sea towards Scandinavia), the maritime routes of the Mediterranean, as well as the ports located on the North Sea coast (Fig. 2).

Benefiting from an attractive location, at the crossroads of two major freight corridors, would not be possible without an efficient and coherent network of passenger transport. Modernization of the infrastructure of the Baltic – Adriatic corridor will also have its role in strengthening the transport, business, and economic integration of settlement strands of Łódź – Warsaw, especially after the implementation of the high-speed line.

6.3. Historical conditions of barriers for the development of Łódź Agglomeration Railway node

The current system of the Łódź railway node, formed at the boundary of two 19th-century railway systems of Russia and Germany, is an example of a coil of subsequent decision-making omissions and lack of final solutions. The first railway line in the Kingdom of Poland, the so-called Iron Road Warsaw-Vienna built in the 1840s, missed Łódź at a distance of about 30 km east of the city center. In the following years, many attempts were made to obtain the consent of the tsarist authorities for the railway connection for Łódź, in various forms, but the proposals did not receive approval.

In 1864, a group of Polish industrialists submitted a license application for the construction of the rail line from Koluszki to Łódź (on the Warsaw – Vienna line) and one year later a permission was granted. Construction of the line from Koluszki to Łódź lasted only three months. It was decided that the terminal station should be located close to Piotrkowska Street – the main axis of the city. In case of consent for the construction of another section to Kalisz, the line was supposed to cut the town in a western direction. Anyway, this option of the line extension never materialized, and the Łódź Fabryczna station remains to this day the final destination (terminal one). During World War I, the Łódź station was destroyed, along with most of the railway infrastructure. Historical development of Łódź node is shown in Fig. 6.3.

In the first decade of the 20th century, there have been shortages of railway lines in the region of Łódź, partly filled by a spontaneous development of tram lines (bearing the name of the Łódź Electric Rail Commuter), creating a vast network covering the towns located around Łódź. It was also planned to expand to cities located within a 50–60 km perimeter from the center of Łódź: Piotrków, Brzeziny, Zduńska Wola, Tomaszów Mazowiecki, and Łęczyca, among others. These plans were unfortunately thwarted by the economic crisis of the 1920s and the development of the automotive industry.

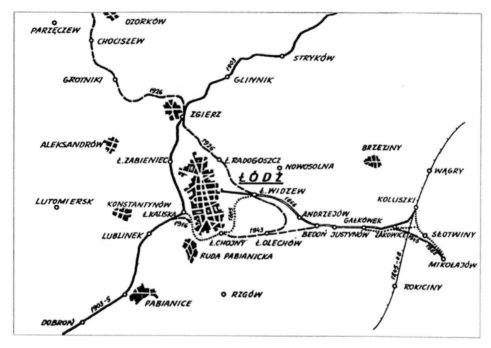

Figure 6.3 Łódź railway node development stages in 1845–1945

Source: [5]

Historically built railway lines within the Łódź railway node system were intended to handle freight traffic, which still has a negative impact on train work organized for passenger traffic.

For the purposes of passenger traffic, the Łódź railway node is characterized by low patency and dysfunctionality. Regional train traffic takes place along two independent network connections. The western part of the city center is not supplied. Centrally located Fabryczna station, before the conversion, principally operated trains to Warsaw, as the long-distance trains overdrove Łódź from the west and changed direction at the Łódź Kaliska station.

The density of the railway network in Poland is the lowest in Central Europe, and in Łódź Province it is even below the national average.[1] In addition, the system of railway lines only in a small degree meets the needs of residents, especially in terms of passenger and rail network, as some structural axes of the Łódź agglomeration are still not supplied with rail communication – this applies to the Łódź – Tuszyn line and the Łódź – Brzeziny line, whose operation is based on road transport.

The Łódź railway node's low infrastructure quality has been for many years and continues to be a negative factor of particular importance, causing restriction for the passenger transport offer – with its low speed, inadequate frequency, and lack of modern transport interchanges.

---

1 Density indicators of the railway network for the countries of Central Europe in 2010: Germany – 120 km/1000 km^2, Czech Republic – 101 km/1000 km^2, Hungary – 85 km/1000 km^2, Slovakia – 75 km/1000 km^2, Poland – 64 km/1000 km^2, and Łódź Region – 58 km/1000 km^2.

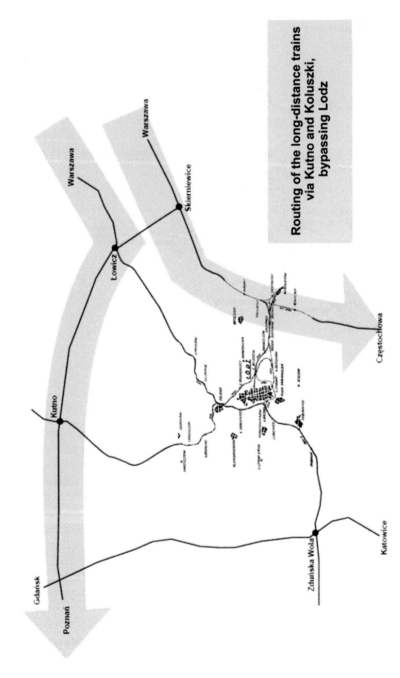

Figure 6.4 Marginalization of Łódź Agglomeration Railway node in routing of long-distance trains

Source: Feasibility Study of Łódź High-speed Railway, PKP PLK S.A.

The Łódź node is not treated at present as a convenient node for the region interchange, due to the breakdown of the system on two poorly connected parts linked to two stations: Łódź Fabryczna and Łódź Kaliska. The Łódź railway node is a relic in the country, and its inconsistency prevents the establishment of a system of national interregional connections covering Poland's central region. Also, it is impossible to create an effective system of regional rail for the Łódź region in relations between the north–south and east–west nodes. In this context, it is obviously impossible to include the railway in the internal communication service system of the Łódź agglomeration. This situation is caused by the lack of rail traffic handling the city center by a part of regional relations and the lack of one central station, which could be a convenient interchange (Fig. 6.4).

6.4. The process of planning and routing of the high-speed line running through Łódź

After Polish accession to the EU in 2004 and the implementation of infrastructural support programs, the improvement of transport availability has been defined as a determinant for further development of the region – in the planning documents of the City of Łódź and the Łódź Region. Therefore, with a view to their implementation, the analysis of the earlier concept of building a high-speed rail in Poland was undertaken. In the study "Directional Program of the High-Speed Line in Poland" [6] published in 1995, construction of a new high-speed line from Warsaw to Poznań, north of Łódź, along the A2 motorway was planned. According to this concept, the Łódź railway node would be connected to a new line in an indirect way, essentially in its present form, which is shown in Figure 6.5.

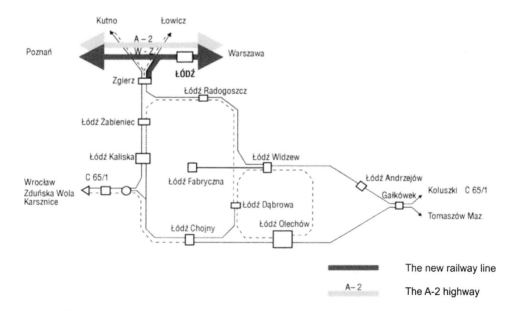

Figure 6.5 Scheme of the Łódź railway node inclusion into the high-speed line projected for 1995–2005

Source: PKP (Polish Railway Company), "Directional Program of the High-Speed Line in Poland" [6]

Such a solution would, however, be ineffective not only for Łódź (not even considering historical barriers as two weakly bound existing systems with two stations) but also for the whole Y line, which rent ability depends on the efficiency and size of the passenger flows on strategic nodes.

Subsequent documents relating to the development of the Łódź railway node were adopted by the Łódź City Council and the Regional Council in 2002 [10] [20]. In these documents, reference is made to the cross-city railway line in tunnel from the 1960s, pointing also to a need for building ramps at Dąbrowa – Stoki and Liściasta – Radogoszcz East. In this situation, it seemed a necessity to adopt the design considerations for the other high-speed Warsaw – Łódź – Wrocław – Prague line, which would run across Łódź city center. For the inclusion of Łódź to the Central Trunk Line (CMK), the document advised constructing a second track on the Tomaszów Mazowiecki – Opoczno line with a link to the CMK line.

In 2002, during the seminar "Fast train connection Łódź – Warsaw railway connections within the network of Poland" [14], organized in Łódź, a proposal to modify the route of the new line in relation to that proposed in previous studies was presented. More economical from an investment point of view, a connection from Warsaw to Poznań and from Warsaw to Wrocław, both using a common path through Łódź via tunnel beneath the city center, should join in the vicinity of Kalisz, where again they would branch out towards Wrocław and Poznań. This concept quickly gained social approval, especially in the concerned regions of Poland, and the connection gained the popular name of "Y line."

Ten years after the first route analysis, in a 2005 preliminary feasibility study for the construction of a new high-speed line in the east–west direction [26], Polish Railway Lines considered two groups of options:

- four variants of the construction of a new high-speed line for passenger traffic V > 300 km/h;
- three variants based on the upgrade of existing lines with newly added short fragments.

As a result of multi-criteria analysis, the economic advantage of options for building a new line running from Warsaw, via Łódź and Kalisz, to Poznań and to Wrocław has been demonstrated.

Analyses of Option I with the Stryków node on the northern side of Łódź and Option II with its course running centrally by Łódź Fabryczna are shown in Figure 6.6. For further analysis, however, two options were recommended, both accordingly to Option I.

The concept of the new high-speed line with its course through the Łódź area was entered in 2005 into the document *Transport Development Strategy for 2007–2013* of the Polish Ministry of Transport. It also received much acclaim during the first meeting of the High-speed Committee UIC in Poland in March 2006. It was a hot time for the crystallization of a variant of the project with its course through the Łódź city node. In 2006, a series of studies [7], including "Study on the course by Łódź of the high-speed V-300 rail line" commissioned by the Board of Roads and Transport in Łódź [18] have been made. They took into account a variant with a railway going through the center of the city. Efforts have also been made to analyze the organization of the construction and the operation of the new line as a part of a separate SPV (single purpose venture) in reference of the public-private partnerships procedure [9].

In 2007, at the request of the Board of Roads and Transport in Łódź, the "Functional analysis of the new central train station in Łódź as a supra-regional intermodal node" [19] was developed. Its task was to analyze existing restructuring of the Łódź railway node projects

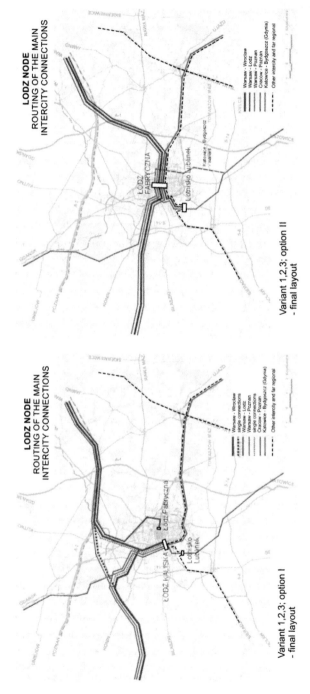

Figure 6.6 Tracking variants on their initial study phases for high-speed lines going on the north from Łódź or through the center of Łódź

Source: Pre-feasibility study for the construction of the high-speed line Wrocław/Poznań – Łódź – Warsaw, 2005 [26].

from the point of view of assessing the possible inclusion of the planned high-speed Y line node in Łódź, as well as the first reported design system concept for the suburban railway network in the region of Łódź.

The principal conclusions were as follows:

1. For a city the size of Łódź, an optimal solution from the point of view of economic exploitation is one station located in the center. For this purpose, the best suited remains currently operating Łódź Fabryczna station providing the redesigning of a system and creating on its basis a supraregional multimodal hub connecting other means of transport. Other locations of the central railway station in the city cause rise in realization costs due to the need for rebuilding the city's transportation system. In addition, a variant of the Central Railway Station in the very center of the city is more functional due to the simplification of the routes for train relations running through the Łódź agglomeration node.

2. Restructuring the Łódź railway node and including it into the national railway network would significantly increase the amount of travelers handled by Łódź stations and railway stops.

3. Construction of the line in a cross-city tunnel is the most justified because of its special importance for the movement of agglomerational and regional passenger traffic. At the diametrical line, intermediate stops should be set up.

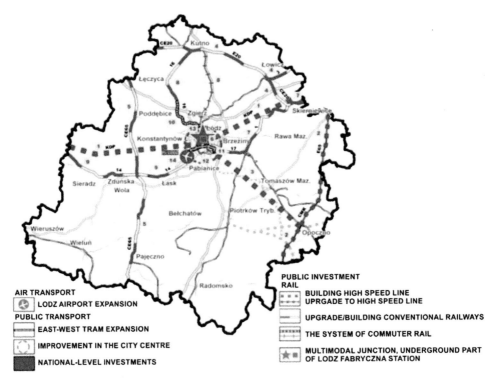

Figure 6.7 Tracing of the projected high-speed rail Y line along with inclusion to the Central Railway Line on the area of the Łódź Voivodship

Source: BPPWŁ, Transport Plan of the Łódź Voivodship

Already in 2009, the Polish high-speed line project was placed in the study of the European high-speed rail commissioned by the Directorate General for Transport and Energy of the European Commission and listed on the railway map of Europe.

In 2010, the Łódź Voivodship Board adopted the Łódź Region Development Plan, which initially traced the course of the new high-speed Y line through the center of Łódź (Fig. 6.7). This is the most favorable option from the point of view of the development of the Łódź node, urban planning of the city, and insurance of the efficiency of the central train station Łódź Fabryczna as a regional and supraregional multimodal hub. Conventional railways would cross in this node, serving the relatively densely populated north–south population area, which could generate for the Y line significant streams of passengers. The Spatial Development Plan introduced a draft merger of the HSR Y line with the existing Central Trunk Line Warsaw – Silesia.

6.5. Description of the modern concept of restructuring of the Łódź Agglomeration Railway node

In 2009–2013, at the request of the Polish Railway Lines, in reference of a series of study projects on the new high-speed line [21] "The feasibility Study for the adaptation of Łódź railway node to handle high-speed rail and to ensure the intermodality with other means of transport" [23] has been developed. This study contained works of studies and ex-ante (preliminary project) phase necessary to undertake the modernization of the infrastructure of railway lines included in the Łódź railway node along with the HSL. The aim was to adapt it to the needs, the scale, and structure of rail transport, as well as to enable an effective integration of different modes of transport on the basis of defined intermodal and interchange nodes.

Thus, the modern concept of the Łódź railway node restructuring was based on the strengthening and development of the external and internal railway system links, i.a. through:

- the implementation of the strategic railway system within the existing and postulated in 2030 TEN-T networks along with its switching to the CMK;
- the development of an interregional and regional system focused on basic incoming directions giving access to the TEN-T networks by rebuilding the system to the speed of 160 km/h (200 km/h) for passenger trains and 120 km/h for freight trains;
- the improvement of the Łódź agglomeration and the Łódź region transport accessibility along with links with the inner city of Łódź by building a Łódź Agglomeration Railway system, developing and integrating a wide range of public transport.

At that time, the city authorities as well as the local government, in cooperation with the manager of the railway infrastructure in Poland, based on their own resources and national and EU funds, had undertaken a package of integrated territorially and functionally transport projects, crucially important for the future of the city and region:

- the construction of the underground Łódź Fabryczna Central Station on a site of a former ground station, which is planned to be operational along with completion of the last stage of the modernization of the railway line from Warsaw to Łódź [24];
- the construction of the tunnel for conventional rail, directly coupling the eastern and western parts of the city, parallelly reducing by a few minutes travel time between the two parts of Łódź [25];

- the switching of the metropolitan area of Łódź and the provincial cities by a high-quality agglomeration rail [22] [3] – in June 2014, the first courses of modern trains within the Łódź Agglomeration Railway began, with their main task to act as the core for public transport in the region, as well as the bonding agent for a whole region;
- the modernization of existing and the construction of new railway stops on the suburban railway lines [1] [4];
- modernization of the railway lines from Łódź to Sieradz and to Kutno;
- routing studies for passing through the region and the introduction on the platforms of Łódź Fabryczna a high-speed rail in relation Warsaw – Łódź – Poznań (Berlin) and Wrocław (Prague), which will reduce the isolation of Łódź from other parts of the country, allowing a quick trip between Poznań, Wrocław, Łódź, and Warsaw and also providing an international rail connection for Łódź to southern and western Europe.

The Łódź railway node, equipped with a diametric tunnel beneath the city and an optimized system of passenger stations adapted for international, interregional, regional, and agglomeration trains, on one hand will finally allow for a full discounting of a privileged central geographical location of the region and agglomeration; on the other hand, it will be a milestone towards the realization of inner sustainable transport policy, which was adopted by Łódź already in 1997. Achieving a railway competitive advantage for passenger transport services, especially in the Łódź region, requires, however, far more ambitious goals for rail than only upgrading and revitalizing degraded infrastructure and replacing obsolete rolling stock. It also needs to provide competitive transit times compared to travel by car, which seems possible to achieve. After implementing HSL through the center of Łódź, establishing a network of domestic routes that also take into account upgraded conventional lines will be possible.

Current and projected arrival times from Łódź to other cities are shown in Table 6.1 and in Figure 6.8.

In recent planning documents, i.e. "The Strategy for the Development of Central Poland up to 2020 with the Prospect of 2030," adopted by a resolution of the Council of Ministers

Table 6.1 Comparison of current and planned travel times for rail connections from Łódź

Connection	Length of the railway	Travel time by rail in 2015	Length of the railway (km)	Travel[2] time after the high-speed line construction
Łódź – Warszawa	126	1 h 07 min[3]	350	0 h 35 min
Łódź – Poznań	286	2 h 51 min[4]	227	1 h 00 min
Łódź – Wrocław	253	3 h 55 min[5]	231	1 h 05 min
Łódź – Kraków	259	2 h 18 min[6]	238	1 h 25 min
Łódź – Katowice	246	2 h 56 min[7]	238	1 h 25 min
Łódź – Szczecin	496	5 h 34 min[8]	440	3 h 40 min

Source: Author's material based on prognostic travel times [11]

2 All departures take place from a central Łódź Fabryczna station.
3 From Łódń-Widzew through Koluszki, Skierniewice to Warszawa Centralna.
4 From Łódń-Widzew through Łowicz, Kutno to Poznań.
5 From Łódń Widzew through Ostrów Wlkp to Wrocław.
6 From Łódń Widzew through Tomaszów Maz.to Kraków.
7 From Łódń Kaliska through Piotrków Tryb., Częstochowa to Katowice.
8 From Łódń Kaliska through Kutno, Poznań to Szczecin.

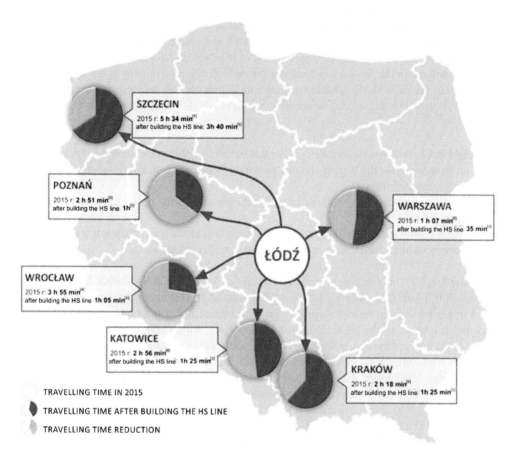

Figure 6.8 Comparison of current and planned after launching of the high-speed rail train travel times from Łódź

Source: Material of the Spatial Planning Bureau of the Łódź Voivodship, 2016.

in July 2015 [16], the realization of development objectives of the strategic macro-region has been defined by creating, through the use of synergetic development potentials of Łódź and Warsaw Voivodships (Mazowieckie), a multimodal transport hub of international importance. Grounding of transport means on which the establishment of a Central Poland Functional Area is based is the line of high-speed rail Y line.

6.6. Implementation of the Łódź Fabryczna station as a central multimodal hub of the Łódź Agglomeration Railway node

In reference of infrastructure projects being undertaken within the Łódź railway node, a strategic project remains the construction of an underground railway station as a multimodal hub on the regional and national level. The next stage will be the construction of the diametrical tunnel beneath the center of Łódź (2022)

Already in the initial conceptual stage, "The functional analysis of the new central railway station in Łódź as a supra-regional intermodal node" was commissioned by the Board of Roads and Transport in Łódź in 2007 [2], which is a benchmarking of foreign solutions and an assessment of their applicability to the case of Łódź. As a result, an assumption of basic conditions for the project were made, considering a planned railway station as an underground train station with the function of multimodal hub for various modes of transport having their impact on supraregional level, in a system in which a high-speed rail is a subsystem along:

- eight tracks (two-edge platforms) at the station; an increase in the number of platforms is not foreseen as for the high-speed rail, provided that its activation will reduce the amount of conventional trains;
- length of platforms up to 400 m that will make possible in the future to dispatch trains of all kinds;
- hollow station at a depth of more than 16 meters will allow the entrance of trains into the diametrical tunnel with respect of the inclinations limit regulations.

Also in 2007, an agreement between the city of Łódź – PKP SA – PKP PLK S.A. (on 31 July 2007) was made, along with an appendix dated 18 December 2007 on the establishment of partnerships in order to regulate the legal land status of the Łódź Fabryczna station and the realization of the New Center of Łódź project, which allowed the implementation phase of the new station construction to begin. Contract for the design and construction of the station as a multimodal hub was signed in August 2011 by all the signatories [12].

Saturday, 15 October 2011, was the last day of operation of the old Łódź Fabryczna station. At 10:40 p.m. a few hundred people boarded the last train departing from the station platforms (Fig. 6.9). The crowd of passengers got off at the Łódź-Widzew station in order to get on board a train heading to Warsaw at 23:12 p.m. – in the last compartment, which reached Łódź Fabryczna station. The atmosphere was that of fun and festivity.

In order to keep the elements of identity designed in 1868 by Warsaw architect Adolf Schmmelpfenin, in accordance with the guidelines of the provincial monument conservator, replicas of the side walls of the old train station should be inserted on the east side of the new train station on the northern and southern elevation in a way that they would be directed towards the interior of the station hall. On the outside façade, there will be a contemporary arrangement (without window openings). In the future, on this side there are expected to appear "outside" walls of commercial buildings accompanying the station.

Planned main stages of the station functioning include:

- The first stage of exploitation of the new underground Łódź Fabryczna station, planned for the end of 2016, will coincide with the commissioning of the modernized railway line from Łódź to Warsaw. The station will provide at this stage a travelers' interchange on the scale of 20 thousand per day. These passengers will be able to smoothly change transport – both between urban transport means, domestic buses and trains of the Regional Transport Company and the Łódź Agglomeration Railway Company, which will provide a good standard for passengers travelling from the center of Łódź towards Koluszki, Skierniewice, and probably also to Warsaw. Agglomeration rail movement from Sieradz, Kutno, and Łowicz will, however, continue to focus mainly on the Łódź Kaliska and Łódź Widzew railway stations.

Figure 6.9 Farewell at the Łódź Fabryczna station at 15 October 2011 before the initiation of its reconvention

Source: http://koloreklodz.blogspot.com/2011/10/odz-fabryczna.html

- The second stage will take place upon completion of construction and commissioning of the diametrical tunnel for conventional rail, connecting the Łódź Fabryczna railway station to the Łódź Kaliska and Łódź Żabieniec railway stations (2021). The terminal station will become a throughout station. This will serve as a breakthrough for improving the efficiency of the Łódź railway node for both the agglomeration and regional lines, as well as for long-distance lines. Removal of historical dysfunctionality, involving the separation of rail services for the eastern and western parts of the city, will create a completely new transport offer. At this stage of planning, the Intercity class IC trains will run from Łódź on connection with Warszawa, Wrocław, and Gdynia Katowice. For the agglomeration traffic, at this stage, there will be a full opening of services to Łódź from Sieradz, Kutno, and Łowicz – it will be possible to take fast trains into the city center. This will allow an opening of new connections in the through traffic – for example from Koluszki to Sieradz and from Łowicz to Koluszki, which will significantly improve the consistency of the agglomeration of Łódź.
- Finally, the third step will be a launching of a high-speed line, which according to previous plans will connect Warsaw, Łódź, Poznań, and Wrocław (approx. in 2030), stopping at the already built within the railway infrastructure station, on built-for-purpose dedicated platforms.

Construction of a diametrical tunnel for a conventional rail outgoing from a western head (turnout head) of the underground station is officially the second phase of the restructuring of the Łódź railway node, which has been recorded in the preamble to the Commission Decision on financing of investment from the Cohesion Fund [8]. In 2016, the studies for the tunnel will be completed [25] and a start of the tender procedure for the implementation phase is planned. The conventional tunnel will have two intermediate stops: in the area of Zielona Street and in the area of Ogrodowa Street. The forecasted volume of passengers at the stops is shown in Figure 6.10. Western turnouts at the head of the station also provide turnouts for the future high-speed rail tunnel, which will direct trains from Łódź Fabryczna station, without stopping, outside the downtown area of Łódź in the direction of Kalisz and further to Poznań and Wrocław.

Table 6.2 Prognosis of traffic volume, as for three functionality stages of the Łódź Fabryczna station

Stages	Planned period	Investment stages	Passenger traffic forecast/day
Stage I	End of 2016	Inauguration for exploitation of the underground Łódź Fabryczna station and completion of Warsaw-Łódź railway modernization. Full operationalization of all Łódź Agglomeration Railway in this stage.	20 ths.
Stage II	2021	Inauguration of exploitation of diametrical tunnel for conventional railway, linking Łódź Fabryczna, Łódź Kaliska, and Łódź Żabieniec stations.	30 ths.
Stage III	2030–2035	Inauguration of exploitation of high-speed rail at sections running through Łódź.	70 ths.

Source: [11] [21]

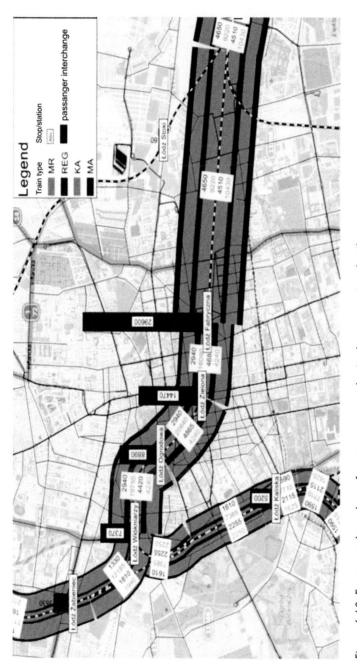

Figure 6.10 Forecasted number of passengers per day in the conventional railway transport (without the high-speed rail) in Łódź – 2035.

Source: [25]

The capacity of the train station and diametrical tunnel between Łódź Fabryczna and Łódź Kaliska (towards the south) as well as to the Łódź Żabieniec station (to the north) was designed (as for 2040) for 22 pairs of trains at rush hours, with 10 of them ending their course at the station [11]. Planned use of the tunnel according to the train operators' programs is shown in Table 6.3.

New quality of passenger handling infrastructure at the station Łódź Fabryczna (Fig. 6.11):

- three levels of underground railway sunlit;
- multimodality including, besides the railway, regional and interregional buses, city buses, trams, cars, and bicycles;
- new functionality of a station enabling integration of the transport function with trade and services;
- pedestrian-friendly space around the station – due to retracting of the car parks and improvement of the environmental quality (green areas and urban small architecture).

Four central platforms with a length of 400 m have been designed, so that they can receive combined trains. The width of platforms of 12 m will allow the free movement of travelers, even when the same platform serve two trains at the same time. A multifunctional bus stop located within the station, dedicated to interregional (long-distance) and regional communication buses, will have a stop point with a capacity designed for 72 departures and 36 arrivals at peak hours. It will have 24 stands located on level − 1.8 m below ground level [12].

Table 6.3 Number of trains planned for diametrical tunnel accordingly to carrier plans

Train specified	Connection	Number of train pass/ day	Max.in a rush hours' time	%	Data source
Interregional	Warszawa – Łódź – Wrocław	4	1	2.7	Train schedule program of the Polish Railway
	Kraków – Łódź – Poznań	4	1	2.7	Train schedule program of the Polish Railway
	Katowice – Łódź – Gdynia	5	1	3.5	Train schedule program of the Polish Railway
	Łódź – Warszawa	30	4	21	Train schedule program of the Polish Railway
Regional	For neighboring Voivodships	20	2	14	Concept of train schedule program for the Łódź Voivodship
Agglomeration	Within the Voivodship area	80	4	56	Concept of train schedule program for the Łódź Voivodship
TOTAL		153	15	100	

Source: J. Raczyński

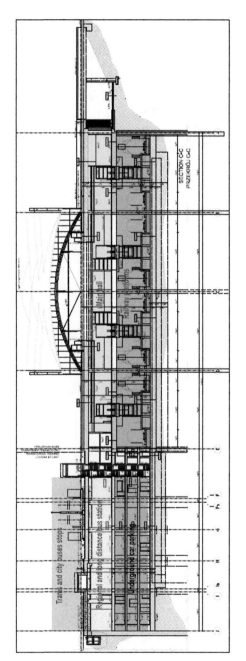

Figure 6.11 Cross-section of planned Łódź Fabryczna station

Source: [12]

Figure 6.12 Broad scope of road investment (marked in red) and modernization/construction of tram infrastructure (marked in green) around Łódź Fabryczna station

Source: Road and Transport Management Board in Łódź

July 2013

November 2013

March 2014

March 2015

October 2015

Figure 6.13 Photos of the underground Łódź Fabryczna station construction stages

Source: EC1 archives

Table 6.4 Share of transport services to get to the Łódź Fabryczna station – prognosis for
2040

Pedestrian and bike traffic 9%	Taxi 6%	Private cars 24%	City buses 18%	Trams 34%	Regional trains 9%
	Cars 30%		Public transport – 61%		

Source: [11]

The urban transport system around the station will also be rebuilt. The main objective is to assure that public transport is in the most possible direct vicinity of the station building. Therefore, within the project a tram route system will be rebuilt. The track on Narutowicza and Kilińskiego Streets will be renovated. Additionally, a new track line along the northern wall of the station will be built. Most of the tram lines will stop at the tram stops located in front of the main entrance to the station. Predictions foresee a fourfold increase in the number of passengers using the tram, therefore assuming full use at the maximum capacity of the infrastructure. For urban traffic buses, three stops are foreseen (Fig. 6.12). The capacity of the underground car parking places, based on the results of benchmarking of existing facilities of its kind in Germany and France and on the fact that access to the station should be ensured mainly through dense urban public transport network, has been calculated for approximately 800 places. Also provided will be areas for short-term parking for taxis and private cars [11].

Expected share of individual means of transport used to get to the Łódź Fabryczna station for 2040 is shown in Table 6.4.

After finishing works and the acceptance procedure the Łódź Fabryczna station was opened in December 2016 (Fig. 6.13).

6.7. Role of the Łódź Agglomeration Railway as a modern distribution system for travelers in the region

Analyses carried out in 2006–2008[9] show that the creation of a suburban railway system around Łódź is technically and organizationally possible, economically justified, and from the urban point of view even necessary.

According to these analyses, current solutions aimed at improving the Łódź Railway Node has been elaborated, the results of which foresee a concept for a diametrical tunnel in later realization stages. For its elaboration, results of previous studies and specific examples of solutions for similar nodes in Western Europe have been used. Major demands that were made for this task are:

- separation of passenger traffic and freight traffic, including the transport of dangerous goods to improve the safety and capacity of the node;
- improvement of the functionality of passenger service throughout the multimodal chain with a view to reduce the total travel time and inconvenient transfers between various means of transport;
- reduction of an operational cost of the passenger service system through rational formation of multimodal nodes in a hierarchical structure;

9 See chapter 5.

• shortening of travel time for passenger trains running through the Łódź node on north–south and east–west axes for all categories of trains. According to the investment provisions, it was foreseen to create a rail system within the agglomeration stages dependent on the development of railway infrastructure in the Łódź region. Implementation of the first phase included only already existing Łódź lines and planned ones to be upgraded in the coming years. In the second stage, once further investments would be operational, including a diametrical tunnel beneath the Łódź city center with two stops underground [25] and the second phase of construction and modernization of passenger stops, it would allow an active incorporation of the agglomeration into the urban area transport handling system [17], creating an agglomeration and a regional backbone for passenger transport.

First stage of the Łódź Agglomeration Railway project included [22]:

• Purchase of 20 electric multiple units having technical parameters for a typical agglomeration traffic utility, with a maximum speed of at least 160 km/h, with starting and stopping acceleration of a minimum of 1.1 m/s^2. Under the signed contract, there has been an obligation to maintain the fleet for 15 years' time, whereas the cost of maintenance was not an eligible project expenditure;

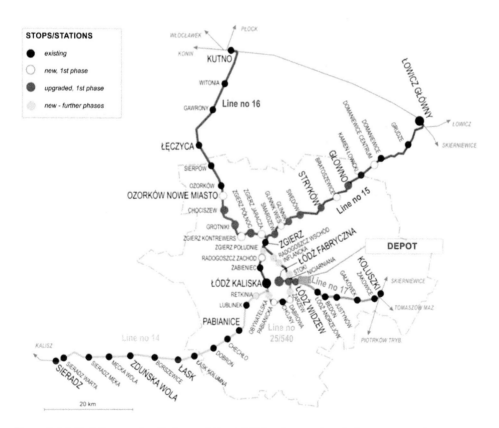

Figure 6.14 Rail lines and rail stops of the Łódź Agglomeration Railway – stage I

Source: Materials of the Infrastructure Department of the Łódń Voivodship Marshall Office

- Construction of facilities for rolling stock maintenance on the Łódź Widzew station, including building structures, stabling tracks, and traction power supply system; tools, diagnostic and information systems, as well as other movable equipment within the scope of the contract for maintenance of rolling stock that were eligible expenditure of the project. Part of the creation of a suburban railway network system was also the construction of a new and reconstruction of existing railway stations as multimodal nodes [4], as well as the establishment of agreements on designed route lines and cooperation with local governments in order to coordinate timetables and tariff agreements.

On 15 June 2014, a passenger service within the Łódź Agglomeration Railway was launched. Starting a freight line from Łódź to Sieradz marked the beginning of a new era in the field of rail transport in the region – regular operations commenced, the public company started featuring an exclusively modern, air-conditioned, and accessible to people with reduced mobility fleet, which is serviced in the in-house technical station – one of the most modern in Europe. The new carrier has achieved commercial success by reactivating a previously unsupported by railway line to Łowicz and obtaining high marks on the surveys measuring passenger satisfaction. The final moment of the first stage of the establishment of the Łódź Agglomeration Railway system will be the opening in December 2016, of an entirely modernized line between Łódź and Warsaw, as well as of the Łódź Fabryczna Railway Station, which will allow better usage of the Łódź Agglomeration Railway's potential (Fig. 6.14).

A second phase of the project of the Łódź Agglomeration Railway is therefore associated with the second stage of the Łódź railway node's restructuring project in the years 2020–21, which will be completed along with construction and commissioning of the diametrical tunnel for conventional rail, connecting the railway station of Łódź Fabryczna – Łódź Kaliska in the south and the Łódź Żabieniec station in the north. Once the terminal station is to become a throughout station, this will be a breakthrough for its functionality. Location of two stops in a tunnel beneath the city center, especially at the node with the main tram agglomeration way along the north–south axis, will significantly increase rail availability and will relieve other means of urban transport.

Figure 6.10 shows the prognosis for the exchange of passengers (passenger flow) at the train stop as approximately 15,000 per day. For regional and agglomeration traffic, this stage will open up the possibility of using new railway express delivery rail connections, offering quick entry to the center of Łódź, Sieradz, Kutno, and Łowicz.

To sum up, urban transportation and the regional Łódź railway node will ultimately be implemented as part of a regional three-axis system:

- Axis 1. Sieradz – Łódź – Skierniewice (east–west).
- Axis 2. Kutno – Łódź – Tomaszów – Opoczno/Radom (northwest–south)
- Axis 3. Łowicz – Łódź – Piotrków and further on to Częstochowa (northeast–south)

Therefore, those three axes will be a strategic element of the urban node of the TEN-T, playing a key role as node integrator in accordance with the provisions of EU Regulation 1315/2013.

The Łódź Agglomeration Railway is already, in 2016, preparing new projects for the purchase of modern rolling stock for new routes with deadlines for their implementation within two years. Expansion plans of the Łódź Agglomeration Railway are based on maintaining a current high standard of passenger service and a high level of operational reliability.

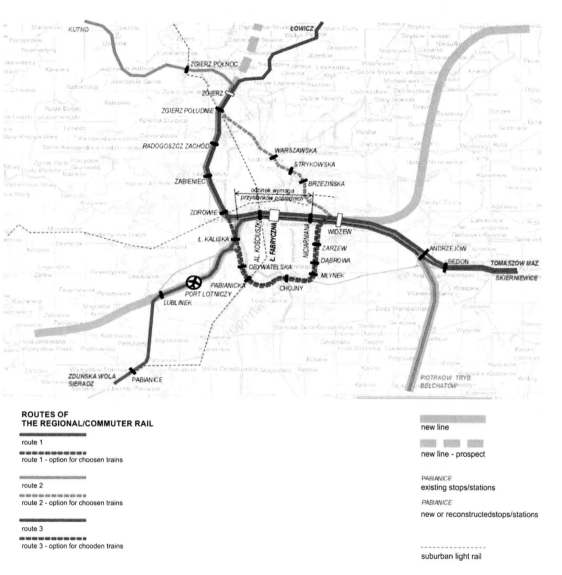

ROUTES OF
THE REGIONAL/COMMUTER RAIL

route 1

route 1 - option for choosen trains

route 2

route 2 - option for choosen trains

route 3

route 3 - option for chooden trains

new line

new line - prospect

PABIANICE
existing stops/stations

PABIANICE
new or reconstructedstops/stations

suburban light rail

Figure 6.15 Łódź railway node restructuring concept with special emphasis on the agglomeration railway and the high-speed lines to Łódź, with use of current and planned infrastructure

Source: [19]

The construction project of the Łódź suburban railway network, was in its assumptions and is still now an essential element enhancing regional development. In the perspective of its implementation, it should contribute to the desired orientation of the process of urbanization within the agglomeration along the main rail routes, thus ensuring greater efficiency of the entire public transport system in the region (Fig. 6.15). The creation of the rail system, which also includes the creation of multimodal interchange [17], will improve the availability of the rail for residents. It was assumed that the simultaneous investment in rail infrastructure would also shorten the travel time from the outskirts of the agglomeration to the Łódź city center, placing it on a level of less than 30 minutes (more than twice as fast compared to the present). As a result, a cheap means of transport with high reliability and punctuality will be provided. What is more, current functioning of the Łódź Agglomeration Railway contributes to an increase of the mobility of citizens and affects the reduction of unemployment. Conditions for commuters travelling to schools and universities will be improved, ensuring equal conditions of access to education for young people from more remote and less urbanized areas of the agglomeration.

Investment in public transport and improvement of its quality will contribute to a reduction of car traffic in the region, thereby reducing CO_2 emissions and air pollution, as well as reducing the cost of congestion, traffic accidents, and the negative impact of road transport on the health and quality of life of the region and its inhabitants.

6.8. Targeted effects of modernization of the Łódź railway node and its surroundings

a. Construction of a diametrical tunnel and multimodal Łódź Fabryczna hub will allow to modernize and better organize traffic within the Łódź railway node through far-reaching separation of passenger and freight traffic. Freight traffic will be routed to the ring roads around Łódź and passenger traffic through the city center. The diametrical line will take over passenger traffic, relieving capacity of actually operational peripheral line for the efficient running of freight traffic. The main TEN-T Zduńska Wola – Warsaw line will run along the southern perimeter rail track using the western and eastern peripheral rail in order to support other lines with freight traffic terminals within the Łódź TEN-T node.

b. An integral part of the modernization of the Łódź railway node will be the shortening of transfer time due to investments in upgrading or at least rehabilitation of the conventional rail line binding Łódź with Sieradz and Kutno. The line from Łódź to Łowicz was subjected for a partial revitalization works in the years 2012–2013. The line from Łódź to Warsaw has been modernized in several stages to a speed of 150–160 km/h. Completion of work on the whole of the Łódź railway node should be coordinated with the final operationality of the diametrical tunnel.

    Lack of coordination in this respect and likely delays in individual projects cause a serious threat for obtaining results to the conventional railway line in the tunnel.

c. Construction of the underground Łódź Fabryczna station as a central hub for region for train and long-distance buses, as well as for trams and city buses. In the future new station will serve also high-speed trains.

d. The Marshal Office of Łódź conducts a consistent transport policy after having introduced in 2010, by the Provincial Spatial Development Plan, the high-speed rail and launching on a basis of sequential stages of the infrastructural investments implementation the Łódź Agglomeration Railway, planned to act as a rail integrator within the Łódź node. In preparation or execution are further projects of investment in rolling stock and expansion of the Łódź Agglomeration Railway on new lines, reconstruction of the Łódź Kaliska station towards rail and urban transport node, and the second stage of construction or modernization of passenger stops on the Łódź Agglomeration Railway (in consultation with the rail infrastructure manager in Poland).

e. The city of Łódź is actively involved in cooperation with the manager of the rail infrastructure in modernizing the municipal part of the Łódź Fabryczna railway station and its surroundings. In parallel, for 10 years works on urban concept and studies for the realization of the great project of the New Center of Łódź are conducted, including comprehensive revitalization of an area of 100 hectares in the quarter of the streets surrounding the station. In a first stage of revitalization, there have been subjected two complexes of old electric power plants that already exist as modern cultural facilities. In development is a bid application for rights of Poland to organization EXPO 2022 in Łódź, with the suggested location for this event in the New Center of Łódź. The advantage is not only the availability of land located in the city center, but also an excellent logistic proximity to the new multimodal central train station Łódź Fabryczna. It is anticipated that this station will be used by passengers and by visitors traveling on regular schedule rail lines as well as with cars left in the system's Park & Ride areas near highways, for which the Łódź Agglomeration Railway will run shuttle bus lines.

f. The organizer of transport in the region, the Marshal's Office, plans to build an interactive knowledge database accumulating data, analysis, and research on the transport behaviors of the inhabitants of Łódź and the Łódź region in order to use them for forecasting traffic and passenger preferences. As part of this project, besides the cooperation of various universities and institutions, a specific cooperation with the Technical University of Łódź in the framework of the Long-Term Research Programme "Railways in the XXI century" is envisaged. In a situation of such radical changes in transport infrastructure and transport operations taking place currently and being planned for the Łódź railway node, implementation of a rational policy of sustainable transport development is needed.

6.9. Summary

After joining the European Union in 2004, a possibility appeared in Poland to use financial aid funds, allowing the completion of previously diagnosed needs and major infrastructure projects within the European transport networks. The Łódź region has received a chance taking advantage of its central location in the country, which was previously not possible because of the historical barriers in the railway infrastructure coming from the political realities of the 19th century and progressive recapitalization of this infrastructure in recent decades. The projected construction in Poland of the high-speed Y line running centrally through the Łódź node was recognized as an obvious chance to change the existing isolation and marginalization of Łódź in the main system of railway lines. The connection of Warsaw, Łódź, Poznań, and Wrocław by a new railway line with modern European standards V > 300 km/h is the most rational, confirmed by feasibility studies and the opinions of international institution to solve effectively the problem of railway transport in the country.

Currently being developed within the Łódź Agglomeration Railway is new network of regional connections that provides the ability to fully exploit the potential of both the high-speed rail Łódź node and the node of TEN-T transport corridors. Prospects are good – according to analysis of current European and world trends, growth on rail transport occurs particularly on the agglomeration transport and on the high-speed rail lines. In recent years, an increase in transport between agglomerations is to be observed as well as progress in the integration of rail with other modes of transport around transportation centers, which are modern and multifunctional railway stations. The phenomenon of passengers returning to rail transport after the introduction of higher standards of travel can also be noticed on the Łódź railway node lines, where a modern fleet operates despite an infrastructure that is still not modernized.

Łódź agglomeration node is fully prepared for the high-speed rail project of the "Y line." The local authorities in cooperation with the rail infrastructure manager in a consistent manner led to the construction and substantial completion of all elements of the model solution integrating high-speed rail with other modes of public transport and the individual Łódź Fabryczna station. The course of the Y line has been entered into development plans of the Łódź, Mazowieckie, and Central Poland macro-regions.

Transport policy of the country requires similar continuity and consistency due to a long-term investment preparation process. The project of the high-speed line Warsaw – Poznan – Łódź/Wrocław is on stage for the execution decision. Its implementation is demanded by the vested socioeconomic interest of regions, including the macro-central region as well as the whole of Poland. The high-speed rail project Y line has been entered in the European transport corridors TEN-T.

Bibliography

[1] Actualisation of the optimal transport system of the City of Łódź – transport system in a perspective of 2025 made by the Office of Development Planning in Warsaw SA on behalf of the Board of Roads and Transport in Łódź, 2014.

[2] Analysis on functional development of the Łódź railway node in terms of the construction of high-speed lines in Poland in a view of meeting the communication needs of Łódź agglomeration, commissioned by the Board of Roads and Transport in Łódź by the SITK RP Branch in Łódź, 2007.

[3] Construction of the Łódź Agglomeration Railway system, Phase I, under the operational program "Infrastructure and Environment" project of the Łódź Agglomeration Railway, 2013.

[4] Concept of organization of the construction and operation of high-speed lines in Poland. Stage I – establishment of a special purpose company CNTK commissioned by PKP SA (Polish Railways) Ltd. Co., Warsaw 2006.

[5] Transport and communication history of the Łódź Region. SITK, Łódź 1974.

[6] Directional Program of High-Speed Lines in Poland. PKP, 1995.

[7] Fast rail Warsaw – Łódź – Wrocław/Poznań. Analysis of possible courses of the Łódź Province in scope of the connection with accessibility of regional train with increased performance, ordered in April 2006 by the Office of Spatial Planning Region of Łódź with headquarter in Łódź.

[8] The European Commission Decision CCI 2014PL161PR018 dated14.07.2015 on the major project implementation "Capacity enhancement of the Łódź Railway Node (TEN-T), phase I section of the Łódź Widzew – Łódź Fabryczna," pointing to the need of implementation by Poland of the project "Capacity enhancement of the Łódź Railway Node (TEN-T), stage II section Łódź Fabryczna, Łódź Kaliska and Łódź Żabieniec" in the perspective of 2016–2020.

[9] Concept of the construction and operation organizing of the high-speed lines in Poland. Stage I – establishment of a special purpose company CNTK commissioned by PKP SA Warsaw, 2006.

[10] The Spatial Management of the Łódź Region – Resolution of the Łódź Regional Assembly of 9 July 2002.

[11] Functional and utilitarian newly designed underground station Łódź Fabryczna, on a commission of PKP SA in cooperation with the City of Łódź and PKP PLK S.A. in frame of the Project OPI & E 7.1–24.2 Modernisation of the Warsaw-Łódź railway line, Phase II, Lot B-2 section Łódź

Widzew – Łódź Fabryczna from Łódź Fabryczna station and construction of the underground part of the Łódź Fabryczna station aimed at handling of departures and arrivals of trains and passenger service.

[12] Draft OPI & E 7.1–24.2 Modernisation of the Warsaw-Łódź railway line, Phase II, Lot B-2 section Łódź Widzew – Łódź Fabryczna and construction of the underground part of the Łódź Fabryczna station adapted for departures and arrivals of trains and passenger service.

[13] Projected of the diametrical railway lines in Łódź as an element of the KDP and conventional rail, made by SENER, commissioned by PKP PLK S.A., the Marshal's Office in Łódź and the Office of the City of Łódź, November 2011.

[14] Raczynski, J.: Łódź – Warsaw High-Speed Rail Network in Poland in comparison with the European solutions. Seminar "Fast train connection Łódź – Warsaw railway connections within the network of Poland, October 2, 2002 organized by the Branch Board of the Railway Transport Engineers Association, the City of Łódź Office and the Office of Rail Projects and Investment Services Lmt Co. in Łódź.

[15] Łódź City Hall and Rail& Investment Services Bureau Lmt. Co. In Łódź. The Regional Transport Plan for the Łódź Province meeting the criteria of the ex – ante conditions for the thematic 7 for the ROP 2014–2020, December 2015.

[16] Development Strategy for the Central Poland for 2020 with the prospect of 2030 adopted by a resolution of the Council of Ministers in July 2015.

[17] Study on construction, reconstruction and modernization of railway stations on the route of the Łódź Agglomeration Railway – Phase I in the framework of the regional operational program of the Łódź Region, by the Railway Institute at the commission of the Marshal Office of Łódź, 2015.

[18] Study on the course by Łódź of the high-speed V-300 rail line, realized in September 2006 by the Development of Cities and Settlements Company Terrain Ltd. Co. in Łódź on the commission of the Board of Roads and Transport in Łódź.

[19] Functional analysis of the new central train station in Łódź as a supra-regional intermodal node commissioned by the Board of Roads and Transport in Łódź by SITK RP Branch in Łódź, 2007.

[20] Study on conditions and directions of spatial development of the city of Łódź, adopted on 3rd April 2002 by the Council of the City of Łódź.

[21] Feasibility study for the construction of the high-speed railway Łódź- Warsaw-Poznań/Wrocław, made by IDOM, commissioned by PKP PLK S.A., 2013.

[22] Feasibility Study for the construction of the Łódź Agglomeration Railway system, phase I. by the SITK RP Branch in Łódź commissioned by the Marshal's Office in Łódź, 2009.

[23] Feasibility Study for the adaptation of the Łódź railway node to handle high-speed rail and to ensure the intermodality with other transport modes, developed by SENER, commissioned by PKP PLK S.A., 2013.

[24] Capacity enhancement of Łódź Railway Node (TEN-T), stage I, section Łódź Widzew – Łódź Fabryczna, which is part of the Operational Programme "Infrastructure and Environment" for structural assistance from the European Regional Development Fund and the Cohesion Fund (2011–2016).

[25] Additional feasibility study along with tender documentation for the variant XIV of the diametrical tunnel implemented for the railway line from Łódź Fabryczna station towards the railway line No. 15 within the project OPI & E 7.1–76, under development by Safaga on behalf of PKP PLK S.A.

[26] Pre-feasibility study for the construction of the high-speed line Wrocław/Poznań – Łódź – Warsaw, made by the Centre for Scientific and Technical in September 2005. Commissioned by PKP PLK S.A.

Chapter 7

European and Polish requirements for a high-speed rail system

Marek Pawlik

The scope of this monographic publication dedicated to high-speed rail system construction and exploitation in Poland covers many important issues, from socioeconomic and economic ones to technical and educational ones, elaborated in 20 chapters prepared by the best Polish experts. The publication presents the deep knowledge in Poland concerning different kinds of circumstances influencing high-speed rail system design, construction, exploitation, and maintenance. It points out and elaborates Polish circumstances, as well as Polish and European regulations, which have to be taken into account during the construction and commissioning of high-speed rail systems in Poland, regardless of the kind of legal entities, Polish, European or global, that are involved.

It is obvious that the Polish high-speed rail system has to be incorporated into the European one to ensure good interconnections with high-speed rail systems in other, neighboring, European Union member states and to ensure fulfilment of the legal regulations in force. Most of the legal requirements influencing, as shown below in the following chapters, operational requirements [5], [6] as well as many detailed technical requirements [8–20], are regulated in European law. However, complementary Polish regulations [23–27] are also obligatory and must not be forgotten. These European and Polish regulations, as well as a general horizontal legal framework [1–4], [7], [23–27], which are directly applicable to high-speed rail systems, are presented below.

In accordance with the "treaty regulating the functioning of the European Union," for practical achievement of European Union objectives, European institutions, particularly the European Parliament in cooperation with the European Union Council and European Commission, set up directives, decisions, regulations, recommendations, and opinions constituting "European derivate law." European directives have to be, and are being, implemented into member state law, including Polish law. Direct application of other legal regulations, including decisions and regulations, is mandatory in the European Union member states. In the range of directives, as well as in the range of decisions and regulations, there are numerous regulations applicable to railway transport. Recommendations dedicated to railway transport are also important; however, their application is not obligatory.

Technical requirements for high-speed rail systems defined in European law [8–20] are relatively well recognized. They are understandable for all railway market stakeholders, including the production industry, contractors, and end users. These requirements are seen, first of all, as a basis for interoperability understood as a set of common solutions, which are defined for achieving seamless rail transport across internal European borders between member states of the European Union. However, regulations in force [1], [3] define the interoperability of the rail system precisely as

> the ability of a rail system to allow the safe and uninterrupted movement of trains which accomplish the required levels of performance for these lines. This ability depends on all the regulatory, technical and operational conditions which must be met in order to satisfy the essential requirements.

Descriptively, but not precisely, creating interoperable railway systems can be seen as a step-by-step incorporation of rail transport into constantly improved fundamental freedoms of the free movement of persons, services, capital, and goods. In particular, it concerns the free movement of persons, including European-wide cross-acceptance of experts' competences, as well as free movement of goods, including European-wide cross-acceptance of products, in accordance with the "once accepted in whichever European member state – proven for the whole European market" principle.

As a consequence of that last principle, European regulations applicable to railway transport are to be incorporated into the national laws of the member states regardless of whether they have railway systems. These regulations form the basis for the creation of the common market for products and services associated with railway transport. The common market entails precisely defined requirements regarding level of the specificity of law, responsibility of producers and contractors, conformity checking, and proving procedures which are applicable to possible solutions [4], [21], [22]. Conformity checking and proving procedures applicable to possible solutions are defined by regulations, which are applicable in all European common markets, i.e. technical areas, within which the free movement of goods is applicable as a fundamental freedom. From the point of view of the railway market, these procedures [4], [21], as defined mainly in horizontal law, are difficult and relatively weakly recognized by railway experts. Simplifying it significantly, it can be stated that European requirements are subdivided into essential requirements and detailed requirements, whereas fulfilment of the detailed ones ensures fulfilment of the essential ones. Fulfilment of the requirements is verified by bodies, which are notified to the Commission as competent by member states thanks to verification of their working procedures against accreditation standards. Uniform interpretation of the requirements is ensured by organizations constituted on the European level, which involve representatives from notified bodies. Verifications of conformity with requirements are conducted by notified bodies applying "modules for EC verification procedures for the subsystems" and "modules for procedures for assessment of conformity and suitability for use of the interoperability constituents." Based on positive results of conformity checking and proving processes, the notified bodies issue EC certificates. Based on EC certificates, producers and contractors issue EC declarations, by which they ensure and declare on their sole responsibility that relevant requirements are satisfied. Declarations have to be issued due to several reasons. As an example, one product of a specific type is tested while the producer sells many products of the same type. Moreover, it should be pointed out that, in the case of highly complex products, notified bodies, due to obvious reasons, do not verify everything, especially as some tests are destructive. EC declarations also have to be issued for products verified with positive results, regardless of their complexity and the scope of tests performed by notified bodies. The responsibility of producers and contractors is not taken on by the notified bodies, however, within checking and proving conformity, the notified bodies create technical files, which allow verification of the conducted processes. Such verifications are conducted by relevant governmental and European authorities. EC declarations, depending on the character of products, allow putting products on the market or applying to governmental authorities of the member states requesting permission to put products in service.

European regulations assume that the railway system in the European Union is safe. This assumption is also applicable individually to railway systems in member states of the European Union, disregarding the fact that the level of safety is differentiated. Nevertheless, on the level of the European legislation, rules are defined for ensuring safety for the new lines and new rolling stock within regulations concerning railway interoperability [1], as well as for the exploitation of existing lines and rolling stock and for interfaces between new and existing solutions within regulations concerning railway safety [2], [3]. High-speed rail systems certainly have to ensure the highest level of safety and have to respect regulations in both areas.

As already stated, requirements derived from the Railway Safety Directive [2], [3] are applicable to the high-speed rail system in the same way as railway interoperability requirements [1], [3]. To ensure railway safety, European law defines common safety methods. These were introduced by six regulations. The most important one for the high-speed rail system in Poland is the regulation defining the common safety method for risk evaluation and assessment [7]. This regulation defines how to perform risk analyses for significant changes, which are implemented into the railway system. Expert judgement on the significance of the change being implemented is performed on the basis of the six criteria: credible worst-case scenario in the event of failure, scale of novelty used in implementing the change, complexity of systems and modules which are foreseen to be implemented, inability to monitor the implemented change influence on safety, change reversibility, and finally accumulation of all recent safety-related changes which were not judged to be significant leading to changes that require detailed analysis. There is no doubt that the construction and start of a high-speed rail system entails significant changes. All risks and associated hazards introduced by such changes must be identified, defined, analyzed, and assessed from the risk acceptance point of view. That requires the setting down of safety requirements and safety measures and putting them together with risks and associated hazards in a hazard record. Such hazard records form an integral part of the safety management systems [2], [3], which are strictly necessary for obtaining safety certificates for railway undertakings and safety authorizations for infrastructure managers. Interconnections between high-speed rail lines and conventional railway lines, as well as the natural supplementary operation of high-speed trains on conventional lines, are good examples of significant changes. In addition, the rules applicable for obtaining safety certificates and safety authorizations are defined in regulations defining common safety methods. Proper understanding requires two important statements. First, safety certificates and safety authorizations are legal documents which confirm the proper organization of legal entities and competences of their staff respectively for railway undertakings and railway infrastructure managers, and therefore do not confirm the correctness of the technical solutions. Second, it has to be stated that implemented safety management systems fulfilling all legally defined requirements are definitely necessary, but not sufficient, for a legal entity to obtain a safety certificate or safety authorization.

Legal documents defining railway operation [5] are also important. Operational requirements cover operational rules and procedures, as well as requirements applicable to safety critical staff. For exploitation of the high-speed rail system, telematic applications [6] must also be taken into account. Telematic applications define the exchange of information between railway undertakings and railway infrastructure managers, as well as between railway undertakings and other legal entities offering public transport with other transport means. In the case of conventional railways, telematic applications cover also the exchange of information between railway undertakings and shippers, or alternatively between railway undertakings and freight forwarders and receivers.

European legal regulations regarding the certification of train drivers operating locomotives and trains on the railway system in the Community, as well as binding European regulations requiring subdivision of the railway system into entities managing infrastructure and competing railway undertakings offering transport services, are also applicable in the case of high-speed rail systems. In light of the safety regulations described above, these regulations have to be taken into account while considering high-speed train movements operated by different railway undertakings, including those that come from other member states connected by the high-speed or even conventional railway infrastructure.

7.1. European technical requirements for HS

It should be stated at the beginning that European technical requirements for the high-speed rail system (HS) are defined up to the speed of 350 km/h. In the case of the construction and exploitation of rolling stock and railway infrastructure for transport services with speeds over 350 km/h, all technical requirements must be defined at the level of national law. As a result, full responsibility for the correctness of the technical requirements lies with individual member states of the European Union. This is why, for construction of the first HS lines and purchase of HS rolling stock, which is foreseen to run with high-speeds on high-speed lines and with adequately lower speeds on conventional lines, a constriction to the speed range up to 350 km/h has been accepted in Poland. The fact that all technical requirements up to 350 km/h are defined in European law is considered evidence that, up to this speed, requirements are well known and proven.

Legal documents are dedicated to railway technical requirements, taking into consideration horizontal regulations defining the common framework for the marketing of products [4], [21], as well as railway specific regulations defining rules for putting in service [22] subdivided railway systems into subsystems and defined technical requirements for individual subsystems [1]. Three subsystems constitute railway lines: infrastructure subsystem [8], [9], track-side control command and signaling subsystem [10], [11], and energy subsystem [12], [13] named respectively as INF, CCT, and ENE. Two subsystems constitute high-speed vehicles: rolling stock subsystem [14], [15] and on-board control command and signaling subsystem [10], [11] named respectively as LOC&PAS (or HS RST) and CCO. The mentioned short names, which are commonly used for subsystems, are not precise enough as they refer only to technical specifications for interoperability (TSI specifications), containing basic requirements applicable to individual subsystems, and omit other requirements which are also applicable but defined in other TSI specifications. Requirements applicable to each subsystem are defined not only in dedicated TSI specifications (e.g. TSI LOC&PAS [15] requirements for the LOC&PAS subsystem), but also in other TSI specifications, including those dedicated to other subsystems due to interfaces between subsystems, as well as those defining requirements applicable to several subsystems [17], [18], 19], [20] (e.g. TSI PRM [17] defining requirements regarding opening railway transport for persons with reduced mobility, which is applicable to the LOC&PAS subsystem and to the INF subsystem) and those defining requirements related to specific issues (e.g. TSI NOIS [16] defining requirements regarding railway noise).

The first set of TSI specifications defining requirements for high-speed rail transport systems was accepted by the European Commission in 2002. From 2006, European Commission decisions put amended TSI specifications into force, defining requirements for high-speed rail transport systems, replacing decisions from 2002. All those specifications were dedicated to the trans-European high-speed rail system [5], [6], [8], [10], [12], [14]. TSI specifications for the trans-European conventional rail system were prepared and put in force independently. Since 2008, TSI specifications have been issued by the European Commission on the basis of legal delegation in the European Parliament and Council directive on railway interoperability – directive 57 from 2008 [1]. From the HS rail system point of view, this directive replaced the directive dedicated solely to the trans-European high-speed rail system and, less importantly from the point of view of our analysis, the directive dedicated solely to the trans-European conventional rail system. It should be pointed out that earlier directives were not dedicated to the whole railway network in the European Union, but only to its so-called trans-European parts. The directive that is presently in force is applicable to the whole railway network, with small exceptions, and therefore to all sections of all HS lines. This is important for HS rail system construction in Poland, as high-speed rail lines are widely defined in respective documents [8], [9].

The directive in force, similarly to earlier ones, subdivides rail systems into five structural subsystems and three functional subsystems:

- INF: track, points, bridges, tunnels, platforms, zones of access, and protective equipment [8], [9];
- CCS (CCT and CCO): track-side and on-board equipment of the European Train Control System (ETCS) as well as track-side and on-board equipment of the Global System for Mobile Communication for Rail (GSM-R), for which all detailed requirements are defined in the European documents, as well as all other active safety equipment, for which requirements remain the responsibility of the member states [10], [11];
- ENE: traction substations, sectioning substations, contact line systems, return circuits, and on-board parts of the electric consumption measuring equipment [12], [13];
- RST: vehicles structures, vehicle control equipment, current-collection devices, traction and energy conversion units, braking, coupling and running gear, suspension, doors, as well as safety devices [14], [15].

It should be mentioned that, according to legal regulations [8], [9], there are three types of high-speed rail line: new lines constructed for a maximum of 350 km/h, lines which are upgraded for speeds on the order of 200 km/h, and railway lines and sections that form connections between such lines and between such lines and stations, even if the maximum speed is significantly reduced due to special features as a result of topographical, relief, or town planning constraints. Additionally, replacing the directive dedicated to the trans-European high-speed rail system with the directive dedicated to the rail system within the Community shows that it is necessary, not only for practical reasons but also because of legal requirements, to ensure cohesion between high-speed rail lines and conventional ones. Requirements regarding infrastructure subsystems are defined with subdivisions for line categories. Requirements regarding control command and signaling subsystems both for track-side and for on-board equipment are defined uniformly for high-speed and for conventional rail without a subdivision for line categories. However, European control command system specifications foresee different configurations, making it possible to equip railway lines according to needs, keeping at the same time conformity between on-board equipment and track-side equipment. Requirements regarding energy subsystems are defined separately for four different traction power supply systems: 1.5 kV DC, 3 kV DC, 15 kV 16.7 Hz AC, and 25 kV 50 Hz AC. To achieve coherence of the railway system, detailed rules have been defined for passing borders between parts of the network, which are equipped with different power supply systems, using multisystem traction units. Requirements regarding rolling stock subsystems are defined in a way that ensures there are no technical barriers for rolling stock fulfilling European requirements while running on the European Union railway network. During the transitional period, when only some railway lines are equipped accordingly (and therefore interoperable), rolling stock must also be compatible with other railway lines of the network on which it is intended to run. Therefore, high-speed trains will be required to accomplish compatibility not only with HS fully interoperable infrastructure, but also with railway lines and sections that form connections between high-speed lines and between such lines and stations. In practice, accomplishing compatibility with additional railway lines is necessary, as such lines may be used for solving operational disturbances. On-board traction power equipment is a good example. The Polish network is equipped with a 3 kV DC traction power supply system. However, such a system does not ensure enough power for achieving speeds of the order 250 km/h and more. High-speed lines have to be equipped with a 25 kV 50 Hz AC traction power supply system, while high-speed trains have to be equipped with 25 kV 50Hz AC on-board traction power equipment. High-speed trains must be capable of running on conventional lines equipped with 3 kV DC, and therefore must also be equipped with 3 kV DC on-board traction power equipment. If trains are intended to run via the Polish-German network, e.g. to Berlin, they have to be additionally equipped with 15 kV 16.7 Hz AC on-board traction power equipment. From the traction point of view, high-speed trains for Poland must be a minimum of double-system and preferably triple-system ones.

It has been recognized in Poland that constructing HS lines for the maximum speed that is being reached by trains now may create a serious limitation in the future. This is because railway lines are constructed for a minimum of 100 years, and new technologies in rolling stock within that time will probably appear several times. HS lines constructed as new ones in relation to the permanent way should fulfil requirements for 350 km/h, e.g. by adequately defining track geometry. At the same time, it has been recognized that traction power supply systems constructed for HS lines should be prepared only for rolling stock that is currently available and foreseen to be purchased. Traction power supply systems have to be calculated for foreseen rolling stock, assuming running with foreseen maximum speeds and foreseen minimum headways, however, taking into account limitations due to the electrical parameters of traction substations and traction units. It is probable that it will not create limitations in the future, thanks to rolling stock technological development. Without such development, speeding up will not be reasonable.

It should be mentioned, when discussing European technical requirements for HS rail, that from January 2015 technical European requirements for high-speed rail and for conventional rail have been combined in common TSI specifications [9], [11], [13], [15], [18], [20]. It is necessary to apply specifications directly applicable to individual subsystems as well as supplementary specifications defining requirements regarding railway noise [16], regarding adaptation of the infrastructure and rolling stock to the needs of persons with reduced mobility [17], [18] and for safety in the railway tunnels, including evacuation routes and flammability of materials used for rolling stock construction.

TSI specifications, besides requirements for subsystems, also define requirements for products that are present on the market individually. These are called "interoperability constituents."

7.2. Polish technical requirements for HS

It is commonly assumed that there is no high-speed rail system in Poland. Such an assumption is imprecise, as in Poland there are railway lines with sections where there is a maximum speed of 200 km/h and lines that are being upgraded at the moment and will shortly have sections with a maximum speed of 200 km/h. According to the European legal definition of high-speed lines, there are high-speed rail lines in Poland, as there are lines upgraded for speeds of the order 200 km/h and railway lines and sections which form connections between them and railway stations, although there are no newly constructed railway lines for 350 km/h.

The first works for upgrading railway lines for 200 km/h were conducted along the Central Trunk Line (CMK line). In view of them, in the years 2000–2001 "technical standards – detailed technical conditions for upgrading CMK line up to 200/250 km/h" were prepared. The scope of the document comprises in particular:

- railway substructure
- railway superstructure
- line crossings and fairing
- engineering structures
- architectural objects
- power supply
- control command and signaling
- rail traffic management

- locomotives up to 200 km/h
- passenger coaches up to 200 km/h
- passenger trains up to 250 km/h
- mobile communication
- telecommunication
- electromagnetic compatibility
- railway traffic
- environmental protection

The document was prepared by a group of experts chaired by the author of this chapter. Around two-thirds of the requirements have been defined to ensure conformity with the rest of the Polish railway network; however, it should be admitted that not only the vast knowledge of the Polish experts but also draft high-speed TSI specifications were used. As a result, the Polish document closed in 2001 is partly based on Commission Decisions published in the EU Official Journal in September 2002. Most of those experts are still working, now with more experience and deeper knowledge, and are handing over that knowledge to their younger colleagues. Other documents were prepared in 2008 in view of the possible construction of a new high-speed line in Poland, namely "Polish Program for HS Construction and Embarking High-Speed Railway Service" and supplementary "Environmental Consequences Forecast." This program, formally accepted by Polish authorities, defines among others: assumptions for HS rail system construction, taking into account complementarity between HS railway lines and existing transport routes, strategic questions regarding HS technology, taking into account solutions accepted in the European Union and solutions used in Poland, as well as different kinds of hazards associated with project realization together with proposed mitigation activities.

A growing scale of upgrading for the railway lines resulted in 2009 in "Technical Standards – Detailed Technical Conditions for Upgrading and Construction of Railway Lines up to Speed Vmax ≤ 200 km/h (for Conventional Rolling Stock)/250 km/h (for Tilting Trains)" as a document complementary to the standards defined in 2001. These requirements are primarily dedicated to conventional rail; however, the scope of application covers railway lines being upgraded for services that, according to European definitions, belong to the high-speed domain. This document has 16 volumes defining requirements, respectively for:

1. Permanent way
2. Structure gauge
3. Railway engineering structures
4. Traction power supply
5. Non-traction electric power engineering
6. Control command and signaling
   Traffic management
7. Telecommunication
8. Running gear malfunctioning detection
9. Electromagnetic compatibility
10. Crossings and parallel roads
11. Platforms
12. Small architecture and color schemes
13. Buildings
14. Line fairing
15. Environmental protection
16. Railway rolling stock

This document was formally accepted in 2010 by Polish Railway Lines S.A., the main railway infrastructure manager in Poland, and is obligatory in tendering processes for the upgrading of railway lines.

It should be emphasized that, besides the technical requirements discussed above, there are Polish legal railway requirements for individual railway products [25], [26], [27], which must be met by such products before they can be put on the market. In the case of the high-speed rail system, it applies only to a limited number of products for which Polish law defines requirements and which are not indicated in EU law as interoperability constituents, e.g. switches. Such products are called "types of constructions" if they belong to the infrastructure subsystem or "types of devices" if they belong to the control command and signaling subsystem or to the energy subsystem.

7.3. European rules for HS product certification

The horizontal rules mentioned above, applicable to the European Union common markets [21] and implemented into Polish law by a parliamentary act [4], are applicable to infrastructure and rolling stock. Procedures defined in the decision of the European Parliament and of the Council, destined to assess whether products comply with applicable European requirements and called conformity assessment modules, have been adapted to railway needs and published in Decision 713 in 2010 [22], which is dedicated solely to railway products. It is important that in the case of rail transport systems not only individual products but also subsystems be considered as products.

For instance, rails, slippers, wheels, and axles are defined as individual products. They are called "interoperability constituents" and have all the requirements necessary to ensure interoperability, defined as European requirements in appropriate TSI specifications put in force by decisions and regulations, as well as in documents indicated by these specifications and, in particular, European standards. This applies to their own features and to all their interfaces. In addition, methods for assessing conformity with applicable requirements are defined for interoperability constituents on the European level [22]. The previously mentioned subsystems are also assessed against European requirements. For them, assessment also covers their features and their interfaces with external ones, including those with other subsystems.

In both cases, the assessment is carried out by notified bodies [1], [3]. For interoperability constituents, positive results allow the notified body to issue an "EC conformity certificate" proving conformity with applicable requirements. For subsystems, positive results allow the notified body to issue an "EC verification certificate" proving conformity with applicable requirements. In both cases, EC certificates constitute the basis for issuing EC declarations, respectively the "EC conformity declaration" and "EC verification declaration" by which producers and contractors declare, on their sole responsibility, that their products satisfy the applicable European requirements. As an example, the notified body assesses rails and, after positive results, issues an EC conformity certificate for rails as an interoperability constituent. The producer manufactures and sells many rails declaring, with the EC conformity declaration, that the rails being produced and sold conform to those for which the certificate was obtained.

7.4. Polish rules for HS product certification

High-speed rail system subsystems will be certified according to the European certification rules. However, some products as "types of construction" or "types of device" will be certified according to the Polish certification rules. These emerge clearly from the Polish Transport Act [3], whose statements in that respect are more precisely stated in a supplementary regulation [25]. This regulation defines a full list of "types of constructions" and "types of devices," as well as the scope of their assessment. Detailed requirements for types are defined by a regulation [26] and stated precisely by a supplementary document issued by the Office of Rail Transport [27]. This applies particularly to railway switches, ballastless superstructures, occupancy checking devices, station interlockings, line block systems, traction return circuits, and even color lights signals.

If, within a subsystem, there are "types of constructions" or "types of devices," then each type needs to have a "type attestation" and each construction and each device as a product needs to have a "declaration of conformity with type." These are the equivalents of "EC conformity certificates" and "EC conformity declarations." The "type attestations" are issued by the Office of Rail Transport based on "type certificates" issued by authorized bodies, while "declarations of conformity with type" are issued by producers based on "certificates of conformity with type." These can be issued based on positive results of tests conducted according to conformity assessment modules. These are not the modules that are defined by the European Commission Decision [22], but the ones defined by the European Parliament and Council Decision [21]. Therefore, checking conformity with requirements defined in Polish legislation is conducted according to assessment modules defined in the European horizontal legislation, which are applicable to all products on the common market except railway ones, as dedicated assessment modules are defined for them.

Independently of product assessment, Polish legislation, especially one regulation [23], requires the verification of conformity with national requirements for subsystems. This is, first of all, applicable in areas indicated in the TSI specifications as those in which national requirements are applicable, such as those indicated as open points and specific cases. This is also applicable in areas that are not covered by the TSI specifications, for instance, those regarding requirements for station interlockings and line block systems. In these cases, assessment with national requirements is conducted according to European assessment modules dedicated to railway transport [22] by bodies notified to the European railway directive.

Thereby, for example, for high-speed rail system control command and signaling subsystems, some products are assessed as interoperability constituents by notified bodies to prove conformity with European requirements using conformity assessment modules dedicated to railways, e.g. radio block centers (RBC) and balises. Some products are assessed as types of devices by authorized bodies to prove their conformity with Polish requirements using general conformity assessment modules, for instance, station interlockings and occupation checking devices. At the same time, control command and signaling subsystems will be assessed by the notified body to prove conformity with European and Polish requirements, using conformity assessment modules dedicated solely to railways, proving the conformity of the complete signaling equipment, which is not covered by the TSI specifications. Additionally, interfaces between Polish and European solutions have to be assessed in accordance with the common safety method for risk evaluation and assessment [7]. Such assessment is also required for national rolling stock, which is modernized non-exhaustively [24] and, as a result, does not fulfil all European requirements, not only before modernization but also afterwards. The acceptability of such national modernization is verified using European procedures for risk assessment.

7.5. European rules for HS rail system operation

The previously mentioned functional subsystems define three groups of requirements. The first group of requirements covers operational rules and procedures defining ways used to operate railway transport, and requirements regarding the health as well as professional and language competences of safety related staff. Language competences are important in the case of train drivers on international routes and for signalmen on railway lines for which infrastructure managers have declared additional languages for train operation [5]. High-speed rail lines frequently connect agglomerations situated in different countries and therefore language competences are important and usually required. On conventional railway lines, the operational rules are significantly differentiated due to interconnections with classic signaling and historically determined differentiation of safety cultures. Minimizing the differentiation of operational rules in conventional railway systems is significantly limited as changing operational rules creates significant risks, especially during transitional periods. On high-speed lines, pure operational requirements are extremely limited thanks to unified control command systems based on track train control data transmission.

The second group of requirements covers signaling rules and procedures linked with operational ones by nature. Relationships between operational situations, states of individual signaling devices, and permission for trains to run, as well as signal aspects, differ so significantly from state to state that changing them into homogeneous ones is impossible, unless it is done on the control command level. Achieving, relatively easily, homogeneous signaling rules is possible on the control command level if the information displayed in the driver's cab provides detailed information about authorized movement distances, speed restrictions, and other information supporting train driving, instead of repeating directly signal aspects displayed on track-side signals. This is the basis on which the European unified control command system together with unified mobile communication system offer unified signaling rules and procedures. These systems require, however, unified operational rules dedicated to them, and such rules are defined in the European legislation. This concerns, for instance, commands given by signalmen to train drivers. Such commands are necessary in degraded situations, such as for trains with malfunctioning on-board control command equipment that has to be removed from main tracks, and for trains which have to be moved through tracks with malfunctioning track-side control command equipment. In that respect, applicable operational rules and procedures can be divided into European-wide and nationwide ones defined according to European rules.

The third group of requirements covers the exchange of information regarding connections between trains [6]. This concerns connections to other high-speed trains and to conventional trains operated by different railway undertakings. This also concerns connections with other public transport means, such as ferries or complementary road connections, for example, from railway stations to airports when there is no direct railway connection to the airport. Information about timetables should be available before the journey. Information about connections and delays should also be available during train journeys.

7.6. Polish rules for HS operation

Creating a high-speed rail system in Poland only using European operational rules and procedures is not possible. Regardless of the arguments mentioned above, it should be stated that high-speed trains will also run on conventional railway lines. Therefore, the complementarity of European and Polish rules and procedures is required. Train drivers in high-speed trains cannot be required to obey different rules depending on the nature of the line on which the train is running – high-speed or conventional railway lines. In addition, signalmen cannot be required to apply different rules depending on the nature of the train for which authority is being given – high-speed or conventional trains. These statements are obvious for high-speed trains running on conventional railway lines, but they are also very important from the point of view of signalmen's training and work arrangements.

While creating the high-speed rail system in Poland, special attention will be required when defining transitions between the national control command system and European unified control command system, as well as between the analog mobile communication system commonly used in Poland and the European unified digital mobile communication system dedicated to railway transport. In addition, in that respect, a common safety method for risk evaluation and assessment will have to be applied.

Bibliography

[1] Directive 2008/57/EC of the European Parliament and of the Council of 17 June 2008 on the interoperability of the rail system within the Community.
[2] Directive 2004/49/WE of the European Parliament and of the Council of 29 April 2004 on safety on the Community's railways.
[3] Polish Railway Transport Act of 28 March 2003 with amendments [*Ustawa o transporcie kolejowym z dnia 28 marca 2003 z późniejszymi zmianami*].
[4] Polish Conformity Assessment Act of 30 August 2002 with amendments [*Ustawa o systemie oceny zgodności z dnia 30 sierpnia 2002 z późniejszymi zmianami*].
[5] 2008/231/EC: Commission Decision of 1 February 2008 concerning the technical specification of interoperability relating to the operation subsystem of the trans-European high-speed rail system adopted referred to in Article 6(1) of Council Directive 96/48/EC and repealing Commission Decision 2002/734/EC of 30 May 2002.
[6] Commission Regulation (EU) No 454/2011 of 5 May 2011 on the technical specification for interoperability relating to the subsystem 'telematics applications for passenger services' of the trans-European rail system.
[7] Commission Implementing Regulation (EU) No 402/2013 of 30 April 2013 on the common safety method for risk evaluation and assessment and repealing Regulation (EC) No 352/2009.
[8] 2008/217/EC: Commission Decision of 20 December 2007 concerning a technical specification for interoperability relating to the infrastructure sub-system of the trans-European high-speed rail system.

[9]   Commission Regulation (EU) No 1299/2014 of 18 November 2014 on the technical specifications for interoperability relating to the 'infrastructure' subsystem of the rail system in the European Union.

[10] 2012/88/EU: Commission Decision of 25 January 2012 on the technical specification for interoperability relating to the control-command and signalling subsystems of the trans-European rail system.

[11] Commission Decision (EU) 2015/14 of 5 January 2015 amending Decision 2012/88/EU on the technical specification for interoperability relating to the control-command and signalling subsystems of the trans-European rail system.

[12] 2008/284/EC: Commission Decision of 6 March 2008 concerning a technical specification for interoperability relating to the energy sub-system of the trans-European high-speed rail system.

[13] Commission Regulation (EU) No 1301/2014 of 18 November 2014 on the technical specifications for interoperability relating to the 'energy' subsystem of the rail system in the Union.

[14] 2008/232/EC: Commission Decision of 21 February 2008 concerning a technical specification for interoperability relating to the rolling stock sub-system of the trans-European high-speed rail system.

[15] Commission Regulation (EU) No 1302/2014 of 18 November 2014 concerning a technical specification for interoperability relating to the 'rolling stock – locomotives and passenger rolling stock' subsystem of the rail system in the European Union.

[16] Commission Regulation (EU) No 1304/2014 of 26 November 2014 on the technical specification for interoperability relating to the subsystem 'rolling stock – noise' amending Decision 2008/232/EC and repealing Decision 2011/229/EU.

[17] 2008/164/EC: Commission Decision of 21 December 2007 concerning the technical specification of interoperability relating to persons with reduced mobility in the trans-European conventional and high-speed rail system.

[18] Commission Regulation (EU) No 1300/2014 of 18 November 2014 on the technical specifications for interoperability relating to accessibility of the Union's rail system for persons with disabilities and persons with reduced mobility.

[19] 2008/163/EC: Commission Decision of 20 December 2007 concerning the technical specification of interoperability relating to safety in railway tunnels in the trans-European conventional and high-speed rail system.

[20] Commission Regulation (EU) No 1303/2014 of 18 November 2014 concerning the technical specification for interoperability relating to 'safety in railway tunnels' of the rail system of the European Union.

[21] Decision No 768/2008/EC of the European Parliament and of the Council of 9 July 2008 on a common framework for the marketing of products, and repealing Council Decision 93/465/EEC.

[22] 2010/713/EU: Commission Decision of 9 November 2010 on modules for the procedures for assessment of conformity, suitability for use and EC verification to be used in the technical specifications for interoperability adopted under Directive 2008/57/EC of the European Parliament and of the Council.

[23] Polish Transport Construction and Maritime Economy Minister Regulation 1297/2013 of 6 November 2013 concerning railway interoperability [Rozporządzenie Ministra Transportu Budownictwa i Gospodarki Morskiej 1297/2013 z dnia 6 listopada 2013 w sprawie interoperacyjności kolei].

[24] Polish Transport Construction and Maritime Economy Minister Regulation 1976/2014 of 15 December 2014 amending the regulation concerning railway interoperability [Rozporządzenie Ministra Infrastruktury i Rozwoju 1976/2014 z dnia 15 grudnia 2014 r. zmieniające rozporządzenie w sprawie interoperacyjności systemu kolei].

[25] Polish Infrastructure and Development Minister Regulation 720/2014 of 13 May 2014 concerning putting in service certain types of railway constructions, devices and vehicles [Rozporządzenie Ministra Infrastruktury i Rozwoju 720/2014 z dnia 13 maja 2014 r. w sprawie dopuszczania do eksploatacji określonych rodzajów budowli, urządzeń i pojazdów kolejowych].

[26] Polish Transport Construction and Maritime Economy Minister Regulation 43/2013 of 27 December 2012 concerning the list of national technical specifications and standardisation documents which allow the accomplishment of essential requirements for rail system interoperability [*Rozporządzenie Ministra Transportu, Budownictwa i Gospodarki Morskiej 43/2013 z dnia 27 grudnia 2012 r. w sprawie wykazu właściwych krajowych specyfikacji technicznych i dokumentów normalizacyjnych, których zastosowanie umożliwia spełnienie zasadniczych wymagań dotyczących interoperacyjności systemu kolei*].

[27] Railway Transport Office President List of national technical specifications and standardisation documents which allow the accomplishment of essential requirements for rail system interoperability dated 26 September 2013 [*Lista Prezesa Urzędu Transportu Kolejowego w sprawie właściwych krajowych specyfikacji technicznych i dokumentów normalizacyjnych, których zastosowanie umożliwia spełnienie zasadniczych wymagań dotyczących interoperacyjności systemu kolei z 26 września 2013 r*].

Chapter 8

High-speed rail passenger services in Poland

Andrzej Żurkowski

8.1. Introduction

High-speed rail systems have been used throughout the world for more than 50 years [2], [7], [8]. In this period, impressive progress was made within technology related to construction of both infrastructure and railway rolling stock. Methods of railway transport organization have developed, especially in the context of a dynamic progress of individual motorization, construction of highway networks, as well as an increase in air transport.

Paradoxically, development of a competitive means of transport doesn't constitute a threat for HSR; nevertheless, it poses a question about principles of a rational transport policy in the scale of the whole country and region, which should lead to such a division of transport tasks so that each of the means of transport execute the ones to which it is predestined the most. The criteria for reaching such a rational division have first of all an economical dimension, but also social and ecological ones.

For obvious reasons, railways should play an important, if not the basic, role in the modern transport system, both in agglomeration traffic and in direct intercity transport. HSR provides in this matter the best transport technology, but because of its profitability, an assumption has to be necessarily made that the serviced passenger flow is high enough.

This chapter presents an outline of problems of transport organization on high-speed lines and in connection with the whole railway network. Initiating from a definition and a brief presentation of the basic issues in this field, presented is a modern way of transport planning, which starts already at the stage of projecting new railway lines or preparing modernization of the existing ones. It concerns both the high-speed lines and the already existing conventional connections.

While concentrating on conditions that are characteristic for Poland, possible variants for organizing passenger railway transport after launching traffic on high-speed lines are proposed and discussed. However, the approach described emphasizes a specific application; the author's objective is to present methods of transport organization by high-speed rails, first in the theoretical dimension.

Separate stages of transport planning are presented according to the generally adopted canon [5]: travel modelling and traffic prediction, investment planning, timetable, rolling stock work, and personnel work.

The chapter begins with a presentation of the size of passenger transport executed nowadays in Poland by land and air transport, in terms of evaluation of railways' share in the market and in comparison with other European Union countries.

8.2. Passenger transport in Poland and in Europe

Size and structure of the passenger transport employed in Poland by means of public transport and by passenger cars have changed radically in the last several dozen of years, both as an effect of political transformation and of technological and socioeconomic development. The structure of the land transport market is presented in Table 8.1.

In the period taken into consideration, a modal split also changed radically. In 1970–90, railways realized 20–25% of all the passenger intercity transport. The share of public communication bus transport was on a similar level, and passenger cars throughout many years remained level, varying from 50% to 70%.

Nowadays, individual motorization in Poland executes more than 90% of the passenger per kilometer (pkm). Share of the public transport is equal then to less than 10%. Railways execute less than 5% of the transport work, therefore its participation in the considered period diminished almost five times (See: Fig. 8.1). Comparing those data with the situation in the European Union countries (Figure 8.2), it may be stated that the role of railway transport in the most developed economically countries of the Community is bigger and is equal to more than 7% in the scale of the whole Community. A question arises to what extent it results from service of a certain part of traffic executed by high-speed rails.

The network of high-speed lines in Europe has been arising since 1981 and is equal nowadays to 7316 km (250 km/h and faster), which constitutes 4.6% of the overall length of the railway lines exploited in the EU-26 countries. The pace of creation of new sections in the years 1990–2013 was on average equal to 255 km per year, and until 2020 there will be put into operation another 1549 km of lines. The passenger per kilometer executed by the high-speed rails on the continent is equal nowadays to 110.37 billion pkm annually, which constitutes 27.13% of the total transport work of railways.

Table 8.1 Size and structure of the passenger transport market in Poland

Year	Railway transport			Bus transport			Passenger cars	
	Number of passengers [thou.]	Transport work [pkm m]	Average distance [km]	Number of passengers [thou.]	Transport work [pkm m]	Average Distance [km]	Number of vehicles [szt.]	Transport work [pkm m]
1970	1,056,470	36,891	34.9	1,373,644	29,140	21.2	279,400	102,378
1975	1,117,959	42,819	38.3	2,237,288	45,792	20.5	1,077,700	104,161
1980	1,100,508	46,331	42.1	2,379,252	49,250	20.7	2,383,000	105,529
1985	1,005,107	51,964	51.7	2,434,423	52,096	21.4	3,682,296	106,682
1990	789,922	50,373	63.8	2,084,708	46,599	22.4	5,260,600	107,698
1995	465,901	26,635	57.2	1,131,593	34,024	30.1	7,517,300	110,700
2000	360,687	24,093	66.8	954,515	31,735	33.2	12,339,400	149,700
2005	258,110	18,157	70.3	782,025	29,314	37.5	12,339,353	197,300
2010	261,314	17,921	68.6	569,652	21,600	37.9	17,239,800	297,900
2011	263,609	18,177	69.0	534,885	20,651	38.6	18,125,490	313,200
2012	273,182	17,826	65.3	497,288	20,012	40.2	18,744,412	327,068
2013	269,814	16,797	62.3	459,947	20,040	43.6	19,389,446	335,378
2014	268,204	16,014	59.7	431,516	21,449	49.7	20,003,863	343,022

Source: Transport. Activity results 2000–2014. GUS, Warszawa 2015

Own calculations by approximation (italic print).

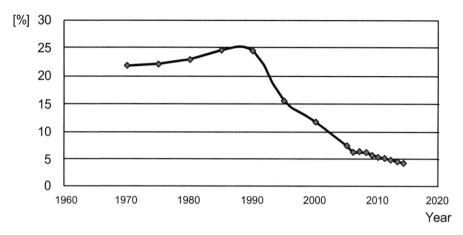

Figure 8.1 Rail part of modal split in land transport in Poland

It is very difficult to estimate to what extent the use of HSR technology enables to increase the share of railways in the passenger transport market. A detailed answer would require a comparison of the transport work executed by the separate means of transport in the intercity transport, both in territories of separate countries and in the international traffic. The statistics in this field are not conducted, especially with respect to road transport.

At the same time, reporting related to the railway transport usually has an accumulated character; therefore, it embraces jointly all types of transport. A certain indicator in this field may be only an average distance of travel of one passenger.

Considering the above limitations, the following conclusions may be drawn. The pace of construction of new sections of high-speed lines and plans in this field for the following years concerning countries having at their disposal a multi-annual experience within HSR exploitation confirm a social and economic efficiency of this system. Despite the fact that in the EU-26 countries the total length of the HSR network constitutes only 4.6% of all the railway lines, transport by high-speed trains is equal to 27.13% of the total paskm of railways. It confirms the thesis about intensive exploitation of such infrastructure.

In the countries having at their disposal the largest HSR systems, the share of this type of transport in the general volume of railway transport is very big: in France nearly 59%, in Spain 29%, in Germany 27%, and in Italy 28%. Comparing the EU-26 countries (among the EU-28 countries, Malta and Cyprus do not have railways), a share of the railway transport in the general volume of land transport is on average equal to:

- 3.2% in the countries not using of high-speed lines;
- 7.4% in seven countries exploiting this type of line.

It can therefore be concluded that high-speed rail transport constitutes nowadays in the analyzed countries of the European Union an important element of transport offer and makes it possible to keep its share in the general traffic volume. At the same time, they have a significant share in passenger transport within the total railway traffic structure.

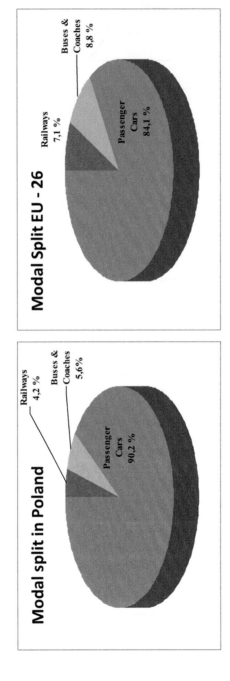

Figure 8.2 Modal split in Poland and EU-27 in 2014

Source: EU Energy and Transport. Statistical pocketbook 2001–2014, Eurostat 2015

The launching in Poland in December 2014 of a series of intercity connections by ED250 (Pendolino) trains reaching speed of 200 km/h on the modernized Central Trunk Line constitutes – in line with Directive 2008/57 on the interoperability of railways – a launching of connections by high-speed rails (HSR). The years to come will enable to evaluate how transport and the share of railways on the market will change, thanks to this first step towards transport with perspective speeds at the level of 300–320 km/h.

8.3. Railway passenger services – some basic issues

8.3.1. Polish systematics of railway passenger services

The railway transport system is constituted by two basic subsystems defined through the purposes they serve: passenger transport and cargo transport. In Poland, a concept of system transport organization was already elaborated in the 1970s [9], [16]. Its structure with regard to passenger transport results from the character of demand for transport services, which is connected to the size, type, or place of settlements, which consist of clusters of different sizes: agglomerations and smaller cities and villages. Accessibility of railways is at the same time limited to stations and railway stops from which passengers enter and exit. From the point of view of railway transport organizers, these are the places that constitute beginnings (so-called sources) and ends (so-called outlets) of passenger flows.

Despite the fact that passenger travels occur between all sources and outlets, it is obviously not possible to create a full system of direct connections. A need therefore arises to standardize travel and ensure connections in the types of relations where the passenger flows justify launching direct connections.

A basic criterion of the classification in use in Poland is the (geographical) accessibility of trains. Four basic subsystems are distinguished:

- (I) intercity services (so-called quality): with accessibility limited only to agglomerations and the largest cities (and during holiday seasons also to particularly popular resorts –e.g. Zakopane);
- (II) interregional services (which in Polish legislation may be identified as "services between voivodships"): with accessibility widened to smaller cities and railway junctions (interchange);
- (III) agglomeration and (IV) regional services: with full accessibility, that is service of all stations and stops (an exception in this field constitutes a so-called zone service).

Intercity and interregional services constitute altogether *long distance services*, whereas the agglomeration and regional ones are called *local services*. Particulars of this systematic are presented in Table 8.2.

Attention should be drawn to the fact that in light of the above systematic, the distance of train service is not important, nor does it matter whether this service is on national territory or in international traffic.

While observing changes happening in the structure of railway passenger services, a basic tendency may be indicated. Size of intercity and agglomeration services and their importance in the system of passenger transport are constantly growing. It results from the fact that changes in the settlement scheme lead to the concentration of population in the largest cities and their suburbs. Railways turn out to be the best means of transport for large flows

Table 8.2 Systematic organization of railway passenger services in Poland [9, 16]

Services	Subsystem	Geographic availability	Railway operator	Products (type of trains)
long distance	I intercity (quality trains)	▪ agglomerations ▪ large cities ▪ holiday resorts ▪ junction stations	Nationwide operators (PKP Intercity, Przewozy Regionalne)	• EIP – Express Intercity Premium (*Pendolino*) • EC – EuroCity • EIC – Express InterCity • EN – EuroNight
	II interregional	▪ agglomerations ▪ large and medium cities ▪ holiday resorts ▪ junction stations		▪ TLK (low-cost trains) ▪ rapid trains ▪ IR – InterRegio ▪ Re – REGIOekspress
local	III agglomeration (urban and suburban) IV regional	all stations and railway stops (except zone services)	Regional operators (local government, private)	local trains

of passengers both in commuting to agglomeration centers and between them. The system of high-speed rail services is doubly important in this field. HSR stock sets are unequivocally adjusted to the realization of intercity services. Their technical-operational characteristics – relatively small movement acceleration and braking delay as well as destination for long-distance course with high speeds – correspond the best to the needs for services between centers of the biggest cities. High-speed lines are projected in a way to enable a crossing-free high-speed train movement, that is with no level crossings on the level of rails and the lines are usually fenced or conducted on bridges.

This basic meaning of the HSR system is supplemented in case of need by additional functions. If the amount of traffic on the high-speed lines requires launching trains with possibly low headways (capacity saturation), in that situation on a line there is movement with the highest speeds exclusively. In case of use of HSL, for example in the scheme of periodic timetable connections, where subsequent trains appear every 30 or 60 minutes, then a capacity margin occurs, enabling to launch interregional and regional trains. A condition lies here in the use of a rolling stock able to operate with speeds of 160 km/h or 200 km/h, and moreover equipped with signaling system devices based on a track–vehicle communication (e.g. ETCS system).

8.3.2. Classification of passenger flows

Passenger flow with respect to *transport network* means the number of passengers moving in a specified time between two considered places of the network: a beginning (a source) and an end (an outlet) in one or both directions [15].

In Polish railway terminology, the basic concept relating to the demand for railway passenger services is called *relational stream* [9, 16]. Considering all stations and railway stops (with joint number Z) as the previously defined sources and outlets $z = 1, \ldots, Z$ may be defined by a square matrix of streams R as follows:

$$R = \begin{bmatrix} r_{11}, r_{12}, \ldots, r_{1Z} \\ r_{21}, r_{22}, \ldots, r_{2Z} \\ \ldots \\ r_{Z1}, r_{Z2}, \ldots, r_{ZZ} \end{bmatrix} = [r_{ij}]_{Z \times Z} \tag{8.1}$$

with elements r_{ij} indicating the number of passengers travelling during 24 hours between two defined stations or stops on the railway network: a beginning i and an end of the journey j. For obvious reasons, for each couple $i = j$ we have $r_{ij} = 0$ (main diagonal). Moreover, aiming to simplify the processes of transport modeling, there is usually an assumption made that $r_{ij} = r_{ji}$.

On separate sections, relational streams become part of so-called section flows p. They indicate the number of passengers travelling through a certain section of a railway line l during 24 hours, which is why their common name is *24-hour flows*.

$$p_l = \sum_{R_l} r_{ij} \tag{8.2}$$

In Equation (8.1), R_l indicates a set of all those relational streams r_{ij}, which pass through the section l. Because of changes in size of streams during different periods, *section flows* may also refer to hours (e.g. peak hours, weeks, or months).

During 24 hours on a certain railway line section operate a series of trains. *Train flows* indicate the number of passengers travelling with a determined train on each agreed on section of its course, usually between stations of commercial stops, on which an exchange of passengers occurs. From the point of view of traffic organization on the high-speed lines in case of a lack of train stops on intermediate stations, a sum of train flows is constituted by *section flows*.

Station flows indicate a number of passengers departing during 24 hours from a certain station.

It is also important to solve a dilemma regarding the size of *relational streams*. According to the traditional rules of passenger transport organization applied in railways, the existence of a strong relational stream connecting (usually) two big cities, even if they are distant from each other, justified launching direct connections. For many passengers, the possibility of travelling without the necessity of a train change was an important argument in favor of choosing the train as a means of transport.

In case of HSR, a typical dilemma is that if between the considered cities only part of the route is constituted by a high-speed line, a question arises whether it is worthwhile to serve the whole connection with an HSR train set.[1] A scheme on Figure 8.3 depicts it.

A series of aspects of this problem exists, which influences the choice of the right solution. An essential one is an evaluation of how big the relational stream will be from C to D thanks to direct service and how many connections per 24 hours shall be predicted in this

---

1 In Poland, there are two denominations regarding HSR train sets depending on the way of traction spacing. In *articulated sets*, it is concentrated in one or both traction vehicles at the beginning and at the end of a set, whereas in *electric multiple-unit sets* traction engines are placed in bogies under coaches throughout the entire set length.

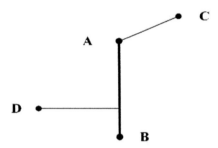

Figure 8.3 Choice of relations for the HSR trains. A – B high-speed line (HSL), C – D considered relational stream

kind of situation. Possible particular solutions will be discussed in later in this chapter, which will concentrate on the scheme of relations of connections related to the HSR. It is only worth noticing at this point that the mere size of the relational stream is not constant and depends on a number of factors, which generally are called passengers preferences.

A knowledge domain which deals with the size of flows is called *modelling transport*, and it is based on econometrical models [1, 8, 11]. Polish literature uses the term *modeling of travel and traffic prognosis* [15, 16]. Issues related to this issue are dealt with in section 8.4.2.

8.3.3. Evolution and types of railway timetables

A timetable is an essential, usually annual, railway work plan. It is therefore a base document, with reference to which railway line exploitation is conducted by an infrastructure manager. In passenger transport, it is also a fundamental element of a transport offer directed by operators to clients.

A base form of timetable is its layout – a train diagram, which imitates train movement driving on a defined section of a railway line in a two-dimensional coordinate system of time and route. A train movement is mapped by a diagonal – a so-called path – connecting a point of departure and of arrival on lines imitating station localization.

Historical forms of train diagram, especially with reference to passenger transport, were evolving. The oldest and the most popular type used until now is a so-called commercial graph, where on specific sections of railway lines appear trains of different categories, arranged from sets of different masses and tractored by different types of locomotives. They have diversified technical-operational parameters: of train acceleration, maximum speeds, braking delays. A construction of such a diagram, aiming to ensure a possibly large railway line capacity, requires efficient computer tools and experienced constructors.

A progressive specialization of railway lines, and in particular a domination of passenger traffic (for example in the day time) as well as need of construction of an attractive transport offer, led to the structuring of traffic diagrams. By these means, the clockface timetables (also called periodic or fixed interval timetables) were created [5]. They are based on fixed time intervals between all trains. There are distinguished three types of such timetables:

- non-symmetrical cyclic timetables;
- symmetrical cyclic timetables;
- integrated cyclic timetables.

A type of train diagram used on the high-speed lines depends on the accepted rules of traffic organization on the whole railway network. At this point, there can be distinguished two basic examples:

- HSL are an element of a railway network, used both for driving at high speeds as well as for traffic execution of other train types,
- HSL is a line specialized only in transport with high speeds.

In the first case, both the commercial graph and one of the three types of the above-mentioned cyclic timetables may be applied. Line capacity depends then on the scheme of train paths on a chart grid and is a consequence of accepted exploitation assumptions.

Particularly interesting is the second example. As previously mentioned, HSL may be close to saturation, and the intention of both an infrastructure manager and a railway operator is to use the system at maximum capacity. Because of the fact of conducting traffic of sets with identical or similar technical parameters, it is then a parallel diagram (of fixed intervals).

Calculation of capacity is trivial, and maximizing it requires a definition and a choice of two basic parameters: headway times of trains and maximum speed. As it turns out there is a speed with which a maximization of capacity is possible and its increase would lead to a deterioration of this parameter [8].

Reaching maximum HSL capacity requires accepting an additional assumption that the railway stations localized in nodes at the ends of such a line are capable of assuring routing and acceptance of the HSR trains with short headway times, which are usually equal to at minimum 3–7 minutes.

The train diagram constitutes a basic element of the timetable. Its construction, however, may be preceded by a series of planning works and organizational preparations.

8.3.4. Speed of passenger trains

In Poland, a speed qualification in the railway transport is based on the differentiation of two essential speed types:

- *maximum speeds* (v_{max}), which are possible to be obtained depending on the construction and state of railway lines, railway signaling devices, and power supply devices and type of railway rolling stock used;
- *average speeds*, which characterize a course of the exploitational process on railways (v_{sr}).

For obvious reasons, $v_{sr} < v_{max}$. A precise list of speed types together with their definitions are presented in Table 8.3. A maximum speed of a train movement v_j for given exploitational conditions is equal to the minor of two speeds: road speed v_d and train speed v_p, which may be noted formally as follows:

$$v_j = \min\{v_d, v_p\} \tag{8.3}$$

Table 8.3 Maximal and average speeds in railway transport [16]

Speed	Symbol	Appellation	Definition
v_{max} maximum	v_d	Line speed	Realizable on the section of railway line
	v_p	Train speed	Possible to get the type of rolling stock
	v	Running speed	Speed of the train
v_{sr} average	v_t	Technical	Speeds reached by train along the entire route from deducting commercial stops at stations and stops
	v^*	Commercial	Speeds reached by train throughout the journey, taking into account commercial stops

An average speed with which the travel of a certain train is realized results fundamentally from a possibility to achieve maximum travel speed v_j and a way of organization of railway traffic, presented as a train diagram and possible to be described thanks to technical speed v_t.

From the point of view of a passenger, the most important is a final commercial speed v^*, prejudging a total travel time on a route L. A connection of this maximum speed of train travel v_i on subsequent sections is described by a value of a so-called factor of maximum speed use w_{max}.

$$w_{max} = \sum_{i=1}^{n} \left(\frac{v_i^*}{v_i} \cdot \frac{l_i}{L} \right) \tag{8.4}$$

where:
n – number of sections i, on which the whole route was divided,
L – route length [km],
l_i – section length i [km].

For long-distance connections, a value w_{max} is usually equal to a number of a minimum of 0.6 on conventional lines to 0.9 on high-speed lines. It results from longer distances between stations on the HSR lines, line specialization, and a homogenous traffic of trains, which bears a relation to its incomparable bigger fluency. The value w_{max} shall be then interpreted as a gauge of evaluation of use of technical possibilities of railway lines in an exploitation process of railways.

The Warsaw – Cracow connection may serve as an example. Before launching of connections served by the ED250 (*Pendolino*) train sets, the conventional Intercity sets were operating with a speed of 160 km/h, which was allowed on 58% of the 292 km route, and on the remaining sections with a speed of 120 km/h. The value w_{max} was equal then to 0.832. The current ED250 train set value w_{ma} is equal to 0.89.

A choice of speed of passenger trains depends on many factors, and in the past it was conditioned mainly by the technical state of the railway lines and by types of an accessible rolling stock. In case of HSR, this choice has a strategic character. A discussion regarding maximum speed on high-speed lines is presented in section 8.4.

8.4. Choice of maximum train speed

8.4.1. General issues

The speed at which the trains operate on high-speed lines is one of the basic technical parameters which characterize an HSR system. In Europe, according to the definitions included in the Attachment I to the 2008/57[2] directive, the high-speed lines comprise:

- specially built high-speed lines equipped for speeds generally equal to or greater than 250 km/h;
- specially upgraded high-speed lines equipped for speeds of the order of 200 km/h;
- specially upgraded high-speed lines in specific topographical conditions.

In Poland from December 2014 on the Central Trunk Line (CMK), train traffic is conducted accordingly to the second of the above-mentioned conditions. In the near future, its increase is planned up to 230 km/h.

As previously stated, the decision regarding the choice of a maximum speed has a strategic character, and its choice is an interdisciplinary issue: first of all a technical one, but also economic, exploitational, or commercial. An issue of the choice of a maximum speed was raised by many authors, e.g. [3], [9], [10], [16]. M. Warlave, dealing for many years with HSR issues, proposes three sets of criteria in this field[3]: *Criteria of Optimization*,

- operational, that is especially important from the point of view of a railway operator;
- infrastructural, that is related to the whole railway system;
- social, embracing among others the external costs related to ecology, etc.

Obviously, speed choice is also determined by the technical limitations related to the infrastructure and the rolling stock.

8.4.2. Service criteria

From the point of view of an operator, an essential element influencing the choice of an optimum speed are the *commercial* conditions, which constitute one of the essential prerequisites while planning high-speed lines.

An essential aim of these investments is to create an efficient transport system that will efficiently concur with other means of transport.

Choices of the means of transport made by passengers are determined by a set of so-called preferences, that is the quality features of a transport offer. In Polish conditions, the set of the most important preferences embraces 11 features, presented in Table 8.4.

While concentrating on high-speed rails, the two most important preferences should be taken into consideration: travel time and its cost. A relatively short travel time has a dominating character, therefore in clients' (passenger) choices it prevails over other preferences.

The second most important reference is undoubtedly the price noticeable by a passenger. In the transport modelling [1], [6], [8], [12], [16] with reference to HSR, a particular use is

---

2 Official Journal of the European Union L. 191/24, 18.7.2008.
3 M. Walrave: "9 main issues about the Optimal Speed," 6th World Congress on High-speed Rail, Amsterdam 2008.

Table 8.4 Revision of the most important clients preferences [16]

Lp.	Preference name	Meaning
1.	price	travel cost noticeable by a passenger
2.	travel time	travel time noticeable by a passenger
3.	directness	travel realization directly or with the lowest possible number of changes
4.	geographical accessibility	distance to the nearest terminal (station, stop, airport)
5.	time accessibility	compatibility of the operating hours with the passenger expectations
6.	communicational safety	probability of safe travel realization
7.	personal safety	evaluation of the threats of criminal type
8.	comfort	travelling comfortableness
9.	timetable accuracy	compatibility of the travel with the timetable
10.	fallibility	probability of a break down during the travel due to technical reasons
11.	elasticity	possibility to modify the travel route while it is in progress

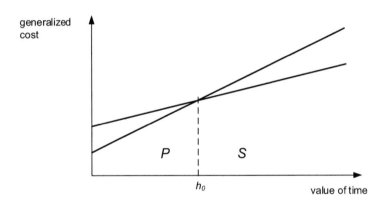

Figure 8.4 Generalized cost of a train G_p and of a plane G_s in a function of the passenger time value [1], [10]

made of a so-called price-time model explaining a modal division between the fast railways and the plane. Use of this model, the key elements of which of the means of transport choice are the values of a unitary passenger time h as well as a generalized cost G_{ij} between cities i and j, is calculated basing on the formula [8], [12]:

$$G_{ij} = c_x + h \cdot t_x \qquad (8.5)$$

where:
c_x – travel cost by a means of transport x between cities i and j,
h – unitary cost of passenger's time,
t_x – travel time by a means of transport x between cities i and j.

Graph of a function $f = G(h)$ enables to explain a share of each of those means of transport in the market, regarding a specific relation (Figure 8.4). The value h_0 in the Figure 8.4 is called neutral for this relation. For values h lower than that, a user chooses a train (field P), whereas for higher – a plane (field S).

Determination of share in the market is based on a graph function of density of passenger times values distribution. Distribution of time values h_r in a population of users r on a relation (a, b) usually is not known. A hypothesis is therefore assumed that it is the same (analogical) as the distribution of incomes in the society.

A limitation of the time-price model is a possibility of its use only for two means of transport and for two preferences. While modeling transport tasks division for example in a transport corridor, where a travel possibility exists – apart from train or plane – also by other means of land transport (car, bus), a following logit model is used [12]:

$$U_q = \frac{e^{v_k}}{\sum\limits_{q=1}^{Q} e^{v_i}} \tag{8.6}$$

where:
U_q – share in traffic q – of this means of transport,
Q – number of transport means functioning in the examined corridor: $q = 1, \ldots, Q$,
v_i – utility function of modes q.

In case of use of the logit model, velocity influencing the travel time should be examined in the context of a series of other preferences, which constitute parts of the utility function. This kind of approach requires a vast data set, obtained mainly thanks to marketing research.

The above-mentioned models depict an influence of velocity on the potential operator's incomes resulting from its share in the transport market under examination. On the costs part, on which a train velocity has impact, there may be marked out the costs related to purchase and exploitation of rolling stock, traction energy costs, as well as personnel costs [3]. It is worth taking into account a relationship between velocity and a number of sets necessary for service. Even the slightest diminishing of travel time (throughout velocity increase) may allow making economies in this field. More precise information will be included in section 8.5.4.

8.4.3. Infrastructural criteria

In relation to a rule functioning in the European Union of separation of operators and infrastructure managers, it may be assumed that the totality of costs connected with infrastructure exploitation is reflected in the charges for access. Depending on local regulations, the amount of charges may be related to the speed offered on the high-speed lines. A choice of such velocity may be therefore a parameter on which an operator balances its potential market revenues with its costs.

Independently from these conditions, it is obvious that an increase of train speed results in increased impact stresses of a track, which accelerates its wear and increases maintenance costs. This dependence is not linear, and for specific solutions there may be indicated threshold speeds, the exceeding of which an increase in speed generates a disproportionally fast increase of line maintenance costs. In Poland, there is a lack of sufficient experience and therefore also of data in this field, particularly regarding speeds above 160 km/h.

An approach described in section 8.3.4, which leads up to a determination of a coefficient of the maximum speed use w_{max}, enables to evaluate collectively the general technical conditions of the railway lines as well as evaluate their practical use in an aggregated way, not analyzing technical possibilities of separate HSR elements, among which a particularly sensitive one is, for example, the power supply system of electric traction (substations, traction network) and a series of others.

However, it is worth analyzing in particular one of these elements with a technical-operational character: a dependency arising between HSL capacity and train speeds. Making an assumption of train traffic management based on the parallel timetable, we obtain a simple equation for calculating capacity:

$$N_d = k \cdot \frac{60}{t_n} \left[\text{train pairs} /24 \text{ h}\right] \tag{8.7}$$

where:
N_d – capacity of high-speed lines,
k – number of hours of HSR train circulation (e.g. 17 between 6 a.m. and 23 p.m.),
t_n – train headways time [min].

Capacity depends therefore mainly from train headways time, which may be calculated from the equation binding velocity with a minimal headway between subsequent trains D_{min}:

$$t_n = \frac{D_{min}}{v} \left[\text{min}\right] \tag{8.8}$$

A function binding the headways time with the velocity is not linear [8] and depend on a series of technical parameters related to the type of signaling system devices (ETCS), technical-operational characteristics of sets, etc. An equation established for movement of sets ED250 on the CMK line in the Polish conditions (automatic block signaling with four block arrangements, ETCS L. 1), is the following [19]:

$$t_n(v) = \frac{L_E}{v} + \frac{v}{2,2 \cdot \exp(-0,007 \cdot v)} \tag{8.9}$$

where:
L_E indicates a route made by a train moving with a defined speed (e.g. 160 km/h or 200 km/h) during reaction time of the line and on-board devices. A function graph is presented in Figure 8.5.

From the conducted calculations, the results show that there exists a determined speed with which the headway time of trains possible to achieve on the high-speed lines is the shortest, and therefore its capacity is the largest. For the conditions of the Polish CMK, this speed is equal to 180 km/h.

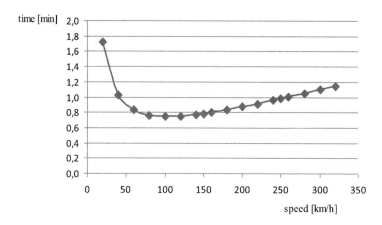

Figure 8.5 A minimum headway time of trains in the speed function

8.4.4. Social criteria

M. Walrave also proposes an analysis of the impact of maximum speed on two key problems related to energy, and in particular with global warming and with noise. Technological development enables a significant reduction of both of these risks. Modern solutions of power transmission, construction of substations, traction network, and on-board devices enable to improve effectively the efficiency of the whole system.

Aerodynamic, constantly improved shapes of bodies of high-speed sets, and in particular the construction of the train head, also serve to diminish energy consumption.

Likewise, the European requirements limiting noise, which refer to new and modernized vehicles, are getting stricter, thanks to which noise both in the vehicles and emitted to the line's environment is quite low. In built-up areas, a solution also lies in sound-absorbing screens.

The above solutions do not change the fact that energy consumption and noise are inseparably connected with speed increase. That is why a postulate of taking these dependencies into account when choosing speed is right and concerns in particular traffic on lines situated in difficult territory conditions, built-up areas, etc.

8.4.5. Economic speed balance

The criteria described in the three essential groups become a basis to formulate four functions, constituting altogether an "economic balance of speed."[4] All of them are not linear and may be presented by means of formulas gathered in Table 8.5.

A set of parameters {A, . . ., J} for the presented functions require precise analyses of a given scheme of connections by the high-speed rails as well as taking into account local technical and economic conditions. Afterwards, an examination of their course enables to determine optimal speed, what was presented schematically in Figure 8.6.

---

4 M. Walrave: "Analyse économique et financière des projets Grande Vitesse," Formation Systèmes Grande Vitesse, UIC 2007.

Table 8.5 Economic balance of speed

No.	Function's name	Function's formula	Type of function	Parameters
1.	Running time	$f_c(v) = A \cdot v^B$	power function	A > 0 and B < 0
2.	Transportation and revenues	$f_p(v) = C \times \ln(v) + D$	logarithmic function	C > 0 and D > 0
3.	Operating costs	$f_k(v) = E \times e^{F \cdot v}$	exponential function	E > 0 and F < 0
4.	Internal benefits	$f_w(v) = G \times v^3 + H \times v^2 + I \times v + J$	polynomial function	G, H, I > 0

Running time function (power function) (Fr. temps de parcours)
Transportation and revenues function (logarithmic function) (Fr. trafic/recette)
Operating costs function f_k (exponential function) (Fr. cout d'exploatation)
Internal benefits function f_c (polynomial function) (Fr. benefice interne)

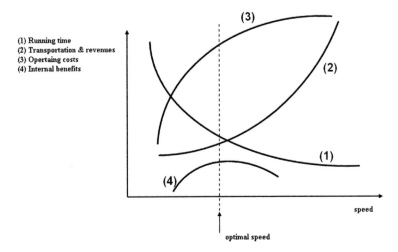

(1) Running time
(2) Transportation & revenues
(3) Operating costs
(4) Internal benefits

(3)

(2)

(1)

(4)

speed

optimal speed

Figure 8.6 Economically optimal HSR speed

Source: M. Walrave

8.4.6. Speed choice

A question arises in which way the choice of an optimal speed may be made, taking into account the above-described four types of conditions with such a differentiated character. In this matter, the following reasoning may be helpful.

According to the multiannual technical tradition, the thresholds of maximum speeds on railways were defined as an integer multiple of number 20, and therefore 80, 100, 120, 140, 160, or 200 km/h. Subsequently, indicated previously in Directive 2008/57 high-speed rails threshold in Europe is equal to 250 km/h (already without preserving the above rule), and further 280, 300, 320, and – as per today – 350 (380) km/h.

Maximum speed choice of trains on a considered HSL may be then initiated from assigning a set of acceptable solutions and simultaneously assume that nowadays the maximally obtained speed is equal to 574.8 km/h.[5] From the theoretical point of view, a speed optimization may lead to a solution of any value v_{max} from within the range, e.g. 200 . . . 570 km/h. The determined maximum speed thresholds used on railways limit actual choice to the values divisible by number 10.

The first step of speed choice should therefore consist in determining a few thresholds predicted to use, 250, 280, 300 or 320 km/h. Passing then to the economic field, it seems justified to analyze for each of those speeds the costs and advantages (revenues) connected with four essential condition groups examined one by one: technical, economic, exploitational, and commercial. Their combination and a critical analysis should facilitate the maximum speed choice that would be the most beneficial in determined conditions of a considered high-speed line.

8.5. Elements of HSR services organization

8.5.1. Modern transport strategies

History of development of the railway transport has strong connections with the development of settlement scheme as well as technological development, which concerns also the other means of transport. The above-mentioned conditions led to the situation in which a traditional model of preparing a railway transport offer has to take into account the limits constituted by the existing transport systems: course of railway lines, localization of stations and stops. At the same time, this results in an imperative to use already existing resources, which is not always in line with the real transport needs existing on the market.

Projecting and constructing new lines for the needs of HSR, often associated with the modernization of previously existing sections, differ from the historical conditioning of the railway system.

With all the objections regarding real possibilities of HSL conducting on a new terrain, their system and course refer to the real, present, and prognosticated needs of the market. Precise econometric models, which are the tools of transport modeling, enable to predict with great precision relational streams and passenger flows on newly projected sections.

Generalizing the above observations on all the transport investments, especially those related to railways, it might be ascertained that the traditional model of transport organization, in which the existing infrastructure notably determines railway traffic capabilities, is subject to a substantial change.

This is depicted by two schemes in Figure 8.7. In the order prevailing until now –case (a) – the timetable is shaped depending on the existing transport systems and on the demand reflected by the previous traffic, also resulting in part from the historically shaped railway lines network.

In the modern approach, which is possible during construction and modernization of railway lines, the sequence is the reverse. This demand, which takes into account also the effect related to the increase of attractiveness of a service offer, decides about parameters of the constructed and modernized infrastructure. What is intended is a line course itself, but also

---

5 This value responds to the record speed obtained by an experimental set V150 produced by Alstom on 3 April 2007.

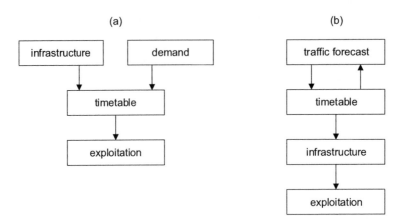

Figure 8.7 Traditional (a) and contemporary (b) rules of railway investment planning

line parameters, including a key parameter for passengers, a route speed, which determines the traveling time.

This contemporary approach is depicted in Figure 8.7(b). Let us notice that this process of investment preparation should embrace not only mere traffic prognoses, but also precise calculations of demand for railway rolling stock, on-board personnel work, etc. It makes it possible to predict precisely the exploitation and economic effectiveness of planned resources, and also – in a broader plan – to place a project in the context of the whole transport plan for the region and the country, as well as the international connections.

8.5.2. Transport offer planning

The process of preparing a transport offer, which is reflected in the train timetable, is presented in Figure 8.8 and constitutes a development of a standard approach to this issue [3]. It constitutes a few stages, starting from the evaluation of transport demand through projecting a transport offer, in which above all of a system and frequency of connections in predefined relations.

Although planning of rolling stock work and of on-board personnel hiring are placed at the last stage of this process, they should also be taken into account at the earlier stages.

Symbolically this was presented through an integrative approach. This attitude results from the fact that sets for high speeds are a particularly expensive element of the whole system [2], [9], therefore their use should be possibly intensive. Paths (relations) as well as departure and arrival terms of trains should be therefore chosen in such a way to ensure the following, while realizing the prognosticated demand:

* traffic of possibly the highest number of passengers in a direct way (with no need of changes);
* opportune connection on node stations;
* the best possible use of passenger sets.

Essential elements of the process of preparation of timetable are described in the following sections, together with a description of methods applied in this field.

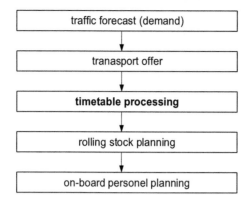

Figure 8.8 Planning process of transport offer (timetable) [5]

8.5.3. Passenger traffic forecasts

The traditional way of preparing a transport offer for the sake of the subsequent timetable bases its evaluation on the recently observed demand in the last few years, analyzing the arising trends and determining on this basis the awaited relational streams and section flows in the following period. On this basis, decisions are made about possible modification of an offer already in use and a timetable is shaped.

The described approach is modified in two cases:

- a decision made by the operator about a fundamental change of a market offer through, for example, rhythmization of services, change of relations scheme of trains, change of served settlement schedule;
- incorporation into use of new or significantly modernized sections of railway lines, which implies new possibilities of construction of connections network, journey time reduction, etc.

In this second case, an evaluation of the traffic size until now and drawing conclusions from the observed trends is not sufficient. It is necessary therefore to use other tools, in particular from within *transport modeling*.[6]

An essential, typical tool in this field is the above-mentioned classical four-step transport model, constituted by the following stages: *Trip Generation* (I), *Trip Distribution* (II), *Modal Split* (III), and *Traffic Assignment* (IV). A transport model is the most frequently applied in case of planning (forecasts) of traffic by different means of transport within an agglomeration. Its elements – subsequent models – may also be helpful, though, in preparing investments of high-speed lines. It refers in particular to the second or third module.

---

6 In Polish literature of the subject, a longer name is used: "travel modeling and traffic forecast."

A classic model used in module II serving to forecast traffic is a so-called gravitational model, which refers to the Newton's law of universal gravitation. It refers to size of flows between determined cities [1], [8], [11], [12]:

$$T_{ij} = K \cdot \frac{(P_i \cdot P_j)^\alpha \cdot (W_i \cdot W_j)^\beta}{G_{ij}^\gamma} \tag{8.10}$$

where:

T_{ij} – traffic between cities (zones) i and j,
K – constant term,
P_i, P_j – population of cities (zones) i and j,
W_i, W_j – income per capita of cities (zones) i and j,
G_{ij} – generalized cost of transport between i and j,
α – elasticity of traffic to the population (usually $\alpha \approx 1$),
β – elasticity of traffic to the income per capita (usually $\beta = 0.9$),
γ – elasticity of traffic demand to the generalized cost of transport (usually $\gamma = 1.6$–1.9).

Having at one's disposal the potential size of passenger traffic between considered cities, it may be determined what part in this market the high-speed rail may have. To achieve this aim, one of the two previously presented formulas may be used (section 8.3.2) that determine modal split depending on a chosen set of preferences. A further step of traffic offer projecting embraces the preparation of the system of connections.

8.5.4. Construction of connection network

Railway transport infrastructure is a coherent system in which trains – as per rule – may move on all of the active railway lines. Four main obstacles exist in this field:

- differentiated track gauge, possible to be bestridden when using bogies of a special construction (with a system of automatic change of wheel gauge) or through raising of wagons or bogies exchange;
- differences in the system of electrical track alimentation (in Poland 3 kV DC), which possibly require multi-system traction vehicles;
- differentiated cabin signaling systems (in Europe there are around 20 different systems – ERTMS is a trial to standardize them);
- line access of rolling stock with crossed gauge.

High-speed lines constitute a complement to the conventional lines,[7] and from a technical point of view they are also adapted to the movement on them of the conventional rolling stock under obvious conditions: compatibility of the system of traction alimentation of a locomotive and their equipment in on-board signaling devices used on HSL.

Not developing further technical issues, we will consider organizational issues. There are four basic models of relationships between conventional lines infrastructure and the one of high-speed movement [13]. There are presented in Table 8.6.

---

7 In order to distinguish high-speed lines and high-speed railways from the other railway lines and trains in the international literature, the terms "conventional lines" and "conventional railways." are used. It is unfortunate that HSR is not an "unconventional railway," which term refers to, for example, magnetic railways (maglev).

Table 8.6 HSR models according to the relationship with conventional services [13]

	High-speed lines	Conventional lines
Model 1: Exclusive exploitation		
High-speed Trains	X	
Conventional Trains		X
Model 2: Mixed high-speed		
High-speed Trains	X	X
Conventional Trains		X
Model 3: Mixed conventional		
High-speed Trains	X	
Conventional Trains	X	X
Model 4: Fully mixed		
High-speed Trains	X	X
Conventional Trains	X	X

Exclusive exploitation model is based on a total separation of HSR movement from the conventional network, which renders independent an organization of fast traffic with regard to other connections.

Mixed high-speed rails model, i.e. access of sets for high speeds on (usually modernized) conventional lines facilitates organization of direct connections with use of high-speed lines between cities situated near conventional lines.

Mixed conventional traffic model is in a certain sense a reverse of the previous model and assumes traffic of railway conventional sets on the high-speed lines.

Fully mixed traffic model assumes both traffic of HSR trains on conventional lines and of conventional sets on high-speed lines. A mixed traffic (including cargo trains) increases significantly maintenance costs of lines intended for speeds 250–300 km/h and causes decrease of their capacity. For obvious reasons (a relatively small number of connections with ED250 sets), this solution was applied in Poland on the Central Trunk Line. Apart from EIP *Pendolino* sets, there are also other passenger trains moving on it: EIC (intercity), TLK (interregional), and regional as well as cargo trains.

As results from the above revision, there are several variants of construction of service offer with the HSR use. Not prejudicing the final shape of the connection network, which will be applied in Poland after launching traffic on the new high-speed line, so-called Y (Warsaw – Łódź – Wrocław/Poznań), there may be presented basic, general rules of shaping of the connection network. They refer to the present rules of passenger traffic organization with use of the CMK line, which means application of a fully mixed model.

Planning of a scheme of relations of HSR trains differs in a certain scope from the work on connections served within a classic set of locomotives and wagons. It results from the fact that the great part of courses of HSR sets should occur on high-speed lines to possibly maximize the use of their technical parameters. Possible extending of trains relations beyond those lines may be justified by adequately large passenger flows indicated for direct service. Lower speeds possible on conventional lines makes for less effective use of HSR sets.

An example justifying the direction of HSR sets is presented in Figure 8.3. In Polish conditions, the current existing line for increased speeds – the Central Trunk Line[8]– enables accelerated travel between Warsaw and Katowice and Cracow (and further Rzeszow), and also Wroclaw. In the past there were plans to extend the CMK line north to reach Tri-city

---

8 CMK – acronym of the Polish name: *Centralna Magistrala Kolejowa*.

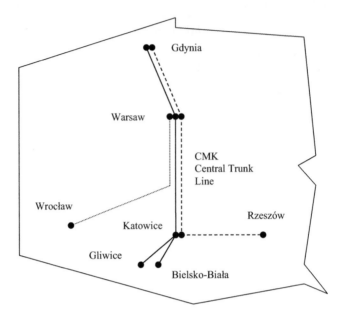

Figure 8.9 An example of use of high-speed line for construction of a network of connections [19]

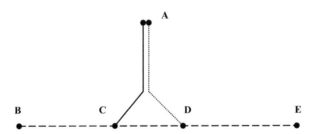

Figure 8.10 An example of intercity train network

which has a population of nearly a million people. Nowadays, trains operate in this relation on more than half of this route (E65 line) with a velocity of 160 km/h.

The connection scheme created with the use of CMK is presented in Figure 8.9. It enables the use of speeds up to 200 km/h in a possibly large number of direct connections between north and south parts of Poland.

The described scheme of connections regards only direct relations. Construction of a network of connections has to take into account changes on node stations. Relations arising in such a case may be analyzed from the following example.

Let us consider a theoretical example of a network of railway connections between five big cities A, B, C, D, and E. Substantial information about a scheme of connections and travel times of trains are presented in Figure 8.10.

Organization of traffic is based on launching of trains with a cycle $T = 60$ minutes in three basic relations: A – C, A – D, and B – E. Because of necessity of a reasonable organization of traffic on the line from A to C and D, the train movement may occur with the headway time of 5 minutes, but in an unrestricted sequence: a train from A to C precedes a train from A to D or the reverse. For the relation B – E, the indicated standing times of trains on the stations C and D aiming to connect trains (in both directions) are equal to e.g. 10 minutes.

Assuming use of a symmetrical timetable, a question arises about a choice of such terms of departures of trains from the initial stations A and B, so that the passenger change on the stations C and D occurs with a possibly minimal loss of time Ω. This time constitutes a product of x number of passengers changing and of times c concentrated for a change in a determined set P of trains i and j on all the S considered node stations.

$$\Omega = \sum_S \prod_P (x_{ij}, c_{ij}) \rightarrow \min \tag{8.11}$$

We assume that the number of passengers changing on the stations C (in the B direction) and D (in the E direction) is known. Search of a solution may be conducted throughout considering basic variants. For the assumed departure times from the station A in the direction of C and D as well as the order of train routing, some basic variants may be calculated regarding routing times from the station B (e.g. *.05, *.10 . . . , *55). Because of the symmetrical timetable used and known travel and stabling on the station times, all the other terms are generated automatically. By using Equation (8.11), we obtain an evaluation of every variant.

The above example is not complicated and easy to calculate. In case of developed connection networks, the construction of periodic railway timetables is algorithmically difficult and has been intensively studied as a periodic scheduling problem (PESP) [15]. Optimization of trains communication in view of time lost by passengers would require using more advanced methods of optimization, for instance within line programming.

While summarizing the rules of creation of connections network, it may be stated that it should fulfill collectively the following criteria:

- ensure traffic of the largest possible number of passengers by trains of direct relations;
- ensure convenient communications on node stations (change stations);
- use at maximum the capacity of high-speed line;
- use the best and accessible HSR sets;
- enable connections according to the cycle of a timetable used on a railway network.

Basic dependencies connecting cyclic timetable with the work plan of rolling stock within optimization of its use is presented in the section 8.5.5.

8.5.5. Planning of rolling stock work and of train crews

Sets destined for KDP transport – because of their value – should possibly be exploited intensively. That is why it is necessary to prepare a reasonable plan of their exploitation.

Important here are two motions used in Polish technical terminology:

- train set's circuit, signifying a set of served subsequently trains $p = 1, \ldots, P$,
- train set's turnaround T, which is the time of circuit's duration.

These motions are connected with the following formula:

$$T = \sum_{p=1}^{P} t_p + \sum_{Z=1}^{Z} t_z \tag{8.12}$$

Train set's turnaround is therefore a sum of travel times of separate trains t_p and of so-called transition times of train sets on terminal stations t_z of a collective number Z. The time t_p in an obvious way depends on the maximum speed and on the achieved coefficient of its use w_{max}. Instead, the times t_p result from the way a timetable was created and are connected with technological requirements in the field of rolling stock maintenance (repairs, cleaning, equipment), both on terminal stations and on the mother station.

The problem especially regards situations when high-speed lines are an element of a network of connections affected by a periodic (fixed) interval timetable. Even the slightest increase of speed may help economize additional time that would prolong the time spent by a train set at a terminal station and will enable to use it for servicing a prior return route.

Projecting of circuits and calculation of need for a number of train sets necessary for servicing HSR connections is relatively easy, especially in case of usage of a cyclic traffic on high-speed lines [5], [12], [17].

$$n = \frac{T}{\tau} = \frac{2 \cdot t_j + t_z^A + t_z^B}{\tau} \tag{8.13}$$

where:
n – number of train sets,
T – train set's turnaround [h],
t_j – travel time with a train in one direction [h],
t_z^A, t_z^B – transition times on terminal stations A and B [h],
τ – headways times of trains [h].

Attention should be paid at this point to the fact that improper planning of a cycle applied on high-speed lines may lead to unnecessary increase of need for rolling stock. It is depicted in the following example, which refers to a situation in which train movement is carried out basing on symmetrical clockface timetable with a zero axis of symmetry.[9] Using real data, let us assume that the travel time of an ED250 train set between stations Warsaw East and Cracow Main Station is equal to 2 h 35 min, and minimal times on terminal stations for service of a train in a return direction are equal to 20 minutes. Essential variants are presented in Table 8.7.

Amid twelve variants of possible service, only four enable service of connections with six train sets. Choice of the most advantageous variant, leaving apart other technical-operational

---

9 That means that arrivals and departures of trains on each station with a determined commercial stabling occur symmetrically with regard to the vertical axle of clock's dial (e.g. 47 and 13, 58 and 2).

Table 8.7 Planning of rolling stock work in the relation Warsaw – Cracow

Variant	Warsaw East		Travel time	Cracow Main Station		Turnaround [minutes]	Timetable period	Number of train sets
	arrival	departure		arrival	departure			
A	*.00	*.00		*.35	*.25	420		7
B	*.55	*.05		*.40	*.20	420		7
C	*.50	*.10		*.45	*.15	360		6
D	*.45	*.15		*.50	*.10	360		6
E	*.40	*.20	2 h 35min one way. Both ways 310 minutes.	*.55	*.05	420	Trains in cycle every 1 hour (60 minutes).	7
F	*.35	*.25		*.00	*.00	420		7
G	*.30	*.30		*.05	*.55	420		7
H	*.25	*.35		*.10	*.50	420		7
I	*.20	*.40		*.15	*.45	360		6
J	*.15	*.45		*.20	*.40	360		6
K	*.10	*.50		*.25	*.35	420		7
L	*.05	*.55		*.30	*.30	420		7

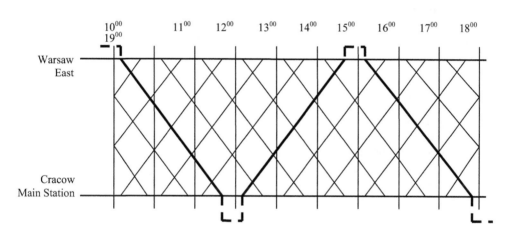

Figure 8.11 Exemplary service of a Warsaw – Cracow connection with ED250 train sets

conditions related to timetables of other remaining trains, may for example consider conditions on terminal stations. Decisively higher traffic on the Warsaw East station leads to the choice of variant C or I related with stabling of 20 minutes. In the example, there were examined only such minute endings that can be easily remembered by passengers. Unfortunately – because of the assumptions made (symmetrical timetable) and of the travel time – it is not possible to establish the most advantageous variant, when departures from both stations occur with identical minute endings.

This kind of variant is possibly in case of a non-symmetrical timetable. An exemplary departure from Warsaw and Cracow would occur 10 minutes after a full hour, therefore an arrival a quarter to the third hour. Stabling time on both stations would be equal to 25 minutes, which travel lasting 155 minutes one way and a cycle of 60 minutes would enable to serve the connection with use of six train sets.

Plan of turnaround of train sets for this particular example is presented in Figure 8.11.

The organization described concerns traffic within one relation only. In real conditions and therefore with a developed network of connections, optimization of used rolling stock is only one of the elements of timetable construction. A fragmentary solution, regarding number of train sets necessary for service of train paths, their communication, etc. may be divergent. Choice of a final, reasonable plan of operations must be made by an operator. It should evaluate, for example, the economic effects of the use of a certain solution.

A crucial element of the timetable is also work of on-board personnel (drivers, train crew, on-board service). Work plans should fundamentally include two aspects: observation of law provisions about work time and minimization of number of necessary crews; that is, minimization of times spent for waiting on terminal stations for undertaking service in another train. It is also recommended to avoid excessively long transitions between served trains, in particular use of train-crew dormitories.

In exploitational practice, the safest solution is an "attachment" of personnel to the served train set, which facilitates assuring service in conditions of operational disturbances and of train delays.

Work of on-board personnel – locomotive crews and train services –is the last element of the described planning process. Finding of optimal (reasonable) solutions requires usually a compromise between partially contradictory criteria.

It is worth emphasizing that before initiation of timetable implementation, its potential variants should be also evaluated according to the methods enabling to evaluate reliability of railway timetable [15].

8.6. Summary

Launching in December 2014 of connections with ED 250 train sets with speeds of up to 200 km/h constitutes a movement for constructing a network of high-speed relations, which will connect in the future the biggest cities in Poland.

The chapter presented the most important transport and organizational aspects connected with the creation of an HSR system. Methods of organization of this kind of services, although they are a continuation of traditional ways used by railway operators, must include the specifics of modern transport technologies.

The outline of issues presented in the chapter does not exhaust all the whole problems. Nevertheless, it may be useful in orienting specialists to the theme of the subject.

Bibliography

[1] Bonnel, P.: *Prévoir la demande de transport* [*Forecasting of transport demands*]. Presses de l'École Nationale des Ponts et Chaussées, Paris, 2004.

[2] Chapulut, J.-N. (ed.): La grande vitesse ferroviaire. La documentation Française [High-speed rail. French documentation]. *Rapports & Documents*, 2011, no 40.

[3] Garcia, A.: *Relationship Between Rail Service Operating Direct Costs and Speed*. UIC, Paris: Fundación Ferrocarriles Españoles, 2010.

[4] Gorlewski, B.: *Kolej dużych prędkości: Uwarunkowania ekonomiczne* [High-speed railways: Economic conditions]. Warszawa: Oficyna Wydawnicza SGH, 2012.

[5] Hansen, A.I., and Pachl, J.: *Railway Timetabling & Operations*. Hamburg: Eurailpress, 2014.

[6] Jacyna, M.: *Modelowanie i ocena systemów transportowych* [Modelling and evaluation of transport]. Warszawa: Oficyna Wydawnicza Politechniki Warszawskiej, 2009.

[7] Kornienko, V.V., and Omellijanenko, V.I.: *Vysokoskorostnoj elektriczeskij transport: Mirovoj opyt* [High-speed electrical transport: World experience]. Kcharkov: Kcharkov University of Technology Publishing House, 2007.

[8] Lebœuf, M.: *Grande vitesse ferroviaire* [High speed rail]. Paris: Le cherche midi, 2013.

[9] Łaszkiewicz, R.: *Organizacja kolejowych przewozów pasażerskich* [Organization of railway passenger services]. Radom: Politechnika Radomska, 1998.

[10] Pita, A.L.: The Effects of High-Speed Rail on the Reduction of Air Traffic Congestion. *Journal of Public Transport*, 2003, 6, 1.

[11] Pita, A.L., and Robuste, F.: Impact of High-Speed Lines in Relation to Very High Frequency Air Services. *Journal of Public Transport*, 2005, 8, 2.

[12] Pita, A.L.: *Explotación de líneas ferrocarril* [Railway lines exploitation]. Barcelona: Ediciones Universito Politécnica de Catalunya, 2008.

[13] Rus de, G. (red.): *Economic Analysis of High-speed Rail in Europe*. Bilbao: Fundación BBVA, 2009.

[14] Towpik, K.: *Koleje Dużych Prędkości: Infrastruktura drogi kolejowej* [High-speed railways: Railway road infrastructure]. Warszawa: OWPW, 2012.

[15] Vromans, M.: *Reliability of Railway System*. Rotterdam: ERIM, 2005.

[16] Żurkowski, A.: Modelowanie przewozów międzyaglomeracyjnych [Modeling of intercity services]. *Problemy Kolejnictwa z.* 148, 2009.

[17] Żurkowski, A., and Pawlik, M.: *Ruch kolejowy i przewozy: Sterowanie ruchem* [Railway traffic and services: Traffic management]. Warszawa: KOW, 2010.

[18] Żurkowski, A.: *Zastosowanie symetrycznego rozkładu jazdy w przewozach międzyaglomeracyjnych* [Use of symmetrical timetable in intercity services]. Prace Naukowe Politechniki Warszawskiej TRANSPORT, zeszyt 97, Warszawa, 2013.

[19] Żurkowski, A.: Central Trunk Line (CMK) 200 km/h: Traffic, Timetables and Operation. *TTS (Technology Transport Systems)* 6/2015.

Chapter 9

Introduction of high-speed rolling stock into operation on the Polish railway network

Marek Czarnecki, Witold Groll, Andrzej Massel and Sławomir Walczak

9.1. Introduction

When identifying factors that have an impact on the operating speed, i.e. the speed achieved by a vehicle in the given conditions, it can be determined that, apart from the vehicle's traction properties (including the design speed), it is also affected by all other railway subsystems, including in particular such elements as railway track, bridges, tunnels, railway traffic control devices, telecommunications systems, catenary, and power supply. An increase of the operating speed is preceded by the validation tests:

- practical verification of correctness of solutions adopted with respect to the surface, facilities, railway subgrade, and the transition zones next to the engineering facilities, as well as the effectiveness of the solutions with respect to noise and vibration suppression;
- verification of safety of the rolling stock-track dynamic interaction and the running stability;
- verification of safety of operation and reliability of the power supply system and various pantograph types, and also evaluation of correctness of parameters adopted and devices selected;
- testing of railway traffic control and communications systems used on the line, with particular attention given to safety of their operation and reliability, and also improving compatibility between on-board and rail track devices.

Positive verification of the solutions adopted does not mean that in case of a significant speed increase other problems will not arise, related, for example, to:

- flying ballast during high-speed train passage;
- increasing noise;
- limiting line throughput forcing operation of highly homogenous traffic.

The main document that defines directions of railway transportation development in Poland is *Master Plan for Rail Transport in Poland until 2030* passed by the Council of Ministers on 19 December 2008. In reference to the railway infrastructure, it sets the primary investment objectives aimed at bringing about, with respect to passenger transportation, first of all, the raising of the maximum speed on the lines as a result of their major upgrade, up to 160 km/h or 200 km/h (depending on the line). With respect to the high-speed rail, it assumes adapting the E65 South (CMK) line to the high-speed rail line requirements.

High-speed rolling stock is expected to be operated on the upgraded line.

Due to the above-mentioned prerequisites, the process of putting high-speed rolling stock into service in the territory of Poland is currently at its initial stage in comparison to Western Europe. In the future, an increase in the quantity and quality of high-speed rolling stock operated in Poland will be fueled by demand for such rolling stock related to the further development of the infrastructure enabling its operation.

9.2. High-speed rolling stock

The term "high-speed rolling stock" is used in reference to the rolling stock with the maximum operating speed higher than 190 km/h. The current technical features of such rolling stock are the result of the advances in rolling stock technology over the last decades.

One of the key issues, the solving of which enabled the advances in the high-speed rolling stock, was the obtaining of appropriate power for traction purposes, allowing for achieving such speed.

As is commonly known for speeds above 200 km/h, aerodynamic resistance (drag) is the main component of the train's resistance. Such drag is proportional to the speed squared and, in case of the operating speed range under consideration, it may represent more than 90% of the train's total resistance.

Therefore it can be assumed, having no fear of making a big mistake, that the power requirement for traction purposes, at the set operating conditions, is proportional to the speed to the power of three (power is the product of multiplying the motion resistance, i.e. the tractive effort and the speed).

Achieving high-power rail vehicles while preserving moderate dimensions and first of all a low weight of the propulsion system (powertrain), allowing for maintaining axle load within the permitted range, has been enabled due to the introduction, more than three decades ago, of alternate current motor (synchronous, asynchronous, and recently using permanent magnets) based propulsion systems, powered and controlled by semiconductor devices. Semiconductor elements used in the rail vehicles' propulsion systems were initially ordinary thyristors, subsequently the GTO (*gate turn off*) thyristors, and in recent years the IGBT (*insulated gate bipolar transistor*) transistors. Each successive change of the semiconductor element type used in the propulsion systems led to the reduction in the number of discrete elements in the system and thus its simplification along with a simultaneous increase of the energy conversion efficiency rate.

An additional, though very important from the point of view of the ability to use high-speed rolling stock in international operations, advantage of the use of the alternating current motor based propulsion system became the ease, in comparison to the legacy solutions, of adapting rolling stock to operate under various catenary power supply systems, since the differences in propulsion systems for various power supply systems exist only in the input module of the propulsion system, whereas the main part of the propulsion system (converter and traction motors) are the same in all systems. In Europe there are four catenary power supply systems: two DC systems, 3 kV and 1.5 kV, and two AC systems, 25 kV, 50 Hz and 15 kV, 16 2/3 Hz; therefore, high-speed rolling stock used in international operations must be prepared to operate under two, three, and even in some cases four power supply systems (e.g. Thalys trains: Paris – Brussels – Cologne – Amsterdam). A certain problem that occurs in multi-system rolling stock in case of four-axle locomotives or trainset propulsion units prepared to operate under AC and DC catenaries is the need to increase weight, and what follows, the maximum axle load. This increase is due to the need to install two relatively heavyweight assemblies on the vehicle, the transformer to operate under AC catenary and the dynamic braking resistors to operate under DC catenary.

As mentioned before, in case of high speeds, the train's aerodynamic motion resistance (drag) is essential. Attempts to achieve the lowest possible values of the aerodynamic resistance coefficient (Cx) are reflected in the shapes of carbodies of high-speed trains (trainsets and multiple units), characterized by the elongated and round front parts of the carbody with very steeply inclined front end (nose) piece (section). Apart from the front end's resistance, airflow resistance over the train's exterior is also important. In order to reduce the resistance, the carbody's external surface is designed so as not to cause even small, air resistance increasing air swirls, i.e. no protruding elements, no window or door recesses, and aerodynamic covers for the lower parts of the carbody and also for the devices installed on the roof, such as pantographs, for example. Particularly important from the point of view of aerodynamic resistance are the zones outside the gangways (passageways) between coaches. They are designed so that any air swirls therein should be as small as possible.

The aerodynamic shape of the front parts of the end units of a trainset or a multiple unit additionally makes it easier to solve the high-speed rolling stock passive safety problem, since it creates space for installing crash energy absorption elements.

Apart from the power and aerodynamics of the high-speed rolling stock, of primary importance in the development of such rolling stock used to be and still are the design and research issues related to the dynamics of the rolling stock-track interaction. Such issues will be discussed further on in this chapter.

High-speed rolling stock can be split into two groups or categories differing by basic parameters, first of all the maximum operating speed and operational use. The first group includes rolling stock with the maximum operating speed within the 250–350 (and higher) km/h range, the second group includes rolling stock with the maximum operating speed within the 190–250 km/h range. Until recently, this split was described in the Annex 1 of Directive 2008/57/EC on the interoperability of the rail system within the Community [1] where these rolling stock groups were defined as the first- and second-class high-speed rolling stock. In the latest edition of TSI, this split is no longer described; nevertheless, from the technical point of view it is still valid.

Rolling stock with the maximum speed within the 250–350 km/h range and higher is to be operated on specially designed lines and built solely for high-speed passenger transportation, on lines characterized by the appropriate configuration and profile, track design and maintenance standard, catenary power supply system, and railway traffic control systems that allow for achieving speeds within the above-mentioned range.

This group of high-speed trains includes trainsets, i.e. trainsets made up of cars that cannot be uncoupled, with the propulsion concentrated on both or one of the end cars, or multiple units, i.e. trains also made up of cars that cannot be uncoupled, but with the main propulsion system and auxiliary devices distributed along the length of the train in the passenger carrying units. Among solutions used in recent years, the most common are multiple units in which the distribution of the propulsion system and auxiliary devices along the length of the train allows for uniform distribution of weight and, as a consequence, maintaining axle loads of 17 tons, which is described in the TSI requirements.

In case of trainsets where the four-axle motor cars (drive units) must include all elements of the main propulsion system and auxiliary devices, i.e. all heavyweight assemblies, the condition of maintaining the weight of the motor car around 68 tons (4×17) is very difficult to meet.

The second group (category) of high-speed rolling stock, i.e. with the maximum operating speed within the 190–250 km/h range, is to be operated both on high-speed lines as well as on the existing and upgraded lines, if the track design and maintenance standard allows for the operation at the speeds within the above range. This high-speed rolling stock group also includes trainsets and multiple units, as well as locomotives with the maximum operating speed from 200 km/h, hauling standard passenger coaches adapted to such speed. An example is Austrian railways' Rail Jet.

In a sense, a separate rolling stock type among the rolling stock with the maximum operating speed within the 190–250 km/h range is the tilting rolling stock.

Achieving high speed, i.e. shortening the travel time on a specific route on the existing and upgraded lines, is frequently hindered by the occurrence on such lines of small radius curves (this is the term used to refer to horizontal track curves with a radius of $R \leq 1000$ m), the alteration of which in order to increase the radius is not possible due to terrain constraints (mountains) or structures built in the direct vicinity of the lines. The parameter that reduces the speed on a curve is the value of the lateral acceleration not eliminated by the track cant and the centrifugal inertia forces arising from it affecting the vehicle and transmitted by the track. The permitted values of the non-compensated lateral acceleration depend on three criteria, namely lateral track stability, safety against derailment (Y/Q), and passenger comfort, to be precise the lateral acceleration felt by passengers.

If on a certain line it is possible to increase the value of the non-compensated lateral acceleration on the rail head plane within the permitted limits defined by the first two criteria (track criterion and Y/Q), and the factor that limits the curve speed is passenger comfort, then the use of the rolling stock with the inwardly tilting carbody enables reducing the lateral acceleration felt by the passenger down to the acceptable level with much higher lateral acceleration on the rail head plane. This is due to the change of the position of the passenger's reference system (coach floor plane) by tilting the coach's carbody by a certain angle (up to approximately 8°) inward towards the track curve's inside. It should be emphasized that the carbody's tilt does not affect the value of the non-compensated lateral acceleration and thus also the centrifugal inertia forces on the rail head plane. The amount that a specific route's travel time is reduced due to the use of the tilting rolling stock depends on many factors [3], first of all on the percentage share of curves with a small radius in the total line length, the share of which, according to the data on the European applications, fluctuates between 5% and as much as 30%. These benefits must be evaluated in the context of increased (by approximately 10% according to the manufacturer's information) costs of purchasing and maintaining the tilting trains in comparison to the similar non-tilting ones. The additional factor affecting the increase of the tilting trains' operating costs is the accelerated wear of outer rails on curves and of wheels caused by lateral wheel-rail forces, which are greater than those of operating conventional trains.

9.3. High-speed rolling stock in Poland

The first vehicles considered, the 190–250 km/h group high-speed rolling stock put into commercial operation by PKP Intercity, were EU44 series locomotives with the maximum operating speed of 200 km/h that could haul standard Z1 type coaches. In the Polish environment, these locomotives may be operated at the maximum speed of 200 km/h solely following the installation of the on-board ETCS system on segments of a line equipped with such a safety system. The tests of dynamic interaction between a locomotive and the track in the Polish railways infrastructure conditions were conducted at the speed of up to 235 km/h.

The ED250 series multiple unit with the maximum operating speed of 250 km/h was put into commercial operation in 2014. In the operational environment of the Polish railways network, its maximum speed is currently limited to 200 km/h. Increasing the maximum operating speed in Poland from the 160 km/h previously in force to 200 km/h was preceded by more than two years of tests of all subsystems that had an impact on the driving safety.

9.3.1. EU44 series locomotive

A versatile EU44 series electric locomotive ES 64 U4 – D (HUSARZ) is to be used to haul passenger and freight trains at the maximum speed of up to 200 km/h. It is a high-power rail vehicle that is member of the Eurosprinter® ("ES") vehicle family manufactured by Siemens.

The design of the bogie frame is a compact fully welded box (cage) type profile (construction) (Fig. 9.1). Its main components are longitudinal beams and the center crossbar (cross-member) as well as end crossbars (transoms). The points where the wheelset bearing guides are connected to the traction motors' suspensions are the steel components welded into the structure and mechanically processed so that they could constitute the fastening for the wheelsets' suspension arms and traction motors' bearings. In the center of the bogie on the longitudinal beams, the secondary suspension springs dampers' pads are placed laterally in relation to the locomotive. In this spot, the center cross bar connects both longitudinal beams. In the middle of the center crossbar, there is an opening in which the kingpin is placed. From the bottom, the kingpin is connected to

Figure 9.1 EU44 series locomotive bogie
Source: [10]

the center crossbar using the lemniscate connectors. Furthermore, on the center crossbar there is a connection point for the fixed traction motor's support. Both end crossbars connect the ends of the longitudinal beams. The powertrain's suspension arms are mounted on the end crossbars.

The primary suspension coil springs (Fig. 9.2) are positioned coaxially in relation to the axle bearings (two on each box). Dampening of the wheelset vibrations is provided by the hydraulic dampers. One hydraulic damper is envisaged for each axle bearing box.

The locomotive carbody rests on each bogie via four coil springs. For the purpose of acoustic insulation, rubber inserts are placed under the spring pads.

Such secondary suspension springs provide vertical and lateral dampening of the locomotive carbody. The hydraulic dampers between the locomotive carbody and the bogie dampen the vertical, lateral, and serpentine motion.

Primary suspension springs

Tractive effort transfer mechanism

Vertical damper

The torque of the traction motor is transmitted to the wheelset via a fully suspended drive pinion and skew gearwheel, and further on via the hollow shaft and star configured coupling links (flexible claw coupling) to the disc-wheel (Fig. 9.3). The traction motor and transmission gear assembly is mounted on the bogie crossbar (member) on a flexible (rotating) support and on the bogie front crossbar (transom) using two suspension arms. A traction motor damper is used to dampen lateral movements. Rotational motion bumpers block the drive unit's lateral shifts.

Gearbox

Braking shaft

Traction motor

Star with claws to rubber bushed links

Figure 9.2 Primary suspension springs

Source: [10]

Figure 9.3 Powertrain overview
Source: [10]

9.3.2. Electric multiple unit ED250

The design of the ED250 multiple unit is derived from the design of the ETR 610 New Pendolino Cisalpino vehicle that was successfully put into operation between Italy and Switzerland in July 2009. The main differences between the two designs were related to:

*   terms of reference;
*   adaptation of the vehicle to interoperate with the Polish railways infrastructure;
*   changes to standards and regulations.

Due to the nature of the lines on which the ED250 electric multiple units were to be operated, such differences included primarily:

*   complete removal of the tilting system and the furnishing related thereto;
*   bogie design changes; and, as a consequence,
*   changes to the weight of the suspended parts.

These differences were physically implemented by removing the tilting system's hydraulic assemblies and the pantograph anti-tilt mechanism as well as modifying the lower cover. Furthermore, pneumatic active lateral suspension assemblies were simplified retaining solely the slip control function. Additionally, the tilting system control unit's functions related to tilting and lateral active suspension were removed, while the functions related to bogie stability control and axle bearings temperature monitoring were retained. In the bogie, instead of

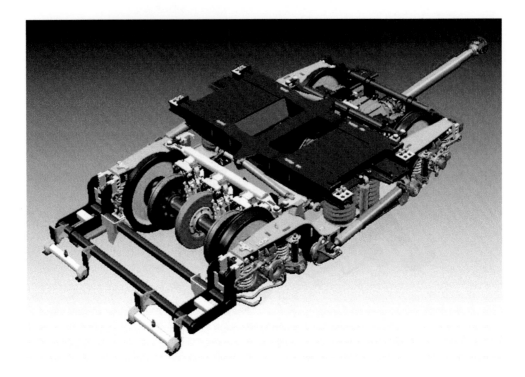

Figure 9.4 Bogie overview

Source: [10]

the removed tilting system a new lateral crossbar (cross-member) was introduced, linking the carbody with the bogie and constituting also an element of the secondary suspension springs.

The design of the bogies is based on the well-known and proven Alstom bogie design, the versions of which, tilting and non-tilting, have been operated worldwide for a number of years (Fig. 9.4).

Trailer bogies and motor bogies are two-axle bogies with the wheelbase of 2700 mm. The bogies are designed for the design speed of 250 km/h.

The main features of the bogie are:

- maximum operating speed of 250 km/h;
- axle load 17 t;
- wheelbase 2700 mm;
- track gauge 1435 mm.

The bogie frame was designed for the axle load of 17 tons, in accordance with the requirements of EN 13749 [8], and is a fully welded frame. It is made up of two longitudinal beams connected with two tube crossbars. The connection between the carbody and the bogie is a Z-type geometry connection. The bogie mechanical braking system is made up of five braking disc assemblies on a motor bogie and six on a trailer bogie: on the powered axle there are two ventilated discs, and on the trailer axle – three.

The primary suspension is made up of:

- axle bearing made of spherical cast iron and one traction rod with rubber-metal couplings (joints) placed in the lower part of the axle bearing and a set of two traction rods with rubber-metal couplings (joints) placed in the upper part of the axle bearing box. All such elements are attached to the bogie frame and allow for the control of lateral and longitudinal movements;
- two groups of two coil springs with a rubber-metal pad, placed on each side of the axle bearing's vertical axis;
- one vertical damper placed between the axle bearing box and the bogie frame, protecting the bogie against running instability.

The secondary suspension is made up of:

- carbody cross bar;
- two sets of coil springs;
- two vertical dampers;
- two lateral dampers;
- two yaw dampers;
- two stabilizers;
- two lateral elastic bumpers (one on each side of the bogie);
- one set of rotation bumpers (rollers).

Two sets of coil springs on each side of the bogie ensure vertical and lateral elasticity. Vertical compression movement is limited using bumpers placed between two secondary suspension springs. A set of two lateral stabilizer rods on each bogie limits lateral rolling (swaying) of the carbody in case of the track cant insufficiency.

The carbody is linked with the bogie via the kingpin placed in the carbody and the Watt's linkage located on the side of the bogie.

9.4. Vehicle dynamics tests

One of the important elements that affect the ability to increase the driving speed is the dynamic interaction between the rolling stock and the track. Tests of the vehicle-track interaction, the description of the planned works, and the methodology applied to conduct the experiment are the subject of this section.

PKP Intercity's ED250 series electric multiple unit with the maximum operating speed of 250 km/h was subjected to the tests.

In order to conduct the tests properly such topics as vehicle design, track condition and test track segment selection, risks related to conducting the experiment, and the appropriate selection of the parameters to be measured were reviewed.

9.4.1. Experiment planning

In order to plan the experiment properly and select the appropriate test track segment, topics related to both the geometry, surface, subgrade of the track as well as vehicle traction and power supply system capabilities were reviewed. Additionally, due to conducting, for the first time in Poland, the rolling stock tests at speeds above 250 km/h, a risk analysis of threats that may occur during the tests was completed.

The basic assumptions for selecting the test track segment were the requirements of the PN-EN 14363 standard [4], defining the minimum length of the track segment to conduct the tests. In accordance with this standard the minimum length of the total of all track sections, both straight as well as with curves with the radius of R > 600 m should be 10 km. This meant that in case of a single run in order to meet the standard's requirements the length of the test track segment should be more than 20 km, excluding the segments needed to accelerate and brake the trainset tested. Inability to meet the test track segment's length requirement during a single run is envisaged in the standard, which allows for the multiple use of the same track sections on curves if cant insufficiency between two test runs changes by more than 7.5 mm (in case of ED250).

The subject of the review was a segment of the Central Rail Line (CMK) running from Olszamowice station to Góra Włodowska station, maintained by the Kielce Railway Line Unit (Zakład Linii Kolejowych w Kielcach), where successful rolling stock tests were conducted at speeds of up to 200 km/h. Figure 9.5 presents track geometric deformations in the form of the lateral rail irregularities for two different wavebands: 3–25 m and 25–70 m. It should be noted that due to the measurement base of the EM120 track geometry, recording car the track irregularities for the 25–70 m waveband are subject to large measurement uncertainty. The track's geometric parameters presented are the parameters measured at the beginning of the experiment planning process. Based on that, a detailed analysis of the scope of track works in order to ensure compliance with the track's technical standards for the speed of up to 200 km/h was conducted. In terms of the required scope of track works, the Olszamowice – Włoszczowa Północ route called for the least amount of effort. However, a number of other factors were also taken into account, such as route profile, in terms of both availability of curves as well as route gradient. Therefore, the Psary – Góra Włodowska route was also included in the analysis.

Analysis of the condition of the CMK line's infrastructure on the Olszamowice – Psary – Góra Włodowska route was also supplemented with the traction calculations that were to give an answer to the question of whether the requirements related to the minimum length of the test track segments could be met.

The calculations were performed with the assumption that that on a straight and level track, for a train with a normal load and 100% of traction power available:

- average acceleration from 0 to 40 km/h is 0.49 m/s^2;
- average acceleration from 0 to 120 km/h is 0.42 m/s^2;
- average acceleration from 0 to 160 km/h is 0.36 m/s^2;
- residual acceleration at the speed of 250 km/h is 0.07 m/s^2.

The results of selected traction calculations are presented in Figures 9.7 and 9.8. Additionally, track parameters such as curve locations, curve cants, curve radii, and the values of non-compensated lateral acceleration on the rail head plane are presented in these figures. The first figure presents an analysis of the situation in which the trainset tested travels from Psary station to Góra Włodowska station assuming that the trainset passes through Psary station at the speed of 160 km/h. In this direction of travel, the trainset reaches the required speed of 275 km/h on kilometer 192,500 which limits the test track segment down to 10 km. Also the relatively large number of curves that the trainset negotiates with the value of non-compensated lateral acceleration of 1.2 m/s^2 should be noted.

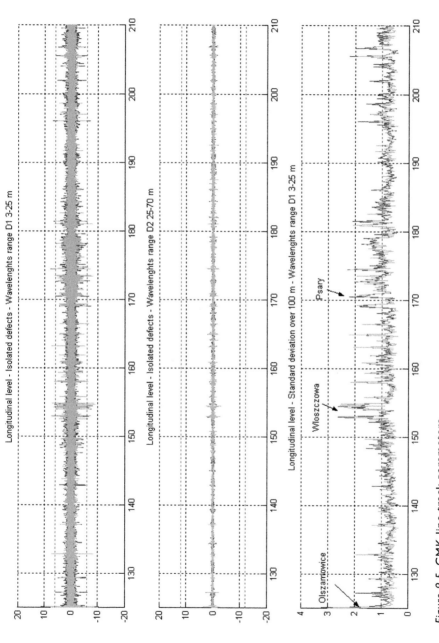

Figure 9.5 CMK line track parameters

Source: Own compilation based on the data from the PKP PLK track geometry recording car Psary – Góra Włodowska route gradient presented in Figure 9.6. It should be noted that the Góra Włodowska – Psary route gradient should help reduce the distance needed to achieve the required test speed.

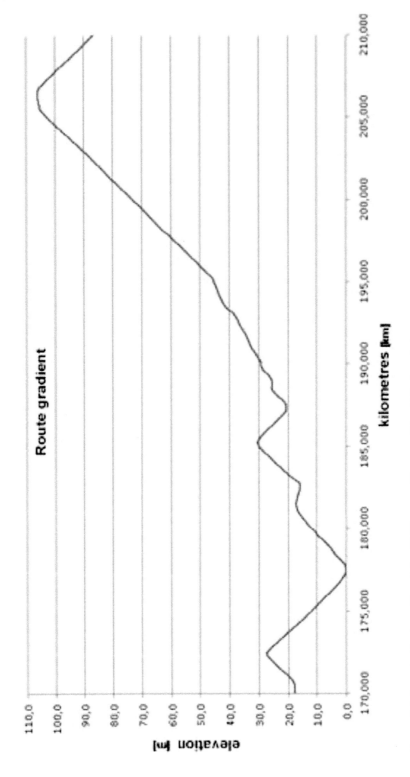

Figure 9.6 Longitudinal profile of the CMK route (Psary – Góra Włodowska)

Source: Own compilation based on the PKP PLK data

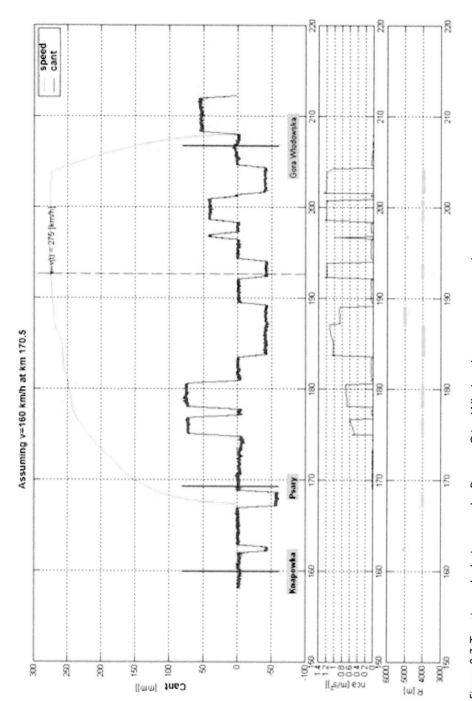

Figure 9.7 Traction calculations on the Psary – Góra Włodowska segment – case I

Source: Own compilation

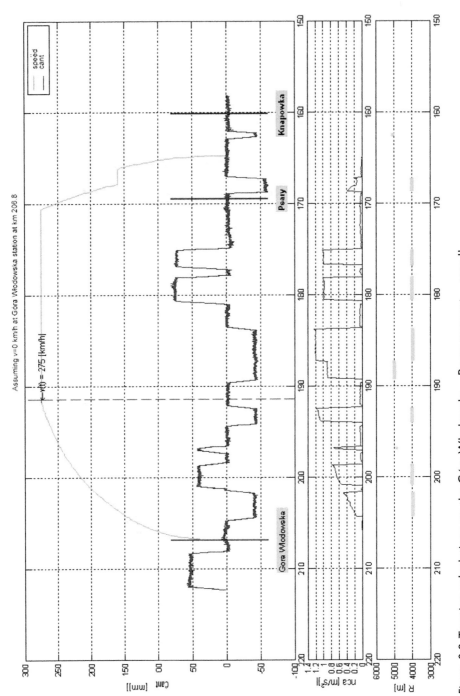

Figure 9.8 Traction calculations on the Góra Włodowska – Psary segment – case II

Source: Own compilation

In case of the opposite direction of driving, reaching the required speed of 275 km/h on kilometer 191,600 will allow for travel at the set speed on the test track segment that is approximately 20 km long.

The calculations and analyses presented allowed for the final choice of the test track segment which was a segment of the Central Rail Line from Góra Włodowska station to Psary station. And on this track segment, PKP PLK completed the required track works that would enable reaching the assumed speeds.

Risk estimate was performed for the selected test track segment and remedies adapted to the local conditions were proposed. Risk estimate related to the test runs, in contrast to the normal operation, was of approximate nature due to the lack of experience with the operation of the trainset tested. The calculations were performed solely based on four areas of threats coming from:

- infrastructure manager;
- carrier that will be operating the trainset tested;
- rolling stock manufacturer;
- testing entity.

Threats identified on the carrier's side included categories "rolling stock," "human factor," and the technical factor. They included:

- lack of correctly translated user manuals for the traction and conductor crews;
- possibility that the traction crew wrongly reads indications of the semaphore (block) signals;
- possibility of the traction crew's errors when driving a new type of train;
- third parties on the track or too close to the passing train – railway enthusiasts (trainspotters);
- death or injury of the traction crew members in case of the trainset's derailment;
- train's technical malfunctions.

Threats identified on the side of the entity performing the test runs included:

- fault of measurement sensors, wires, or other testing equipment;
- electric shock of the crew performing the tests;
- fire of the measurement installation, testing equipment;
- third-party involvement;
- panic;
- fainting;
- wrong choice of the test route and configuration of the trainset tested;
- test runs at night time;
- changes to the track's geometric parameters caused by operation of regular trains between test sessions;
- brake system's failure;
- exceeding of allowable rail stress limits or a threat of lateral track deformation;
- wrong interpretation of the test results.

FMEA method, allowing for risk estimation, was applied in all cases. To estimate risk using the FMEA method risk estimation, tables that take into account three elements – number of events, threat significance, and threat detectability – were applied.

Threat significance was defined using the formula:

$$RPN = W \times Z \times O \qquad (9.1)$$

where:
RPN – estimated risk (risk priority number),
W – number of events,
Z – threat significance,
O – threat detectability.

Each of the estimated risk components can assume integer numbers within the <1,10> range, and the estimated risk product (multiplication result) can assume integer numbers within the <1,1000> range.

Table 9.1 presents the risk matrix based on which the estimated risk level was determined. The following accepted risk level was assumed:

- RPN < 120 admissible risk;
- 120 < RPN < 150 – tolerable risk, but steps aimed at eliminating it should be taken;
- RPN > 150 – inadmissible risk.

Table 9.2 presents risk analysis for threats posed by the entity testing the train.

Table 9.1 Risk matrix

Unavoidable	10	100	200	300	400	500	600	700	800	900	1000	10
Very often	9	81	162	243	324	405	486	567	648	729	810	9
	8	64	128	192	256	320	384	448	512	576	640	8
Often	7	49	98	147	196	245	294	343	392	441	490	7
	6	36	72	108	144	180	216	252	288	324	360	6
On the average	5	25	50	75	100	125	150	175	200	225	250	5
	4	16	32	48	64	80	96	112	128	144	160	4
Rarely	3	9	18	27	36	45	54	63	72	81	90	3
	2	4	8	12	16	20	24	28	32	36	40	2
Practically never	1	1	2	3	4	5	6	7	8	9	10	1
Frequency	w/z	1	2	3	4	5	6	7	8	9	10	Z/O
Qualitative term	Effect	Minimal	Small		Average		Big		Very big		Disastrous	

Key	Admissible risk
	Tolerable risk
	Inadmissible risk

Source: Based on [1]

Table 9.2 Risk analysis

Group	Threat	W	Z	O	RPN	Accepted risk evaluation
Test measurements	fault of measurement sensors, wires, or other testing equipment	2	1	1	2	admissible risk
	electric shock of the crew performing the tests	1	4	5	20	admissible risk
	fire of measurement installation, testing equipment	1	6	2	12	admissible risk
	third party involvement	3	6	6	108	admissible risk
	panic	2	2	9	36	admissible risk
	fainting	2	2	8	32	admissible risk
	wrong choice of the test route and configuration of the trainset tested	2	8	4	64	admissible risk
	test runs at night time[*]	3	8	6	144	tolerable risk
	changes to the track's geometric parameters caused by operation of regular trains between test sessions	4	4	3	48	admissible risk
	brake system's failure[**]	3	7	6	126	tolerable risk
	exceeding of allowable rail stress or a threat of lateral track deformation[***]	3	6	7	126	tolerable risk

Source: Based on [1]

Proposed remedies:
 [*] performing runs at late night hours with local track closings;
 [**] preparation of a buffer zone to Knapówka;
 [***] preparation of the testing station on the curve with the largest cant deficiency.

9.4.2. Organization of the test runs

The test run venue was a regular railway network track no. 1 Psary – Góra Włodowska; however, the test runs, at speeds from 160 km/h to 250 km/h, were performed from Góra Włodowska station to Psary station. To ensure safety, the runs were performed on the track segment closed for regular traffic, with both tracks closed so as to eliminate aerodynamic interaction in case of two trains running in opposite directions on adjacent tracks. To ensure safety, if the brake systems were to fail, to perform the runs also track no. 1 from Psary station to Knapówka station, that formed the "buffer zone," was closed in case the trainset would not come to a stop in front of the Psary station's exit signals. The test runs involved multiple runs of the ED250 train on the above-mentioned route, starting at the speed of 160 km/h and then increasing successively to 275 km/h or higher. The train operating company contracted by the Railway Institute (Instytut Kolejnictwa) to perform the test runs was PKP Intercity S.A. All the test runs were coordinated by the Railway Institute.

The test runs preparation work was divided into three stages.

The first stage included installing the testing equipment to measure the catenary lift at km 180,200, constructing two testing stations on the track to measure excessive rail dynamic stress, rail dislocations, land vibrations caused by the passing trainset tested, and the noise generated.

The second stage included the track and catenary diagnostics prior to the commencement of the test runs.

Following the closing of track no. 1 on the Psary – Góra Włodowska route, the train dispatcher (rail traffic controller) at Psary station allowed the entry on the above-mentioned track of the track geometry recording car in order to carry out the track geometry measurements. Following the entry of the track geometry recording car to Góra Włodowska station, an IZ Kielce engineer or track surface and subgrade inspector conducted an analysis of the measurements completed and, in case of a positive result, provided the train dispatcher (rail traffic controller) at Góra Włodowska station with the written consent permitting driving on track no. 1 Psary – Góra Włodowska at the speed of up to 200 km/h.

Upon the arrival of the EM120 track geometry recording car to Góra Włodowska station, the train dispatcher (rail traffic controller) at Psary station allowed the entry on track no. 1 Psary – Góra Włodowska of EZ Kielce's diagnostics coach in order to test the catenary. Following the entry of the diagnostics coach to Góra Włodowska station, the maintenance train's diagnostics specialist notified Góra Włodowska station's ISDR in writing that there were no obstacles to perform test runs at the speed of 200 km/h.

Data obtained from the track and catenary measurements performed by the EM120 track geometry recording car and the diagnostics coach were provided to a representative of the Railway Institute present at Góra Włodowska station who, following the review thereof, would forward them by phone to the Railway Institute's coordinator of the test runs in order to make the decision on commencing the test runs.

The third stage included the test runs. During the measurements, a special procedure for test runs aimed at achieving the speed of 275 km/h was implemented.

The basic assumptions of this procedure can be characterized as follows:

- Driving at the maximum speed of 270–280 km/h (V_{dop} + 10% ± 5 km/h). Maximum non-compensated lateral acceleration on the rail head plane of approximately 1.2 m/s².
- Driving with the entire route (both tracks) closed. Test runs are performed in one direction, the trainset is not reversed;
- Test runs are performed at day time.
- Force components Q and Y at the wheel-rail contact point on two of the vehicle's bogies, lateral accelerations on the frames of other selected bogies, vertical and lateral accelerations on the vehicle's carbody, and the driving speed are recorded.
- Railway Institute's Power Unit (Zakład Elektroenergetyki IK) is monitoring the performance of the catenary and the pantograph/catenary interactions.
- Railway Institute's Railway Lines and Transportation Unit (Zakład Dróg Kolejowych i Przewozów IK) is monitoring the track surface and the vehicle/track interactions.
- Test runs start at the speed of 160 km/h (the highest normal speed on the PKP network).
- Following the completion of a test run, based on the processing and analysis of the recorded measurement signals, the commission makes a decision to increase speed.
- If the recorded and calculated values do not exceed 85% of the permitted value*, then in the next test run the speed is increased by 20 km/h (180 km/h, 200 km/h, etc.). If the recorded and calculated values exceed 85% of the permitted value, then in the next test run the speed is increased by 10 km/h (220 km/h, 230 km/h, etc.).
- If the recorded and calculated values do not exceed 95% of the permitted value**, then in the next test run the speed is increased by 10 km/h. If the recorded and calculated values exceed 95% of the permitted value, then in the next test run the speed is increased by 5 km/h.

- Steps defined in the previous item – are repeated until the maximum speed of 270–280 km/h is reached or the maximum measured and calculated parameters that are critical for the safety of driving are reached.
- In case the maximum speed of 275 km/h is not reached in the given test run series (on the given day), on the following day we commence with the control run at the speed of 200 km/h, and then at the highest speed at which the measurements were performed on the previous day.
- Before restoring regular trains operation the catenary is checked by PKP Energetyka and the evaluation of the condition thereof on the basis of the measurements performed using the track geometry recording car is made.

* calculated as 20 km/h: 160 km/h = 12.5%, i.e. maximum 87.5% of the permitted value, assuming linear increase of the measured parameters in the function of speed,

** calculated as 10 km/h: 250 km/h = 4%, i.e. maximum 96% of the permitted value, assuming linear increase of the measured parameters in the function of speed.

9.4.3. Test runs

The tests were performed on 16–17 and 23–24 November 2013 for a partially loaded vehicle (the first unit in the driving direction was loaded with steel weights with the total weight of 6600 kg, whereas the other units were empty). The diagram showing the vehicle load during the tests is presented in Figure 9.11.

For the needs of the dynamic tests, 43 signals were measured (16 forces, 24 accelerations, linear velocity and angular velocity around Z axis, and non-compensated lateral acceleration on the rail head plane) from the vehicle's side in total. Measurement sensors were installed in the following units:

- first unit in the driving direction – forces and accelerations;
- third unit in the driving direction – accelerations;
- fifth unit in the driving direction – forces and accelerations.

The forces are measured using the specially prepared by the Railway Institute measuring (instrumented) wheelsets that enable measuring vertical interaction forces Q and lateral interaction forces Y. The measurement of these forces was performed using the wheelset axle bending (section) moment measurement method along with the simultaneous online solving of the set of equations describing the balance of forces and torques (moments) interacting at the wheel-rail contact point. The example of the trailer measuring set used is presented in Figure 9.9. Four measuring sets were used in the tests (three trailer and one powered). The powered measuring set was able to measure the longitudinal force T arising from the tractive torque applied. The powered measuring set was installed as the first leading wheelset in the first bogie in the driving direction. A trailer measuring set was also installed on the same bogie. The other trailer sets were installed on unit 5's leading bogie. Accelerometers were installed on the vehicle's bogies and carbody to measure vibrations both in the vertical as well as in the lateral planes. An example of the ED250 vehicle's instrumented trailer bogie is presented in Figure 9.10.

Figure 9.9 Measuring set used to measure Q and Y forces

Source: Andrzej Zbieć

Figure 9.10 Instrumented measuring bogie

Source: Grzegorz Wysocki

Inside the passenger section

100 kg weight 6.6 t in total

During the tests, all the driving speeds envisaged according to the plan were achieved. Because of the vehicle's very good dynamic performance, the decision was made to carry out a test run at the maximum achievable speed. During the run, the speed of 293 km/h

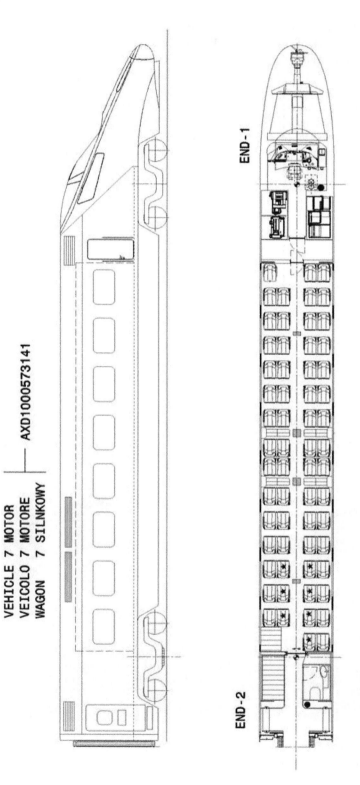

VEHICLE 7 MOTOR
VEICOLO 7 MOTORE ———— AXD1000573141
WAGON 7 SILNKOWY

END-1

END-2

Figure 9.11 Unit no. 7's load diagram
Source: Own compilation

was achieved between kilometer 177,000 and kilometer 176,820. The achieved speed of 293 km/h is the highest, record-breaking speed reached up to now on the Polish railway lines. It is worth emphasizing that this speed is also a world record for this type of a railway vehicle, since never before had any trainsets that would be similar to ED250 achieved such speed. The speed diagram in the function of the test time is shown in Figure 9.12.

The flat speed curve between the 150th and the 260th second at approximately 270 km/h was due to the restrictions imposed by the analysis of the rolling stock's dynamic interaction with the track observed during earlier test runs.

Figure 9.13 presents the indications of the speedometer installed additionally in the driver's cabin during the record breaking test run.

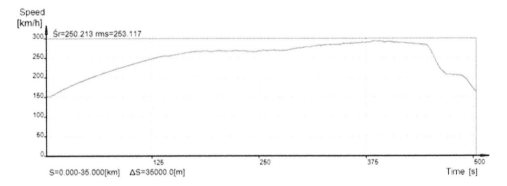

Figure 9.12 Vehicle speed during the fastest test run

Source: Own compilation

Figure 9.13 Speedometer

Source: Witold Groll

9.5. Operation of trains with the maximum speed of 200 km/h

Commercial operation of passenger trains with the maximum speed of 200 km/h was commenced in Poland in December 2014. In contrast to the test runs described in the previous sub-section, these are the trains included in the timetable and carrying fare paying passengers. The description of these trains' operational performance is provided below.

9.5.1. Speed characteristics

The maximum speed is a characteristic of the infrastructure related to the technical equipment of a railway line and its technical condition. It should be emphasized that maximum speeds usually vary along the length of the line; therefore, in order to characterize them it is necessary to adopt synthetic measures. Such measure could be the highest maximum speed on the entire line or on the train's route. It is very important since it defines the requirements for the rolling stock that would take full advantage of the railway infrastructure's capabilities.

In order to reflect the impact of the maximum speed differentiation on the travel duration, and as a consequence also on the line throughput, it is justified to apply the harmonic mean as the weighted average maximum speed capacity. It is calculated according to the formula:

$$V_{0\max} = \frac{\sum_{i=1}^{n} l_i}{\sum_{i=1}^{n} \dfrac{l_i}{V_{i\max}}} \qquad (9.2)$$

where:
l_i – length of the i segment of the route with constant speed,
$V_{i\max}$ – maximum speed on the i segment,
n – number of segments.

Physical interpretation of the formula is as follows: weighted average (harmonic) maximum speed represents the ratio of the total line (route) length to the total of the theoretical travel times on individual segments at constant speed.

Train speed is also an operational characteristic. Typical operational measures are speeds: technical v_t and commercial v_h. These speeds are calculated as follows:

$$v_t = \frac{l}{\sum t_j} \qquad (9.3)$$

$$v_h = \frac{l}{\sum t_j + \sum t_p} \qquad (9.4)$$

where:
l – route length,
t_j – travel times on individual segments,
t_p – stop durations.

Technical speed is based solely on the travel time, while commercial speed also takes into account stop (dwell) times. Apart from these speeds, average speed is also extremely important, calculated for a train on segments between successive stops (the so-called *start-to-stop*). Average segment speed has been the basis for classifying the fastest trains for years. World Speed Survey has been published bi-annually since 1975 by the Railway Gazette International. The most recent World Speed Survey appeared in July 2017 [6].

9.5.2. Speed utilization ratio

Based on the data of PKP Polskie Linie Kolejowe (Polish Railway Lines), the speed utilization ratio by the qualified trains servicing the connections between Poland's metropolitan areas that use the Central Rail Line was examined. The runs between successive stops of Ekspres Intercity Premium (EIP) category trains serviced by ED250 series electric multiple units were subjected to analysis. The analysis covered the data from the initial period after the 2015/2016 timetable had been put in force, from 14 December 2015 until 12 March 2016. For each of the segments under review, the following data was defined:

- segment length based on appendix 2.6 to the Network Statement of PKP PLK;
- weighted average maximum speed calculated as the harmonic mean on the basis of appendix 2.1 to the Network Statement of PKP PLK;
- travel time of the fastest train according to the timetable for the given period;
- fastest train's average speed (*start-to-stop segment speed*);
- weighted average maximum speed utilization ratio.

The analysis covered segments, partly or fully using the Central Rail Line, on which EIP category trains run. In total, five segments in both directions were reviewed, i.e. 10 runs of trains between successive commercial stops. The longest of such segments is 290.2 km long (Warszawa Zachodnia – Kraków Główny), while the shortest one is 69.7 km (Włoszczowa Północ – Zawiercie). The results of the analysis are provided in Table 9.3.

The highest weighted average maximum speed – 191.1 km/h in both directions – is in force on the Włoszczowa Północ – Zawiercie segment. It is a part of the CMK on which the

Table 9.3 EIP train start-to-stop runs on segments that use CMK (as of Dec. 2015)

From	To	distance [km]	Vo max [km/h]	Vmax [km/h]	t [min]	Vavg [km/h]	utilization ratio of Vo max
Warszawa Zach.	Kraków Główny	290.2	148.2	200	130.0	133.9	0.904
Kraków Główny	Warszawa Zach.	290.2	148.4	200	132.5	131.4	0.885
Warszawa Zach.	Włoszczowa Płn.	180.9	162.8	200	77.0	140.9	0.866
Włoszczowa Płn.	Warszawa Zach.	180.9	162.1	200	75.0	144.7	0.892
Włoszczowa Płn.	Zawiercie	69.7	191.1	200	26.0	160.8	0.841
Zawiercie	Włoszczowa Płn.	69.7	191.1	200	25.0	167.2	0.875
Włoszczowa Płn.	Sosnowiec Gł.	105.0	156.6	200	45.5	138.4	0.884
Sosnowiec Gł.	Włoszczowa Płn.	105.0	156.5	200	46.0	136.9	0.875
Warszawa Zach.	Częstochowa Str.	246.8	148.2	200	110.5	134.0	0.904
Częstochowa Str.	Warszawa Zach.	246.8	147.8	200	113.0	131.0	0.887

Table 9.4 Train runs on selected upgraded lines in Europe (2015/2016 timetable)

From	To	distance [km]	Vo max [km/h]	Vmax [km/h]	t [min]	Vavg [km/h]	utilization ratio of Vo max
Berlin-Spandau	Hamburg Hbf	276.3	210.0	230	90	184.2	0.877
Berlin-Suedkreuz	Luth. Wittenberg	91.3	192.9	200	33	166.0	0.860
Lutherstadt Wittenberg	Leipzig Hbf	69.6	168.2	200	29	144.0	0.856
Paris Austerlitz	Les Aubrais-Orleans	118.9	164.2	200	58	123.0	0.749
Paris Austerlitz	Vierzon Ville	201.1	176.0	200	84	143.7	0.816
Saint Petersburg	Moscow	649.3	190.3	230	220	177.1	0.931
Bologoje	Twer	163.6	195.9	230	57	172.2	0.879
Twer	Moscow	166.5	178.9	200	67	149.1	0.834

speed of 200 km/h generally applies (except for the approach to Zawiercie station). The highest average speeds of the fastest train (*start-to-stop* speeds) are also achieved on the same segment, reaching, depending on direction, 160.8 km/h and 167.2 km/h.

A very high weighted average maximum speed utilization ratio should also be emphasized, reaching even 0.9, which demonstrates a very smooth train ride.

It seems useful to compare the results achieved recently by the Polish railways on the CMK with the segment speeds (*start-to-stop*) on similar lines in other European countries.

The comparative analysis included segments on the lines upgraded to the speed of 200 km/h or higher, located in Germany, France, and Russia (Table 9.4).

It can be stated that on upgraded lines it is possible to achieve average segment (*start-to-stop*) speeds that represent 80–90% of the weighted average maximum speed. The best results are achieved on long line segments on which maximum speed fluctuations are relatively low. Classic examples of such lines are the Berlin – Hamburg route in Germany and the Saint Petersburg – Moscow railway line in Russia.

9.6. Conclusion

In December 2014, for the first time in Poland, commercial operation of trains at the speed of 200 km/h was launched on the CMK. It was possible as a result of the introduction of the new rolling stock equipped with the ETCS on-board safety system, the upgrade of the CMK with respect to the track's geometric parameters, catenary, power supply system, and equipping of the line with the ETCS system level 1. The goal of the tests performed by the Railway Institute (Instytut Kolejnictwa) was to confirm that all the subsystems met the high-speed rail requirements in terms of safety, dynamic interaction with the track and catenary, passenger comfort, and power supply. The tests have proven that the appropriate track quality and the right relationship between rigidity and vibrations generated during the train's run are maintained. Good geometric quality of the track surface with respect to horizontal and vertical irregularities, and also the vehicle's very good dynamic parameters, were confirmed. The interaction between the rolling stock and the catenary as well as functioning of the ETCS safety system met the applicable requirements. No vehicle running instability was detected, even at speeds exceeding the design speed. Test runs performed at speeds of up to 293 km/h were a part of the homologation procedure of the new vehicle type and its compatibility with the Polish railway infrastructure tests.

The experience from the first year of the ED250 series trains regular operation indicates a very high ratio of utilizing the maximum speed permitted on the line.

Bibliography

[1] Kukulski, J.: Identyfikacja i ocena ryzyka przy jazdach próbnych składu zespołowego ED250 w testach dynamicznych na linii CMK z prędkościami powyżej 200 km/h [Identification and evaluation of risk during test runs of the ED250 trainset in dynamic tests on the Central Rail Line (CMK) at the speeds above 200 km/h]. Compilation of the Railway Research Institute no. 2743/21/ LW/66.01/13.

[2] Kukulski, J., and Groll, W.: *Nowoczesny* Tabor do przewozów aglomeracyjnych [Modern rolling stock for metropolitan area transportation]. *Problemy Kolejnictwa* [Railway problems], 148, pp. 82–118.

[3] Massel, A.: Koleje dużych prędkości w Chinach – wybrane zagadnienia techniczne i eksploatacyjne [High-speed railways in china – Selected technical and operational topics]. *TTS* no. 11/12/2010, pp. 55–62.

[4] PN-EN 14363:2007 – Tests of rail vehicles' dynamic properties before admission for operation. Tests of driving properties and stationary tests.

[5] Wolfram, T.: Tabor trakcyjny dla pociągów dużej prędkości (wzorce dla PKP) [Rolling stock for high-speed trains (reference solutions for PKP)]. *Problemy Kolejnictwa* [Railway problems], 151, pp. 5–18.

[6] World Speed Survey. Railway Gazette International. July 2017, pp. 28–31.

[7] Wyszyński, R., and Malepszak, P.: Koleje dużych prędkości – najnowsze zagadnienia [High-speed railways – latest topics]. *Rynek Kolejowy* [Railways market], nos. 7–8/2009, pp. 74–78.

[8] EN 13749 Railway Applications – Wheelsets and Bogies – Method of Specifying the Structural Requirements of Bogie Frames.

[9] Directive 2008/57/EC of the European Parliament and of the Council of 17 June 2008 on the interoperability of the rail system within the Community.

[10] Sergiejczyk, M. (ed) – Koleje dużych prędkości w Polsce [High-Speed Rail in Poland], *Instytut Kolejnictwa* [Railway research Institute], Warszawa, 2015, pp. 219–221.

Chapter 10

High-speed rail power supply systems

Adam Szeląg and Tadeusz Maciołek

10.1. Power supply systems for electrified railway lines

10.1.1. Contact line supply for rail vehicles

An electric railway system (ERS), in which transport demand (TD) requires a supply of sufficient quantities of electric energy (EE) to obtain suitable transport output (TO), is shown in Figure 10.1. The system comprises following subsystems: vehicles (ETV) operating based on a specific timetable (TT), traction power supply system (TPSS), to which electric energy is supplied from a power supply network (PSN).

Specific conditions of energy supply to vehicles (ETV) apply in the ERS system:

- high variability of loads, depending on the vehicle's operating conditions, number of vehicles, and their weight, output, and operating speed between stations;
- effect of contact line voltage to energy intake capacity of a vehicle and achieved traction force, acceleration, and speed;
- variability of power supply source (traction substation – TS) parameters vs. the distance between a vehicle and a substation, and power intake of a vehicle;
- non-linearity of characteristics;
- differences in electromechanical (seconds) and electromagnetic (milliseconds) time-constants;
- comprehensive control functions, especially an ETV – with converters and AC motors;
- difficulty in receiving detailed data concerning equipment operation under the conditions of voltage fluctuations in a catenary.

All these factors affect simplified assumptions taken, methods applied, and results obtained in a number of analyses referring to contact system voltage as a vital quality parameter of energy supplied to the ETV. Parametric or variant analyses are often required.

The subsystem of an electric power transmission grid supplying electric energy used by the ETV comprises a contact system, a return circuit, and supply points (substations). Electric power supply contact systems: AC – 25 kV or 50 kV, 50 Hz; 15 kV, 16 2/3 Hz AC and DC: 3 kV and 1.5 kV DC, used worldwide basically do not differ in terms of a circuit structure. In each contact system, electric energy is taken from the source system (electric power grid), then it is converted to transmission voltage, and then converted to the type (DC or AC at specific frequency) and voltage level suitable to supply the ETV within a given contact system. The ETVs are mobile energy consumers supplied from the contact system – return system

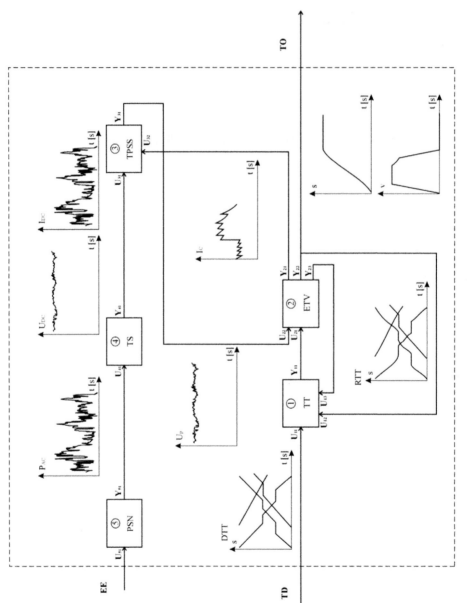

Figure 10.1 Functional diagram of an ERS system after decomposition into subsystems and presentation of exemplary time runs (time axis scaled in seconds) of input and output values

Source: Author's own study [33]

circuit. In railway applications, a typical contact system is a multiple wire overhead system, whereas the return line is a railway track; Polish railways uses a 3 kV DC power supply system in such configuration.

The sources of electric energy for electric vehicles are AC public electric power systems with power plants and high-voltage (HV) and highest-voltage (THV) transmission grids, system switching stations and high-voltage and medium-voltage distribution stations. Electric power systems, also referred to as public grids, are operated at industry-grade frequency (50 Hz or 60 Hz). PSN, connected to power systems, supply both DC (equipped with rectifier units) and AC traction substations, operated at industry-grade frequency (50 Hz). AC traction substations with reduced frequency of 16 2/3 (16.7) or 25 Hz are supplied from public grids (via conversion substations) or from separate railway power lines and power plants generating electric energy with reduced voltage frequency. Public power system (PSN) and TPSS system links are shown in Figure 10.1, while TPSS schemes are described in subsections: 10.1.2–3 kV DC power supply, 10.1.3–15 kV 16 2/3 Hz AC, and 10.1.4–25 kV 50 Hz AC.

Power supply system – a configuration of a power supply circuit with energy sources (traction substations), contact system circuit infrastructure (catenary sections, electrical connections of the contact system), commutation system (high-speed breakers, power switches, load switches and dis-connecting switches, isolating gaps of the contact system) – constitutes the practical application of an electric vehicle power supply system.

Commonly used structural solutions of the TPSS (Figure 10.2, Table 10.1) include:

- one-sided power supply (Figure 10.2a),
- two-sided power supply (Figure 10.2b),
- one-sided power supply systems with sectioning cabins or transversal connections of systems (25 kV 50 Hz AC system),
- two-sided power supply systems with sectioning cabins or transversal connections of systems (systems: 15 kV 16 2/3 Hz AC, 3 kV DC).

A rule of thumb in the TPSS contact line system is the reciprocal transversal isolation of contact line systems of individual tracks on lines with two or more tracks.

A contact system divided into supply sections is connected to the contact system substation (contact system switching station) via cable or overhead lines. Each such line (contact power feeders) is fitted with a metering circuit, overvoltage protection, disconnecting switches, and a circuit-breaker. In DC systems, these are current-limiting air-gap current breakers (CBs), called "high-speed breakers" (HSB) due to their quick operation, and in AC systems one-phase oil, gas, or vacuum circuit breakers. Both the traction voltage switchgears in the substation and connections of feeders to the contact line system are fitted with operating disconnectors. CBs and power breakers are used to disconnect short-circuit currents and overload currents.

CBs of substation and cabin feeders are usually coupled via control and automation circuits in order to allow breaking of short-circuit currents and switch-off voltage in a contact system under emergency conditions in two-sided power supply systems.

Approximately 25% of all railway lines in the world are electrified, with various AC systems (Table 10.2) and DC systems (Table 10.3) used. Development of 750 V, 1.5 kV and 3 kV DC systems and 15 kV 16 2/3 Hz AC systems started during the period of intense electrification before the World War II. The 50(60) Hz 25 kV system has been introduced since 1950s and length of railway lines with this system is increasing year by year, specifically in Asia.

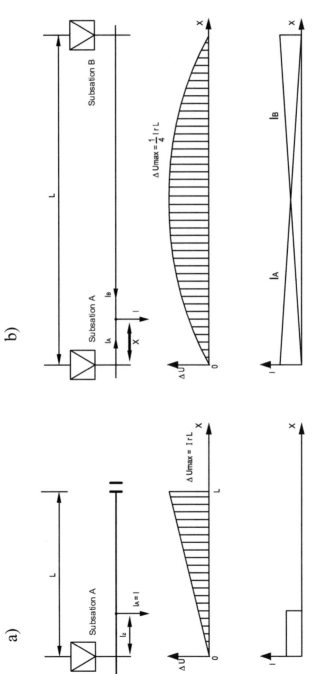

Figure 10.2 Distribution of loads and voltage ∆U drops in the power supply system at one large input current I, at one-sided (a) and two-sided (b) power supply

Source: Author's own study [31]

Table 10.1 Typical parameters of power supply systems

Item	Traction system	Length of section supplied from one substation [km]	Power supply
1	0.6 kV DC	0.5–2.0	one-sided
2	0.75 kV DC	1.0–4.0	two-sided
3	1.5 kV DC	10–15	two-sided, two-sided with cabin
4	3.0 kV DC	12–30	two-sided, two-sided with cabin
5	15 kV 16 2/3 Hz AC	50–70	one-sided/two-sided
6	25 kV 50 Hz AC	50–90	one-sided
7	2×25 kV 50 Hz AC	60–120	one-sided

Source: Author's own study

Table 10.2 AC power supply systems in use

Voltage [kV]	Frequency [Hz]	Country
6.5	25	Austria
11		USA
20	50	Japan
25*	50	Bosnia, Bulgaria, China, CIS, Croatia, Czech Republic,
50* kV	50	Denmark, Finland, France, Greece, , Holland, Hungary, India, Italy Japan, Luxembourg, Marocco, New Zealand, Pakistan, Portugal, Romania, South Africa, Slovakia, Spain, Turkey, Serbia, South Africa, United Kingdom Zaire, Zimbabwe,
20	60	Japan
25*		Japan, South Korea
50*		USA, Canada
11	16.7 (16 2/3)	Switzerland
15		Austria, Germany, Norway, Sweden, Switzerland

* systems considered as suitable for newly constructed lines

Own study based on multiple sources, including web materials and [34]

Table 10.3 DC power supply systems in use

Voltage [kV]	Transport system type	Country
3	Railways	Algeria, Belgium, Brazil, Chile, CIS, Czech Republic, Italy, Luxembourg, Morocco, Poland, Slovakia, South Africa, Slovenia, Spain, USA,
1.5	Railways	Australia, Czech Republic, Denmark, France, United Kingdom, Netherlands, India, Japan, New Zealand, Portugal, Spain, USA
0.63–0.75–1.2 1.5*	Commuter railways	UK, Spain, Ireland
0.75–0.9*	Metro, commuter railways	
0.6	Tramways, trolleybuses	
< 0.6	Mining and industrial railways	

* systems considered as suitable for newly constructed lines

Source: Own study based on multiple sources, including web materials and [34]

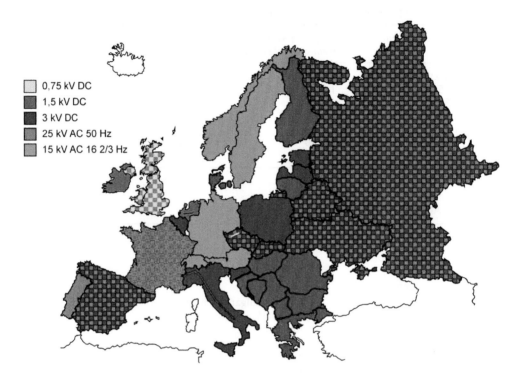

Figure 10.3 Diversification of railway powers supply systems in Europe

Source: Author's compilation based on various sources

In Europe, railway power supply systems are highly diversified (Figure 10.3), which makes difficulties for interoperability and resulted in introduction of Technical Specifications of Interoperability (TSIs) for the Energy subsystem [25].

When selecting and designing contact system power supply, the basic criterion is to ensure the supply of energy of adequate quality, suitable for electric vehicles with specific output:

- with sufficient available power;
- with a required voltage value (useful voltage);
- free from prohibited negative impact to a contact system and line infrastructure;
- highly reliable.

In addition, engineering and financial aspects are considered (number of substations, cost of their construction, costs of a contact system and electric power lines connected to electric power system) and local aspects (suitable connections to a power grid, existing railway power supply system).

Currently, the systems recommended for use on newly electrified lines include the following: 25 kV or 50 kV 50/60 Hz and 15 kV 16 2/3 Hz (in countries where this system is in use). This refers in particular to lines designed for operating speeds that exceed 250 km/h, due to the requirement to supply trains with high power (above 10 MW) [22], [25] (see table 10.4).

In non-conventional high-speed rail systems, which implement a magnetic levitation (maglev) principle, the issue of high vehicle power intake was solved by using a linear induction motor, allowing for supply of a track (where the stator of an induction motor is set), instead of a vehicle, thus eliminating an overhead contact system. Japanese maglev set a new speed record on 21 April 2015, travelling at the speed of 603 km/h for 11 s, at the 1.8 km-long section.

10.1.2. DC power supply system

DC power supply system has been used since the dawn of an electric traction contact system (Figure 10.4). The variant 3 kV DC adopted in Italy or Poland, allows train operation in the range from 220 km/h to 250 km/h, power output of 6–8 MW [4,12,14]. Larger loads (vehicle output and traffic density as in Italy), however, cause significant voltage drops and large load currents, which require an increase in the cross-section of a contact line conductor (up to over 600 mm² copper), smaller distances between substations (as little as 10–12 km), higher installed power in substations (above 10 MW). This forces an increase in investment outlays and is costly in operation. In addition, increased load and generation of harmonics result in the necessity to move the point of common coupling of a substation to the electric power grid at the high voltage level. So a single-stage transformation (as 110/1.3/1.3 kV in Poland) is applied to feed the rectifier set in a traction substation. Attempts were taken to increase

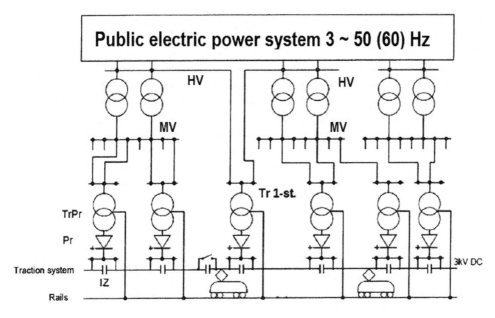

Figure 10.4 Example solutions for railway power supply system with 3 kV DC electric contact system

Source: Author's own study [33]

voltage to 6–18 kV DC in a contact line; the major issue was of breaking the fault currents, which caused resignation of practical usability of such voltages.

Legend to Figure 10.4
HV – high voltage;
MV – medium voltage;
TrPr – transformer of rectifier unit;
Pr – rectifier,
Tr 1st – power supply with single-stage HV transformation.

DC contact system for the HSR causes a number of issues, both concerned infrastructure (large cross-section of catenary up to 620 mm² Cu, low distances between traction substations) and operational (high energy losses, high costs). The good example of the HSR line, with an operating speed 250 km/h, supplied at 3 kV DC makes 315 km long Direttissima line between Rome and Florence in Italy. Experience with this line caused that 3 kV DC system was analyzed as a preliminary variant for Rome – Naples HSR line project for operating speed of 300 km/h [4]. Outputs of rectifier units designed for installation on this line reached 2×10 MVA with 12-pulse rectifier units and a redundant one, LC type filters and HSBs with breaking capacity above 50 kA. Two-sided power supply from substations constructed every 12 km without sectioning cabins was assumed. AC feeding with a 132 kV 50 Hz line was supposed to ensure a high level of reliability in the n-1 arrangement of supplying DC traction substations (excluding one substation from operation allows maintaining the assumed volume of traffic), simultaneously eliminating negative effect of a traction substation harmonic currents on a power grid. However finally 2×25 kV 50 Hz system was selected for electrification of the line Rome-Naples, which has been operating since 2005.

10.1.3. 15 kV 16 2/3 Hz AC power supply system

Such a system (Figure 10.5) requires voltage to be generated with a reduced frequency of 16 2/3 Hz (in some systems currently 16.7 Hz) in power plants, or by means of single-phase electromechanical or static converters. Substations are supplied directly from a dedicated centralized system of 110 kV one-phase lines with reduced frequency or individually (via converters), from a public three-phase system. For the electric power grid, a contact system load is therefore a 3-phase symmetric consumption, and disturbances may be caused by starting rotating machinery with outputs in the order of 10 MVA (old solutions), or by harmonics introduced by the static converters 3~ 50Hz/1~16.7 Hz with outputs in order of several dozens of MVA (modern solutions).

Thanks to reduced frequency, the fault impedance of a contact system is lower than in the case of 50 Hz frequency, which reduces voltage drops and improves breaking conditions of minimum short-circuit currents. In addition, as compared with the 25 kV 50 Hz system, there is a possibility of a longitudinal connection of individual contact system sections with various substations in the event of interrupted operation of one substation. This allows power supply in emergency conditions. Power supply redundancy for adjacent line sections, which is necessary with 25 kV 50 Hz systems, can be eliminated. This is because transformers of 25 kV system substations are connected to various voltage sources and to various phases. HSR lines supplied with voltage at the frequency of 16 2/3 Hz are mostly the ICE lines in Germany, but also lines in Switzerland and Sweden. Contact systems in AC supply systems

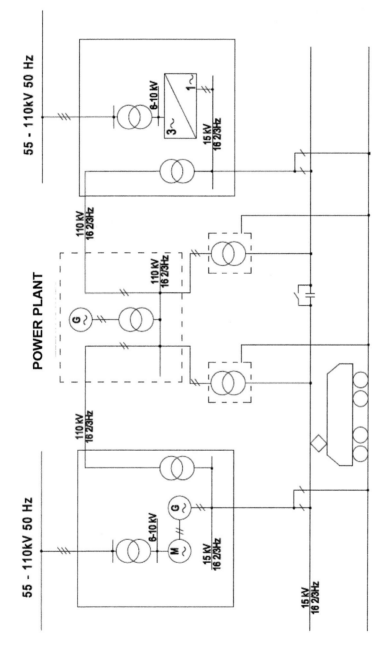

Figure 10.5 Diagram of electric power supply in the system 15 kV 16 2/3 Hz AC

Source: Author's own study based on [8]

are much lighter (smaller cross-section of wires) than in DC systems due to smaller values of currents.

10.1.4. 50/60 Hz AC power supply system

Currently, various solutions of AC systems with system frequency of 50 Hz or 60 Hz (Figure 10.6) are being used. In order to identify these different solutions, symbols are used as for a 1×25 kV 1 AC 25 kV power supply system. The 2×25 kV system may have the same symbol –2 AC 50/25 kV 50 Hz, 2 AC 43.3/25 kV 50 Hz or 2 AC 35.4/25 kV 50 Hz, which is due to an angle shift (180°, 120°, or 90°) between phases of the secondary side of main transformers used (acc. to [8],[23], [37]).

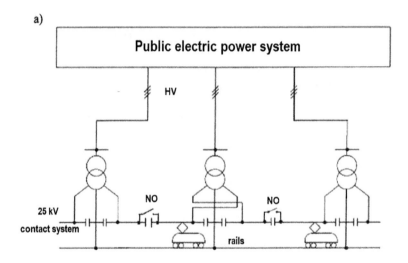

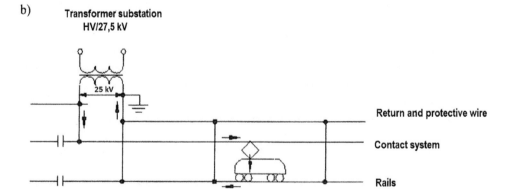

Figure 10.6 50 Hz 25 kV power supply system: (a) general diagram, (b) feeding using dedicated transformers, (c) variant with booster transformers (BT), and (d) 2×25 kV system with autotransformers AT; NO – normally open

Source: Author's own study based on [4],[8],[15], [37]

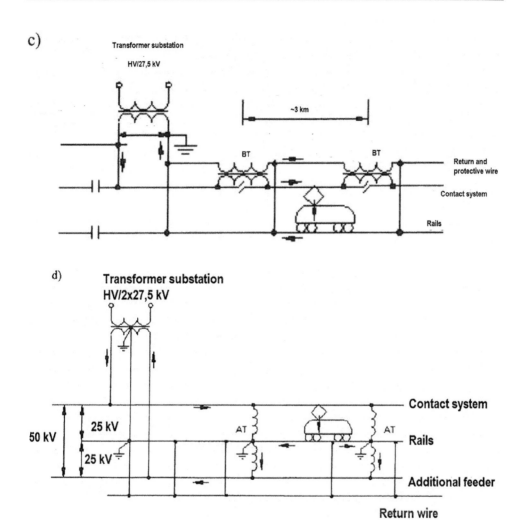

Figure 10.6 (Continued)

The 25 kV power supply system of a contact line is a power supply system with con-nected to rails and earthed one terminal of a transformer in a substation while the second terminal is connected to a contact line (Figure 10.6a). One rail of each track is used as a return rail for traction currents with an additional return and protection wire on a supporting structures, comprising an traction earthing system in order to reduce voltage drops in rails and equalize the potential in a catenary impact zone.

The simplest diagram of power supply in this system is shown in Figure 10.6b, but due to the flow of significant current in earth and interferences in telecommunication circuits, it is practically not used. The system is used in a modified variant, with booster transformers BT connected every few kilometers between a track system and additional return wire (Fig-ure 10.6c). Their function is forcing current flow from running rails to the return wire. The latest solution is the so-called 2×25 kV system (Figure 10.6d). In this system, there is a transformer with two secondary windings, with the center earthed and connected via a return wire to rails. One terminal of one secondary winding is connected to a contact system-catenary, and the other one to an additional feeder. With such a connection, voltage between the contact system and rails is 25 kV, and at the transformer output the voltage is 50 kV. The contact system conductors and additional feeder are suspended on the same posts, comprising a 2-phase 50 kV system (Figure 10.7). At the section between consecutive substations, there are autotransformers installed every 7÷25 km, connected between a contact system and an additional feeder, with centers led out and connected to running rails. In the course of a vehicle moving alongside the section, autotransformers, one by one, take over the task of supplying a train, and the current taken by the train from the contact system flows in rails only in the section between adjacent autotransformers. This reduces voltage drops and extends supply sections, with substations located in 2–2.5 times larger distances than in the 1×25 kV system, ensuring identical operating conditions. This system is also an alternative to the 50 kV system used in some countries, but it has to be supplied from a HV line 220 kV or higher. However, it allows for construction of railway lines in areas where poor development of HV lines forces high distances between substations. As compared with 1×25 kV, the 2×25 kV system reduces earth leak of return current and interferences with low-current circuits. System's higher complexity is its weakness. One needs additional equipment to provide adequate protection. Differences in equipment used with the 2×25 kV system, as compared with the 1×25 kV system, include mostly the types of transformers at HV substations (2×25 kV) and, additionally, autotransformers on the line (50/25 kV), significantly increasing project costs. Transformers supply voltage depends on their output, e.g. for a TGV line, using 60 MVA oil transformers supply voltage is 225 kV or 400 kV (similar values adopted in the *Studium* . . . [27]). Autotransformers have rated power from 5 MVA on conventional lines up to 10 MVA on the TGV lines and 15 MVA on the Rome – Naples line.

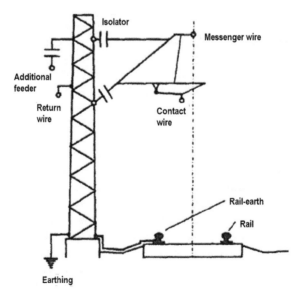

Figure 10.7 Layout of contact system wires in the 2x25 kV system with additional feeder (typical in the 2×25 kV system)

Source: Author's own study based on [2],[4],[8],[37]

Substations of the system 1×25 kV, located every 40–60 km (in the 2×25 kV system even every 80–120 km), usually comprise two 25 kV feeders, each of double and independent type, capable of transferring overall contact system load under normal operating conditions all by themselves, ensuring high reliability (100% redundancy). Each feeder has its own circuit breaker on the 25 kV side. 110/25 kV transformers are supplied from two independent sources of a HV grid.

Complex protection systems [8],[37] are used in AC systems, whose task is to ensure the safety of people and reduce risks due to fault or overload.

From the viewpoint of impact to electric power grid, the 25 kV 50 Hz power supply system makes an asymmetrical load. In order to minimize such negative influence, 110 kV supply voltage or higher is required to have sufficient short-circuit capacity. Adjacent substations are usually connected to different phases, so that the load in a public electric power system system may be more or less balanced equally among three phases (Figure 10.6a). Another solution requires using special transfers at substations (e.g. Scott, Woodrige, Le Blanc type), reducing asymmetry fed to the electric power system, or, when necessary, using symmetrizers. In some solutions of modern 25 kV 50 Hz substations static frequency converters (SFC) 3~ 50Hz/1~50 Hz, assuring symmetrical load of public power grids are applied.

In order to calculate the level of asymmetry determination of the vector of negative sequence voltage and limits of this value is used. One of the formulas below can be used to calculate the value of an asymmetry coefficient [1], [2], [4], [17],[37]:

$$nps = \frac{U\max - U_{\acute{s}r}}{U_{\acute{s}r}} *100\% \qquad (10.1)$$

$$[\%]$$

or

$$nps = \frac{U\max\% - U\min\%}{1,7} \qquad (10.2)$$

$$[\%]$$

where: $U_{max\%}$, $U_{min\%}$ – maximum and minimum RMS value of phase voltage, expressed in % rated value, U_{max}, $U_{\acute{s}r}$ – maximum and average value of phase voltage.

In case of contact system loads connected to two phases of the supply transformer, nps value can be estimated as

$$nps = \frac{S_{obc.tr.}}{kS_{zw3-faz}} *100\% \qquad (10.3)$$

$$[\%]$$

where:
$S_{obc.tr.}$ – output of contact system load connected to two phases [MVA]. Such load is calculated for 10 s or 1 min peak load conditions, at the highest impedance of the supply system; in practice 30 min peak load is often adopted.
$S_{zw3-faz}$ – 3-phase fault capacity at the feeding point of contact system substation, k = 1.2 – coefficient considering the possibility of exceeding the symmetry.

Table 10.4 Example HSR trains and power supply systems

Train	Country	Maximum speed [km/h]	Maximum operating speed [km/h]	Power supply system
TGV*	France	574.8 (2007)	320	AC 25 kV/50 Hz
Shinkansen	Japan (also operated in Taiwan)	405.0	320	AC 25 kV 50/60 Hz
ICE	Germany	406.9 (1988)	330	AC 15 kV, 16 2/3 Hz
Pendolino	Italy	293.0 (Poland, 2013)	250	3 kV DC
ETR500	Italy	319/360	300	3 kV DC /25 kV,50 Hz
KTX	South Korea	421.4 (2013)	305	25 kV 60 Hz
CRH3	China	380	350	25 kV 50 Hz

*TGV, France – the system 2×25 kV was first used on a 163 km section of TGV South – East line in 1981 to supply trains 6.6 MW, and TGV Atlantic line to supply trains 8.8 MW (sequence every 4 min)

Source: Own study based on multiple sources [34], including web materials

Due to the contact system substation introducing to the power grid both asymmetry and higher harmonics, both interferences should be considered jointly, as their impact to electric consumers and generators of power plants is cumulative. Other interferences from the same system are voltage fluctuations caused by traction substation load quick changes, impact to telecommunication lines and low-current circuits, and consumption of reactive power.

10.2. Requirements concerning HSR railway power supply systems

10.2.1. Requirements specified by Technical Specifications of Interoperability (TSI)

Commission Regulation (EU) No. 1301/2014 of 18 November 2014 on Technical Specifications for Interoperability for Energy subsystem [25] (defined by Annex II (2.2) to Directive 2008/57/EC) of the EU railway system is a document combining requirements from the previously implemented directives referring to TSIs for Energy subsystem for *conventional railways*, as specified pursuant to Annex I, Section 1.1 of Directive 2008/57/EC and for *the trans-European high-speed rail system network* (TEN), as described in Annex I, section 2.1 of Directive 2008/57/EC. Simultaneously, the previously mandatory TSIs for conventional railways, and separate for high-speed rail, were cancelled. Requirements of particular importance are listed in the standard [22].

Energy subsystem comprises contact system substations, sectioning cabins, phase separation sections, system separation sections, a contact system, and a return system. This section includes requirements referring to general parameters of the supply whose task is to provide sufficient energy of adequate quality to trains operating based on a preset timetable.

Operating parameters of the energy subsystem depend on:

- maximum operating speed of the line;
- type and output of trains;
- timetables.

In HSR lines this applies to:

- category I: dedicated HSR lines, operating speed of at least 250 km/h;
- category II: lines upgraded to operating speed of approximately 200 km/h;
- category III: upgraded HSR lines with individual speed limits due to, for example, geographical conditions and restrictions due to town planning.

In the infrastructure registry, an infrastructure manager declares the operating speed for a given line and a maximum current input of a train. The structure of the electric power system should guarantee suitable operating parameters.

Recommended (allowed) power supply systems for HSR lines in the above categories are listed in Table 10.5, and their typical parameters are listed in Tables 10.6 and 10.7.

General parameters of power supply include [22]:

- voltage and frequency (Table 10.8);
- efficiency (ability to transfer energy to all trains consuming specific power), defined by average utility voltage (Table 10.9) and maximum power consumption of train (Table 10.10) and train output coefficient [22.]. Trains with output below 2 MW should be subject to no restrictions in current/power intake;

Table 10.5 High-speed rail power supply systems

Speed ranges [km/h]	Categories of HSR acc. to TSI	
	I	II and III
$v \geq 300$	AC 25 kV 50 Hz, AC 15 kV 16.7 Hz	NA
$250 \leq v <300$	AC 25 kV 50 Hz, AC 15 kV 16.7 Hz, DC 3 kV*	NA
$200 \leq v <250$	NA	AC 25 kV 50 Hz, AC 15 kV 16.7 Hz, DC 3 kV, DC 1.5 kV

* DC 3 kV is not used with new lines

Source: [22]

Table 10.6 Characteristics of AC power supply systems for HSR lines

Type of line	Categories of lines, HSR	
	I	II and III
Installed power at substations [MW]	20–60	15–45
Single vehicle output [MW]	8–20	5–15
Length of section supplied by one substation [km] (with double-sided power supply this is half of the distance to the nearest substation)		
AC 25 kV 50 Hz	15–30	20–30
AC 2×25 kV 50 Hz	20–45	20–50
AC 15 kV 16.7 Hz	15–30	20–35
AC 2×15 kV 16.7 Hz	30–60	30–60

Source: [22]

Table 10.7 Characteristics of DC power supply systems for HSR lines

Type of line	Categories of lines, HSR	
	I	II and III
Installed power at substations [MW]		
3 kV DC	10–20	6–12
1.5 kV DC	–	8–15
Single vehicle output [MW]	–	7.5
3 kV DC	8–12	6
1.5 kV DC	–	7.5
Length of section supplied by one substation [km] (with double-sided power supply this is half of the distance to the nearest substation)		
3 kV DC	6–12	7.5–12.5
1.5 kV DC	–	2–8

Source: [22]

Table 10.8 Rated voltages and their allowed limits of traction voltage

Electrification system	U_{min2} [V]	U_{min1} [V]	U_n [V]	U_{max1} [V]	U_{max2} [V]
DC (average value)	400	400	600	720	770 (800)*
	500	500	750	900	950 (1000)*
	1000	1000	1500	1800	1950
	2000	2000	3000	3600	3900*
AC (root mean square value)	11,000	12,000	15,000	17,250	18,000
	17,500	19,000	25,000	27,500	29,000

* applicable to vehicles with recuperation braking system

Source: [18]

Table 10.9 Minimum value of mean useful voltage Umean_usefulon a current collector under normal operating conditions

Power supply system	Minimum useful voltage U_{mean_useful} [V] of a power supply zone and train	
	Line with speed v > 200 [km/h]	Line with speed v ≤ 200 [km/h]
25 kV 50 Hz	22,500	22,000
15 kV 16.7 Hz	14,200	13,500
3 kV DC	2800	2700
1500 V DC	1300	1300

Source: [22]

- coordination of protections (Table 10.11 and 10.12);
- ensuring stability of substation – vehicle operation, in particular with AC systems – reduction in overvoltage caused by transient and harmonics (Table 10.13); for details see standard [22].

Table 10.10 Maximum train current [A]

Power supply system	High-speed rail category I	High-speed rail category II	High-speed rail category III
25 kV 50 Hz	1500	600	500
15 kV 16.7 Hz	1700	1000	900
3 kV DC	4000	4000 (3200*)	4000 (2500*)
1500 V DC	–	5000	5000

* in Polish railway network

Source: [22],[26]

Table 10.11 Maximum short-circuit current contact line – rails

Power supply system	Two-sided power supply from a substation	Maximum short-circuit current [kA]
25 kV 50 Hz	NO	15
15 kV 16.7 Hz	YES	40
3 kV DC	YES	50
1500 V DC	YES	100

Source: [22]

Table 10.12 Current-limiting CB (HSB in DC systems) response upon short-circuit on-board of vehicle

Power supply system	CB tripping sequence required upon an internal fault within a vehicle	
	Traction substation CB – tripping	Vehicle's on-board CB – tripping
25 kV 50 Hz	immediate tripping (duration ca. 80 ms)	immediate tripping
15 kV 16.7 Hz	immediate tripping (duration ca. 100 ms)	Primary side of the transformer – staged trip Secondary side of the transformer – tripping immediate (ca. 100 ms)
3 kV DC	immediate tripping (duration 20–60 ms)	immediate tripping (duration 20–60 ms)
1500 V DC	immediate tripping (duration 20–60 ms)	immediate tripping (duration 20–60 ms)

Source: [22]

Note 1: Use of a CB tripping device on-board vehicles that responds within 3 s from the contact line system voltage decay and which reactivates the CB at least 3 s after restoring contact system voltage is required.

Note 2: Use reactivation automation systems with CBs supplying sections of a contact line system with (or without) line test (test of contact system insulation).

Note 3: In DC systems HSBs are applied as CBs and reduction of current derivate di/dt to values below 20 A/ms during 20 ms, until a di/dt value is 60 A/ms (in order to reduce the possibility of a substation HSB triggering upon switching on HSB on-board of a vehicle fitted with an input filter) at minimum inductance of a contact line system and substation choke of 2 mH.

Table 10.13 Allowable momentary unique peak voltage, by power supply system

Power supply system	Momentary peak voltage (unique) [kV]
25 kV 50 Hz	50.0
15 kV 16.7 Hz	30.0
3 kV DC	5.1
1500 V DC	2.6

Source: [18],[22],IEC-60364-4-443,

The design of power supply systems has to meet the requirements concerning allowable values and electric parameters, a minimum and maximum voltage value (long-term, Table 10.8, and short-term, Table 10.13) at the current collector of an electric vehicle, and values of average utility voltage on a current collector and in a contact system within the supply zone.

Permitted voltages (pursuant to TSI) on output terminals of contact system substations are listed in Table 10.8, with following symbols:

U_n – rated voltage;
U_{min1} – minimum continuous voltage;
U_{min2} – minimum transient voltage, duration up to 2 min;
U_{max1} – maximum continuous voltage;
U_{max2} – maximum voltage, duration up to 5 min.

Table 10.8 indicates that a permitted continuous value of off-load voltage of a contact system substation may not exceed 3.6 kV in the DC 3 kV system, 17.5 kV in the AC 15 kV 16 2/3 Hz system, and 27.5 kV in the AC 25 kV 50 Hz system. Pursuant to the *Operation and Power Transmission Grid Maintenance Manual*, voltage at common tapping points of the electric power grid in Poland, in the order of 110 kV and 220 kV, may fluctuate within ±10% of a rated value, while voltage of 400 kV within +10/−5% of a rated value. In order to maintain beneficial supply conditions, it is recommended that a contact system substation be supplied from an HV transformer station, in which transformers are fitted with on-load tap voltage control, or that rectifier transformers are retrofitted with such a control. Due to the nature of contact system consumption (supply category I), requiring the level of high reliability, individual elements of a power supply circuit and substation equipment are redundant. Only the contact system supplying power directly to a vehicle cannot be made redundant, which imposes special requirements with regard to its correct design and maintenance.

In addition, requirements are imposed on the design of power supply systems that ensure the possibility of using recuperative braking systems – providing smooth exchange of power with other trains, also by other means, in AC systems, and enabling at least exchange of power with other trains, in DC systems.

Voltage criteria used in connection with an electric contact system allow for the technical evaluation of the power supply system and may be presented synthetically, mostly as the value of $U_{meanuseful\,z}$ on a current collector and in a contact line system.

The standard [22] formulates the criterion referring to evaluation of power supply quality using the formula:

$$U_{usruz} = \frac{\sum_{I=1}^{n} \frac{1}{T_i} \int_0^{T_i} U(t)_{pi} I_{ci}(t) dt}{\sum_{I=1}^{n} \frac{1}{T_i} \int_0^{T_i} I(t)_{ci} dt} \qquad (10.4)$$

where:
$I_{ci}(t)$ – current of i-th *EPT*,
$U_{pi}(t)$ – voltage on current collector of i-th *EPT*,
T_i – duration of i-th train operation.

This criterion varies in the voltage calculation method from the previously used classical methods of determining the average value in that it uses an integer value after instantaneous power time divided by the average value of current, and obtaining such values is possible practically only when simulation techniques are used. Such an approach increases the weight of a voltage level during power consumption, which is justified by the effect of a voltage value in the system to contact system and operating parameters of vehicles [32], especially high-output vehicles. Calculated mean useful voltage U_{mean_useful} "at pantograph" (previously referred to as UIC voltage) should be not less than values indicated in Table 10.9.

Calculations of vehicles, substations, and contact system working currents (maximum values of train working currents depending on a line category are listed in Table 10.10) should be made using suitable methods and simulation software for the assumed traffic under peak loads, with simultaneous determination of U_{mean_useful} values for the analyzed line section.

Ensuring breaking of maximum short-circuit currents by current breakers (CB) (Table 10.11) and coordinating protections between a substation and vehicle, ensuring selectivity of fault breaking (Table 10.12) and overvoltage protection (Table 10.13) in a power supply system is a significant problem.

Electric railway line is a complex system, comprising a series of subsystems that are to be EMC compatible, i.e. work together without interferences.

As regards impact on technical infrastructure, the effect of an electric railway line on electric power grids and equipment and systems (such as traffic control, signaling, safety, and telecommunication systems) in the railway line vicinity (impact zone) has to be considered. This requires coordination of various services and companies operating in the area of railway line infrastructure and rolling stock.

Emission of interferences (harmonics, asymmetry, fluctuations) to the public grid is the responsibility of an infrastructure manager, who should consider applicable European and domestic standards and requirements concerning a power plant. External electromagnetic compatibility for HSR TEN-T is not described in detail.

Within the impact zone of the electric railway line, voltages may occur in systems, conductors, and structures made of conductive materials that may not exceed permitted values, as specified in applicable regulations, due to the safety of staff, passengers, and third parties, and the risk of damage or incorrect operation of devices and systems. For this very reason, the study of railway line impact on infrastructure (electromagnetic and electrostatic induction, stray currents) has to be conducted as early as at the design stage, possibility of interferences occurrence checked, including levels, and possible remedial measures developed, pursuant to applicable regulations.

All the equipment used in the railway infrastructure of the power supply system has to have a relevant certificate and admissions to use on electric railway line attached, covering:

- minimization of stray currents phenomenon influence [20];
- high energy efficiency and effectiveness of recuperative braking;
- allowing the use by stationary systems and their protections of recuperative braking, except for situations as specified in the standard [22]. Stationary system compatibility assessment is conducted as per the standard [22];
- safe operation and electric shock protection [19];
- selection of power supply sources (electric power system) with suitable parameters – this applies most to the value of fault capacity of the source, the measure of source internal impedance and voltage level (sufficiently high) in the node of common tapping of a contact system and other public consumers;
- location of a contact system substation, so that the lengths of power supply lines do not result in excessive voltage drop;
- selection of maximum distances between substations due to permitted voltage drops, load capacities of contact system substations, and fault identification;
- values of installed power of contact system substations ensuring transfer of assumed loads;
- use of devices compensating voltage drops in the power supply system circuit, such as sectioning cabins;
- selection of cross-sections of electric conductors in major current lines, ensuring transfer of required loads;
- selection of switching equipment with suitable parameters;
- use of filters and other appliances required for correct operation of a substation;
- use of insulation with suitable parameters, protection earthing, protection devices and systems.

In addition, the sources of contact system energy and power supply should be designed in such a manner as to allow uninterrupted supply in case of disturbances. This can be achieved by dividing the contact system into sections and installing redundant equipment at substations. Compatibility assessment should include inspection of circuit diagrams; such inspection should demonstrate that devices and systems installed ensure uninterrupted power supply. In the 3 kV DC system, two-sided power supply has to be used.

10.2.2. Standards applicable in Poland regarding 3 kV DC high-speed rail power supply systems

Currently, HSRs are lines with operating speed that exceeds 250 km/h, but in Poland, until new lines of 25 kV 50 Hz are built, they are the lines with operating speed in the range of 200–250 km/h, classified as upgraded lines, and including the HSR (categories II and III – Table 10.5). This refers, in particular, to the line E65, Grodzisk Mazowiecki – Zawiercie (CMK-Central Rail Line), planned for operating speed of 200–220 km/h, and the section Warszawa – Gdańsk for up to 200 km/h (Section 10.3).

The network of PKP PLK S.A. is subject to standards developed by CNTK-Railway Institute [25], based on regulations and standards applicable to lines depending on maximum operating speed. Required parameters specified in "Standards" [25] are listed below (symbols: P250 – passenger operations, at v_{max} = 250 km/h; M200 – mixed operations, at v_{max} = 200 km/h; T120 – freight operations, at v_{max} = 120 km/h).

10.2.2.1. Operating currents and short-circuit currents

Maximum current intake of a train on the already upgraded line P250, M200, and P200 is 3200 A, according to PN-EN 50388.

Maximum current intake of a train on the already upgraded line P250, M200, and P200 is 4000 A, according to [22].

Main circuits of all power supply and contact line system equipment should feature immunity to short-circuit current 50 kA, according to PN-EN 50388 [22].

Minimum short-circuit currents at a supply section should be at least 10% higher than setpoints of trips in a HSB of a feeder supplying this section, when the last one value is above 3000 A.

HSB trip setpoint current should be at least higher by 200 A than the maximum current flowing through HSB during its normal operation, as results from instantaneous loads of given supply section during peak hours.

Subject to consent of an infrastructure manager, breaking of short-circuit currents may be based on criteria other than the content of minimum short-circuit currents, such as steepness of di/dt rise, current rise in specific time $\Delta I/\Delta T$ or current I maintaining on the specific level for specific time $T – I(T)$.

10.2.2.2. Contact system voltage

The value of and changes in contact system voltage should comply with requirements of the standard [18]. 3 kV DC system of contact line power supply system has to be used with all electric line types.

10.2.2.3. Stray currents

In order to reduce stray currents and electric corrosion caused by stray currents, earthing and protection bus bar systems (bonding to rails) used as short-circuit protection in DC systems and shock protection should be separated and isolated. As regards earthed elements or elements connected to earth, the use of open bonding to rails is recommended.

10.2.2.4. AC supply of traction substations

One 110 kV line is allowed for feeding a substation.

110 kV AC feeding lines should be connected to the power grid at main and point of common coupling (PCC) to grid points or to the 110 kV line.

10.2.2.5. Location of a contact system substation and sectioning cabin

Contact system substation or sectioning cabin should be so located as to ensure:

1. access by tractor trailers with a low-loading trailer or that access road with required parameters can be erected;

2.   routes of feeders and return cables are possibly short and straight;
3.   distance between a traction substation or a cabin band earthing and the external rail of the electrified line track is at least 20 m – in particularly difficult conditions, the distance may be 16 m; the distance from a non-electrified line track may be smaller, provided that insulating inserts are installed in the track.

10.2.2.6. Feeder lines

When calculating AC supplying lines, one should consider the location of a traction substation and a PCC switching station, short-circuit power at the station output, 15 min power and maximum instantaneous power of the traction substation and power redundancy for auxiliary purposes.

Line cross-section should be selected according to three criteria:

1.   thermal load capacity;
2.   acceptable voltage drops;
3.   short-circuit current capacity.

10.2.2.7. 110 kV HV switchgear

1.   rated voltage – 123 kV;
2.   insulation level – 550 kV;
3.   rated current – 1600 A;
4.   rated breaking current – 31.5 kA.

10.2.2.8. MV AC switchgear

MV AC switchgear at the substation supplied with 110 kV provides power supply to non-traction system consumers and auxiliary power supply. The switchgear should have single arrangement of common bus bars, sectioned with a disconnector or a CB with disconnectors.

Required parameters of the MV AC switchgear:

1.   working voltage – 15 (20) kV, 50 Hz;
2.   rated continuous current for feed bays, bay coupling, common bus bars – 630 A;
3.   rated continuous current for other bays – 400 A;
4.   peak short circuit current – 31.5 kA;
5.   rated 1 s current – 12.5 kA.

10.2.2.9. Rectifier units

12-pulse rectifier units have to be used. Rectifier units should be rated in class III overload capacity, with the following output voltages:

1.   rated rectified voltage – 3300 V;
2.   off-load voltage of the unit U_{d0} 3600 V.

In substations supplied by 110 kV AC line rectifier units (Figure 10.14a) with minimum rated DC side current 1700 A should be used, while in substations supplied by 15 kV or 20 kV AC line – 1600 A.

10.2.2.10. 3 kV DC switchgear

Working voltages of the switchgear should comply with the standard [18] and current parameters suitable for the power supply system. Rated voltage in auxiliary circuits should be 220 V DC.

Required parameters of DC switchgear and HSBs:

1. rated voltage – 3.3 kV;
2. maximum operating voltage – 3.6 kV;
3. steady-state short-circuit current (value expected at time constant 20 ms) – 50 kA;
4. auxiliary circuits voltage – 220 V DC;
5. rated current of a primary common bus bar and section disconnectors:

    a. minimum 2 kA for P80 line;
    b. minimum 4 kA for lines P160, M160, P120, M120, T120, M80, T80, and T40, and for the upgraded lines P250, M200, and P200.

10.2.2.11. 3 kV feeder lines

Feeders have to be designed as cable lines.

Cross-section of cable feeders is selected based on the equivalent value of 15 min current, however not less than:

1. 500 mm^2 for the line P80;
2. 1000 mm^2 for other types of lines.

10.3. Increasing speed in Poland on railway lines with 3 kV DC power supply

The network of electrified lines in Poland is 12,000 km (Figure 10.8).

The vast majority of currently operated 3 kV DC voltage traction substations in Poland is supplied by medium voltage (MV) AC lines (15 kV or 20 kV, seldom by 30 kV). In the 1990s, attempts were made using 110 kV voltage in direct feeding of transformers of 3 kV DC rectifiers in traction substations (one-stage transformer without intermediary transformer 110 kV/15 kV arrangement). General Directorate and Division of Principal Power Engineer of PKP initiated the upgrade of the Huta Zawadzka contact system substation (CMK line), including rebuilding of a 110 kV 50 Hz feeding system, and a prototype construction of two rectifier units PD17 (Table 10.14, Figure 10.9). Each PD17 unit is composed of a 4-winding transformer with windings arrangement of 110 kV/1.3 kV/1.3 kV/15 kV (Figure 10.9), intended for feeding two connected in-series bridge rectifiers creating a 12-pulse rectifier unit equipped with an aperiodic (LC) filter as well as 15 kV switchgear to feed non-traction consumers and auxiliary supply.

The following entities have participated in the project: Division of Principal Power Engineer of PKP (currently private company PKP Energetyka S.A.), Electric Traction Division of Warsaw University of Technology [14], "Elektroprojekt" Design Office, Central Bureau of Studies and Projects of Railway Construction Sector "Kolprojekt," PKRE company (currently: Trakcja PRKiI S.A.), CNTK (currently: Railway Institute). The rectifier unit was constructed by EMIT Żychlin (transformer) and Instytut Elektrotechniki-Warsaw (rectifier). Company APENA Bielsko-Biała developed and implemented the construction process of

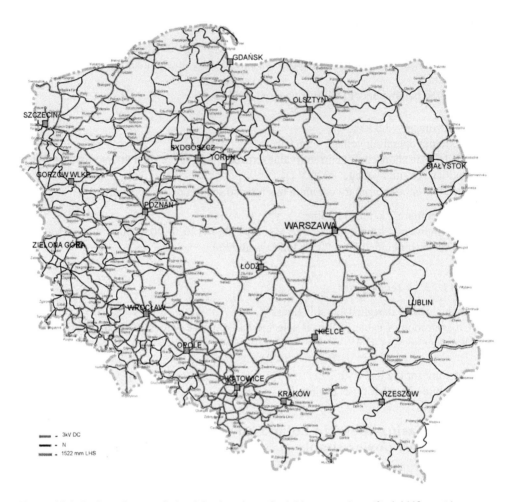

Figure 10.8 Railway lines in Poland (red – electrified, N – non-electrified, LHS – wide-gauge, non-electrified)

Source: PKP – Polish State Railways

Table 10.14 Technical specifications of the 12-pulse rectifier unit type PD-17

Rectifier transformer	–
Power source	110 kV
Rated output	7.4 MVA
Rectifier unit characteristics	
Rated output of AC rectifier windings	6.4 MVA
DC rated output	5.61 MW
Rated rectified current I_{dn}	1700 A
Rated rectified voltage U_{dn}	3300 V
Overcurrent class	III
Allowable overloads	150% I_{dn} for 2 min (2550 A)
	200% I_{dn} for 10 s (3400 A)

Figure 10.9 Rectifier PD17 in a 3 kV DC traction substation

Source: Authors' own source

Figure 10.10 110/1.3/1.5/15 kV traction rectifier set transformer in a 3 kV DC traction substation of line E65

Source: Authors' own source

high-speed breaker BWS50. Then, the construction of a Barłogi substation (container 3 kV DC traction substation supplied by 110 kV AC on the crossing of E20 and 131 lines) was commenced, as well as the construction of Przecza (E30 line), Mienia, Sosnowe, and Sabinka (E20 line) with ABB company manufactured rectifier units. Upgrade projects for lines E20, E30, and E65 provided for construction of more 3 kV DC traction substations supplied by 110 kV 50 Hz lines, which are presently completed (Figure 10.10).

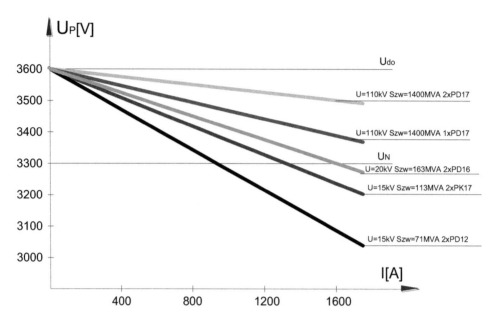

Figure 10.11 External characteristics Up(I) of a 3 kV DC traction substation according to the type of AC supply, its short-circuit level (S$_{zw}$) and rectifiers installed; U$_{do}$ – DC side idle voltage

There are fundamental differences in the shape of voltage/current U_d (I_d) output characteristics of substation supply voltage of 15 kV, 20 kV, or 30 kV and 110 kV (Figure 10.11). Voltage drops in substations with supply voltage of 110 kV are two or three times smaller than the similar voltage drops at supply voltage of 15 kV or 20 kV. Using a one-stage transformer allows obtaining a system efficiency ratio of 95–96% for central lines and reducing costs of energy purchase from energy distribution companies. It also significantly reduces the impact of a substation with 110 kV supply voltage on the supplied electric power system (harmonics, voltage fluctuations) (Figures 10.12 and 10.13).

When upgrading a main line to operating speeds of 160–200 km/h, a typical solution is to add a contact system substation between the existing substations, usually substituting the existing sectioning cabin (Figure 10.14). Such a solution was used in the upgrade of lines E20, E30, and E65, section Warszawa – Zawiercie and Warszawa – Gdańsk.

Unfortunately, the possibility of feeding such a substation from a near 110/15(20) kV PCC or a nearby existing line of 110 kV is not always possible. In such a situation, the only practical solution is feeding electric energy (by cables only) from the MV bus bar of an adjacent substation, usually at the distance of 12–15 km. Thus, supporting substations are created, typically fitted with a one rectifier unit, with a significant gradient of external characteristics resulting in their contribution to feeding the section being smaller than that of adjacent substations, usually with two or more rectifier units, with a smaller gradient of external characteristics (Figure 10.11).

Upgrade of a CMK feeding system for *Pendolino* trainsets (without tilting body, projected maximum speed in Poland is 200–220 km/h), and for locomotive trains – 200 km/h,

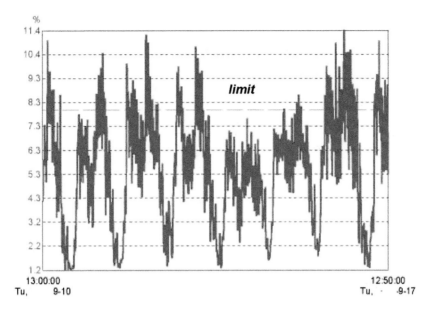

Figure 10.12 Results of one-week measurement of THD U (total harmonic distortion of voltage) coefficient at 15 kV AC line supplying 3 kV DC traction substation with 6-pulse rectifier units (limit 8% – broken line), the limit has been exceeded

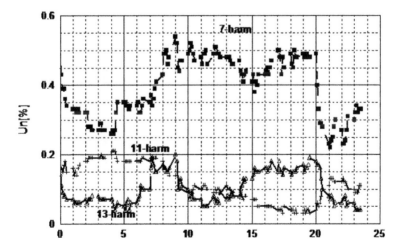

Figure 10.13 Results of voltage harmonics measurements at the 110 kV 50 Hz AC side of the 3 kV DC traction substation with 12-pulse rectifiers

including the upgrade of the existing traction substations (TS) and erection of new TSs (Figure 10.15) at most critical sections (longest substation-to-substation distance and largest expected loads), while maintaining the used cross-section of the contact system 440 mm² Cu. High level of specific electric loading of installed power, over 1.1 MW/km, was obtained.

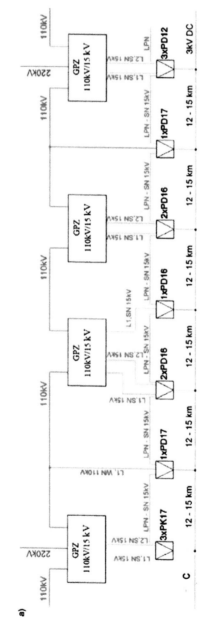

Figure 10.14 Example diagram of typical feeding system for (a) main lines (upgraded, with 12-pulse rectifiers) and (b) secondary lines (not upgraded, with 6-pulse rectifiers)

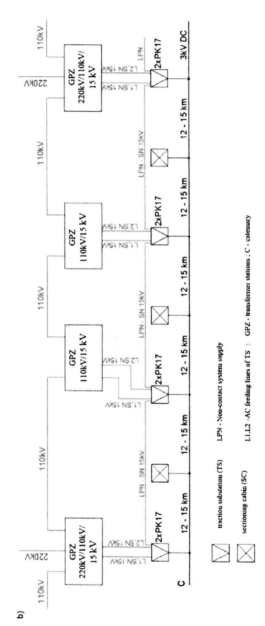

b)

Figure 10.14 (Continued)

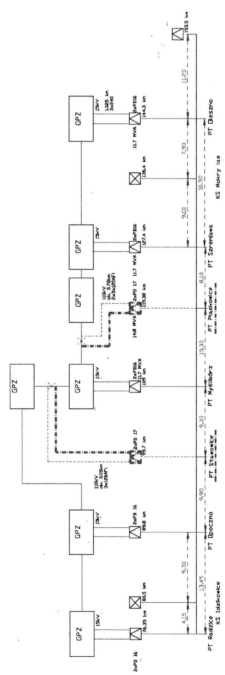

Figure 10.15 Scheme of objects following the upgrade of a 3 kV DC power supply system of CMK line – adding seven new and upgrading the remaining TS (except TSS Huta Zawadzka, upgraded earlier), broken lines – new objects, bold line – proposals for priority objects. GPZ – Point of Common Coupling-PCC (110/15 kV)

Source: Author's own study based on [45]

Similar upgrade works were undertaken on the line E65 North (Warszawa – Gdańsk), with a comple45tion date in 2015.

Due to such speeds, the CMK line feeding system has to meet the most restrictive requirements concerning a 3 kV DC system, that is mean useful voltage U_{mean_useful} should be no less than 2.8 kV (Table 10.9), and suitable for supplying trains with currents of 3.2 kA (Table 10.10). It needs to be stressed that rated voltage of this type of trainsets is over 3.2 kV, and in locomotive trains 2.8 kV. This might indicate, under some conditions, the need for reduction in power intake of trainsets below the rated value of 5.5 MW. Ensuring the value of utility voltage of 2.8 kV (which guarantees the intake of little over 75% of installed power) should, however, have no significant effect on maintaining maximum operating speed due to relatively low resistances to motion, and only minor effect on dynamics of trainset operation. Only further reduction in voltage and power below 75% of the rated power should result in perceivable reduction in operating dynamics and speed [46]. This is shown in Figures 10.16 and 10.17. Type of operation and increase in operating speed significantly affect the nature of line substation load (Figure 10.18).

10.4. Preparations for implementing a new 25 kV 50 Hz power supply system in Polish railways

Preparations undertaken by PKP PLK S.A. to implement a 25 kV 50 Hz power supply system that has not so far been used in Poland were discussed by PKP PLK S.A. representatives in the series of materials and publications. This section mostly focuses on discussions of those works as specified in the article [3].

In 2005, PKP PLK S.A. commissioned Scientific and Technical Railway Centre (CNTK - currently Railway Institute to prepare the study [38], in which electrification of the line using the 2×25 kV 50 Hz power supply system was agreed. Pursuant to *Master plan for railway transport in Poland by 2030* [11], the 2×25 kV 50 Hz traction power supply system has to be used at newly built high-speed lines. Grounds for selecting this voltage were specified in *Program budowy i uruchomienia przewozów Kolejami Dużych Prędkości w Polsce [High-speed rail in Poland – Construction and Operation Programme]* [24], as follows:

> Since due to technical reasons it is not possible to use only 3 kV DC supply voltage for the HSR, one has to assume that supply voltage between HSR nodes shall be 25 kV 50 Hz, but in railway junctions 3 kV DC supply voltage shall be maintained. HSR traction vehicles have to allow electric power intake from at least those two power supply systems.

The article also specified the grounds for rejecting the 15 kV AC 16.7 Hz frequency system.

Further preparatory actions shall include studies [5],[38],[40] and in particular *Studium KDP . . .* [HSR study . . .] prepared by the Spanish company IDOM [27] with its continuation on extension of the HSR towards a German border and the HSR line route through railway junctions: Wrocław [28], Poznań [29] and Łódź [30]. The PKP PLK S.A. plans also includes pilot implementation of the 25 kV 50 Hz system on a non-electrified (so far) section near the border with Lithuania, where this system is used [39],[40],[42]. Preliminary views, declarations, and interest of Polish manufacturers for this type of traction power supply

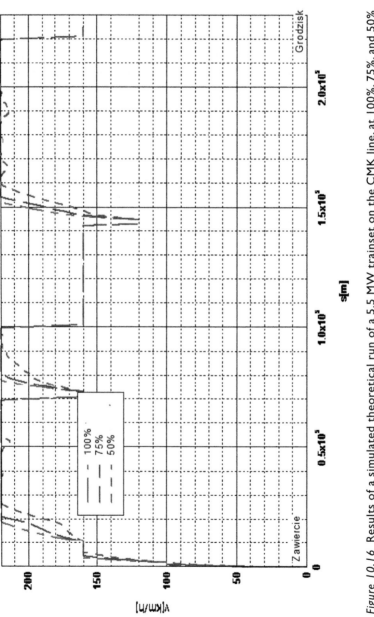

Figure 10.16 Results of a simulated theoretical run of a 5.5 MW trainset on the CMK line, at 100%, 75%, and 50% of output power

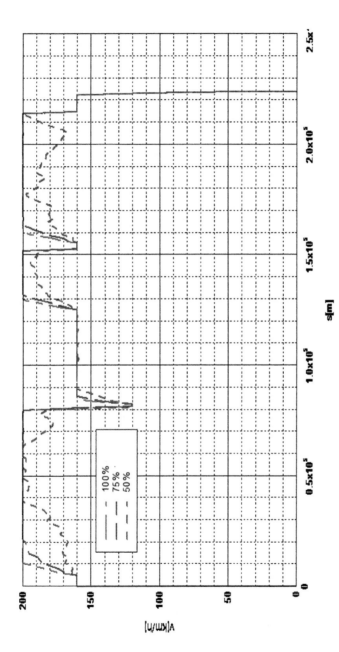

Figure 10.17 Results of a theoretical run of a train with 500 t wagons and a 6 MW locomotive at 100%, 75%, and 50% of output power, respectively

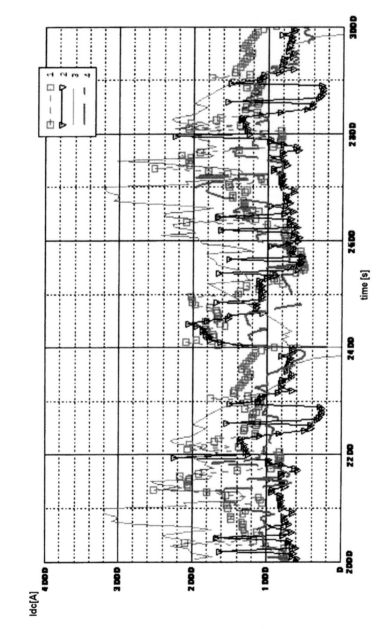

Figure 10.18 Time curves of a 3 kV DC TS load current for various types of trains (maximum speed from 160 km/h to 250 km/h (1–350 t trainset, 6 MW, vmax = 200 km/h; 2–350 t trainset, 6 MW, vmax = 160 km/h; 3–500 t trainset, 6 MW, vmax = 250 km/h; 4–300 t trainset, 4 MW, vmax = 220 km/h)

were also gained. Furthermore, the possibility of obtaining power supply from a power grid for the HSR in Poland by 2014 was investigated, together with requirements and energy demands [7].

Conclusions from such preparatory works, as presented in [3], allow for stating that change in supply voltage alone shall not cause any significant troubles. Regulations in force specify precisely:

- minimum permitted distances, heights of guards, protective fences and railings, as well as dimensions of protection zones in indoor switching stations with open cells;
- minimum permitted distances, heights of fences and protection zones in overhead substations with internal protection fencing;
- minimum permitted distances and protection zones in overhead substations at locations without internal protection fencing;
- minimum permitted distances in the air between uninsulated live parts of devices and various structural components of indoor and overhead substations and switching stations.

In a case of a traction substation section under HV potential (110, 220 kV), no issues related to interpreting and adopting correct distances and heights for a given voltage level are anticipated. The situation becomes more complicated for the section of a traction substation and contact line under 25 kV 50 Hz.

A serious issue for operation shall be a significant increase, for voltage exceeding 30 kV, of distances defining safety limits for works near the voltage and at the voltage. Such distances are specified in the Regulation by Minister of Economy of 17 September 1999, on occupational health and safety for power machinery and installations [**Błąd! Nie można odnaleźć źródła odwołania.**]. This, however, shall not be the only reason for changing the safety regulations for works on railway electric power equipment, currently in force. Coexistence of interfaces between systems – an existing one 3 kV DC and a new one 25 kV 50 Hz (Section 10.5) shall be important.

Below presented is synthetic information on solutions for a 25 kV 50 Hz power supply system resulting from a concluded study.

10.4.1. HSR line feasibility study

The study, prepared by the IDOM company [27] for HSR line, so-called Y line Warszawa – Łódź – Poznań/Wrocław for passenger operation at the speed of up to 350 km/h, weight 483 t, and net output of 8.8 MW, auxiliary power output of 1 MW, proposed a contact system power supply of 2×25 kV 50 Hz from eight substations fitted with two transformers, 60 MVA (three substations), and 30 MVA (five substations). Autotransformers, 10 MVA each, shall be installed about every 10 km. Required overload capacity of substation transformers was the result of projected loads: 20% for 2 h, 50% for 15 min, 100% for 10 min, and autotransformers: 20% for 2 h, 50% for 15 min, and 100% for 5 min. Due to projected loads and asymmetric nature of energy consumption, feeding of substations shall require tapping to nodes 220 V (two substations) and 400 kV. One of substations could be fed with 110 kV supply voltage. It was needed, under certain conditions (low short-circuit capacity of substation power supply), to install a filtering and symmetrizing device. There is proposed a link designed between the Y line and CMK line, via upgraded line no. 25. Also, the prognosis for energy demands were presented, allowing estimation of investments required in the Public Utility System. Studies are continued with regard to extension of the Y line to the German border.

10.4.2. Plans for integrating an HSR line with the existing railway junctions

As regards development of an HSR line with 25 kV supply voltage, design works were undertaken to integrate an HSR line with the following junctions: Wrocław [28], Poznań [29], and Łódź [30]. Due to the existing extensive network of a 3 kV DC power supply system in those junctions, it was decided unjustified to introduce 25 kV supply voltage to junctions [24], so that the change of voltage in a traction power supply system occurs within the approach area ahead of the junction. Problems related to mutual impact of the 3 kV DC and 25 kV 50 Hz systems, in particular at interface areas at junctions, are discussed in Section 10.5.

10.4.3. Projects for applying 25 kV supply voltage on the E65 South line

Prior to upgrading the 3 kV power supply system of the E65 South line for the purpose of supplying a *Pendolino* train [44],[46] studies were conducted with regard to possible change in a power supply system from 3 kV DC to 25 kV 50 Hz, to achieve operating speed of trains to 300 km/h. Studies included not only preparations, but also development of contact system support structures for supply voltage of 3 kV DC and 25 kV 50 Hz [5] (similar works were conducted earlier, i.a. on the line planned for change in supply voltage from 3 kV DC to 25 kV 50 Hz, an Italian line called Direttissima), as well as the study on power supply [4]. Prepared concepts of railway line power supply (Figure 10.19) indicated the need for construction of four contact system substations of 220 (110)/2×25 kV 50 Hz for the CMK line, and a fifth one to supply the new section of a CMK line extension towards Kraków and Katowice, via Olkusz. Unfortunately, the CMK line is routed through the area without any energy nodes with sufficient short-circuit capacity and power transmission capacity to which such new substations of a 25 kV 50 Hz power supply system could be tapped. Thus, the need to construct over 200 km of new HV lines and extend the existing ones. It would be much simpler (due to better availability of electric energy) to connect the substation in the Olkusz area (Figure 10.19), plus two substations planned for feeding the new HSR line section Katowice/Zebrzydowice/Zwardoń (Figure 10.20). Due to the extensive upgrade and reinforcement of a 3 kV DC power supply system on the E65 line, completed between the years of 2012 and 2015, the considered plans are not scheduled for implementation any time soon.

Figures 10.21 and 10.22 represent forecasted, based on simulation, load with momentary power of TSs shown in Figure 10.19 and voltage at current collectors of trains vs. their positions, while Figures 10.23 and 10.24 represent substation load and voltage on train current collectors, respectively, upon complete failure of the TS Opoczno.

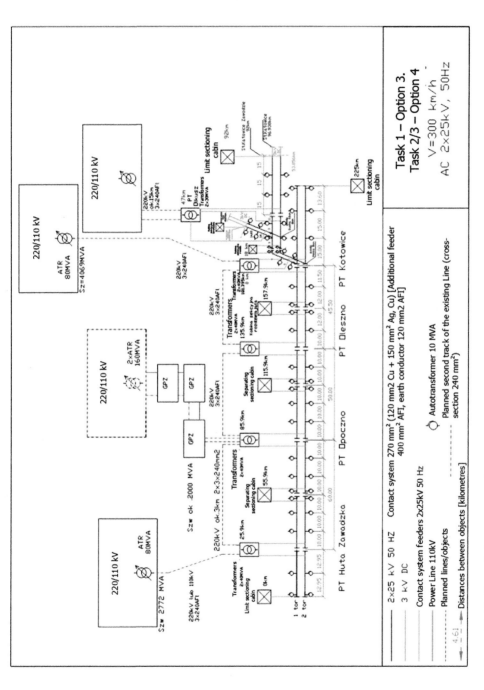

Figure 10.19 One of the variants of the planned 2×25 kV power supply system on the existing CMK line and on a new railway line section to Kraków/Katowice

Source: [43]

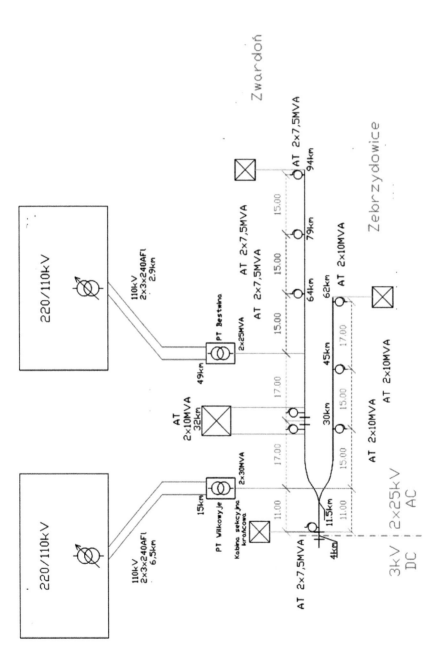

Figure 10.20 One of the variants of planned power supply system 2×25 kV for the new railway line section Katowice-Zebrzydowice/Zwardoń

Source: [43]

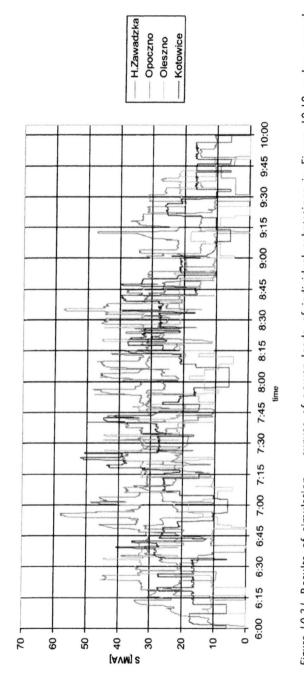

Figure 10.21 Results of simulation — curves of power loads of individual substations in Figure 10.19 under normal conditions

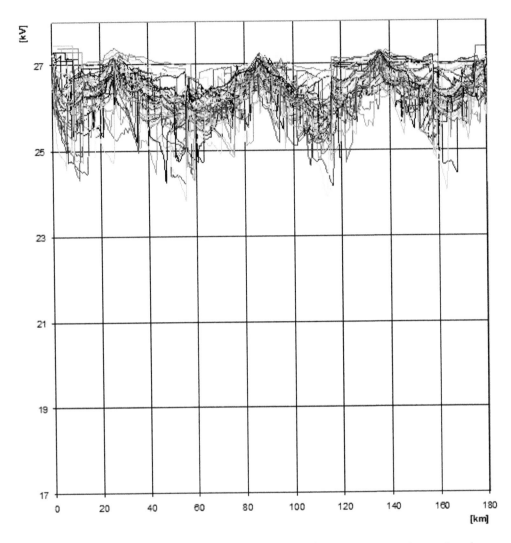

Figure 10.22 Voltage on current collectors of trains vs. their position, on the section shown in Figure 10.19, under normal conditions

10.5. Issues related to integration of a 25 kV AC supply voltage railway line with the existing 3 kV DC infrastructure

Works in this area were conducted by Electric Traction Division Warsaw University of Technology [5],[14],[39],[40],[42],[43],[44],[45],46] and the subject was summarized in a series of publications [1],[6],[15],[16],[31], [32],[36] as well as in a Ph.D. thesis by M. Patoka [17]. In the event of introducing an AC TPSS to the electrified railway junction with 3 kV DC supply voltage, due to the limited spaces for railway infrastructure, a need for the AC and DC

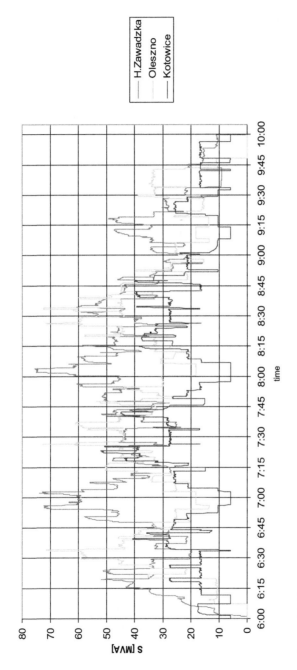

Figure 10.23 Results of simulation – shape of power loads of individual substations in Figure 10.19 under emergency conditions (TS Opoczno out of service)

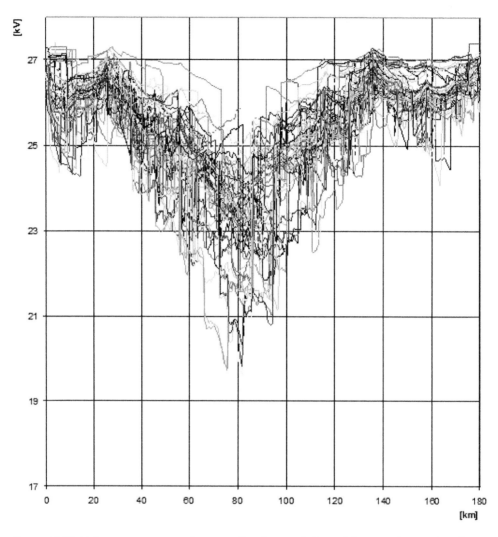

Figure 10.24 Voltage at current collectors of trains vs. their positions, at the section shown in Figure 10.19, under emergency conditions (TS Opoczno out of service)

TPSS installation close to one another has to be foreseen; it is important, however, due to induction voltage, to ensure adequate, safe potential in rails. Suggested solution employs the 2×25 kV AC 50 Hz TPSS, with an additional return conductor at interfaces with the DC TPSS (low resistivity of ground makes interaction more significant). Parallel routing of the HSR along the existing DC line is only possible with relatively short parallel sections and with relatively (depending on the length of a parallel section) small outputs. Such an arrangement often allows routing the AC and DC PSS practically everywhere within city limits, although it increases the occurrence of described interactions due to interferences conducted between the AC power supply and DC power supply [15],[21]. What is recommended is a possibly good separation (considering safe step voltage of DC rail–earth) of the DC TPSS towards earth in order to minimize stray currents at interfaces and effect of AC return current to DC rails and DC current from DC TPSS to AC TPSS rails. In addition to increasing the risk of dangerous voltages, AC current, especially in a not upgraded DC TPSS, may also interfere with operation of track circuits. DC current going into AC rails may interfere with operation of traction energy systems.

The broad spectrum of issues occurring at AC TPSS and DC TPSS interfaces results in the best solution being withholding from introduction of a 25 kV AC railway power supply system to junctions fitted with a 3 kV DC power supply system (Figure 10.25), as provided for in plans for HSR implementation in Poland [24]. Such a solution, however, would not eliminate all the issues related to parallel operation.

Implementation of a new railway traction power supply system in Poland shall lead to the appearance of supply voltage interfaces within individual TPSS routes. Presently, such interfaces are only observed near a state border, wherever the adjacent contact system supply voltage is different (e.g. German border). Thus, changes in supply voltage and related interfaces have never constituted a significant issue before. However, integrating a new HSR line with a 2×25 kV AC TPSS with the existing 3 kV DC TPSS (or change in voltage or expansion of CMK line with sections of 25 kV 50 Hz TPSS) shall result in aggravation of the situation, which has to be foreseen and considered at the stage of feasibility study or planning [6],[8],[37].

In the event of longitudinal separation, the AC TPSS and DC TPSS are not parallel to one another, but the AC TPSS is the extension of the DC TPSS, and other way round. In such arrangements parameters of the earth near the TPSS are also very important.

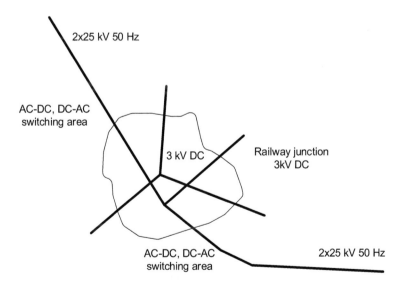

Figure 10.25 Railway junction electrified with a 3 kV DC power supply system, with change of supply voltage of an HSR line from 2×25 kV to 3 kV DC ahead of the junction

Source: Ref. [16]

There is a probability of vehicles in one system transferring potentials via rails and ground to the other system, resulting in more stray currents and the presence of voltage with AC and DC components in rails. Presence of a DC component in AC systems similar to the one described above may lead to malfunction of railway infrastructure. Such phenomena shall intensify with larger currents in the system, which conditions critical cases for, i.a. short-circuit states.

There exist numerous solutions for the TPSS change area DC/AC [6],[8], [16],[17], [37], [40], with different TPSSs separation limited to implementing a so-called neutral section, allowing safe transit of a multisystem traction vehicle between different TPSS areas. Basically, voltage change sections mutually isolate both TPSSs; however galvanic coupling may occur in these areas, with the following phenomena intensified:

- DC system stray currents;
- voltages with AC and DC components in rails;
- saturation of transformers, autotransformers, and booster transformers;
- malfunctions of traffic command and control systems;
- presence of one of the system components in another system on-board return circuits.

The above phenomena depend mostly on such parameters as longitudinal resistance of rail, earth resistivity, conductance of rail-to-earth transition in a DC system, resistance of earth electrodes, and resistance of inter-car passages in long trainsets.

Creating interfaces of two TPSSs with different supply voltage, especially within city areas, is not a trivial matter [22], and requires, i.a.

- detailed analysis of electric power loads (anticipated traction loads, power supply circuits schemes, TPSS parameters);
- analyses of environmental parameters – natural conditions and technical infrastructure (determination of earth resistivity and a degree of area urbanization);
- suitable location of the voltage change area and selection of suitable technical solutions.

Decision of not introducing a new 25 kV 50 Hz voltage supply system to railway junctions with other supply voltage (3 kV DC in Poland) [24] allows reducing the number (although not complete elimination) of issues at interfaces between different TPSSs as early as at the stage of project planning (e.g. selection of shock protection measures, stray current protection measures, etc.), although it shall require continued research on the longitudinal separation of TPSSs at interfaces.

10.6. Conclusions

Each currently used HSR TPSS has its specific strengths, weaknesses, and limitations. The 3 kV system has the lowest power transmission capacity among all discussed systems, and its capacity to provide suitable energy for high-speed rails is limited by the speed of 250 km/h. Construction of new lines or preparation for a change of the system at the Direttissima line in Italy [4] results also from operational factors – higher wear of the contact line and high energy losses with heavy traffic. Also, construction of the HSR supplied by a 25 kV 50 Hz TPSS in France and in Holland was justified, as for the TPSS with low efficiency, such as the system 1.5 kV, operating speeds higher than 160 km/h were the challenge. Due to voltage drops and introduction of higher current harmonics to the public electric power system, a DC system requires shorter distances between substations, which should be supplied from grid points of high short-circuit capacity (high voltage grids). The strength of a DC TPSS is the symmetry of loads.

Due to introduced asymmetry, the 25 kV 50 Hz system requires connection to high-voltage power grids via an separated power supply lines. As requirements concerning short-circuit capacity may be satisfied only with grid voltages exceeding 110 kV, with its own connection to a power electric system network, sometimes there is a need to use additional costly equipment for symmetrization of traction loads in transformer substations. Due to rapid development of high-power electronic equipment reduction in costs of symmetriza-tion systems can be expected with the best, however the most costly solution as application of SFC 3~50 Hz/1~50 Hz. This may increase technical and economic benefits of using this system (longer distances between substations, lighter network, smaller energy consumption, investment outlays usually smaller by approximately 20–30%). In order to maintain long distances between substations at high power consumption, systems of induction voltage drop compensation are used or supply voltage of 2×25 kV (50 kV).

Reduced frequency AC TPSS (16 2/3 Hz) requires creating its own isolated supply net-work, and except for countries where it has been used for years and is well developed, such a system is not very likely to be expanded. Although its impact on a power grid (when tapped to the public grid) is symmetrical, voltage drops can be observed when starting up rotating machinery, or higher harmonics, when static converters are used in traction substations.

Considering specific features of discussed systems (their properties related to supply of high-speed rail lines are listed in Table 10.15 for comparison), they can be related to railway lines in Poland.

Distances between Warsaw and major metropolitan areas not exceeding 300–400 km indicate that the network of railway lines with operating speed of 200–250 km/h should improve competitive advantage of railway transport over road transport and air transport. To supply operation of trains at these speeds the existing TPSS in Poland, 3 kV DC, selected before WW II, is sufficient. Please note that in those times, this system was modern, and, as can be seen now, sufficient in the long term. Selection of then well-developed 1.5 kV DC system would force, as in France or Holland, the need for switching or complementing it with railway lines supplied by the AC TPSS. In all fairness, we have to admit, after the 80th anniversary of electrification of Warsaw Rail Junction, that the decision that allowed Poland to maintain a single system for electrification of railway lines, with resulting benefits, which is rare in present Europe, was right.

To sum up, introduction in 2014 of rapid trains on the Central Railway Line required an upgrade of the existing 3 kV DC system. Measures taken to meet the criteria included in TSIs and other relevant regulations (including standards) were (see Section 10.3): reduction in distances between substations by erecting some new substations; applying one-stage voltage transformation in new substations, wherever possible; retrofitting existing substations with 12-pulse rectifiers with higher output then the existing ones; and (as target) increasing the cross-section of a contact system on sections that require so, where no new substations were built. The CMK line was built in 1970s, originally for freight operations, and then for pas-senger operations at the speed of 160 km/h, which was achieved in 1980s. Extensive upgrade of the CMK line, including the existing 3 kV DC TPSS section Grodzisk Mazowiecki – Zawiercie (after postponing the implementation of HSR Project beyond year 2030), allowed for achieving a regular operating speed exceeding 200 km/h. Bringing Poland to the club

Table 10.15 Comparative evaluation of traction power supply systems used at HSR lines

System type	Maximum operating speed [km/h]/ vehicle output	Asymmetry of loads	Estimated cost of electrification [%]	Harmonics introduced into power supply network	Contact system	Interferences in telecommunication circuits	Voltage drops in contact system	Stray currents	Number of substations	Fault capacities and requires voltage level
3 kV DC	250 6 MW	+	100	−	− maximum cross-section, high consumption	+/−	−	−	high	>1000 MVA 110 kV
1×25 kV 50 Hz	250 10 MW	−	80	−	+	−	−	+	average	50× substation transformer output >110 kV
2×25 kV 50/60 Hz	350	−	70–80	−	+	+	+	+	Low	50× substation transformer output. >110 kV
50 kV and more, 50/60 Hz	350	−	65–75	−	+	+/−	+	+	Minimum	50× substation transformer output. >110 kV
15 kV 16 2/3 Hz	350	+	?	+/−	+	+	+	+	Low	own transmission system or static converters supplied from 50 Hz network

+ not applicable, +/− applicable, − applicable to a limited extent.

"200+ km/h" railways was possible thanks to the vision of CMK planners working over 40 years ago and making full use of the capacity of the 3 kV DC TPSS selected in 1930s by Roman Podoski, Professor of the Warsaw University of Technology for electrification of Warsaw Rail Junction.

The TPSS that is used in Poland, 3 kV DC, still has unexploited potential for delivery of suitable energy for high-speed rails in Poland, limited to the speed of 250 km/h. Energy supply for trains operating at that speed is sufficient (although not too energy efficient) in the 3 kV DC TPSS, provided that sufficiently high level of voltage is ensured at pantographs of a modern ETV (in a contact system). In the case of the CMK line, a supply voltage of 3 kV DC is feasible within the next 15–25 years, as there are plans to switch to 25 kV 50 Hz upon an increase in operating speed to 300 km/h. This, however, requires time-consuming and costly investments in energy infrastructure (Section 10.4). Such a migration may be advisable because of energy efficiency obtained after completion of present upgrade works on the 3 kV DC TPSS, and parameters achieved by the HSR trains on such a line.

Many countries have recently commenced the construction of HSR lines supplied with voltage at the frequency of a national power grid. They are not only highly developed countries, but also countries like Turkey or Morocco. It is expected that also in Poland the decision shall be made, as soon as justified by traffic and transport policy, to construct the HSR line with operating speed in the range of 300–350 km/h, which would require a 25 kV 50 Hz power supply system. Similarly, in the event of electrification of the LHS line (between the Ukrainian border and Sławków in the Silesia area – Fig. 10.8), the alternative 25 kV 50 Hz power supply for this line, isolated from the power supply system due to a wide-gauge track, might be considered. First, however, Polish industry and companies in the railway sector have to be prepared to build and operate such a system, e.g. by erecting a pilot section as per prepared plans [40],[42].

Bibliography

[1] Altus, J., Novak, M., Otcenasova, A., Pokorny, M., and Szeląg, A.: *Quality Parameters of Electricity Supplied to Electric Railways*. Scientific Letters of the University of Żilina-Communications 2–3/2001.

[2] Barnes, R., and Wong, K.T.: *Unbalance and Harmonic Studies for the Channel Tunnel Railway System*. IEE Proc. B, 138, 2, March 1991.

[3] Burak-Romanowski, R.: *Zmiana systemu zasilania widziana z perspektywy zarządcy infrastruktury kolejowej* [Change of power supply system as seen from a railway infrastracture operator]. Międzynarodowa Konferencja Naukowo-Techniczna Problemy projektowania i budowy systemu zasilania sieci trakcyjnej KDP. Krzyżowa, 19–21 November 2009 r.

[4] Capasso, A., and Morelli, V.: *Elektryfikacja nowych linii kolejowych wysokich prędkości we Włoszech* [Electrification of new high-speed rail lines in Italy]. *Technika Transportu Szynowego*, 1996.

[5] Ekspertyza naukowo-techniczna dla projektu rozmieszczenia konstrukcji wsporczych na odcinku Góra Włodowska – Zawiercie linii CMK dla podwieszenia sieci jezdnej zasilanej napięciem, 3 kV DC, a w przyszłości do wykorzystania ich do wywieszenia sieci jezdnej 2×25 kV AC przystosowanej do prędkości do 350km/h. [Expertise for a design of a support structure for 3 kV DC catenary with possible transition to 2x25 kV 50 Hz with maximum speed 350 km/h]. A study commissioned by PKP PLK S.A. 2007 carried out by Zakład Trakcji Elektrycznej Politechniki Warszawskiej and CBPBPBK Kolprojekt.

[6] Jordan, N., and Palmer, M.: *AC and DC Electric Railway Interfaces*. 3rd IET Professional Development Course on Railway Electrification Infrastructure and Systems, 2007

[7] *Kierunkowy program rozwoju Kolei Dużych Prędkości w Polsce do roku 2040*. Etap III: Potrzeby energetyczne wynikające z systemu Kolei Dużych Prędkości. [Strategic programme for development of HSR in Poland up to 2040. Power supply demand]. CNTK study, Warszawa, III/2010.

[8] Kiessling, F., Puschmann, R., and Schmieder, A.: *Contact Lines for Electrical Railways*. Planning, Design, Implementation. Publicis, 2001.

[9] Kroczak, M.: *Symulacja funkcjonowania zelektryfikowanej linii kolejowej o złożonej strukturze sieci zasilającej*. [Simulation of functioning of electrified railway line with complex structure of power supply network] Rozprawa doktorska [Ph.D. Thesis], Wydział Elektryczny Politechnika Warszawska, 2008.

[10] Lewandowski, M.: Trakcje w Kolejach Dużych Prędkości [Traction in HSR]. *Logistyka* 3/2012.

[11] Master plan dla transportu kolejowego w Polsce do roku 2030 [Masterplan for development of railway transport in Poland up to year 2030]. Ministerstwo Infrastruktury [Ministry of Infrastructure], Warszawa, August 2008

[12] Mierzejewski, L., and Szeląg, A.: Infrastruktura elektroenergetyczna układów zasilania systemu 3 kV DC linii magistralnych o znaczeniu międzynarodowym. [Power supply infrastructure of 3 kV DC system for main lines of international importance], *TTS* nos.1–2/2004.

[13] Mierzejewski, L., and Szeląg, A.: Infrastruktura elektroenergetyczna układów zasilania systemu 3 kV DC linii magistralnych o znaczeniu międzynarodowym (2) – projektowanie efektywnego układu zasilania zlk. [Power supply infrastructure of 3 kV DC system for main lines of international importance(2)- design of effective power supply], *TTS* no. 3/2004.

[14] Mierzejewski, L., Szeląg, A., and Matusiak, R.: *Ocena układu zasilania linii CMK pod kątem wprowadzenia prędkości jazdy pociągów 250 km/h i wstępna analiza wprowadzenia jednostopniowej transformacji napięcia w podstacjach trakcyjnych tej linii* [Assessment of CMK line power supply due to introduction of trains with maximum speed 250 km/h and a preliminary analysis of introduction one-step voltage transformation in traction substations of this line]. A study commissioned by DG PKP, Warszawa, 1996–1997.

[15] Ogunsola, A., and Mariscotti, A.: *Electromagnetic Compatibility in Railways: Analysis and Management*. Springer, 2013.

[16] Patoka, M., and Szeląg, A.: Kolejowe obszary stykowe systemów 25 kV AC i 3 kV DC – wybrane zagadnienia [Zones of close operation 25 kV 50 Hz and 3 kV DC power supply systems – selected problems]. Zeszyty Naukowo-Techniczne Stowarzyszenia Inżynierów i Techników Komunikacji w Krakowie. Conference materials, 2014.

[17] Patoka, M.: *Analiza oddziaływań zakłócających w strefie styku systemów trakcji elektrycznej 3 kV DC i 25 kV 50 Hz*. [Analysis of disturbing influence of 3kV DC and 25 kV AC railway power supply systems in a contact zones]. Rozprawa doktorska [Ph.D. Thesis], Politechnika Warszawska, Wydział Elektryczny, 2014.

[18] PN-EN 50163: 2006 – Zastosowania kolejowe. Napięcia zasilające systemów trakcyjnych [Railway applications. Supply voltages of traction systems].

[19] PN EN550122–1 – Zastosowania kolejowe – Urządzenia stacjonarne – Bezpieczeństwo elektryczne, uziemianie i sieć powrotna – Część 1: Środki ochrony przed porażeniem elektrycznym [Railway applications – fixed installations – electrical safety, earthing and the return circuit – Part 1: Protective provisions against electric shock].

[20] PN EN 50122–2 – Zastosowania kolejowe – Urządzenia stacjonarne – Bezpieczeństwo elektryczne, uziemianie i sieć powrotna – Część 2: Środki ochrony przed skutkami prądów błądzących powodowanych przez systemy trakcji prądu stałego [Railway applications – fixed installations – electrical safety, earthing and the return circuit. Part 2: Provisions against the effects of stray currents caused by d.c. traction systems].

[21] PN EN50122–3 – Zastosowania kolejowe – Urządzenia stacjonarne – Bezpieczeństwo elektryczne, uziemianie i sieć powrotna – Część 3: Oddziaływanie wzajemne systemów trakcji prądu

przemiennego i stałego [Railway applications – fixed installations – electrical safety, earthing and the return circuit. Part 3: Mutual interaction of a.c. and d.c. traction systems].

[22] PN-EN 50388: 2012 – Zastosowania kolejowe – System zasilania i tabor – Warunki techniczne koordynacji pomiędzy systemem zasilania (podstacja) i taborem w celu osiągnięcia interoperacyjności. [Railway applications – power supply and rolling stock – technical criteria for the coordination between power supply (substation and rolling stock to achieve interoperability].

[23] PN-EN 61293:2000 – Znakowanie urządzeń elektrycznych danymi znamionowymi dotyczącymi zasilania elektrycznego – Wymagania bezpieczeństwa [Marking of electrical equipment with ratings relating to electrical supply – safety requirements].

[24] *Program budowy i uruchomienia przewozów KDP w Polsce.* [Agenda for construction and starting of high-speed rail traffic in Poland], Ministry of Infrastructure, Warszawa, August 2008.

[25] Commission Regulation (EU) No. 1301/2014 of 18 November 2014 on Technical Specifications for Interoperability for Energy subsystem

[26] Standardy techniczne. Szczegółowe warunki techniczne dla modernizacji lub budowy linii kolejowych do prędkości Vmax ≤ 200 km/h (dla taboru konwencjonalnego)/250 km/h (dla taboru z wychylnym pudłem) [Technical standards. Specific technical requirements for modernization or construction of railway lines for maximum speed 200 km/h and 250 km/h (for titing trains)] – adopted to use by Resolution no. 263/2010 of PKP Polskie Linie Kolejowe S.A. Management Board of 14 June 2010.

[27] Studium Wykonalności dla budowy linii kolejowej dużych prędkości "Warszawa – Łódź – Poznań/Wrocław" [Feasibility study for construction of HSL Łódź – Poznań/Wrocław]. IDOM. A study for PKP PLK S.A., 2012.

[28] Studium wykonalności dla przystosowania Wrocławskiego Węzła Kolejowego do obsługi kolei dużych prędkości oraz zapewnienia jego intermodalności z innymi środkami transportu. Etap IV – Analizy techniczne opcji modernizacji linii, wraz z oszacowaniem kosztów TOM 4.8 – Układ zasilania sieci trakcyjnej [Feasibility study for Wrocław Railway Junction modernization for HSR. Power Supply] Sener Sp. z o.o., Warszawa, 2013.

[29] Studium wykonalności dla przystosowania Poznańskiego Węzła Kolejowego do obsługi kolei dużych prędkości oraz zapewnienia jego intermodalności z innymi środkami transportu. Układ zasilania sieci trakcyjnej [Feasibility study for Poznan Railway Junction modernization for HSR. Power Supply] Sener Sp. z o.o., Warszawa, 2013.

[30] Studium wykonalności dla przystosowania Łódzkiego Węzła Kolejowego do obsługi kolei dużych prędkości oraz zapewnienia jego intermodalności z innymi środkami transportu Etap IV – Analizy techniczne opcji modernizacji linii, wraz z oszacowaniem kosztów TOM 4.8 – Układ zasilania sieci trakcyjnej [Feasibility study for Wrocław Railway Junction modernization for HSR. Power Supply] Sener Sp. z o.o., Warszawa, 2013.

[31] Szeląg, A., and Patoka, M.: *Some Aspects of Impact Analysis of a Planned New 25 kV AC Railway Lines System on the Existing 3 kV DC Railway System in a Traction Supply Transition Zone.* SPEEDAM 2014 International Symposium on Power Electronics, Electrical Drives, Automation and Motion Ischia, Italy VI 2014.

[32] Szeląg, A.: *Wpływ napięcia w sieci trakcyjnej 3 kV DC na parametry energetyczno-trakcyjne zasilanych pojazdów.* [Influence of voltage in 3 kV DC catenary on energy-traction parameters of supplied vehicles], pp. 1–158, Instytut Naukowo-Wydawniczy SPATIUM, Radom, 2013.

[33] Szeląg, A., and Mierzejewski, L.: *Ground Transportation Systems.* Monographic article in 22-volumed "Wiley Encyclopedia of Electrical and Electronic Engineering," Nowy Jork, Supplement I, 1999.

[34] Szeląg, A., and Mierzejewski, L.: Systemy zasilania linii kolejowych dużych prędkości jazdy [Power supply systems of HSR]. *Technika Transportu Szynowego,* nos. 11–12 2006.

[35] Szeląg, A., and Maciołek, T.: A 3 kV DC Electric Traction System Modernisation for Increased Speed and Trains Power Demand- Problems of Analysis and Synthesis. *Przegląd Elektrotechniczny* no. 3a/2013.

[36] Szeląg, A. – Application of Modeling and Simulation Techniques as Methods for Feasibility Studies and Design in Electric Traction Systems. Elektryfikacja transportu no. 8/2014, pp. 56–65, ISSN 2307–4221, http://etr.diit.edu.ua/article/view/42894+1

[37] White, R.D.: AC/DC Railway Electrification and Protection. Electric Traction Systems, Conference paper. London, November 2014.

[38] Wstępne studium wykonalności budowy linii dużych prędkości Wrocław/Poznań – Łódź – Warszawa [Preliminary feasibility study of HSL Wrocław/Poznań – Łódź – Warszawa], CNTK, 2005.

[39] Wytyczne projektowania, budowy i odbioru sieci trakcyjnej oraz układów zasilania 2×25 kV AC dla linii kolejowych o prędkości do 350 km/h [Guidelines for designing, construction and receipt of work of catenary and power supply 2x25 kV 50 Hz for railway lines with speed up to 350 km/h]. A study carried out by CBPBBK Kolprojekt Sp. z o.o. commissioned byPKP PLK S.A., 2007–2008.

[40] Zakład Trakcji Elektrycznej PW/CBSiPBK Kolprojekt (Konsorcjum). Przygotowanie pilotażowego wdrożenia w Polsce systemu zasilania trakcji 25 kV prądu przemiennego [Preparation of a pilot implementation in Poland 25 kV 50 Hz power supply system], Warszawa, 2008.

[41] Zakład Trakcji Elektrycznej PW. Seminarium naukowo-techniczne. System 25 kV 50 Hz – wymagania i wytyczne [Seminar – 25 kV 50 Hz system – requirements and recommendations], Warsaw University of Technology, Warsaw, 2009.

[42] Zakład Trakcji Elektrycznej PW. Przygotowanie pilotażowego wdrożenia w Polsce systemu zasilania trakcji 25 kV prądu przemiennego. Etap II – projekt prototypowej podstacji trakcyjnej i sieci trakcyjnej [Preparation of a pilot implementation in Poland 25 kV 50 Hz power supply system – phase II – a design of a prototype traction substation and catenary]. A study commissioned by PKP PLK S.A., Warszawa, 2012.

[43] Zakład Trakcji Elektrycznej PW. Modernizacja linii kolejowej E65-Południe odcinek Grodzisk Mazowiecki – Kraków – Katowice – Zwardoń/Zebrzydowice – granica państwa [Modernization of E-65 South railway line section Grodzisk Maz – Cracow-Katowice-Zwardon/Zebrzydowice-state border]. Commissioned by: Scott Wilson Sp. z o.o. Warszawa, Halcrow Sp. z o.o. 2010–2013.

[44] Zakład Trakcji Elektrycznej PW. Ekspertyza dotycząca układu zasilania sieci trakcyjnej linii CMK, etap II – opracowanie zakresu niezbędnych działań do przeprowadzenia jazd testowych pociągów zespołowych dużych prędkości na linii CMK [Expertise concerned power supply of CMK railway line – phase II – preparation for high-speed trains tests]. A study commissioned by PKP Energetyka, 2013.

[45] Zakład Trakcji Elektrycznej PW. Wykonanie ekspertyzy dotyczącej układu zasilania sieci trakcyjnej linii CMK [Expertise concerned power supply of CMK railway lines]. A study commissioned by PKP Energetyka, 2012.

[46] Zakład Trakcji Elektrycznej PW. Ekspertyza dotycząca parametrów eksploatacyjnych pociągów zespołowych na odcinku W-wa Wsch. – Gdańsk Gł. ciągu linii E-65. [Expertise concerned exploitation parameters of electric multiple unit trains at section Warsaw East – Gdansk Main Station power supply of E-65 line]. A study commissioned by PKP Energetyka, 2014.

Chapter 11

Overhead contact line systems for high-speed rails

Michał Głowacz, Marek Kaniewski,
and Artur Rojek

11.1. Introduction

This chapter presents general design and construction principles of overhead contact lines (OCLs) used on Polish railways that to be part of a future high-speed system. Whole description is based on TSI "energy" approach and its requirements. The TSI approach provides the "energy" subsystem containing:

- substations, regarding devices transforming high-voltage energy drawn from the power grid into energy suitable for traction stock;
- section cabins, regarding all electrical equipment placed between substations in order to transmit the traction energy using the overhead contact lines connected in parallel (on double-track lines) while allowing to break the circuit or disconnect a section when needed;
- overhead contact line, regarding all overhead wiring necessary to transmit the power along the line from the substations and section cabins to moving electric vehicles, with respect to any environmental conditions;
- return circuit, regarding all track-level equipment necessary to transmit the energy from the traction vehicles back to the substations with respect to personal safety, uninterrupted operation of signaling and control systems, and protection of any underground civil structures.

Although pantograph is considered an integral part of a vehicle, its interaction with the overhead contact line is being assessed according to the TSI "energy" requirements.

This chapter focuses on the need of upgrade of all the "links" mentioned above in a "chain" of power supply system in order to make it suitable for high-speed operation.

The paper presents basic OCL parameters and test results of current collection. One of the basic parameters limiting maximum service speed on a given line is the pantograph-OCL cooperation quality. The survey presented in this chapter confirmed that Polish overhead line equipment type 2C120–2C-3 and YC150–2CS150 can be safely used on lines with speed up to 250 km/h, which seems to be a technical limit for 3 kV DC system.

The chapter presents the following tendencies in overhead line equipment design in Poland: common use of copper alloys (CuAg, CuMg, CuSn), increased wire cross-section up to 150 mm^2, and increased mechanical tension (for CuMg and CuSn). For AC overhead contact lines, it is anticipated that bronze messenger wires will be used.

This chapter presents the thesis that development of high-speed railways in Poland should be linked with the introduction of the 2×25 kV 50 Hz system, as recommended in TSI "Energy."

11.2. Overhead contact line designs used in Poland

1   All OCL types used in Poland have been designed according to following climate conditions:

- minimum temperature (frost): −25 °C;
- frost accumulation temperature: −5 °C;
- normal temperature: +10 °C;
- temperature for blow-off calculations: +15 °C;
- maximum temperature (heat): +40 °C.

11.2.1. Overhead contact line of type 2C120–2C and 2C120–2C-3

New EU regulations about the "energy" and "rolling stock – locomotives and passenger rolling stock" subsystems [2], [3] relating to Directive 2008/57/EC of 17 June 2008 no longer divide the railway system into "high-speed (HS)" and "conventional." Lack of such a division can be justified by the thesis that high-speed trains should be free to transit from the HS network to a conventional one, so most requirements must be the same.

However, Directive 2008/57/EC divides railway lines into newly built lines for speeds 250 km/h or higher and into conventional lines upgraded for operation up to 200 km/h.

In Poland, there is one upgraded line providing 200 km/h operation in commercial service. This is the national line no. 4, commonly called "Central Rail Line" (CMK), linking Grodzisk Mazowiecki (near Warsaw) with Zawiercie (near Katowice).

The CMK line was built in 1971–1977 with track geometry intended for future 250 km/h passenger operation: minimum curve radius of 4000 m and a maximum cant set at that time at 100 mm [6]. The line was intended for mixed traffic. Tracks were electrified using a special type of overhead contact line marked as 2C120–2C.

The upgrade of the line is still in progress, enhancing the power supply and implementing an ETCS L1 system (without infill) [13]. At present, the line is operated with speeds of up to 200 km/h using ED250 Pendolino trainsets branded as EIP services on the following routes:

- Gdynia – Warsaw – Kraków – Rzeszów;
- Gdynia – Warsaw – Katowice – Bielsko-Biała/Gliwice;
- Warsaw – Wrocław;
- Kraków – Kołobrzeg;
- Warsaw – Jelenia Góra.

Real dynamic speed profile recorded on 08 January 2016 for EIP 1401 gives an objective perception about present performance of the line (Figure 11.1).

OCL designs type 2C120–2C and its further modification 2C120–2C-3[1] are DC compound catenary OCL systems [4]. A distinctive feature of this design is that the main and

---

1 Meaning of the Polish OCL marking is the following: first digit "2"stands for the number of messenger wires and "C120" for 120 mm² copper messenger wire (each). After the first dash, "2" stands for the number of contact wires and "C" stands for 100 mm² copper contact wire (default cross-section "100" is omitted for clarity). Suffix "3" placed after the hyphen stands for the third version of the basic OCL type.

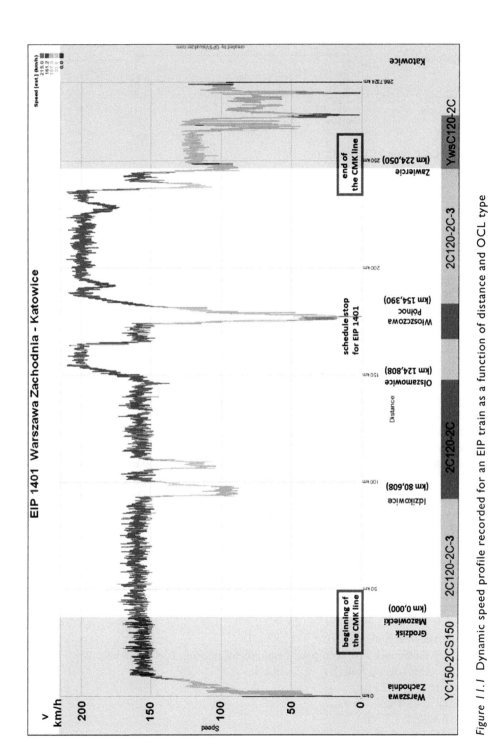

Figure 11.1 Dynamic speed profile recorded for an EIP train as a function of distance and OCL type

Source: Michał Głowacz – own measurement (2016)

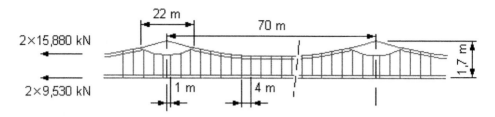

Figure 11.2 OCL type 2C120–2C

Source: [4]

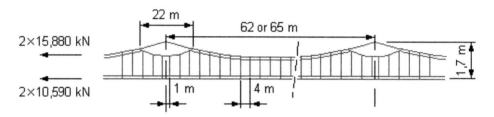

Figure 11.3 OCL type 2C120–2C-3

Source: [4]

auxiliary messenger wires are laid together along a span, but they split up at every support. The main messenger wire is fixed to the cantilever and the auxiliary messenger wire acts as symmetrical stitch wire 2×11 meters long (Figures 11.2 and 11.3).

Every messenger wire has a cross-section equal 120 mm² and every contact wire has 100 mm² respectively. Total OCL cross-section is therefore equal 440 mm² of copper.

When comparing technical parameters of the first (basic) version 2C120–2C with recommendations stated in the UIC Leaflet [5], it becomes clear that its Doppler coefficient (α), reflection coefficient of interfering wave (r), and amplification coefficient (χ) are not suitable for operation with speed 250 km/h. Therefore, the pantograph interaction with 2C120–2C contact line was supposed to prove unsatisfactory in practice. To improve its parameters, the OCL was redesigned. The tension of contact wires was increased by 10%. For that reason, the construction of the five-span overlap was changed by separating the auto-tension device into two independent ones, installed on two consecutive supporting structures for each direction.

After those design changes, the basic mechanical parameters are the following [4]:

- main messenger wire tension: F_a = 15.88 kN;
- auxiliary messenger wire (compound equipment) tension: F_c = 15.88 kN;
- contact wires tension (total): F_b = 21.18 kN;
- split wire length: 22 m;
- nominal span length: 70 m;
- system height: 1.7 mm;
- distance between first dropper and support: 1 m;
- minimum elasticity: e_{min} = 3,14 mm/daN;

- maximum elasticity (in the middle of a 62 m span): e_{max} = 4.30 mm/daN;
- coefficient of elasticity non-uniformity: u = 15,5%;
- mechanical wave propagation speed (v_c): 393 km/h;
- Doppler coefficient α calculated for line speed v_b = 200 km/h is equal to 0.35 and for v_b = 250 km/h is equal to 0.27 respectively;
- reflection coefficient of interfering wave: r = 0.58;
- amplification coefficient calculated for line speed v_b = 200 km/h is equal to χ = 1.65 and for v_b = 250 km/h: χ = 2.16 respectively;
- contact wire sag: 0 mm;
- distance between two consecutive droppers placed on the same contact wire: 8 m and on adjacent contact wires 4 m respectively;
- current capacity: I_{zn} = 2500 A.

11.2.2. Overhead contact line of type YC120–2CS150 and YC150-2CS150

OCL of type YC150–2CS150 and its modification YC120–2CS150 were constructed and implemented in 2004–2008. Design of the OCL assumed line speed between 200 km/h and 250 km/h.

OCLs of type YC120–2CS150 and YC150–2CS150 are DC catenary systems consisting of three wires. OCL of type YC150–2CS150 (Figure 11.4) has one 150 mm² copper messenger wire and its modification YC120–2CS150 (Figure 11.5) has copper messenger wire of smaller cross-section equal 120 mm². Both types contain two contact wires of 150 mm² each made of copper silver alloy. Therefore, total metallic cross-section for type YC120–2CS150 is equal to 420 mm² and for type YC150–2CS150, it is 450 mm².

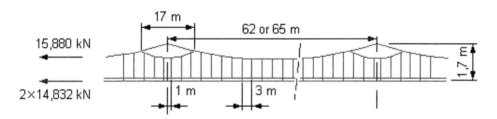

Figure 11.4 OCL type YC120–2CS150

Source: [12]

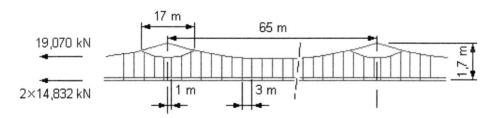

Figure 11.5 OCL type YC150–2CS150

Source: [12]

Mechanical and electrical design parameters of the OCLs are as following:

- messenger wire tension: F_a = 15.88 kN (for 120 mm² wire [YC120–2CS150]) F_a = 19.07 kN (for 150 mm² wire [YC150–2CS150]);
- stitch wire tension: F_c = 2.50 kN;
- contact wire tension: F_b = 29.66 kN;
- stitch wire length: 17 m;
- nominal span length: 65 or 62 m, according to adopted blow-off (wind zone);
- system height: 1.7 m;
- distance between first dropper and support: 1 m;
- minimum elasticity: e_{min} = 2.84 mm/daN (for type YC120–2CS150) and e_{min} = 2.56 mm/daN (for type YC150–2CS150);
- maximum elasticity: e_{max} = 3.83 mm/daN (for type YC120–2CS150) and e_{max} = 3.63 mm/daN (for type YC150–2CS150);
- coefficient of elasticity non-uniformity: u = 15% (for type YC120–2CS150) and u = 16% (for type YC150–2CS150);
- mechanical wave propagation speed: v_c = 369 km/h;
- Doppler coefficient α calculated for line speed v_b = 200 km/h is equal to 0.31 and for v_b = 250 km/h is equal to 0.21 respectively;
- reflection coefficient of interfering wave is equal to r = 0.4 (for type YC120–2CS150) and r = 0.42 (for type YC150–2CS150);
- amplification coefficient for line speed v_b = 200 km/h is equal to χ = 1.3 (for type YC120–2CS150) and χ = 1.4 (for type YC150–2CS150), although for line speed 250 km/h the coefficient is equal to χ = 1.89 and 1.99 respectively;
- contact wire sag: 0 mm;
- current capacity: for type YC120–2CS150 is equal to I_{zn} = 2500 A, and for type YC150–2CS150 is equal to 2730 A, assuming line speed of 200 km/h with 10-minute headway and at wind speed 0.6 m/s.

11.2.3. Overhead contact line of type YwsC120–2C-M

OCL of type YwsC120–2C-M is a DC catenary system consisting of three wires. The OCL of has one 120 mm² messenger wire and two contact wires of 150 mm² each made of copper silver alloy. Therefore, total metallic cross-section for type is equal to 320 mm². Stitch wire 16 m long is provided under every support (Figure 11.6).

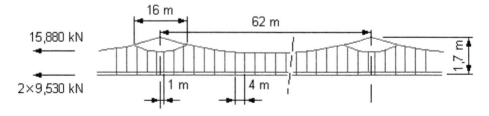

Figure 11.6 OCL type YwsC120–2C-M

Source: [16]

Mechanical and electrical design parameters of the OCLs are as following:

- messenger wire tension: 15.76 kN;
- contact wire tension: 19.06 kN;
- stitch wire tension: 16 m;
- system height: 1,700 m;
- nominal stagger: ±300 mm;
- overlap: supported by six structures
- Doppler coefficient α calculated for line speed v_b = 200 km/h: 0.33;
- reflection coefficient of interfering wave r: 0.419;
- current capacity: 1700 A.

11.3. Tests of high-speed overhead contact lines in Poland

11.3.1. 2C120–2C-3 tests

Instytut Kolejnictwa (at the time CNTK) performed tests of a modified Italian EMU class ETR 460 *Pendolino* [8]. During these tests, the overhead contact line was interacting with a SBD 89 3 kV pantograph manufactured by Schunk Bahn und Industrietechnik. Static force of the pantograph was dependent on train speed. Precisely, up to 140 km/h the force was 90 N and for higher speeds 140 N.

These tests were performed on 11 May 1994 on track no. 2 between Strzałki and Biała Rawska on the CMK line. The tests resulted in setting a new speed record for Central Europe railways of 250.1 km/h. Before the test run, the height position of the OCL had been adjusted along the route and set on 5465 mm with standard deviation of 40 mm.

Basic test parameter was contact wire uplift, measured at two consecutive supports and with 2/3 of distance between them. The tests results are shown in Figure 11.7. Another important recorded parameter was pantograph voltage. The voltage recordings proved that no loss of contact occurred during the 250 km/h run. Maximum recorded uplift was equal to 94 mm and was recorded at 235 km/h, which was the maximum speed reached at the track-side measuring point.

In 2009, other tests were performed on track no. 1 between Psary and Góra Włodowska. The test subject was a pantograph of type DSA 250 installed on a Siemens ES64U4 electric locomotive. During these tests, contact wire uplift at a support and contact point vertical displacement were recorded. Moreover, contact force was estimated. Maximum speed obtained during the tests was 235 km/h.

An interesting waveform of contact wire uplift is shown in Figure 11.8. Interference of a wave generated by the pantograph passing at 220 km/h and a reflected wave resulted in creating a standing wave of amplitude 35 mm and duration of 28 s.

Figure 11.9 shows estimated force between the contact wire and the pantograph passing at 235 km/h. Mean static contact force measured right before the test was equal to 114 N, and the difference between measured lowering and raising force (caused by friction) was equal to 12 N. Mean contact wire height was 5203 mm with a standard deviation of 12 mm.

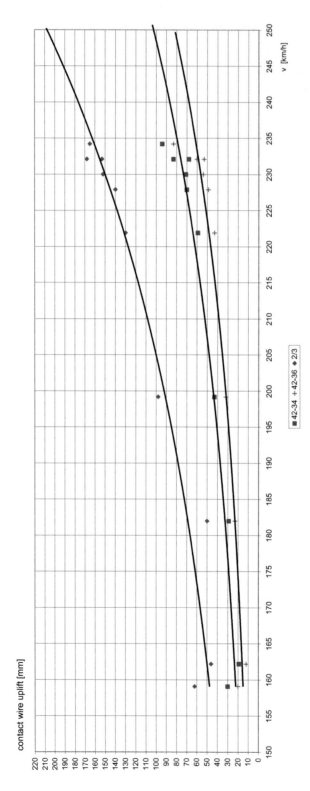

Figure 11.7 Contact wire uplift as a function of speed

Source: [8]

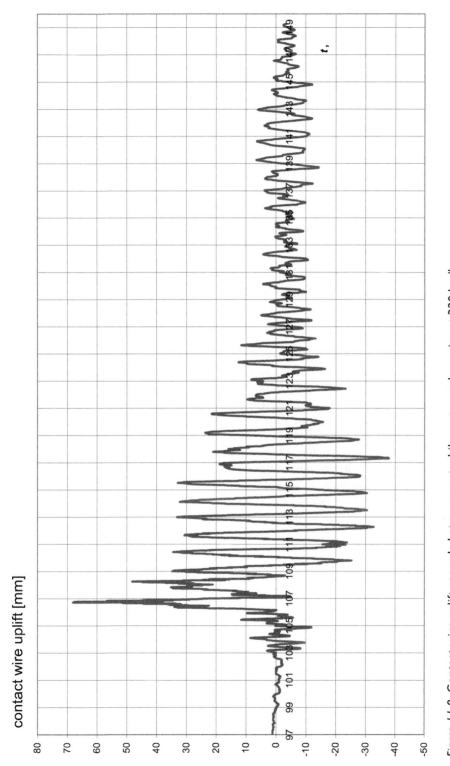

Figure 11.8 Contact wire uplift recorded at a support while pantograph passing at 220 km/h

Source: [11]

Assessment of contact force value F between pantograph and OCL has been performed using the following formula:

$$F = F_{st} \pm F_t \pm m_z \frac{d^2 y}{dt^2} + F_{ae}$$

Each element of the formula mentioned above has been measured separately, where:
F – contact force,
F_{st} – static force,
F_t – dry friction force,
m_z – effective mass of the OCL and the pantograph,
$\frac{d^2 y}{dt^2}$ – contact point acceleration calculated on basis of measurement of the contact point vertical displacement F_{ae} – aerodynamic force measured according to EN 50317:2012, 6.

As a part of the dynamic tests of ETR 610 (PKP class ED250) conducted in November 2013 between Psary and Góra Włodowska, the interaction between pantograph of type DSA 250.14_PKP and the overhead contact line was checked. Several test runs performed at this time resulted in attaining new speed records: a record for EMU in Central Europe and a record for the *Pendolino* trainset family. This speed record of 293 km/h was set on 24 November 2013.

Another survey conducted on the Psary – Góra Włodowska section concerned the DSA 200 pantograph with copper contact strips. During this survey, the temperature rise inside the contact strip was checked while the train was running at 200 km/h. The maximum current drawn by the locomotive was 2679 A. The temperature was measured by thermoelements placed in small holes drilled in the strip. Eight holes were drilled along the strip, placed at 100 mm, 200 mm, 300 mm, and 400 mm from the track center on both sides. The highest temperature was recorded at 400 mm from the track center and was equal to about 90 °C. The temperature rise was achieved almost immediately after the current reached its maximum. Temperature recordings are shown on in Figure 11.11. According to EN 50119:2009, 5.1.2, permanent temperature rise should not exceed 80 °C and temperature rise above 120 °C should not last longer than 30 minutes.

11.3.2. YC120–2CS150 and YC150–2CS150 tests

During 2005–2007, as a part of Project No. 6T08 2004C/06482 entitled *Opracowanie i wdrożenie technologii wytwarzania z miedzi stopowej przewodzących elementów górnej sieci trakcyjnej o znamionowej obciążalności prądowej powyżej 2,5 kA i podwyższonej wytrzymałości mechanicznej*, (Technology development and implementation of alloy copper conductive elements of OCL production with a rated current capacity above 2.5 kA and increased mechanical strength), OCLs of new types YC120–2CS150 and YC150–2CS150 were designed. The project involved designing and constructing a new test section of OCL with enhanced current capacity through using wires made of copper silver alloy. The first section of the OCL was built between Stara Wieś and Żychlin. In 2006 and 2007, type tests of the new OCL were conducted [10]. The first stage of the tests involved test runs with speed up to 176 km/h and the second up to 210 km/h. Using higher speed for the tests was not possible due to traction parameters of available motive stock.

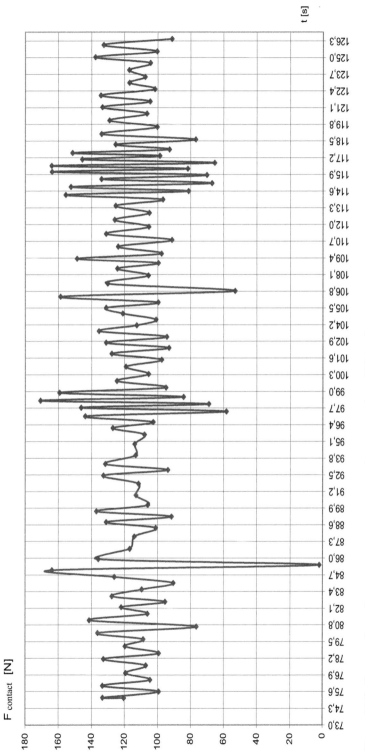

Figure 11.9 Contact force between contact wires and pantograph at 235 km/h

Source: [11]

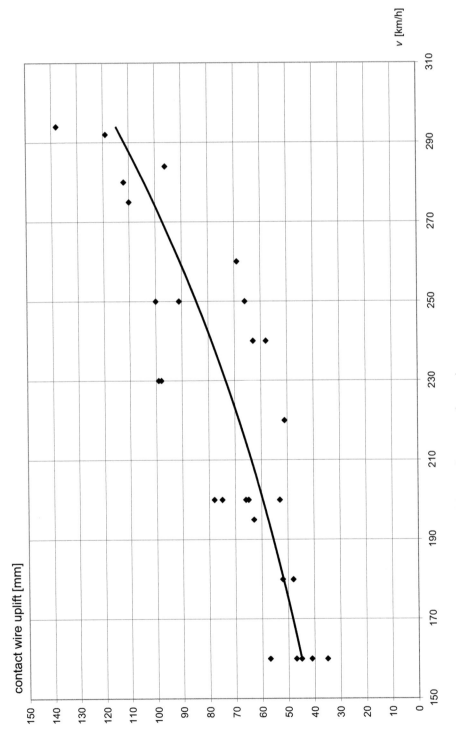

Figure 11.10 Measured contact wire uplift as a function of speed

Source: [14]

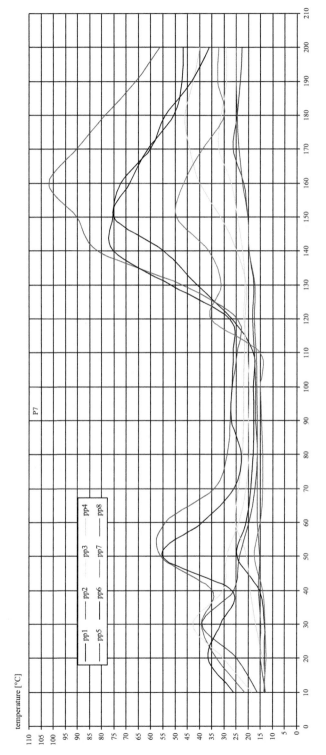

Figure 11.11 Waveforms of temperature rise measured inside the copper contact strip of DSA 200 pantograph. Measuring points pp1 to pp8 were placed along the strip.

Source: [11]

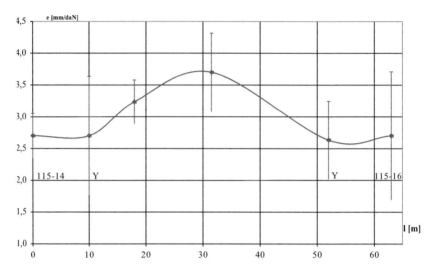

Figure 11.12 Elasticity distribution of YC120–2CS150

Source: [10]

Measurement results of static elasticity confirmed theoretical calculations performed at the design stage. Elasticity distribution along a 63-meter span is shown in Figure 11.12. During the tests, the following parameters were measured or calculated: contact wire uplift at a support, points of contact loss, and contact force between the wire and collector head. Pantograph used for the tests was diamond-shaped, spring-loaded, and equipped with a collector head with two carbon strips.

During the first series of tests when the train speed was about 180 km/h, the maximum observed uplift caused by a pantograph with static force 109 N was 58 mm. The uplift caused by the same pantograph with lower static force 91 N was 52 mm. No contact loss was found while collecting 100% of nominal traction current.

During the second batch of tests, the same pantograph was interacting with OCLs with speeds up to 210 km/h. At 210 km/h, the highest recorded contact wire uplift was 63 mm, caused by the pantograph with mean static force of 109 N. Contact wire uplift record is shown in Figure 11.13. A standing wave of amplitude 20 mm was created from the interference of the primary and reflected waves. The amplitude is lower than that recorded for 2C120–2C-3 on the CMK line. Comparison of contact wire uplift against speed that had been recorded for both types of overhead line equipment is shown in Figure 11.14.

Contact force between the pantograph and both types of overhead contact lines is shown in Figure 11.15. Mean contact force is equal to 143 N with a standard deviation of 13 N. The obtained results comply with current requirements concerning dynamic behavior and quality of current collection specified in TSI for "energy" subsystem.

11.3.3. YwsC120–2C-M tests

Tests of YwsC120–2C were conducted in 2009 during ES64U4 certification process [11]. OCL was interacting with a DSA 250 pantograph. A tensioning section of the OCL that was

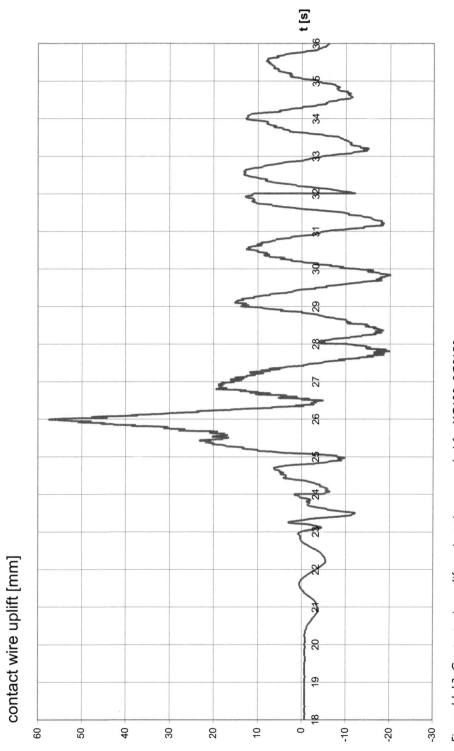

Figure 11.13 Contact wire uplift against time recorded for YC120–2CS150

Source: [10]

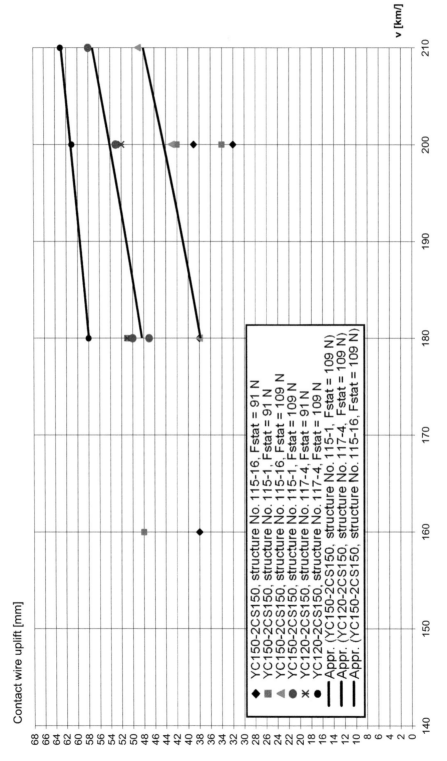

Figure 11.14 Comparison of contact wire uplift against speed recorded for both types of OCL

Source: [10]

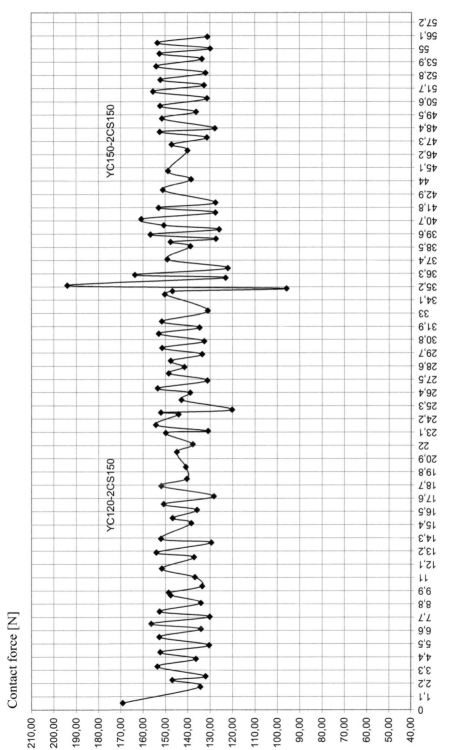

Figure 11.15 Comparison of contact force recorded for both types of OCL at 210 km/h

Source: [10]

used during the tests was hung up at 5473 mm ± 38 mm over the top of the rail. The maximum difference between two consecutive supporting structures was proved to be not greater than 81 mm (1.4‰). Therefore, the setup of the OCL section did not comply with UIC Code 799–1 requirements stating the maximum permitted value of the parameter was 30 mm. Stagger of the OCL section was equal 30 mm.

Uplift of the contact wire at a support was measured. Then, contact force was estimated on the basis of the measurement. The uplift of the contact wire is shown in Figure 11.16, and the contact force at 200 km/h in Figure 11.17.

The contact force profile calculated between the OCL and the pantograph is shown in Figure 11.17. Mean value of the force is 113 N and its standard deviation 30 N. The results obtained during the survey confirmed that the parameters of the OCL comply with TSI "energy" requirements regarding current collection quality and dynamic characteristics.

11.4. Newly constructed high-speed lines in Poland

The Polish Ministry of Transport, Construction and Marine Economy is planning to build a new high-speed line. The line is to connect Warsaw with Łódź, Poznań, and Wrocław. The project is branded the "Y" line.

A feasibility study prepared by CNTK in 2005 provides for the line electrification with 2×25 kV 50 Hz power supply system. The power supply system is to use substations spaced by 40 km to 60 km and fed by separate 220/400 kV transmission lines. Between the substations, autotransformer stations are to be provided. A concept drawing of such system is shown in Figure 11.18.

One of the three following OCL types were proposed to be used [15]: YBz95-CMg150, YBz70-CS120, or YBz70-CSn120. Within city junctions (i.e. in the Łódź junction area), it would be possible to retain the 3 kV DC system, but it would be enhanced by spacing the substations by 12 km to 15 km, providing single-stage 110/3 kV transformation, and using an OCL with 600 mm² cross-section.

11.5. Summary

Since the beginning of 1990s, we can observe a rapid growth of the high-speed network in Europe. The commercial speed of 300 km/h becomes a standard. The most common electrification system for the lines is 2×25 kV 50 Hz with additional feeder cable and autotransformers. A traction unit is therefore powered from a mono-phase contact line but with quasi-bilateral feeding – which allows to transmit high power at relatively low cost and at full compatibility with a classic 25 kV system (Figure 11.18).

The survey confirmed that overhead line equipment type 2C120–2C-3 and YC150–2CS150 can be safely used on lines with speeds up to 250 km/h while operating 7.5 MW traction vehicles, which seems to be a technical limit for a 3 kV DC system. Moreover, an overhead contact line of type YwsC120–2C-M can be used to supply motive stock with power limited to 5.1 MW.

We can observe the following tendencies in overhead line equipment design in general in Europe: common use of copper alloys (CuAg, CuMg, CuSn), increased wire cross-section up to 150 mm², and increased mechanical tension (for CuMg and CuSn). It is also popular to use bronze messenger wires for AC overhead contact lines. The concept of using additional

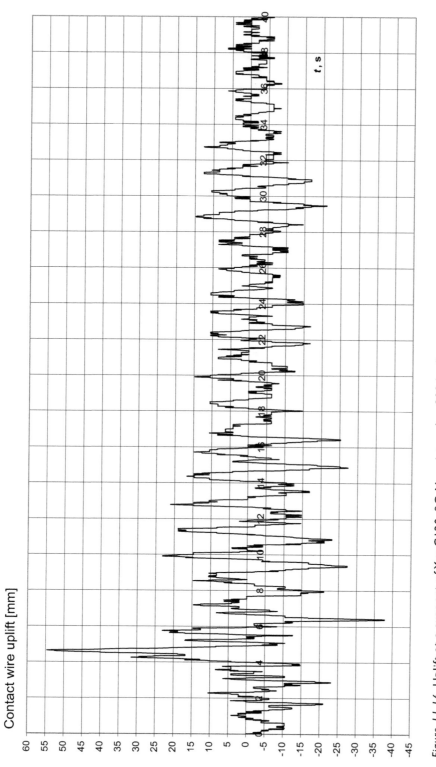

Contact wire uplift [mm]

t, s

Figure 11.16 Uplift at a support of YwsC120–2C–M registered at 200 km/h

Source: [11]

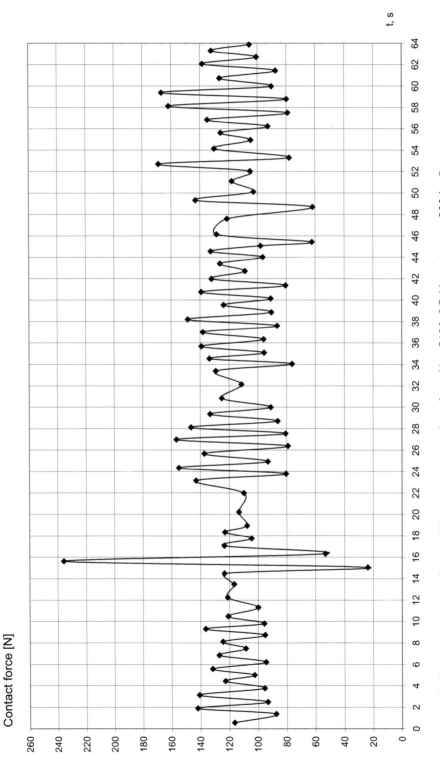

Figure 11.17 Calculated contact force while passing an overlap of two YwsC120–2C-M sections at 200 km/h

Source: [11]

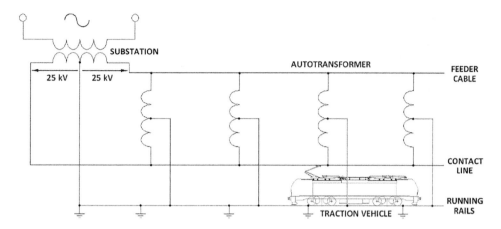

Figure 11.18 2 × 25 kV 50 Hz power supply system
Source: Michał Głowacz – own work

stitch equipment in order to improve uniformity of elasticity distribution also comes back into engineers' favor.

Bibliography

[1] Directive 2008/57/EC of the European Parliament and of the Council of 17 June 2008 on the interoperability of the rail system within the Community. OJ L191, 18.7.2008, pp.1–45

[2] Commission Regulation (EU) No 1302/2014 of 18 November 2014 concerning a technical specification for interoperability relating to the 'rolling stock – locomotives and passenger rolling stock' subsystem of the rail system in the European Union. OJ L356, 12.12.2014, pp. 1–393

[3] Commission Regulation (EU) No 1301/2014 of 18 November 2014 on the technical specifications for interoperability relating to the 'energy' subsystem of the rail system in the Union. OJ L356, 12.12.2014, pp. 1–227

[4] Kaniewski, M., and Maciołek, T.: *Sieci trakcyjnej dużych prędkości jazdy na PKP* (*High Speed of Overhead Contact Line on PKP*), 6th International Conference "Modern Electric Traction in Integrated XXIst Century Europe" MET'2003, Warsaw University of Technology, pp. 212–216.

[5] UIC Code 799–1 OR *Characteristics of Direct-Current Overhead Contact Systems for Lines Worked at Speeds of Over 160 km/h and Up to 250 km/h*. UIC 799-1ED.1, 01.01.2001, pp. 1–25.

[6] 30 lat Centralnej Magistrali Kolejowej (*30 years of the Central Railway Line*), Świat Kolei, 2004, 12, pp. 22–27.

[7] Kaniewski, M.: *Opracowanie kryteriów technicznych projektowania górnej sieci trakcyjnej o znamionowej obciążalności prądowej powyżej 2,5 kA z zastosowaniem elementów przewodzących ze stopu miedzi. (Elaboration of Technical Criteria for Designing the Overhead Contact Line with a rated Current Capacity above 2.5 kA using Conductive Elements made of Copper alloy).* Wykonane w ramach projektu celowego 6T08 2004C/06482 pt. *Opracowanie i wdrożenie technologii wytwarzania z miedzi stopowej przewodzących elementów górnej sieci trakcyjnej o znamionowej obciążalności prądowej powyżej 2,5 kA i podwyższonej wytrzymałości mechanicznej (Targeted Project 6T08 2004C/06482. Development and Implementation of Technology for the conductive elements of the Overhead Contact Line production with a rated current capacity above 2.5 kA and increased mechanical strength).* Warszawa: Instytut Kolejnictwa, 2005, pp. 1–16.

[8] Jarosz, T.: *Sprawdzenie sieci jezdnej na wybranych odcinkach linii Warszawa Gdańsk i CMK w związku z problemami pociągu ETR460 kolei FS na PKP (Veryfication of the Overhead Contact Line on selected sections of the Warsaw – Gdańsk and CMK lines due to train problems of the ETR460 train on the FS railways at PKP)*. Warszawa: Instytut Kolejnictwa, 1994, pp. 1–21

[9] Kaniewski, M.: *Badania sieci YC150–2CS150 i YC120–2CS150 przy prędkości jazdy 200 km/h. (Tests of the YC150-2CS150 and YC120-2CS150 Overhead Contact Line at a speed of 200 km/h.)* Warszawa: Instytut Kolejnictwa, 2007, pp. 1–65.

[10] Kaniewski, M.: *Badania eksploatacyjne doświadczalnego odcinka sieci trakcyjnej zbudowanej z doświadczalnej partii elementów przewodzących z miedzi srebrowej (Operational Tests of the Experimental section of Overhead Contact Line constructed from an experimental part of conductive elements made of silver copper)*. Warszawa: Instytut Kolejnictwa, 2006, pp. 1–57.

[11] Kaniewski, M.: *Przeprowadzenie badań technicznych lokomotywy ES64U4 produkcji Siemens (Siemens ES64U4 locomotive Technical Research)*. Warszawa: Instytut Kolejnictwa, 2009, pp. 1–68

[12] Kaniewski, M., and Burak Romanowski, R.: *Dokumentacja techniczno-ruchowa sieci trakcyjnej typu YC120–2CS150 i typu YC150–2CS150 (The Overhead Contact Line Type YC120-2CS150 and type YC150-2CS150 Technical and Operational Documentation)*. Warszawa: Instytut Kolejnictwa, 2011, pp. 1–13.

[13] Massel, A.: *Przygotowanie infrastruktury kolei dużych prędkości w Polsce. Koleje dużych prędkości w Polsce* [Preparation of high speed rail infrastructure in Poland. High-speed railways in Poland], joint publication edited by Mirosław Sergiejczyk, Warszawa: Instytut Kolejnictwa, 2015

[14] Kaniewski, Marek: *Proces weryfikacji EMU250 PKP (współpraca odbieraka prądu z siecią trakcyjną). (EMU 250 PKP verification process (cooperation of the Current Collector with Overhead Contact Line)*. 12/2743/21, Warszawa: Instytut Kolejnictwa, 2014, pp. 1–121

[15] Szeląg, A., Maciołek, T., Kamiński, B., Kaniewski, M., Knych, T., Kawecki, Niemyski W., Siennicki, Z., Zgiep, J., and Freliszka, J., *Przygotowanie pilotażowego wdrożenia w Polsce systemu zasilania trakcji 25 kV prądu przemiennego – Etap II Projekt prototypowej podstacji trakcyjnej i sieci trakcyjnej (Preparation of a pilot implementation in Poland of a 25 kV AC traction power supply system – Stage II prototype traction substation and Overhead Contact Line Design)*, Warszawa: Instytut Kolejnictwa, 2012, pp. 1–32.

[16] Kaniewski, Marek: *Dokumentacja techniczno-ruchowa sieci trakcyjnej typu YwsC120–2C-M (The Overhead Contact Line Type YwsC120-2C-M Technical and Operational Documentation)*, Warszawa: Instytut Kolejnictwa, 2008, pp. 1–12.

Chapter 12

High-speed lines control command and signaling

Marek Pawlik

12.1. Introduction

Control command and signaling from the point of view of a high-speed rail system should be subdivided into four parts: primary signaling, track-side control command, on-board control command, and communication. This chapter describes the first three of them. Communication systems should not be forgotten, of course, as they form an integral part of control command and signaling; however, they are described in another chapter. Primary signaling comprises systems checking occupancy of tracks and switches and systems that use information about occupancy – station interlockings, line block systems, and level crossing protection systems. Primary signaling is used on most of the railway lines – not only for high-speed rails. It is not used on some low-density lines, on which traffic is managed according to operational rules using significantly minimized technical support, e.g. based on advance notices passed by telephones. Primary signaling is closely interlocked with operational rules applicable on railway lines in individual countries. These rules cannot be changed in a substantive way in a short period of time. Possible step-by-step change means there is a long implementation period, within which the risk level in railway transport increases so dramatically that it becomes unacceptable. Therefore, it is assumed that such change would cause a loss of safety in railway transport and therefore unification of the technical solutions used in primary signaling was not and is not foreseen on the level of European law. Differentiation of the operational rules, and the resulting differentiation of the technical solutions in primary signaling, creates an important barrier for high-speed rails. This is important, especially as connecting agglomerations, which are in different member states of the European Union, is one of the objectives of implementing high-speed services. The situation is different in the case of control command. National, country-specific control command solutions were used and are still used in different countries; however, the control command technical solution is currently fully defined in European Union law. Control command is based on safe digital data transmission [1], [6]. Data taken from primary signaling systems are transmitted to vehicles. In practice, the only vehicles running on high-speed lines are trains; therefore, the term "train" is frequently used below. The received data are used in trains by on-board equipment, which checks train movement against speed and distance limits, which are known thanks to data transmission from primary signaling. Such limits are called below movement authorities (MA). Full knowledge of the running limits, which is available for the on-board equipment from the received MA authorities, is used additionally to display, on the driver's desk in the drivers' cab, information about the limits [1]. The equipment used for displaying MA on the driver's desk is called cab signaling or Driver Machine Interface (DMI).

12.2. Primary signaling

As already mentioned, primary signaling is used not only on high-speed rail lines. However, on high-speed lines, primary signaling is also required as a source of data necessary for compiling MA authorities. Primary signaling for high-speed lines practically does not include level crossing protection functionality. Level crossings are strictly forbidden on lines with a maximum speed higher than 160 km/h. Therefore, on all sections where the maximum speed exceeds 160 km/h, all roads that cross the line must go below or above the railway line. Level crossings are allowed on sections with a value lower than the maximum speed. Such sections are frequently required, because of topographical, relief, or town-planning constraints, to reach main stations in agglomerations and big cities, where passenger service is due to take place. However, in such places multilevel crossings are also preferable for other reasons, e.g. for achieving fluent road traffic within the cities.

The main goal of the station interlockings is to completely exclude any possibility to set conflicting routes on the railway stations. Setting conflicting routes would create an unacceptable safety risk [6]. High-speed trains are, therefore, subject to the same rules as conventional trains. Station interlockings have to be common, as routes set for trains via stations for high-speed trains and for conventional trains cannot conflict. No track, switch, overlap, or side protection can be used at the same time for MA authorities for different trains, regardless of whether the trains are high-speed ones or conventional ones.

On the sections between stations, trains need to keep safe distances between each other while running [3], [6]. This is ensured by line block systems that subdivide sections between stations into block sections. Such systems are used on more important conventional railway lines between main stations. At intermediate stations, passing the station under the control of the block system is allowed, but only when running through on the main tracks and when the station is equipped accordingly. In the case of high-speed lines, the number of block sections between stations is commonly growing. Block sections are becoming shorter. The number of sections required for braking from maximum speed to standstill is rising. Together with the increase in speed, the potential consequences of accidents are also increasing, while the time available for drivers to make decisions and act is decreasing, e.g. for initiating brake intervention. The time window in which drivers are able to observe signal aspects displayed on the track-side color light signals is getting shorter. Therefore, frequently only some block sections are protected by track-side signals. Other block sections are only visible for control command systems and are therefore called virtual block sections. From the point of view of the trains not equipped with on-board control command, such virtual blocks do not exist and, as a result, headways between conventional trains are longer than headways between high-speed trains. Availability requirements are increasing. The highest level of safety has to be maintained, regardless of speed. Occupation of the virtual block sections, in the same way as the occupation of the block sections protected by track-side color light signals, is checked for each block section separately. The line block system therefore has information about the occupation of all individual block sections. Such knowledge about track occupancy is taken into account by track-side control command equipment when computing MA authorities based on data from the line block system.

Devices checking the occupancy of track sections and switches are used as a source of data by station interlockings and by line block systems. On high-speed lines, axle counters are used for that purpose. On conventional lines, track circuits, which were used for that purpose for a long time, are still being used. Track occupancy is determined thanks to an axle counter composed of a pair of senders and a pair of receivers installed together on the rail and generating electrical signals, which enable axles passing such a point on the track in both directions to be counted. The presence of any vehicle on the individual track section or switch is determined by linking axle counters in groups.

12.3. Control command

12.3.1. Relationships between primary signaling and control command

As already mentioned, control command was introduced and is still used in many countries as a national, country-specific technical solution. Currently, control command is fully defined at the level of European Union legislation. A transition period is currently pending, and national solutions are being gradually replaced by a unified solution defining track-side control command and on-board control command equipment. Both functional and technical requirements are defined and imposed by European law. The imminent subdivision of the control command systems into track-side and on-board equipment, in combination with high-speed trans-border railway traffic, requires deeply unified requirements for fully coherent track-train data transmission, as well as adequate functional and technical unification of the on-board and track-side equipment as such. This does not mean a single producer, as all the documents defining the European control command system belong to the public domain. Nearly 50 documents, precisely defining the European solution, are pointed out as obligatory in the annex to the European Commission Regulation that defines requirements for track-side control command subsystems and for on-board control command subsystems. Subdivision of the control command into two subsystems arises, first of all, from the separate realization of implementations, as there are many separate implementation projects, which are tendered separately and frequently by different contracting entities, and as equipment is constructed by different providers and, finally, verified within separate processes before being put into service.

Relationships between primary signaling and control command, including the subdivision into track-side and on-board equipment, are shown in Figure 12.1. Primary signaling, which is installed track-side by its nature, and track-side control command equipment are shown schematically on the left side. Internal primary signaling arrows show that, on the one hand, interlockings, block systems, and level crossing protection systems are connected with track occupancy checking devices, as they need safe information about track occupancy. On the other hand, interlockings, block systems, and level crossing protection systems are usually connected with remote control realized by local control centers and operational management realized by regional operational centers. On the left side of Figure 12.1, track-side control command is also shown, distinguishing control command equipment and its interfaces with primary signaling. It can be seen that taking data from primary signaling can take place locally directly from interlockings, block systems, and level crossing protection systems, and can also take place from the remote control systems. Data from primary signaling cannot be taken from operational management systems, as safety of the data acquisition cannot be ensured in such a case. On the right, arrows are used to mark the main data flow directions in on-board control command equipment. Before a journey, a set of train data has to be entered by the driver (weight, length, brake type, etc.). Without such data, MA authority cannot be precisely displayed on the DMI. The data in question

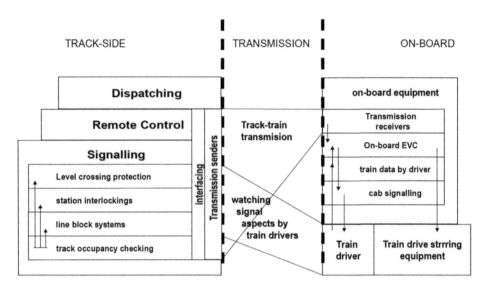

Figure 12.1 Relationships between primary signaling, track-side control command, and on-board control command [4], [9]

have to be entered at standstill. The train then receives MA authorities while running and, after their verification in fractions of milliseconds, the on-board equipment checks whether the train has violated limits defined in the MA authority. If so, emergency braking is initiated automatically. Precise data concerning MA authority, detailed by taking into account the train characteristics, allow the driver to drive the train in a safe way without the need to observe track-side color light signals displaying signal aspects, whose visibility significantly decreases with increased speed. It is believed, not only in Poland, that driving trains based on track-side signals is possible only up to 160 km/h. Such a restriction arises, among others, from the rules, according to which track-side color light signals are installed along the tracks, in a way that ensures the visibility of signal aspects. Such a restriction in Poland also arises from Polish law.

The general characteristics of the control command and signaling described above also apply to the European Train Control System (ETCS), which uses the unified European communication system, called Global System for Mobile Communication for Railway (GSM-R), as one of the data transmission channels. ETCS and GSM-R combine to make the unified European Railway Traffic Management System (ERTMS). ETCS is described below in this chapter. GSM-R is described separately in the chapter dedicated to communication. The general characteristics described above also apply to many national control command systems, although not all functions shown in Figure 12.1 are utilized by all of them. For instance, using train data entered by the train drivers via DMI to make MA authorities more precise by taking train characteristics into account is not common. The most important differences between the systems designed, constructed, and used in different countries, regardless of the technical solutions, are described by control command classes defined based on the scope of embedded functionalities.

12.3.2. Control command systems subdivision into classes

Systems based on data transmission between equipment installed on tracks and equipment installed in trains, frequently called track-train transmission systems (TTT), are subdivided into three groups: vital train movement control systems called communication based train control (CBTC), systems imposing braking in all trains present in a predefined area and systems tracking and tracing vehicles and loads called telematic applications (TA), which are used for logistic data exchange necessary for including railway transport in logistic networks [7].

For high-speed rails, a key role is played by the CBTC systems. They are shown in cooperation with primary signaling in Figure 12.1. Systems imposing braking in all trains present in a predefined area, important from an operational point of view, are described separately in the chapter dealing with communication systems. From the control command and signaling point of view in that respect, linking the technical solution called RADIOSTOP, which is commonly used in Poland, with the European solution embedded in GSM-R for emergencies, called Radio Emergency Call (REC), is crucial.

Within communication based train control systems: AWS, ATP, ATC, and ATO systems are distinguished, as well as ATS systems used together with the ATO ones [4], [9].

Automatic warning systems (AWS) provide audible and/or visual indications to alert the driver when approaching potentially dangerous places, e.g. track-side signals able to show stop aspect, level crossings, permanent, or temporary speed restrictions. The way the AWS systems work does not depend on signal aspects shown on track-side signals, as they do not take into account present operational distance and speed restrictions and do not use transmission of the MA authorities. Because of the lack of information about signal aspects shown on the track-side signal, it should be admitted that AWS class systems do monitor driver vigilance, but do not check the way the driver drives the train against MA authority. AWS class systems cannot be accepted as the only CBTC solution for high-speed rails; however, they can be used as supplementary solutions together with higher-class systems, e.g. ATP class or ATC class systems.

Automatic train protection (ATP) systems check the way the driver drives the train against restrictions represented by the signal aspects and increasingly directly against MA authorities, while track-side signals are omitted. Checking takes place at defined points along the railway line, including particular points where speed restrictions start. Omitting track-side signals is only possible when all trains running on the line are equipped with the same ATP class system that is installed on the line. ATP class systems transmit information about the present signal aspect shown on the track-side color light signal (more broadly speaking, about operational restrictions, thereby about MA authority) and check the way the driver drives the train, comparing train speed with the appropriate allowed value. ATP class systems compare the speed and location of the train with a stepwise speed function defined in relation to the distance travelled, as they do not take into account train movement dynamic effects. ATP class systems are frequently, but not always, equipped with cab signaling, which displays, in the drivers cab, signal aspects equivalent to those being shown on track-side color light signals. Systems that belong to this class are no longer used on high-speed lines, as they are being replaced by ATC class systems.

Automatic train control (ATC) systems check the way the driver drives the train against MA authority in real time. ATC class systems not only possess information about movement restrictions imposed by primary signaling, but also take into account train movement dynamic effects. Train speed is thereby checked not only against speed limits, but also braking to the beginning of the speed limit and, in the same way, not only against passing the end of MA authority, but also braking intervention and braking to the end of authority. Continuous checking is complemented by the continuous display on the DMI of up-to-date information supporting drivers in driving trains near the speed limit during ceiling speed monitoring and driving trains near braking intervention curve during target speed monitoring. The European unified control command system imposed by European law called the European Train Control System is an ATC class system.

Automatic train operation (ATO) systems automatically operate devices controlling train movements, in accordance with all data received from track-side control command and signaling, replacing drivers in driving the trains. ATO class systems provide a procedure which ensures required changes of the train speed without any driver action, called automatic train driving. ATO class systems frequently complement ATC class systems, which filter commands given by ATO, providing train movement safety. ATO class systems require some additional functionality track-side, which is ensured by automatic train supervision (ATS) systems for monitoring the movements of a number of trains at the same time. Together, ATO class systems and ATS class systems are called automatic people movers (APM). ATO class systems are currently used only for rail transport operated using vehicles with similar or even identical dynamic parameters, where the maximum speed does not exceed 100 km/h. Thereby, at present, ATO class systems are not used for high-speed trains. Such systems would enable driverless operation, but it is not currently acceptable for the railways or for the passengers. However, intensive works aimed at including the ATO functionalities in the unified European control command system are ongoing. It has to be expected that ATO functionality will be used on long sections between main stations, but not while entering stations, driving through the stations, and leaving the stations. The ATO systems will therefore support the drivers in driving trains on sections with the highest maximum speeds, where the time left for driver reaction is the shortest, and will not take over the work of the train drivers. The ATS class systems are also currently used only for supervising groups of vehicles with similar or even identical dynamic parameters, running on separated infrastructure with a maximum speed not exceeding 100 km/h. The ATS functionalities in the case of high-speed rails can be implemented partly by enhancing the functionality of the interlockings on the level of local control centers and of the operational management systems on the level of regional operational centers.

12.4. European unified control command

12.4.1. European Railway Traffic Management System ERTMS

The European Railway Traffic Management System (ERTMS) has been developed to create a unified system for managing railway traffic in the European Union [6], [7], [8], [9]. It has to ensure railway transport interoperability, and thereby the capability of safe and fluent train movement on the railway networks of the individual European Union member states, and for crossing borders without the need for stopping, changing traction units, or changing drivers. Far-reaching European conformity is required to achieve such capability, consisting of:

- infrastructure (tracks, engineering structures, platforms, etc.);
- power supply (overhead contact line, on-board energy measuring systems, etc.);

- control command and signaling (control command, radio communication, driver vigilance checking, etc.);
- operation (operational rules, train audibility and visibility, safety-related staff competences, etc.);
- rolling stock (structure gauge, axle load, mechanical strength parameters, etc.);
- maintenance (maintenance works); and
- telematics (telematic applications for passenger and for freight services).

The ERTMS system is a necessity due to the differentiation of national control command systems used in different countries and their significant incompatibility. As a result of such differentiation, trains running over the borders between member states (or between railway networks managed by different infrastructure managers) have to be equipped with multiple on-board control command installations, all separately certified in individual countries based on verifications on track-side control command installations that are already certified in those countries [1], [3], [8]. Because of introducing unified requirements in relation to the topics quoted above, infrastructure managers will offer routes with interoperable permanent way, traction, and control command for many competing railway undertakings using interoperable rolling stock for transport services. From the European Union transport policy point of view, this means the creation of a common European railway system able to compete with other transport modes. It also creates a common market for railway products and services, thanks to which European products will become even more competitive in the global market; however, this is not important for the high-speed rail in Poland.

The ERTMS system comprises:

- European Train Control System (ETCS) –communication based train control (CBTC) ATC class system enabling railway operation with speeds up to 500 km/h; and
- Global System for Mobile Communication for Rail (GSM-R) – a railway radio-communication system for operational purposes based on GSM system working in the 900 MHz band, functionally generally equivalent to GSM 2+, ensuring radio-based data transmission channels for voice communication and for data transmission via radio for the ETCS system.

In the past, ERTMS was intended to comprise not only ETCS and GSM-R but also the European Traffic Management Layer (ETML). In 1996, the European Commission decision setting the requirements for ERTMS finally excluded ETML from ERTMS, leaving traffic management to a separate legal decision, which created the so-called telematic services later.

ETCS is a unified European solution, which should replace many different incompatible national CBTC control command systems. Similarly, the GSM-R system should replace many different national radio communication systems used for operational purposes. During the transition period, which has already begun and will continue, due to the scale of the railway system in the European Union, national systems are used and will still be used. The unified European systems that are being implemented are called "class A" systems (one CBTC "class A" system – ETCS and one mobile communication "class A" system – GSM-R). In contrast to single "class A" systems, there are many "class B" systems (many still used national systems). The CBTC "class B" system used in Poland is the train self-braking system (*System Hamowania Pociągu* – SHP). The operational radio communication "class B" system used in Poland is a broadcasting analog radio, working in the 150 MHz band, together with embedded RADIOSTOP functionality.

The ETCS system ensures:

- information for the driver about MA authority given via the DMI; and
- checking the conformity of the train movement with the MA authority.

The first functionality enables the drivers to drive the trains without watching signal aspects shown on track-side signals. By assumption, information on the cab signaling always takes precedence over the aspects on the track-side signals. This is especially important for high-speed trains when correct interpretation of the information derived from the signals is more difficult or even impossible. According to Polish rules, driving a train with a speed higher than 160 km/h requires cab signaling.

12.4.2. European control command – support for drivers

The ETCS system transmits the maximum speed permitted on a section of the railway line from track-side equipment to on-board equipment in a train. Instead of direct repetition of the signal aspect, the permitted speed is shown on the display of the Driver Machine Interface. The DMI interface was proposed for both ETCS and GSM-R by the International Union of Railways (*Union Internationale de Chemins de fer* – UIC). This is not a standard imposed by EU law, but ERTMS suppliers use it as a standard.

The proposed visualization is based on touchscreen technology with the display subdivided into areas, on which icons are displayed. The main areas of the DMI, shown in Figure 12.2a, are:

- A – braking data;
- B – speedometer;
- C – exact speed value;
- D – planning area – route description;
- E – on-board equipment monitoring;
- F – driver keyboard.

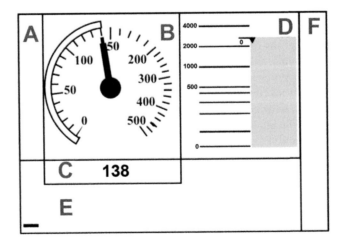

Figure 12.2a Main areas of the DMI common for ETCS and GSM-R [2], [4], [9]

The A, B, and C areas are dedicated to ETCS, while D, E, and F are used by both ETCS and GSM-R; for instance, area F on the right edge constitutes the driver keyboard. The keyboard comprises ETCS buttons, such as:

* on-board mode of operation;
* passing track-side signal at danger;
* entering train data;

as well as GSM-R buttons assigned to on-board radio functionalities, such as:

* connect with primary signalman;
* connect with auxiliary signalman, with dispatcher;
* connect with power supply dispatcher;
* emergency call to all trains on a predefined area.

The last one can be seen as a button on the edge between ETCS and GSM-R.

Each icon appears in a well-defined place on the display and is displayed in a color adequate to the situation:

* the white icon informs about events which do not require any driver action;
* the yellow icon informs about events which require driver action;
* the orange icon informs about exceptional events which require immediate driver action;
* the red icon informs about system intervention which took place – the driving did not conform with authority or driver action was not correct and therefore the system took over driving and imposed braking, replacing the train driver.

Colors are used for icons understood in a classic way and also for the speedometer arrow and speed bar at the circumference of the speedometer.

The main areas are subdivided into sub-areas as individual icons are displayed in predefined places. The generally used recommended detailed subdivision of the DMI into sub-areas is shown in Figure 12.2b.

As "class A" systems are developing, an obligatory DMI specification has already been elaborated and accepted. The recommended DMI specification is used for ETCS version 2.3.0, while a new obligatory DMI specification is due to be used for ETCS version 3.4.0, respectively called ETCS baseline 2 and ETCS baseline 3. The detailed subdivision of the DMI into sub-areas for touchscreen technology is shown in Figure 12.2c and for softkeys technology in Figure 12.2d.

12.4.3. European control command – implementation levels

The ETCS system is adjustable. It can be adjusted to different railway line characteristics by choosing the appropriate implementation level. Levels 1, 2, and 3 are downwards compatible. This means that a vehicle equipped with a higher level can run not only on lines equipped with the same ETCS level, but also on lines with lower-level equipment.

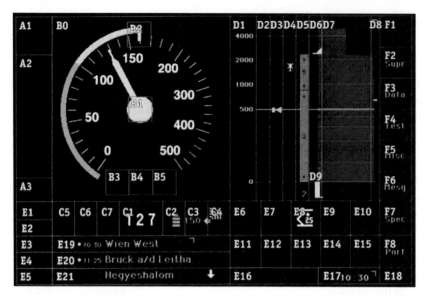

Figure 12.2b Sub-areas of the DMI in the case of touchscreen technology – baseline 2 [10]

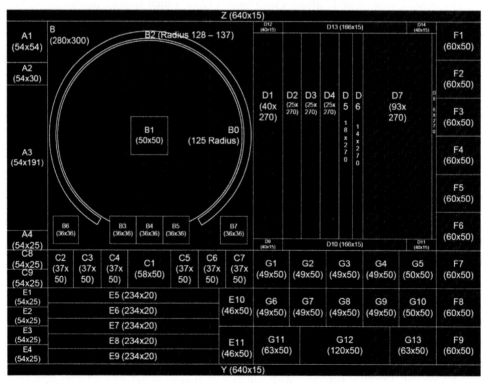

Figure 12.2c Sub-areas of the DMI in the case of touchscreen technology – baseline 3 [10]

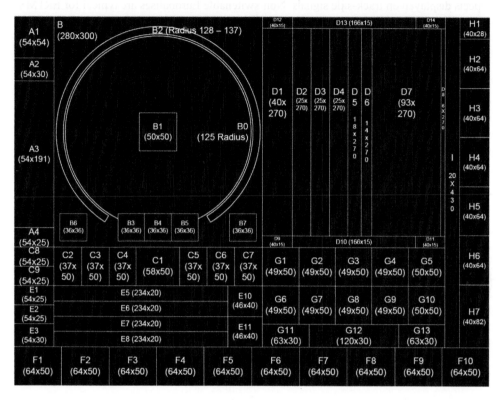

Figure 12.2d Sub-areas of the DMI in the case of softkeys technology – baseline 3 [10]

12.4.3.1. ERTMS/ETCS level 1

ERTMS/ETCS level 1 configuration is the simplest one. It is also the easiest from the implementation point of view, as such configuration does not require upgrading of the primary signaling. In practice, such a solution forms an add-on to the already existing signaling. Track-side signals are used and therefore mixed traffic is possible without any constraints. Operation of the ETCS equipped trains can be mixed with operation of trains without ETCS, which are equipped with a national control command "class B" system or which have no such "class B" equipment. The latest ones without ETCS and without "class B" equipment can run if the rules applicable in a certain country allow such operation. This solution is important for the early stage of ERTMS/ETCS implementation, when a limited number of traction vehicles are equipped with unified on-board control command.

ERTMS/ETCS level 1 configuration is a distributed solution based on transmission of the electronic MA movement authorities via Eurobalises connected to the primary signaling system. Usually, primary signaling track-side signals are used as a source of data. Signaling circuits used for displaying signal aspects are used for connecting lineside electronic units (LEU), which are defining messages containing MA authorities, transmitted to the trains via Eurobalises. Data received by the on-board balise antenna, which is part of the unified on-board control command equipment, is used for displaying MA authority on the DMI and for supervising train movement.

Eurobalises are frequently called balises. A switchable Eurobalise is typical for ERTMS/ETCS level 1 and sends messages that depend on the operational situation, i.e. on signal aspects displayed on track-side signals. Non-switchable Eurobalises are typical for ERTMS/ETCS levels 2/3 and send fixed messages to on-board equipment, regardless of the operational situation. To ensure quick and easy identification of the train running direction, level 1 installations use non-switchable balises together with switchable ones in groups.

There are different ETCS level 1 configurations:

- without infill (shown in Figure 12.3);
- with infill via additional Eurobalises;
- with infill via Euroloop (leaking cable) (shown in Figure 12.4);
- with infill via the GSM-R system.

12.4.3.1. ERTMS/ETCS level 2

ERTMS/ETCS level 2 configuration provides wider functionality compared with ERTMS/ETCS level 1. ERTMS/ETCS level 2 configuration is a centralized solution based on radio block centers (RBC), which transmit the electronic MA movement authorities via GSM-R to on-board equipment. Operational data are collected from primary signaling from all the line sections and stations covered by the RBC area and used centrally in RBC for computing the MA authorities for all the trains running in the RBC area. Data required for ERTMS/ETCS, reflecting in particular the occupation of the track sections and routes set for the train movements, are collected from the primary signaling via vital interfaces. Interfaces are called vital, as their functioning directly influences the safety of the whole control command and signaling system. Using mobile transmission for MA authorities makes the use of track-side signals optional. ERTMS/ETCS level 2 is shown schematically in Figure 12.5.

12.4.3.1. ERTMS/ETCS level 3

ERTMS/ETCS level 3 configuration is an amplification of the ERTMS/ETCS level 2 obtained by moving the checking occupancy of tracks and switches from track-side to on-board functionalities. Level 3 configuration enables the managing of train spacing according to the moving-block principle, and resignation from using track circuits and axle counters for checking occupancy, as this functionality is replaced by the on-board ERTMS/ETCS equipment supplemented with the train integrity unit. Electronic MA movement authorities are calculated, taking into account the position of the end of the train in front, transmitted via GSM-R to RBC, and transmitted via GSM-R from RBC to the on-board equipment of the train which follows.

ERTMS/ETCS level 3 is shown schematically in Figure 12.6.

12.4.4. European radio-communication GSM-R – interfaces

In the case of ERTMS/ETCS level 2 and level 3, a key role is played by data transmission via ERTMS/GSM-R. Although the GSM-R system is described in a separate chapter, points of contact between ETCS and GSM-R have to be shown when discussing ERTMS/ETCS level 2 and level 3. There are two such interfaces shown in Figure 12.7, namely:

- fixed network interface I_{FIX} – interface between the ETCS radio block center and fixed transmission network, described in the ETS 300 553 standard; and
- radio network interface I_{GSM} – interface between the GSM-R mobile terminals and on-board ETCS equipment, described in the ETS 300 011 standard.

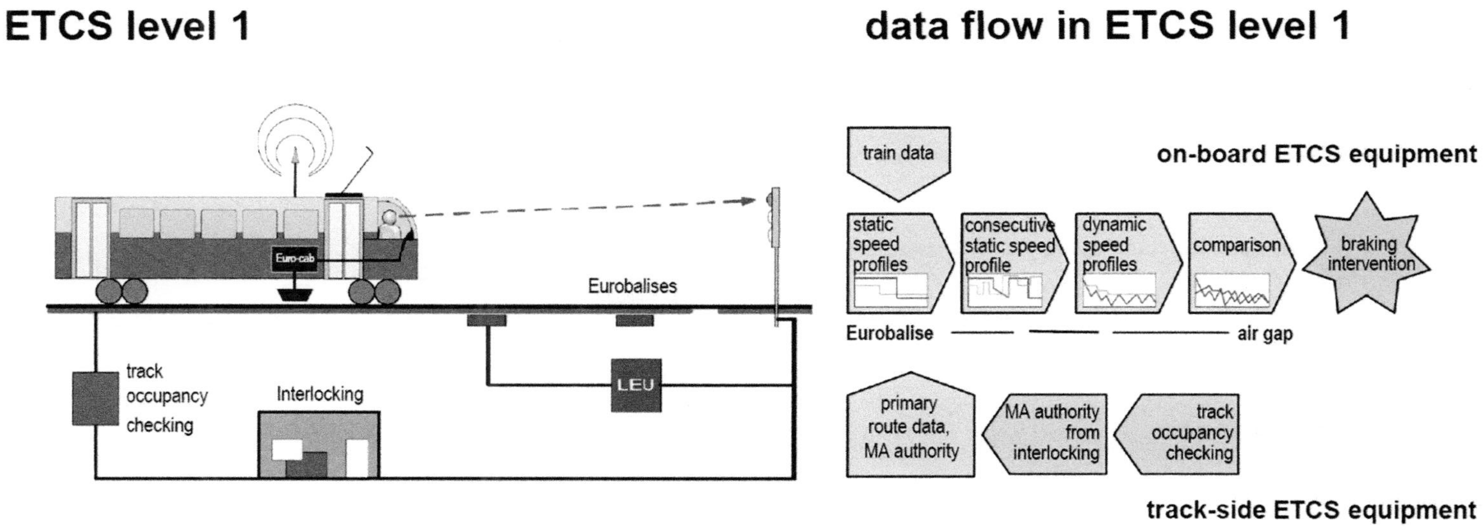

ETCS level 1

data flow in ETCS level 1

Figure I2.3 ETCS level I without infill [2], [5]

ETCS level 1 with Euroloop infill

data flow in ETCS level 1

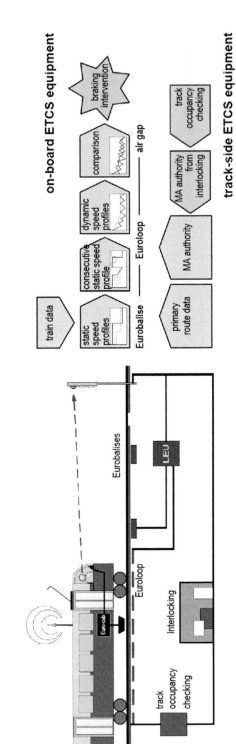

Figure I2.4 ETCS level I with infill via Euroloop [2], [5]

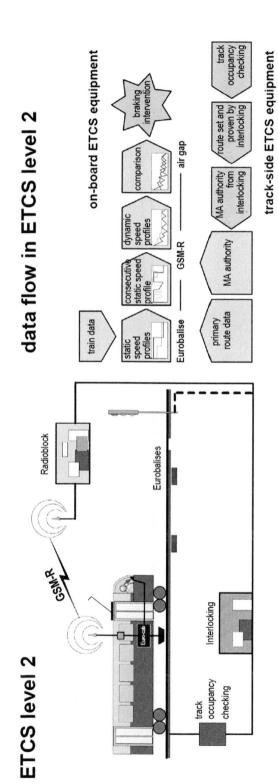

Figure 12.5 ETCS level 2 [2], [5]

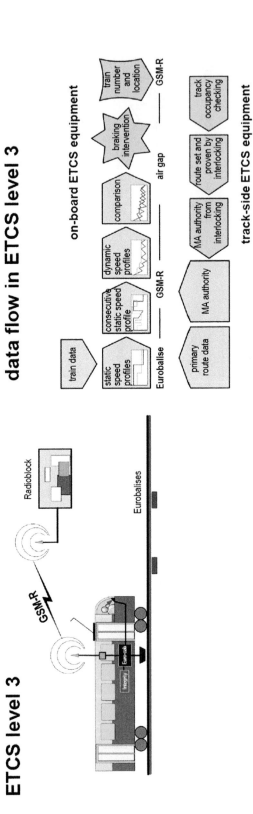

Figure 12.6 ETCS level 3 [2], [5]

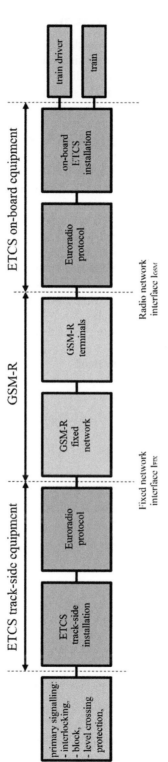

Figure 12.7 Interfaces between ETCS and GSM-R within ERTMS [2], [5]

12.5. Polish control command system

The train self-braking SHP system is an AWS class system commonly used in Poland for trains running up to 160 km/h. The SHP track-side equipment is installed on 17 thousand kilometers of double track lines, including many lines which are not foreseen to be equipped with the unified European control command. The system is based on vigilance CA equipment providing SIFA functionality and ensures:

- drawing driver attention to signal aspects shown on block signals and station entry signals located 200 m in front of the train;
- checking driver vigilance at the station exit signals and at the common signals; and
- initiation of emergency braking when there is a lack of driver vigilance.

SHP track-side equipment is a passive resonance circuit attuned to the frequency 1000 Hz, called the track-side electromagnet. The location where it is installed is called the impact point. SHP track-side installation is shown schematically in Figure 12.8.

In the SHP system, data transmission from track to train uses inductive impact between resonance circuits installed track-side and on-board. Passive resonance circuits called track-side electromagnets are installed on tracks, while active resonance circuits called on-board electromagnets are installed in traction vehicles.

On-board SHP system equipment, shown schematically in Figure 12.9, includes:

- the SHP main apparatus, called the generator;
- on-board active electromagnets;
- reduction resistors in power supply circuits;
- warning lamps;
- howlers;
- vigilance buttons;
- electro-pneumatic valve;
- the main switch.

The SHP train self-braking system cooperates with:

- the driving direction selection lever;
- the braking system;
- the recording speedometer.

Driver vigilance checking is initiated by SHP when the traction unit on-board electromagnet passes the track-side electromagnet at the impact point. Initiation switches on

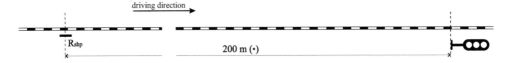

Figure 12.8 SHP system track-side equipment overview Rshp – SHP system track-side electromagnet

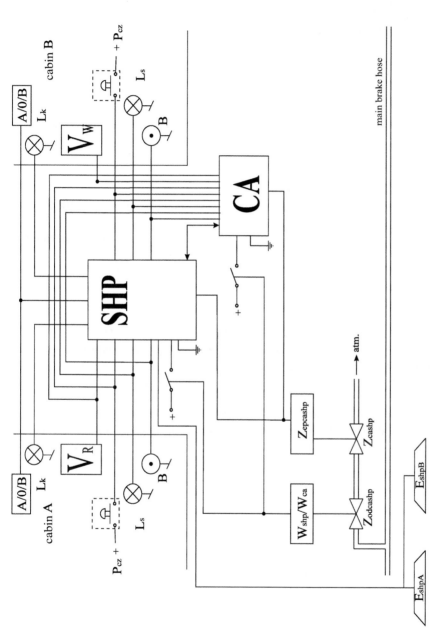

Figure 12.9 SHP system on-board equipment overview

Pcz – vigilance button, Ls – warning lamp, B – howler, Lk – control lamp, VR – recording speedometer, VW – speedometer without recording functionality, A/0/B – driving direction selection lever, Wca – CA main switch, Wshp – SHP main switch, Zepcashp – CA&SHP electro-pneumatic valve, Zodcashp – CA&SHP cut-off valve, Zcashp – CA&SHP valve, atm. – main brake pipe mouth into atmosphere, EshpA – SHP electromagnet active while driving forward from cab A, EshpB – SHP electromagnet active while driving forward from cab B continuous lighting of the warning lamp. When there is a lack of driver vigilance (vigilance button not operated) after approximately 3 seconds, the howler is switched on. If driver vigilance is still not acknowledged, after 2 seconds SHP initiates emergency braking. When the vigilance button is operated before 5 seconds elapses from the moment when the warning lamp was switched on, the generator is activated and the vigilance checking cycle of the CA is reset. CA vigilance checking takes place approximately in 60 seconds.

12.6. Control command on HS lines in Poland

12.6.1. Polish national ERTMS deployment plan

In 2006, all European Union member states were obliged to elaborate national ERTMS deployment plans to create a base for the elaboration of the European ERTMS Implementation Masterplan. Plans were obliged to define:

- railway line sections to be equipped with ETCS and GSM-R, taking into account in particular the ETCS network defined in Annex H to the Commission Decision 2006/679/ WE;
- the main characteristics of the foreseen implementations (ETCS levels, data acquisition from primary signaling, national values, etc.);
- outline of the implementation plan (order in which lines will be equipped);
- migration strategy for infrastructure and for traction rolling stock;
- potential risks, which may influence plan realization.

The National Polish European Railway Traffic Management System Deployment Plan (*Narodowy Plan Wdrażania Europejskiego Systemu Zarządzania Ruchem Kolejowym w Polsce*) ERTMS NPW was formally accepted by the Polish government on 6 March 2007 (communique no 41(128) dated 6 March 2007). This document defined the ERTMS implementation plan for Poland, as well as the strategy for migration to the final solution [7]. Moreover, some national requirements were defined to ensure, on the one hand, usage of the same solutions in different implementation contracts in Poland and, on the other hand, to preserve European interoperability of the control command and communication. It should be emphasized that Polish preconditions for ERTMS implementations do not step out of the functions defined in European "class A" system specifications.

ERTMS NPW requires the following functionalities to be guaranteed for each railway line equipped with GSM-R, as long as the line is equipped with a 150 MHz radio:

- automatic generation of the GSM-R radio emergency call when the RADIOSTOP signal is generated from track-side equipment;
- automatic generation of the GSM-R radio emergency call when the RADIOSTOP signal is received by the track-side equipment (e.g. RADIOSTOP signal generated by the train driver);
- automatic generation of the RADIOSTOP signal when the GSM-R radio emergency call is generated from track-side equipment;
- automatic generation of the RADIOSTOP signal when the GSM-R radio emergency call is received by the track-side equipment (e.g. generated by the train driver or by staff walking along the tracks).

Functionality for effective receiving and proper utilization of the RADIOSTOP signal is ensured by the specific transmission module (STM) for the SHP system. Trains equipped with ETCS and SHP STM running on lines equipped with a 150 MHz radio are capable of receiving RADIOSTOP signals and reacting accordingly.

Additional Polish requirements for GSM-R implementations are as follows:

- GSM-R dispatcher terminals, cab radios, and mobile terminals have to communicate with users in Polish. It is required to use words and expressions accepted by the infrastructure manager in the framework of the pilot ERTMS implementation on the E30 line;
- The possibility to transfer areas of responsibility (roles) between dispatchers must be ensured. The role transfer has to be organized in a way that ensures that only one dispatcher is responsible for each area (role) at any moment.

Additional Polish requirements for ETCS implementations are as follows:

- Level crossings equipped with autonomous protection systems have to be directly visible for ETCS. Information about an approaching not-closed level crossing has to be given in ETCS language by a temporary speed restriction to 20 km/h and a text message;
- Introducing temporary speed restrictions into the ETCS system must be possible from the RBC operator level, as well as by installation on the track of two movable balises;
- The DMI driver interface and RBC operator interface have to communicate in Polish. It is required to use words and expressions accepted by the infrastructure manager in the framework of the pilot ERTMS implementation on the E30 line;
- It is required to use values of the national variables that are defined in the ERTMS NPW.

12.6.2. SHP train self-braking within HS railways in Poland

There are no doubts from the safety point of view regarding the further utilization of the SHP system on railway lines, which are equipped with SHP and not equipped with ETCS, and frequently even not foreseen to be equipped with ETCS. An appropriate record is included in the ERTMS NPW deployment plan. However, a question arises if and how the SHP system should be utilized in the high-speed rail system in Poland.

Using SHP in high-speed trains creates no doubts, as high-speed trains should be prepared to run on other lines. This will take place, without any doubt, in the case of traffic disturbances, but also in the case of scheduled movements. Using SHP in high-speed trains, while running on high-speed lines equipped with ETCS, is an open question. High-speed trains currently running on the Polish railway network, namely Intercity Premium trains, which obtain the speed of 200 km/h in scheduled movements, are equipped with SHP on-board equipment, which is, however, switched on only when running outside the ETCS equipped area. Switching on and off the SHP on-board equipment has been achieved by implementing the specific transmission module. Automatic SHP switching on and off would not be possible in the case of using SHP on-board equipment independent from the ETCS on-board equipment.

Such a configuration is also allowed according to ERTMS NPW. Using such a configuration in high-speed trains will, however, require the abandonment of SHP track-side equipment installation on, or disassembling track-side electromagnets from, high-speed lines. Otherwise, in the case of high-speed movements, the SHP system will require continuous vigilance acknowledgement from drivers, hindering train driving.

No track-side SHP equipment on high-speed lines, in the case of conventional trains running on high-speed lines, means a safety level for conventional trains running on high-speed lines lower than the safety level for conventional trains running on conventional lines. Therefore, resignation from installing SHP track-side electromagnets on high-speed lines is not recommended. This leads to two statements. First – high-speed lines should be equipped with SHP electromagnets. Second – high-speed trains should be equipped with STM providing SHP and RADIOSTOP functionality.

In the SHP context, it is also necessary to ensure automatic switching on of the SHP on-board functionality by on-board ETCS equipment when leaving the ETCS equipped area. Such a function is foreseen in ETCS and simply requires the inclusion of appropriate information in data transmitted in the ETCS system from track to train.

12.6.3. Control command and signaling safety chain verification for HS railway systems in Poland

The use of many safety devices linked with each other to ensure operational safety can easily generate so-called dangerous failures in control command and signaling. Systems checking the occupancy of tracks and switches, station interlockings, line block systems, track-side and on-board ETCS equipment, track-side and on-board SHP equipment, as well as all kinds of interfaces, have to fulfil technical safety requirements based on the fail-safe principle and assessment of the safety integrity level.

Safety verification performed for individual devices and systems does not ensure safety of the whole control command and signaling. Safety verification of the whole control command and signaling requires assessment of the whole data processing chain from the safety point of view, taking into account different technical configurations, and different operational situations, including degraded modes of operation. The completeness of the analyses has to be verified using the common safety method for risk evaluation and assessment defined by European Commission regulation supplementing the Railway Safety Directive.

Bibliography

[1] Białoń, A., and Pawlik, M.: Safety and Risk in Control Command and Signaling [*Bezpieczeństwo i ryzyko na przykładzie urządzeń sterowania ruchem kolejowym*], Railway Problems [*Problemy Kolejnictwa*], Railway Institute [Instytut Kolejnictwa], Warsaw, 2014;

[2] Billin, D.R., Bourges, S., Deutsch, P., Frøsig, P., Godziejewski, B., Jones, D., Jonsson, Ö., Kurniawan, H., Leadbeater, Ch., Lochman, L., McPherson, I., Molinero, V., Pawlik, M., Rijpkema, H., Saalbach, H., Schulz-Klingner, A., Stamm, B., Sterner, B.J., Sturmack, C., Tricker, R.L., Vedelaar, B., and Walter, K.L.: *System Requirements Specification for the European Train Control System*, filing no. A200/SRS.04-A5499A5–03.01–960809, European Rail Research Institute, Utrecht, 1996;

[3] Dąbrowa-Bajon, M.: *Control Command and Signaling Basics* [*Podstawy Sterowania Ruchem Kolejowym*]. OWPW, Warsaw, 2005;

[4] Dyduch, J., Pawlik, M.: *Automatic Train Movement Control* [*Systemy Automatycznej Kontroli Jazdy Pociągu*]. Radom Technical University [*Wydawnictwo Politechniki Radomskiej*], Radom, 2002, ISBN: 978-83-7351-438-6;

[5] European Railway Agency, ETCS System Requirements Specification, subset 026–1, issue 3.4.0, 2014;

[6] Pawlik, M.: *Control Command Systems Impact on the Railway Operational Safety*. SEMTRAK Conference proceedings, Ceacow University of Technology, Cracow, 2014;

[7] Pawlik, M.: *Polish National European Railway Traffic Management System Deployment Plan* [*Polski Narodowy Plan Wdrażania Europejskiego Systemu Zarządzania Ruchem Kolejowym ERTMS*], Rail Transport Technic [*Technika Transportu Szynowego*] no 1, Warsaw, 2007;

[8] Winter, P.: Compendium on ERTMS, ISBN: 978-3-7771-0396-9, Eurailpress, Hamburg, 2009;

[9] Żurkowski, A., and Pawlik, M.: *Railway Operation and Transport. Control Command and Signalling* [*Ruch i przewozy kolejowe. Sterowanie ruchem*], ISBN: 978-83-930600-5-4, PKP Polskie Linie Kolejowe S.A., Warsaw, 2010;

[10] Pawlik, M.: *European Railway Traffic Management System, Functions and Technical Solutions Overview – from an Idea to Implementation and Exploitation* [*Europejski System Zarządzania Ruchem Kolejowym, przegląd funkcji i rozwiązań technicznych – od idei do wdrożeń i eksploatacji*], ISBN: 978-83-943085-1-3, KOW, Warsaw, 2015.

Chapter 13

Digital radio communication system for high-speed rail lines in Poland

Mirosław Siergiejczyk

13.1. Introduction

Pursuant to European Union rail transport policy, establishment of the transeuropean rail network in which the intermodal rail infrastructure, maintained by infrastructure managers and made available to railway operators providing passenger and cargo transport services using intermodal rolling stock, is a priority. The need to implement the European Rail Traffic Management System (ERTMS), combining the European Train Control System (ETCS) and Global System for Mobile Communications-Railways (GSM-R), stems from both technical premises (technical developments in telecommunications and control of rail transit safety) and legislation (implementation of interoperability is mandatory, subject to provisions of Community law and Polish law). Interoperability also refers to data exchange, and thus the GSM-R system was accepted at the European level as an interoperable means of radio communications.

GSM-R was created to facilitate communication between dispatch operator, train driver, and all operation-related services on the entire train route. Thanks to its structure, the system allows prioritizing given transfer, so that it is sent first. Such operations lead to optimum planning of traffic density on the route, which allows more efficient use and significantly improves safety. The GSM-R system is the transmission medium of ETCS, mediating between transmission of information to drivers and railway services [2]. The implementation of aforementioned systems improves the safety of railway traffic, allows real-time train diagnostics, and monitors shipments and wagons/cars. In addition, it may significantly increase the throughput on individual railway lines by allowing accurate determination of distances between trains [8], [17].

GSM-R is the European standard for radio communication, developed for railway purposes and used in the railway environment. GSM-R is based on the GMS system, presently the most popular standard of mobile telephony, and uses a 900 MHz band. The system allows services combining voice and data transmission, as well as texts and multimedia. As it uses GPRS, package data transmission, it allows speedy data transmission. This is very important for quick and reliable data exchange in "track-traction vehicle" communication over GSM-R for purposes of ETCS level 2 system.

GSM-R is part of the European Rail Traffic Management System (ERTMS). ERTMS has been designed for the purpose of railway transport integration. The system objective is to ensure interoperability, resulting in free movement of trains within railway networks of individual member states without the need for stopping and changing locomotives/drivers.

Presently, the GSM-R system is being implemented in France, Italy, Sweden, Switzerland, United Kingdom, and Germany. In Poland, the telecommunication network in this standard has been implemented by PKP Polskie Linie Kolejowe S.A. since 2009, and is the component of European Rail Traffic Management System. The project is co-financed from EU funds.

13.2. General characteristics of GSM-R system

When the GSM system was first implemented in 1992, the decision was made at the European level to adopt GSM as a standard for railway radio communication purposes. In the same year, the International Railway Union (UIC) (*Union Internationale des Chemins de Fer*), with European Union and European railway organizations, initiated the project titled the European Integrated Railway Enhanced Network (EIRENE). The objective was to specify functional and technical requirements for a mobile radio communication network that would match railway requirements and ensure interoperability. Development of mandatory requirements for operational communication services was foreseen, in order to ensure uninterrupted access to the network in the future. Two work groups were founded under the EIRENE project:

- Functional Group, whose task was to develop Functional Requirements Specification (FSR), defining obligatory system parameters, necessary to ensure interoperability, and optional system parameters, accounting for specific requirements of national systems;
- Project Team, whose task was to develop System Requirements Specification (SRS) based on functional requirements. This document was the first one to define specific technical parameters related to rail traffic control and specification of Advanced Speech Call Items (ASCI) properties. As problems related to ASCI were part of the GSM system, it became necessary to carry out works jointly with specialized ETSI Work Group in order to define such problems in GSM-R standard correctly. A first draft SRS was prepared in 1995.

In order to coordinate technical verification measures of EIRENE specification, the MORANE (Mobile Oriented Radio Network) group was founded in 1995, associating three major European railway organizations, the European Commission, and GSM system providers, ready to support development of GSM-R system. The team developed, launched, and tested prototype components of GSM-R system[1] on pilot sections.

Relationships between GSM-R standardization organizations are shown in Figure 13.1.

In 1997, UIC prepared a memorandum of understanding obligating European railway organizations to stop investing in analog radio communication and focus on implementing GSM-R. The document included a declaration that railway organizations that implemented the system earlier should assist organizations yet to implement the system with their knowledge and experience. The memorandum was signed at that time by 32 European railway organizations. Presently, the memorandum has 37 signatories – including railway organizations from outside of Europe.

---

1 Pilot implementation for GSM-R system was completed on the Florence – Arezzo and Stuttgart – Mannheim sections and on the metropolitan railway line in Paris.

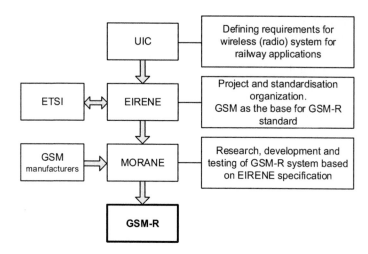

Figure 13.1 Relationships between GSM-R standardization organizations
Source: Own study based on [27]

As regards the declaration in aforementioned memorandum, UIC drafted an agreement of implementation obligating signatories to commence the implementation process in 2003 at the latest. This agreement had been signed by 17 railway organizations.

Projects EIRENE and MORANE were closed in late 2000. Their efforts produced FRS 4.0, SRS 12.0, MORANE FFFIS (Form Fit Function Interface Specification), MORANE FIS (Functional Interface Specification) and a final report with conclusions from tests conducted on pilot sections.

UIC started a new project, ERTMS/GSM-R, in order to complete works presented by EIRENE and MORANE. Applying such a name confirmed that GSM-R, as a new digital mobile radio communication system and transmission medium for the ETCS system, is one of its major components. The project combines experiences and knowledge of railway organizations obtained on pilot sections by railway organizations who implemented GSM-R system earlier on, and it is still being implemented this way. The ERTMS/GSM-R group comprises work groups that mostly took over the competence of EIRENE and MORANE projects. Permanent work groups are:

- European Radio Implementation Group (ERIG) – a forum of railway organizations that signed the memorandum of understanding in 1997. In this group, i.a. problems related to system implementation, implementation progress in individual countries are discussed, and information is exchanged on the work of other work groups (including new solutions and requirements concerning GSM-R). Since mid-2008, ERIG is a body verifying formal change proposals, representing UIC in the ERTMS department of the European Railway Agency (ERA);
- GSM-R Functional Group – charged with the task of continuing works on EIRENE FRS. This is the major body verifying implementation reports and issue reports affecting functional requirements and raising formal FRS change proposals. The group is also

responsible for migrating new and future technologies to requirements, in order to maintain uniform rules of rail traffic control.

- GSM-R Operators Group – the group continuous works on EIRENE SRS, defining technical solutions for functional requirements; it is also responsible for updating MORANE documents. Among the group's tasks is also raising formal claims concerning changes [34] based on implementation reports, and working with technical part of the Technical Industry Group (TIG) in seeking solutions and possibilities for improvement. Whenever work is required on detailed problems, the Observer Group calls ad hoc groups of experts – representatives from GSM-R suppliers and railway organizations.

In addition, under the ERTMS/GSM-R project, groups of specialists from aforementioned teams are called ad hoc to analyze particular problems. Two groups, called in late 2007, whose task is to supervise the work of the GSM-R system in all countries that implemented it, can serve as example:

- Network Management Group (NMG), dealing with operational problems;
- European Networks Integration for Railways (ENIR), dealing with technical aspects of network interoperability.

Pursuant to UIC data, in late 2009, 65,800 km of railway lines were fitted with GSM-R communication equipment. This is 43% of the target GSM-R network in Europe (150,650 km) [35].

Figure 13.2 represents the hierarchy of legal and normative documents applicable to GSM-R.

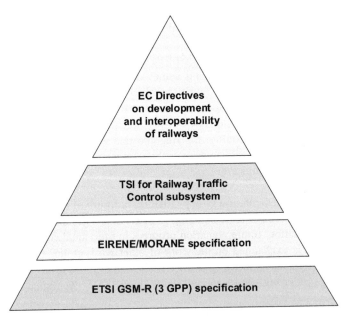

Figure 13.2 Hierarchy of legal and normative documents for GSM-R system

Source: Own study based on [11]

In the Community, the Legislation Directive is the major act of law, second only to Regulation. Function of that document is the harmonization of legislation, which is binding as regards its consequences – member states are obligated to approximate their legislation to assumptions and provisions in a directive (transposition period is usually 1–3 years), although they are free to choose the form and measure of its implementation.[2] As regards interoperability of Community railways, presently in force is the Directive of the European Parliament and the Council 2008/57/EC of 17 June 2008. The directive refers to the interoperability components, interfaces, and procedures, as well as conditions for absolute compliance of each railway subsystem, which is required to ensure interoperability of railway systems. The railway system is divided into the following subsystems:

- structural:

  - infrastructure;
  - energy;
  - control;
  - rolling stock;

- operational:

  - rail traffic;
  - maintenance;
  - telematic applications for passenger operations and freight operations.

Basically, each subsystem is covered by Technical Specification for Interoperability (TSI) defining essential requirements concerning the subsystem. TSIs are published as annexes to decisions by the European Commission related to relevant subsystems. System GSM-R is part of CCS (Control-Command and Signaling) subsystem. The GSM-R system is subject to decision by European Commission Decision 2006/679/EC of 28 March 2006 on Technical Specification for Interoperability for Rail Traffic Control subsystem of transeuropean conventional railway network, and:

- Decision by European Commission 2007/153/EC of 6 March 2007 amending Annex A to Decision 2006/679/EC on Technical Specification for Interoperability for Rail Traffic Control subsystem of transeuropean conventional railway network, and Annex A to Decision 2006/860/C on Technical Specification for Interoperability for Control subsystem for transeuropean high-speed rail network;
- Decision by European Commission 2008/386/EC of 23 April 2008 amending Annex A to Decision 2006/679/EC on Technical Specification for Interoperability for Rail Traffic Control subsystem of transeuropean conventional railway network, and Annex A to Decision 2006/860/C on Technical Specification for Interoperability for Control subsystem for transeuropean high-speed rail network;

---

2 In contrast with Regulations, which harmonize legislation and come into force "straight-up", on the date of publishing, without national authorities of Member States having any regulative capacity (obliged however to cancel any provisions not consistent with the content of regulation and prohibited to issue acts of low contrary to such content).

- Decision by European Commission 2009/561/EC of 22 July 2009 amending Decision 2006/679/EC on implementation of Technical Specification for Interoperability for Rail Traffic Control subsystem of transeuropean conventional railway network;
- Decision by European Commission 2010/79/EC of 19 October 2009 amending Decision 2006/679/EC and 2006/860/EC as regards Technical Specification for Interoperability for subsystems of transeuropean conventional railway network and transeuropean high-speed rail network.

Annex to Decision 2010/79/EC contains an up-to-date list of specifications mandatory for the GSM-R system:

- Index 32: EIRENE FRS. GSM-R Functional Requirements Specification, rev. 7;
- Index 33: EIRENE SRS GSM-R System Requirements Specification, rev. 15;
- Index 34: A11T6001 12. (MORANE) Radio Transmission FFFIS for EuroRadio, rev. 12;
- Index 48: Test specification for mobile equipment GSM-R (in the process);
- Index 61: GSM-R version management (in process).

The standard defined in EIRENE/MORANE specifications (indices 32–34) refers to 3GPP (Third Generation Partnership Project) specifications included in the standard EN 301 515, based on requirements of Technical Report TR 102 281. These documents have been updated on an ongoing basis, ever since establishing the GSM-R standard in 1999. Current mandatory revisions are as follows:

- ETSI EN 301 515 (V2.3.0): "Global System for Mobile communication (GSM); Requirements for GSM operation on railways" of February 2005;
- ETSI TR 121 900: "Digital cellular telecommunications system (Phase 2+); Universal Mobile Telecommunications System (UMTS); LTE; Technical Specification Group working methods (3GPP TR 21.900)" of September 2004

13.3. System GSM-R, architecture and services

13.3.1. Architecture of the system GSM-R

GSM-R is a system of digital cellular telephony used for railway purposes. The system provides digital voice communication and digital data transmission functionality. It offers extended functionality of the GSM system. System infrastructure is located only near railway lines. The 900 MHz frequency was adopted, in order to prevent electromagnetic interferences. GSM-R role is to assist systems implemented in Europe: European Rail Traffic Management System and European Train Control System, collecting and exchanging rail vehicle data, such as speed or position, on an ongoing basis. GSM-R system is the transmission medium of ETCS, mediating between transmission of information to drivers and railway services. Implementation of the aforementioned systems improves the safety of railway traffic, allows real-time train diagnostics, and monitoring of shipments and wagons/cars. In addition, it may significantly increase the throughput on individual railway lines by allowing accurate determination of distances between trains.

Operation of ETCS using GSM-R for data transmission does not require lateral signalization, which reduces the cost of investment and maintenance of currently used equipment. Using the satellite navigation system GPS (global positioning system) allows even further reduction in quantity of land equipment and increase in line throughput.

QoS (quality of service) requirements were imposed on the GSM-R system. Availability of services is among the major quality parameters of GSM-R, established at 99.95%. Other parameters include the maximum time of establishing normal priority connection, which should not exceed 7.5 s, and the probability of failure when establishing connection, which should be less than 10^{-3}. Major quality parameters for the established connection are, i.a. bit error rate, defining the quantity of error bits to all transmitted bits ratio. For a channel with data rate 2.4 kbit/s BER value should not exceed 10^{-4} throughout 90% of connection duration. The probability of disconnection was set at 0.0001.

GSM-R system handles terminals operating at the speed up to 500 km/h. The higher the operating speed the more frequent switching of cells, and therefore the maximum handover time is very important. Adopted handover mechanisms should ensure correct handover of cells within 300 ms in 99.5% of cases.

The essential GSM-R infrastructure is very similar to GSM system infrastructure. However, for functional reasons, some components are added that are not present in public GSM systems. This is typically due to railway services such as functional addressing – the classification of individual subscribers by priority group; therefore a special database was used, referred to as functional addressing registry. GSM-R infrastructure is also expanded by components mated with the automatic train control system and control center. The system is required to handle group connections, control connections, and high-priority connections, in which time of establishment should not exceed one second. In addition to mobile terminals used in GSM systems, GSM-R also utilizes an on-board terminal installed in locomotives or railway vehicles. When designing the GSM-R system, the decision was made to implement from the GSM system also algorithms assigning radio channels and algorithms for network management and maintenance.

Distribution of base stations in the GSM-R system may be selected in four ways, depending on required safety level.

Selection of distribution and connection of base stations should be dictated by class and purpose of railway line, throughput, and required safety level.

There are three types of cells in GSM-R systems: They are shown in Figure 13.3. Type one cells (1) are cells that, on principle, cover only the area of railway line. They are long and narrow. Type two cells (2) cover station area and part of the railway line. They are usually circular or elliptical. Type three cells (3) are cells covering other railway areas, such as sidings, buildings, etc. Each type of cell is handled by all types of radio telephones. Sizes and shapes of cells can be modified by adjusting the power level and use of omnidirectional antennas, wide-angle antennas, or linear antennas. Since the GSM-R system is designed for professional services, no radio coverage of areas other than railway areas is ensured (Figure 13.3.).

In addition to mobile terminals used in GSM systems, GSM-R also utilizes on-board terminal installed in locomotives or railway vehicles. In order to ensure efficient location of mobile subscribers, a complex hierarchical structure was adopted in GSM/GSM-R network (Figure 13.4).

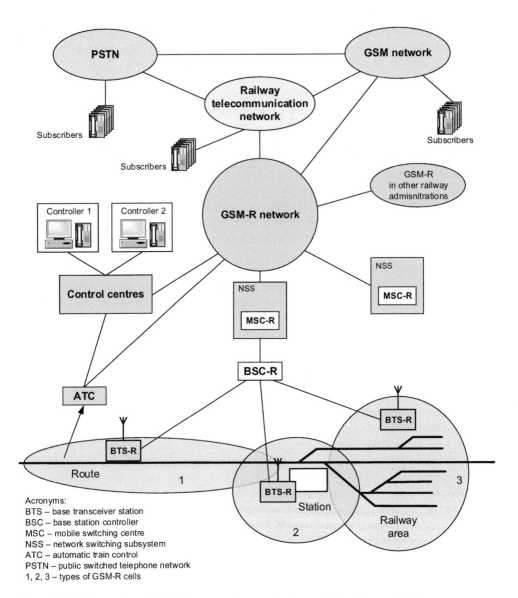

Figure 13.3 Illustration of GSM-R network organization within railway areas

Source: Own study

There are following layers identified within GSM-R network:

- GSM for Railways Network or GSM for Railways Service Area, the area covered by GSM-R services. Geographically, it represents all member states (operators) using the GSM-R system. Connections between the GSM-R network and the public phone network are effected via one of GSMC transit exchanges.
- PLMN service area, the area of GSM-R operation managed by single operator. In case there are more operators, there are a few parallel GSM-R systems communicating via the public cellular network.

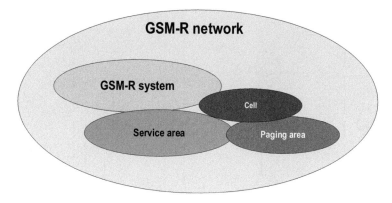

Figure 13.4 Hierarchical structure of GSM-R system
Source: Own study

- MSC service area – part of the GSM-R system, serviced by one of the MSC centers. Incoming calls to the mobile station are directed to the MSC center working in the service area in which given mobile station is located; HLR registry of own stations stores information on current position of mobile station, with accuracy to one service area.
- Location Area (LA) – section of service area in which moving mobile station does not have to transmit to the system the updated position data. In the event of incoming call to the moving mobile station, it is the area within which paging message is transmitted to call the desired subscriber. Location area usually covers a small number of cells (of the same service area), handled by one or more base controllers. Definition of location area size is related with optimum (efficient) use of radio channels – a larger location area ensures better range of call signals to mobile base and a lower frequency of mobile station position updates. A mobile station moving from one location area to another, within the same MSC center, causes an update in the record of VLR registry of a given station. Changing one location area to another, belonging to the other MSC center, causes the update of records in the HLR registry.
- Cell – the smallest fragment of the GSM-R system. A group of cells constitutes a location area. Each cell means one base station.

Based on the hierarchical structure of the network and considering functions to be fulfilled by GSM for railway applications, the structure of GSM-R can be shown as in Figure 13.4. Jointly included shunting communication, train communication, and area communication correspond with a logical level of the GSM system (Figure 13.5).

There are following layers identified within GSM-R architecture [11]:

- Base station subsystem layer;
- Network switching subsystem layer;
- Operation maintenance center layer;
- Application control layer;
- Monitoring layer;
- Mobile devices layer.

Architecture of GSM-R system is shown in Figure 13.5.

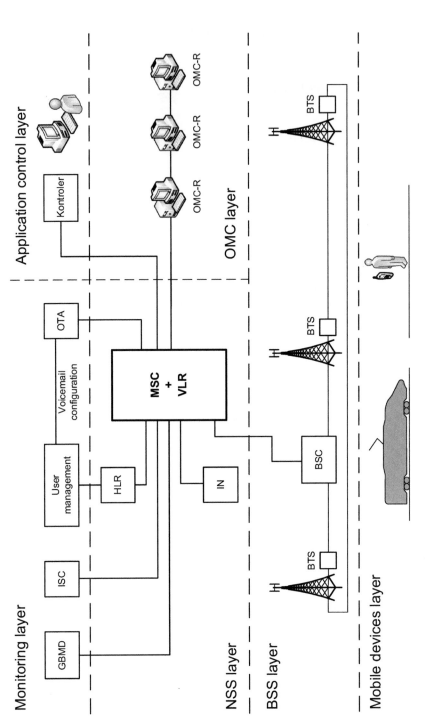

Figure 13.5 Architecture of GSM-R system

Source: Own study based on [30]

The mobile devices layer is the GSM-R subsystem comprising following elements:

*   controller stations;
*   mobile stations;
*   digital radio phones.

Controller terminals in the GSM-R network have to be integrated with the railway tele-communication network, in order to meet their functional requirements. They are connected, along with other telecommunication equipment (such as track-side phones), to the GMSC center.

The mobile station is used by the subscriber to communicate with other users of the tel-ecommunication network. It consists of a subscriber terminal (mobile phone, GSM phone), comprising two parts – radio device and subscriber identity module (SIM card). The SIM card contains subscriber identification data and services provided to subscriber.

The mobile devices layer is the GSM-R subsystem comprising following elements:

*   controller stations;
*   mobile stations;
*   digital radio phones.

The BSS (Base Station Subsystem) layer is the subsystem of GSM-R architecture com-prising three elements:

*   Base Station Controller(s) (BSC), whose role is to manage all radio functions in the GSM network;
*   Transcoding Units (TCU), whose task is to code and decode voice;
*   Base Transceiver Station (BTS), whose task is to control radio interfaces of mobile stations.

Such stations are controlled by the BSC system directly using BSF module. Single BTS is able to provide up to 30 transmission channels with a 2 Mbit data rate. Correct planning of this layer has great significance to ensuring availability and continuity of data and voice transmission services in GSM-R network.

The BSC should also allow individual (individual base stations), serial (many base sta-tions), or loop serial connection (a ring of base stations) connection of base stations, pursuant to description of interface A-bis of the GSM-R standard. Network topology has to correspond with the required 0.00005 availability of the GSM-R system.

The Network Switching Subsystem(NSS) layer is the subsystem of GSM-R architecture comprising the following elements [30]:

*   MSC (Mobile Switching Center);
*   VLR (Visitor Location Register);
*   HLR (Home Location Register);
*   GPRS (General Pocket Radio Service) Support Nodes Module;
*   GMSC (Gateway Mobile Switching Center);
*   AuC (Authentication Center) Module;
*   EiR (Equipment Identity Register) Module;

- SCP/IN (Service Control Point/Intelligent Network) Module;
- STP (Signaling Transfer Point) Module;
- MGW (Media GateWay) Module;
- CBM (Core Billing Manager) Module;
- SMSC (Short Message Service Center) Module.

MSC center is the most important element of NSS subsystem. This is a radio center whose task is to establish connections within the network and to allow users connecting with external networks, e.g. railway telecommunication networks. For purposes of GSM-R, MSC is implemented in technology 3GPP R4. Due to the critical importance of this component, it is implemented in georedundant system, including mated elements. All calls, ingoing and outgoing, are controlled by MSC. In addition to standard functionalities of telecommunication center, MSC also offers functionalities to mobile devices in the network, such as recording subscribers' position or calling subscribers.

OMC layer is the subsystem of GSM-R architecture with following components:

- OMC-R (Operations and Maintenance Center – Radio), a tool for management, control, and monitoring of all radio network elements (BSS layer);
- OMC-R (Operations and Maintenance Center – Switching), a tool for management, control, and monitoring of all functions related to circuit switching and calls (NSS layer);
- NMC (Network Management Center), master tool for management, control, and monitoring of the GSM-R network, sending information to specific OMC modules, able to take over their functions in the system if need be.

The application control layer is a GSM-R specific (nonexistent in GSM) subsystem, where management and supervision over critical system services, such as Location Area (FN), Railway Emergency Call (REC), or Location Dependent Addressing (LDA), is effected.

The monitoring layer is the subsystem of GSM-R architecture with the following components:

- GBMD (GSM Billing Mediation Device) module, mediating between customer service and billing system of the NSS layer, collecting data on subscribers' payments (separately for each service, e.g. voice transmission and data transmission over GPRS);
- OTA (Over The Air) module, working with type OTA SIM cards, which allow dynamic download of new services and configurations over the radio transfer;
- ISC (International Switching Center) module, whose role is routing international connections;
- applications responsible for specific functions, such as voicemail;
- user management centers, directly connected with customer service.

Interfaces between components used in GSM-R are similar to those used in the GSM network
(Figure 13.6). These are:

- Interface Um – between mobile stations and BTS;
- Interface Abis – between BTS and BSC;
- Interface A – between BSC and MSC;

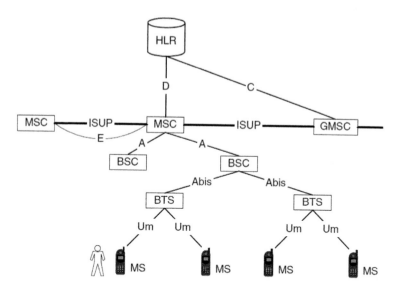

Figure 13.6 Interfaces used in GSM/GSM-R network

Source: Own study based on [15]

- Interface B – between MSC and VLR;
- Interface C – between GMSC and HLR;
- Interface D – between MSC and HLR, and also between VLR and HLR;
- Interface E – between two MSCs;
- Interface G – between two VLRs;
- Interface Gr – between SGSN and HLR;
- Interface Gn/Gp – between *Serving GPRS Support Node* (SGSN) and *Gateway GPRS Support Node* (GGSN);
- Interface Gc – between GGSN and HLR;
- Interface Gs – between MSC and GGSN;
- Interface H – between Authentication Center (AuC) and HLR;
- Interface ISUP – between MSC and TCP/IP devices;
- Interface T – between MSC and SCP.

13.3.2. Services and functions of GSM-R system

GSM-R system is based on Phase 2 public GSM standard, implementing all essential services plus additional services, complemented with GSM Phase 2+ standard (voice broadcast service, group call, GPRS, call priority). This allows for implementing the following functionalities:

- Voice Broadcast Service (VBS) – consisting in broadcasting voice information to indicate a predefined group of recipients, without the option of confirmation of reception by voice. Mobile stations confirm reception of information with a brief message, so that it could be verified who received a given message. Message recipients may be a group of recipients identified by an address list on a sender's SIM card, specialized staff (e.g. shunting groups), or all users within the coverage area of the specific base station. Group addresses may be changed automatically at given hours or modified manually under emergency conditions.

- Voice Group Call Service (VGCS) – allows simultaneous mutual communication with a predefined users group (SIM card), and each user may activate or deactivate his/her connection individually. Call initiator or controller supervises group formation, selects active users, and monitors the whole service. Completion of the teleconference is up to the initiator, system controller, or expiration of connection duration, as predefined on the SIM card.
- Enhanced Multi-Level Precedence and Pre-emption – service consisting in granting priority of connections to some network users, used in emergency. A higher priority user may interrupt the call of a lower priority user. Lower priority user answers such a call automatically. Time of establishing priority connection shall be short (up to 1 s, including interrupting the existing connection).
- Functional Addressing (FA) – consists in assigning to railway workers addresses related to functions performed by them. This allows the calling user using an ID of a particular function, and not the physical terminal. Connection can be established with the driver, for example, not via his unique address but also via train ID or locomotive ID. Addresses are assigned to corresponding functions by a regional control station or controller, who may change or cancel them, or assign them to another terminal.
- Location Dependent Addressing (LDA) – ensures communication with a worker performing a given function, depending on the location of train and paging area connected with the given function. Such type of addressing allows, among others, connections between the train driver and traffic controller by pressing only a single button. Basically, the location is determined based on the cell identifier (GSM-R cell number, i.e. Cell ID) in which the train is currently located. However, due to cell size differences, in order to improve the level of positioning accuracy, other sources may be used, such as on-board positioning systems, track-side balises, or information from systems based on fixed infrastructure.
- General Packet Radio Service (GPRS) – mostly used in track-vehicle transmission (ETCS);
- Enhanced Railway Emergency Call (eREC) – emergency connection notifying drivers, traffic controllers, and other staff of dangers that require stopping rail operations, or taking other measures, within a specific area. Two modes of emergency railway connections are defined: train operation emergency mode (non-shunting) and shunting operation emergency mode. The mode of initiated connection is determined automatically based on the operating mode of initiating terminal. Emergency connection to trains has to be sent to all drivers and traffic controllers within the predefined operation area.
- Shunting mode – ensures communication between staff involved in in shunting operations.
- Direct mode – supported by mobile terminals, refers to situations when they communicate without using the GSM-R network. This function is provided for situations such as network failure or no GSM-R coverage. The user may activate this mode only when the base station "decides" that no GSM-R service is available. This is an optional functionality of the GSM-R network.

13.4. Selected aspects of GSM-R system implementation in Poland

13.4.1. Dimensioning of GSM-R radio network

13.4.1.1. Preparing the energy balance of radio link

Based on available materials, it may be assessed that there are no recommendations concerning the construction of link budget. This allows for great freedom in creating completely different link budgets for same coverage conditions, from extremely pessimistic, with useful station ranges of a little over a few kilometers to a coverage range of 10 km. Presented below is a discussion based on experiences when planning public GSM networks, and in particular cases of planning GSM-R network; such guidelines (in particular values of specific parameters) may slightly vary. The idea of calculating radio link budget remains, however, unchanged, as do the majority of radio links.

The following parameters should be considered in radio link budget drafted for specific conditions and in calculations for individual station ranges [2], [5], [3], [14]:

- required minimum level of signal and probability of signal coverage;
- suspension height of base station antenna and mobile station antenna;
- handover margin and signal standard deviation;
- safety margin;
- losses in radio link;
- antennas' gains and separation gain;
- projected ranges of base stations;
- planning of radio channels.

The margin of handover, that is, handing over radio connection from one base station to the other while the connection remains active, is a significant parameter from the viewpoint of radio planning, as its value affects both link budget and handover speed. A margin that is too high might result in unnecessary concentration of base stations as well as delays in the handover of connections. Too little or no handover margin may cause unnecessary handover. The practice of radio planning indicates that the standard deviation for signal in flat and open countryside is approximately 5 dB. However, in the case of planned GSM-R, values higher than e.g. 7–10 dB should be adopted, due to diverse land characteristics. The safety margin (this is the minimum standard deviation for the signal) in accordance with EIRENE recommendations is 3dB and takes into account the aging of the GSM-R system infrastructure.

As is the case with public networks, losses in radio link should include overall attenuation of radio signal between the base station signal output and antenna input. Such losses comprise attenuation in flexible cables and couplings and attenuation in concentric cable.

Antenna gain in a mobile device to be adopted for the link budget should be 0 dBi, and as regards antennas of base stations, standard values adopted in the budget should be between 10dBi and 20 dBi, as antenna gain.

Diversity is the gain related to multichannel propagation and depolarization of signal. The value of this parameter adopted in public networks usually is 3 dB. However, because diversity gain is smaller in open areas (and the GSM-R system is erected to a great extent in such areas), smaller values have to be adopted for the budget.

Prior to drafting the link budget, base station configuration types suitable for use within the planned area have to be determined. Line budget calculation requires information on transmitter and receiver, i.e. transmitter power, receiver sensitivity, cable attenuation, cable length, antenna gain, and distances between points and frequency. The result is EIRP (effective antenna power) of the transmitter and path attenuation.

$$EIRP(dBm) = transmitter\ power\ (dBm) - cable\ attenuation\ (dB)$$
$$+ antenna\ amplification\ (dBi)$$

(13.1)

Using the above-mentioned data and calculations, one can determine the approximate number of base stations and their ranges and locations.

13.4.1.2. Sources of telecommunication traffic in GSM-R system

The GSM-R network has to ensure handling all types of services that generate all kinds of network traffic. Therefore, all possible traffic profiles should be analyzed at the stage of network dimensioning. The volume of traffic offered may be determined as the product of the number of notifications in time unit (usually related to GNR peak hour) and average duration of connection. Calculations of generated traffic should account for traffic generated by voice calls and data transmission with circuit switching [6], [7].

In this phase, the peak hour (GNR) also has to be determined. This will help determine the volume of traffic in a given area. Calculations should account for the presence of route maintenance personnel and shunting personnel, also generating traffic.

In order to dimension the network properly for the purpose of determining the volume of telecommunication traffic, classification of major traffic sources resulting from EIRENE isolated applications of GSM-R network, concerning the typical modern railway network, should be accounted for [14], [15]. Sources of telecommunication traffic in GSM-R network are

- Control and signalling:

  - train automation and control;
  - remote control of railway network infrastructure.

- Process communications:

  - controller – driver communication;
  - emergency communication;
  - shunting communication;
  - driver's communication;
  - route maintenance communication;
  - service staff communication.

- Local communication on routes and at junctions;
- Non-operational area communication;
- Passenger communication, including passenger information.

The GSM-R network has to ensure handling all types of services generating all kinds of network traffic. Therefore, all possible traffic profiles should be analyzed at the stage of network dimensioning. Volume of traffic offered may be determined as the product of the number of notifications in time unit (usually related to GNR peak hour) and average duration of connection. Generated traffic and data transmission with circuit switching are expressed in Erlangs [Erl].

$$A = \frac{n_{GNR} \times \tau_{sr}}{3600} \qquad\qquad (13.2)$$

where:
A – intensity of telecommunication traffic [Erl],
n_{GNR} – number of connections during peak hour,
τ_{sr} – average duration of one connection(s).

13.4.1.2.1. VOICE COMMUNICATION

Voice communication can be divided into:

- point-to-point communication (PTP);
- point-to-multipoint connection (PTMP);
- voice group communication system (VGCS);
- voice broadcasting system (VBS).

13.4.1.2.2. DATA TRANSMISSION

Data transmission in the GSM-R system can be divided by transmission procedures used. Generally, traffic generated by data can be determined based on the quantity of messages sent in time unit and the size of a single message. In addition, transmission intensity has to be considered – whether continuous or interrupted. The following mechanisms ensuring data transmission in GSM-R system can be isolated:

- text – the number of texts should be used to dimension the system, irrespective of average size of one text (maximum 140 bytes);
- package transmission – average size of the message of 64 byte can be used;
- additional data transmission – used in registering and deregistering of mobile station for function addressing purpose – the unit here is the number of messages;
- data transmission using circuit switching – may be dimensioned the same way as voice connections;
- data transmission for ETCS.

The main use of data transmission in GSM-R is ETCS. Implementation of ETCS level 2/3 requires a transmission medium that ensures communication between train and track-side equipment. GSM-R, as a part of ERTMS/ETCS, allows bidirectional exchange of information between on-board equipment, radio broadcast center (RBC), and radio interface unit (RIU). It requires a relatively small band (up to 9600 bps). RBC periodically sends small messages (up to 200 bytes) to the train, every 30 seconds. Train reports to RBC by 32-byte (user data) messages, sent every 10 seconds. It is worth noting that from the ETCS perspective, the most important is ensuring minimum transmission delays in both directions, as well as ensuring reliability. Presently, the transmission is effected via circuit switching, which means than each ETCS transmission channel occupies one timeslot. Each train with ETCS occupies such a timeslot throughout the whole train operation, which might cause capacity problems at junction stations.

Hence, GPRS transmission is planned, reducing the required capacity by multiplexing many users in one timeslot. Reliability of package transmission can be improved by adopting suitable mechanisms, such as ARQ (Automatic Repeat Request) mechanism or support from TCP/IP transport layer. In the case of GPRS, the effect of switching between stations on radio interface and delays in data transmission is visible. When switching time in DCSS transmission was approximately 0.5 seconds, in package transmission it may take as much as 8 seconds. In addition, in GSM network's voice transmission has priority over package data transmission, which means that in the absence of sufficient resources on radio interface, data transmission is stopped and the freed timeslot is allocated to the incoming voice connection. The solution here is timeslot allocation permanently dedicated to GPRS. This in turn may reduce capacity gain thanks to multiplexing, as blocking one or more timeslots should reduce the number of resources available to voice communication and consequently might result in the need to increase capacity by installing an additional TRX module. That may be the case with a small-capacity station (sectors with one TRX module), although in the case of base stations with high-intensity telecommunication traffic, significant gain from data transmission over GPRS is anticipated. Correct dimensioning of ETCS-GPRS traffic capacity developing of traffic profile is necessary, which would serve as the base for traffic measurements in the actual network.

Another aspect of GPRS use for ETCS data transmission that the handling of voice communication and ETCS-GPRS data transmission should be separated. This may mean the modification of terminals by implementing separate transmission and reception channels for each domain. Dimensioning and optimizing a radio network for ETCS-GPRS has to account also for an estimation of maximum delays for the assumed data stream and available resources, as well as the impact of radio network and handover times.

13.4.1.3. Profiles of telecommunication traffic in GSM-R system

Depending on the type of traffic and geographic location of users and their functions within the organization, individual groups of GSM-R network can be isolated. First, however, we have to determine what types of traffic can be observed in the GSM-R network. The following types can be distinguished:

- point-to-point voice communication and data transmission with circuit switching, expressed in telecommunication traffic intensity units [Erl];
- voice group communication [Erl], point-multipoint [Erl];
- text messages SMS [number of messages/h];
- circuit switching [number of messages/h];
- package messages [number of messages/h];
- point-to-point communication using location dependent addressing [Erl], expressed in Erlangs;
- data transmission with circuit switching using location dependent addressing [Erl].

Offered traffic refers to specific areas and includes traffic generated from the train, both by staff terminals and data transmission systems. Such traffic should also account for other system users, both landline and mobile, defined as the number of active terminals. From the viewpoint of GSM-R system functionality, four user groups may be isolated, including their traffic profiles.

Telecommunication traffic is generated by train staff. Train-related traffic is generated by driver–traffic controller connections, driver–driver connections, or driver–train staff connections. In addition, the driver is a party to VGCS connections and VBS connections. Assuming the standard traffic model, we may expect that traffic generated by driver is in the range 0.02–0.01 Erl. Each train using ETCS allocates one timeslot to data transmission for the duration of train operation, meaning that 1 Erl should be reserved for each train. Communication is effected mostly at stations and, to a smaller extent, during operation.

Shunting teams. Groups located within the shunting area or signal tower are responsible for the organization of train composition. Voice connections carried out by group members feature long duration and include VGCS group connections between group members and VGCS communication with traffic controller. We can assume that traffic generated by that staff is in the range of 0.1–0.5 Erl.

Route maintenance staff. Traffic is generated by workers of route maintenance staff. Average generated traffic is 0.05 Erl per user. It includes maintenance staff as well as construction teams, security services, and drivers of construction equipment.

Station staff. Traffic generated at stations is initiated by workers who occupy themselves with ensuring correct functioning of the railway station. They are, among others, control tower staff, logistic staff, porters, and security. Such traffic is mostly local in nature, within the station area (95%), point to point (90%).

13.4.1.4. Model of telecommunication traffic in GSM-R system

Traffic model defines the volume of voice traffic and data traffic related to railway operator applications, as well as the geographic distribution of traffic in the railway network. Since generated traffic is an individual feature of each user, it is hard to design a universal model. It should therefore result from network analysis based on busy hour call attempts (BHCA). Considering the characteristics and intensity of traffic, when determining network capacity GSM-R network organizational variants should be included, namely

- Railway network junctions, in proportion to their size;
- Railway stations – the analysis of GSM-R telecommunication traffic generated at railway stations should consider the number of railway lines at the station, the number of platforms used at the same time, the number of through traffic lines, the number of workers on platforms, and the number of workers outside of platforms (shunting within station limits);
- Shunting areas, considering the number of simultaneous trainset operations, the number of shunting locomotives, the number of workers in shunting teams, and other users within the shunting area;
- Railway lines – heavy traffic lines and light traffic lines. The analysis of GSM-R telecommunication traffic generated on railway lines should consider the number of lines, the frequency of train operation, the maximum speed of trains, and the distribution of trainsets (closer to one another near the station, farther apart between stations).

Telecommunication traffic in railway network junctions comprises traffic generated by trains within the station and local traffic generated by workers. In the case of railway terminals, the number of trains in peak hours should be equivalent to the number of simultaneously used platforms. It can be assumed that at the majority of stations, the number of trainsets at the station is related to the occupancy of platforms. Then the obtained number of trainsets should be multiplied by average telecommunication traffic generated by single trainset. This should help in determining traffic generated by trains at stations. Next, traffic generated locally, at the station – by station staff and shunting groups – has to be determined.

For determining the intensity of telecommunication traffic on railway lines, one can use the ratio between the number of trainsets per hour and the maximum speed of the train. The result is the number of trainsets per kilometer, which in the case of GSM-R system is important for determination.

The following is an example determination of GSM-R network capacity:

Assumed maximum speed near the station is approximately 60 km/h, maximum speed between stations is 160 km/h, and in the case of HSR trains, as much as 300 km/h. Offered traffic refers to specific areas and includes traffic generated from the train, as the sum of traffic generated by staff and data transmission systems, as well as traffic generated by other system users, both landline and mobile, is defined as the number of active terminals.

Estimation of required capacity of GSM-R system radio network was prepared by dimensioning, in which the value of traffic described in Section 3.3 was adopted, and it was assumed that on all lines fitted with an ETCS level 2 system, there is no package data transmission, and group and broadcast calls are considered, with a block coefficient 0.01.

During dimensioning, it was assumed that telecommunication traffic is generated by trains fitted with ETCS and initiated by shunting teams and station staff. Erlang B [5] model was used in calculations [34].

The results of the example estimation of railway traffic in GSM-R radio network are shown in Table 13.1.

Another problem is determining the number of users (terminals) of GSM-R system, by coverage area. It can be assumed that at a big station, the number of users is 60, at a medium station 30 users, at a small station 10 users, with the shunting area having 30 users. Then, at the phase of the dimensioning of radio network capacity, the average number of operating trains served by a given base station has to be determined.

Table 13.1 Results of estimation of railway traffic in GSM-R radio network

Source	User	Service	Average duration of service [s]	Number of calls in GNR	Traffic [Erl/terminal/GNR]
Train	ETCS	data	1800	2.0	1.000
	Cabin radio	voice	30	5.0	0.042
	Train staff	voice	50	1.0	0.120
	VBS	voice	50	0.1	0.002
	Train diagnostics	data	1	2.0	0.006
Shunting teams	VGCS	voice	20	1.0	0.056
	VBS	voice	40	0.1	0.001
Station staff	PTP	voice	60	5.0	0.083
	VBS	voice	40	0.1	0.001

The last phase is determining the number of transmitting and receiving devices (TRX), responsible for sending and receiving signals. Pursuant to PKP PLK S.A. guidelines [5], their number should be two.

This can be calculated for each station be determining, in the following sequence [12]:

- overall number of GSM 900 system station sectors;
- average traffic handles by one network sector;
- required sector capacity in individual types of geographic areas;
- number of channels necessary to handle required traffic per sector, using B Erlang calculator.

Having estimated the offered traffic based on the B Erlang model for the assumed block coefficient 1%, the required number of digital channels for the analyzed area was calculated. Based on the values presented earlier and assuming that this is a medium-sized station (average daily traffic is 60 trains), the number of channels necessary to handle telecommunication traffic can be calculated. Such number is 18 channels, with block probability 0.0092.

13.4.2. Environmental aspects of planning and implementing GSM-R network in Poland

BSS layer (Figure 13.7) is the subsystem of GSM-R architecture comprising three elements:

- Base Station Controller(s), whose role is to manage all radio functions in the GSM network;
- Transcoding Unit (TCU), whose task is to code and decode voice;
- Base Transceiver Station (BTS), whose task is to control radio interfaces of mobile stations. Such stations are controlled by BSC system directly using BSF module. Single BTS is able to provide up to 30 transmission channels with 2 Mbit data rate.

Base station (site) consists of following components (Figure 13.7):

- standalone radio tower with GSM-R antennas and possible radio line antenna used to send signals to BSC;
- telecommunication container as the space with BTS equipment, teletransmission equipment for sending signals to BSC (backhaul), and power supply equipment (batteries, UPS).

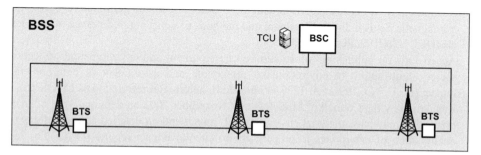

Figure 13.7 BSS layer of GSM-R architecture

Source: Own study based on [11]

In Polish conditions, one should also consider that on some antenna towers and in some telecommunication containers (probably in every second container), suitable space should be provided for antennas and transmission and reception equipment of SZS 150 MHz equipment (if implemented).

ERTMS/GSM-R network should be implemented in the "design and develop" process, and the design should be prepared by the project contractor. For that reason, there are only general guidelines available for such projects, included in tender documentation in the description of the subject matter (DSM), as per the principal's requirements. Among major requirements with regard to design, the following recommendations and guidelines can be listed:

- GSM-R subsystem has to ensure implementation by train radio communication all mandatory requirements, as specified in applicable documents of EIRENE standard;
- In the event of using the ERTMS/GSM-R network to handle ETCS level 2, the network has to be so designed as to allow data transmission for the ETCS subsystem with the required quality level (QoS);
- Ensure connection with other GSM-R railway operators and public phone network.

Base stations of cellular telephony (masts and equipment) are practically everywhere. Based on various sources, presently in Poland there are tens of thousands of base stations and counting. They are erected in metropolises, villages, at airports, along roads and railway tracks, in the mountains, and on the coast. They arouse strong emotions and anxiety in people living nearby. Undoubtedly, they also spoil the beauty of the landscape. People's fears concerning masts refer most of all to harmful radiation emitted by equipment installed on masts and the reduction in the value of real property in direct vicinity of such masts. Such fears are often expressed in the form of protests, taking various forms, which often inhibit or prevent completion of a given project. It is worth noting that presently there are no places on earth free from magnetic fields, and differences are only in the intensity and frequency of waves comprising such fields. It has to be stressed that Poland has one of the lowest allowed limits of public exposure to electromagnetic fields, meaning that Polish regulations referring to cellular telephony base stations are among the strictest in the world. Erection of cellular telephony masts in Poland is subject to a series of acts of law, with varying rank. Among the most important are [18]:

- Environmental Protection Law Act of 27 April 2001 (uniform text Dz. U. 2008 No. 25, item 150, as amended) – hereinafter referred to as POŚ;
- Regulation by Minister of Environment of 30 October 2003 on permitted levels of electromagnetic fields in the environment and methods of verifying observance of such levels (Dz.U. No. 192, item 1883);
- Act of Law on publishing information on environment and environmental protection, society contribution to environmental protection, and assessment of environmental impact (Dz.U. No. 199, item 1227 as amended), hereinafter referred to as UOOŚ.
- Regulation by the Council of Ministers of 9 November 2003 on defining types of enterprises that may have material environmental impact and on detailed considerations for qualification of enterprises to prepare environmental impact reports (Dz.U. No. 257, item 2573, as amended), hereinafter referred to as RRPOŚ;
- Supporting Development of Telecommunication Services and Networks Act of 7 May 2010 (Dz.U. No. 106, item 675);

- Regulation by Minister of Infrastructure of 22 October 2010 amending the regulation on technical requirements to be met by telecommunication structures and their locations (Dz.U. No.115, item 773);
- Regulation by Minister of Environment of 12 November 2007 on the extent and manner of performing periodic examination of electromagnetic field levels in natural environment (Dz.U. No. 221, item 1645);
- Regulation by the Council of Ministers of 9 November 2010 on defining types of enterprises that may have material environmental impact and on detailed considerations for qualification of enterprises to prepare environmental impact reports (Dz.U. No. 213, item 1397).

As shown by the above, the process of cellular telephony mast erection is regulated by a series of different acts of law. Detailed discussion of this process is outside the scope of this analysis. It is worth noting, however, the environmental aspects, in particular the erection/construction of such masts requires obtaining an environmental impact assessment and, as a consequence, drafting an environmental impact report. Analyzing obligations of an entrepreneur who intends to erect, start, and operate the cellular telephony mast, first he/she has to obtain the decision on environmental considerations for project implementation ("Environmental impact decision"), pursuant to Art. 71 UOOŚ. An environmental permit defines environmental considerations of project implementation. Obtaining environmental permit is required for planned:

- enterprises that might have certain significant impact on the environment;
- enterprises that might have potential significant impact on the environment.

The environmental decision is issued prior to obtaining a building permit (including the erection of a cellular telephony mast), approval of detailed design, restoration of construction works permit, and the permit for change in use of the structure or part thereof.

Commune head, mayor or president of the city, respectively, is the body competent to issue environmental impact decision with regard to construction and operation of gas stations.

Pursuant to RRPOŚ, projects related to cellular telephony masts are qualified as projects with significant environmental impact. Pursuant to § 2 (1) item 7 RRPOŚ, drafting the environmental impact report is required each time for the following projects: radio communication, radio navigation and radiolocation systems, excluding radio line, emitting electromagnetic fields at frequency range 300 kHz to 300 GHz, in which equivalent radiated power determined isotropically for a single antenna is [1], [18]:

- not less than 2000 W, and general access locations are in the distance not more than 100 m from the electric center along the main axis of radiation beam of such antenna;
- not less than 5000 W, and general access locations are in the distance not more than 150 m from the electric center along the main axis of radiation beam of such antenna;
- not less than 10,000 W, and general access locations are in the distance not more than 200 m from the electric center along the main axis of radiation beam of such antenna.

The essential components of ERTMS, which require obtaining a building permit, are GSM-R radio communication objects (ROs). Each RO comprises tower/mast, usually two antennas, TRX base station transmitter-receiver, and supply and control equipment, installed in the container/cabinet at the foot of the tower/mast.

BTS equipment is operated in the band GSM-R 900 MHz, in two band sections of 4 MHz wide each. This band is defined by following frequency values: 876–880 MHz (uplink – transmission from mobile terminal to BTS) and 921–925 MHz (downlink – transmission from BTS to mobile terminal).

Due to the cost of land purchase, costs of use and operation, as well as property rights, radio communication objects (RO) are usually erected on railway land.

Mast height and antenna configuration varies depending on the type of landscape in which an RO shall be installed and the coverage area. Masts with a height exceeding 40 m are rarely used due to the "uplink" (mobile terminal MS to base station) power budget and limited transmitting power of mobile terminal MS. General working configurations of GSM-R antennas are shown in Figure 13.8. Radio communication objects alongside the railway line are usually fitted with two bidirectional antennas facing opposite directions (Figure 13.8 BI). At railway stations, mostly omnidirectional antennas are installed (Figure 13.8 OMNI). An alternative solution is the multi-sector base station, e.g. tri-sector sector base station (Figure 13.8 TRI) [18].

The majority (over 90%) of GSM-R radio communication objects in Poland should be type BI systems located in the countryside and in suburban areas, in which propagation along the track does not collide with human settlements. An example distribution of GSM-R signal field intensity distribution for the first system (planning phase: section of the line E30, Legnica – Bielawa Dolna) is shown in Figure 13.6. Radio coverage was prepared for the Epstein Peterson propagation model. Requirements concerning radio coverage are as follows:

- minimum signal level 95 dBm with 95% probability;
- overlapping of coverage areas of adjacent cells;
- using base stations that can ensure so-called "double coverage."

Detailed location of radio communication objects and parameters of GSM-R system equipment, such as mast height, transmitter power, and the direction and inclination of

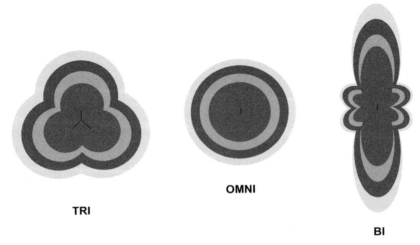

Figure 13.8 General configurations of GSM-R antennas

antennas should be specified in detail in the drafting detailed design phase, with values of permitted levels of electromagnetic fields verified at the stage of commissioning (pursuant to Article 122a of Environmental Protection Law Act of 27 April 2001). Measurement shall be carried out up to 2 m above the ground level, or above any on general access surface.

As already mentioned, contrary to the public GSM system, the GSM-R system in over 90% of cases (suburban and countryside areas) is more environmentally friendly, as its radiation propagates directionally alongside railway track, without colliding with any settlements, and if there any settlements, they comprise low buildings. Also at railway stations, one can assume the nonexistence of high residential buildings (over three floors).

It has to be stressed at the same time that as regards systems related to cellular telephony and wireless technologies, decades of research have not proved their negative effect to human health, provided that environmental protection standards are maintained in general access locations.

Due to the nature and specificity of GSM-R system, classification of a project of building GSM-R radio communication objects (ROs) is carried out pursuant to regulation by the Council of Ministers of 9 November 2010 on projects with potential environmental impact, is classified in the group of radio communication, radio navigation and radiolocation systems, excluding radio line, emitting electromagnetic fields at frequency range 0.03 MHz to 300 GHz. Projects are classified by two parameters:

- effective isotropic radiated power (EIRP) determined for single antenna;
- distance from the electrical center of such antenna to general access locations along the main axis of antenna radiation.

When determining the possible range of signal emitted from the base station, it is important to consider watt density of radiation in specific direction, and not overall power radiated from the antenna.

For those reasons it is convenient to characterize the source of signal emission by P_{EIRP} (EIRP –effective isotropic radiated power), which defines the power that should be radiated using hypothetical isotropic antenna, in order to obtain such watt density in desired direction as radiated by antenna with gain G_i supplied with power P via antenna channel with losses L_f [5]:

$$P_{ERIP} [dBm] = P[dBm] + G_i[dBi] - L_f[dB] \qquad (13.3)$$

where:
P_{ERIP} – effective power of isotropic radiation in [dBm],
P – antenna supply power,
G_i – antenna gain,
L_f – antenna line (feeder) losses.

Figure 13.9 represents an example of a radio communication system with antenna radiation axes indicated.

When determining P_{EIRP}, a couple of parameters are considered – most of all transmitter power, antenna gain, and antenna line attenuation. As standard, the power obtained at the output of TRX transmitter is in the order of 65–70 W (near-maximum value for GSM transmitting equipment). Adopting values of individual component parameters, such as antenna

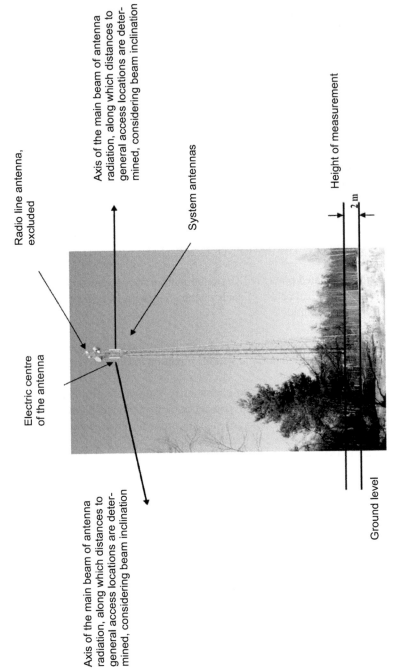

Radio line antenna,
excluded

Axis of the main beam of antenna
radiation, along which distances to
general access locations are determined, considering beam inclination

System antennas

Height of measurement

Electric centre
of the antenna

Axis of the main beam of antenna
radiation, along which distances to
general access locations are determined, considering beam inclination

2 m

Ground level

Figure 13.9 Example radio communication system with antenna radiation axes indicated

Source: [1].

Table 13.2 Qualification of radio communication, radio navigation and radiolocation systems, excluding radio lines, as projects with significant environmental impact

Projects with significant potential environmental impact, requiring completion of environmental impact assessment procedure			Projects not requiring completion of environmental impact assessment procedure
Effective isotropic radiated power W	Projects for which the report is not required	Projects for which the report is required or may be required	
	Distances between general access locations and the electrical center of the antenna, along the main axis of radiation beam of such antenna [m]	Distances between general access locations and the electrical center of the antenna, along the main axis of radiation beam of such antenna [m]	Distances between general access locations and the electrical center of the antenna, along the main axis of radiation beam of such antenna [m]
15 to 100	–	less or equal to 5	more than 5
100 to 500	–	less or equal to 20	more than 20
500 to 1000	–	less or equal to 40	more than 40
1000 to 2000	–	less or equal to 70	more than 70
2000 to 5000	less or equal to 100	100 to 150	more than 150
5000 to 10,000	less or equal to 150	150 to 200	more than 200
10,000 to 20,000	less or equal to 200	200 to 300	more than 300
20,000 and more		irrespective of the distance	

Source: [1]

gain and antenna channel attenuation, P_{EIRP} W can be determined at 1700 W. This value is in the range high or equal to 1000 W and lower than 2000 W, for which 70 m is the limit value defining distance to general access locations, determined from the electric center of the antenna and along the radiation beam axis (Table 13.2).

Qualification of radio communication, radio navigation, and radio location systems, except for radio lines used in projects with a possible high environmental impact, is subject to Regulation by Council of Ministers of 9 November 2003 on defining types of enterprises that may have material environmental impact and on detailed considerations of qualifying enterprises to prepare environmental impact reports (Dz.U. No. 213, item 1397).

Implementation of radio communication, radio navigation, and radiolocation systems (including radio communication objects of GSM-R system), which pursuant to the letter of the aforementioned regulation are not qualified as projects with significant environmental impact, does not require obtaining decision on environmental considerations for the project.

Pursuant to Article 143 of the Environmental Protection Law Act, a technology used in newly opened or significantly modified systems and appliances should meet the requirements defined in consideration of below:

- use of substances with small risk potential;
- efficient use of energy;
- ensuring rational use of water and other materials and fuels;
- using no-waste and little-waste technologies, recycling of waste produced;
- type, range, and volume of emissions;

- using comparable processes and methods that were successfully used in industry (on railway lines outside of Poland);
- state of the art.

GSM-R technology meets all aforementioned requirements, and is the best available technology as regards railway traffic management, including safety of railway traffic control.

13.5. Summary

GSM-R is the system of wireless railway communication designated for the maintenance of railway traffic control, which is scheduled to substitute the analog 150 MHz system presently used by PKP Polskie Linie Kolejowe S.A. GSM-R is a component of ERTMS, ensuring safe operation of trains. Thus, GSM-R has to be a reliable system of (voice and data) transmission, in which reliability and dependability should exceed the public GSM system, which is ensured by additional measures (hardware redundancy, suitable coverage of electromagnetic fields).

Due to aggravating condition of analog radio communication equipment and costs incurred by the infrastructure manager for their maintenance and upgrade, the fastest possible implementation of GSM-R on critical railway lines in Poland is of paramount importance. GSM-R implementation rate is to a great extent dependent on obtaining administrative permits, including a decision on environmental considerations of the project (if required) and building permit. Since each upgraded railway line requires custom design, EC Verification Certificate is issued by a notified body (NoBo) separately for each line.

Capacity requirements allowing estimation of the required number of transmission and reception equipment at each base station directly result from the number of appliances that would be in use in a given station simultaneously. Hence, GPRS transmission is planned, reducing the required capacity by multiplexing many users in one timeslot. Reliability of package transmission can be improved by adopting suitable mechanisms, such as support from a TCP/IP transport layer. In the case of GPRS, the effect of switching between stations on radio interface and delays in data transmission is visible. When switching time in DCSS transmission was approximately 0.5 s, in package transmission it may take as much as 8 s. In addition, in GSM networks voice transmission has priority over package data transmission, which means that in the absence of sufficient resources on radio interface, data transmission is stopped and the freed timeslot allocated to incoming voice connection. The solution here is timeslot allocation permanently dedicated to GPRS. This in turn may reduce capacity gain thanks to multiplexing, as blocking one or more timeslots should reduce the number of resources available to voice communication and consequently might result in the need to increase capacity by installing an additional TRX module. That may be the case with a small-capacity station (sectors with one TRX module), although in the case of base stations with high intensity of telecommunication traffic, significant gain from data transmission over GPRS is anticipated. Correct dimensioning of ETCS-GPRS traffic capacity developing of traffic profile is necessary, which shall serve as the base for traffic measurements in the actual network.

Above brief characteristics of GSM-R system, including the description of radio communication objects (ROs), allows to classify the system in the group of radio communication, radio navigation, and radiolocation systems, excluding radio line, emitting electromagnetic fields at frequency range 0.03MHz to 300 GHz. Based on the value of power P_{EIRP} which for

GSM-R ROs in the majority of cases should be in the range of 1000 W to 2000 W, and based on relevant applicable regulations, it may be concluded that in places accessible to the public, there will be no unacceptable Polish laws, values of electromagnetic fields. Making the above assumptions, it may be concluded that the erection and operation of radio GSM-R communication objects does not qualify as a project with a possible high environmental impact (pursuant to Regulation by Council of Ministers of 9 November 2003 on defining types of enterprises that may have material environmental impact) and does not require obtaining a decision on environmental considerations for the project.

Implementation of the GSM-R system should improve the attractiveness of Polish railway network. Thanks to such a project, European standards will be implemented in transeuropean railway corridors crossing the territory of Poland. Thanks to ensuring the interoperability of railways, railway operators from other member states may access the Polish railway infrastructure, facilitating the movement of people and goods on domestic and international routes, as well a transit transport between EU and neighbor countries [16].

A direct consequence should be the provision of digital voice communication (mostly between traffic controllers and drivers) and digital data transmission, in particular for ETCS. Implementation of railway-specific services, such as REC and eREC, should improve the safety of railway operations. From the viewpoint of operators, improvement of service portfolio is the most important change. Competitive advantage of railways over other means of transport should improve, especially in terms of speed and punctuality, consequently reducing transit time.

In the near future, the GSM-R system will be implemented on railway line E20 and E65, finally to include the majority of railway lines in Poland. Migration from the analog system of radio communication, using RADIOSTOP , to the GSM-R system will be a challenge for Poland. The transition period, during which both systems will be used simultaneously, should also provide the challenge of ensuring required availability of both systems on railway lines in which they are used. It is worth mentioning that the use of RADIOSTOP in analog VHF radio communication is a unique solution, adopted only in Poland. Years of operation of such configured VHF system proved the need for controlled use of the RADIOSTOP system in an emergency, in many cases helping to avoid incidents or disasters. This is why substituting the emergency (RADIOSTOP) signal, automatically stopping all trains on given section of railway line, with information emergency signal (REC) should be thoroughly analyzed. The result of such analysis should be the development of procedures ensuring the same (or nearly the same) level of safety as provided by the RADIOSTOP signal.

Bibliography

[1] Bałkowiec, P.: *Wpływ systemu GSM-R na środowisko w ujęciu procesu inwestycyjnego*. Prezentacja PKP Polskie Linie Kolejowe S.A. [The impact of the GSM-R system on the environment in terms of the investment process. Presentation of PKP Polskie Linie Kolejowe S.A.]. Warsaw, 2013 (in Polish).

[2] Białoń, A.: *Masterplan wdrażania ERTMS w perspektywie krajowej i wspólnotowej* [Masterplan of ERTMS implementation in the national and community perspective]. Transport i Komunikacja, 2010, nr 2 (in Polish).

[3] Decyzja Komisji Europejskiej 2008/386/WE z dnia 23 kwietnia 2008 r. zmieniająca załącznik A do decyzji 2006/679/WE dotyczącej technicznej specyfikacji dla interoperacyjności odnoszącej się do podsystemu sterowania ruchem kolejowym transeuropejskiego systemu kolei konwencjonalnych oraz załącznik A do decyzji 2006/860/WE dotyczącej specyfikacji technicznej

interoperacyjności podsystemu "Sterowanie" transeuropejskiego systemu kolei dużych prędkości [European Commission Decision 2008/386 / EC of 23 April 2008 amending Annex A to Decision 2006/679 / EC concerning the technical specification for interoperability relating to the rail traffic control subsystem of the trans-European conventional rail system and Annex A to Decision 2006/860 / EC concerning the technical specification for interoperability of the control-command and signaling subsystem of the trans-European high-speed rail system] (in Polish).

[4]   GSM-R Radio Planning Guidelines, 02–2006, JERNBANEVERKET UTBYGGING Document number 3A-GSM-036.

[5]   Hammuda, H.: *Cellular Mobile Radio Systems: Designing Systems for Capacity Optimization.* John Wiley & Sons Ltd, 1997.

[6]   www.uke.gov.pl

[7]   Jajszczyk, A.: *Wstęp do telekomutacji* [Introduction to telecommuting]. Warsaw: Publ. WNT, 2009 (in Polish).

[8]   Kabaciński, W., and Żal, M.: *Sieci telekomunikacyjne* [Telecommunications networks]. Warsaw: Publ.Wydawnictwa Komunikacji i Łączności, 2008 (in Polish).

[9]   Katulski, R.J: [Radio waves propagation in wireless telecommunications]. Warsaw: Publ. WKiŁ, 2010 (in Polish).

[10] Mandoc, D.: *GSM-R in 2009 – International Operations Take off.* European Railway Review, 2010, nr 1.

[11] Mandoc, D., Konrad, K., and Winter, P.: *ERTMS Training Programme 2008 – Handbook.* UIC, Paryż, 2008.

[12] Markowski, R.: *Aspekty łączności GSM-R w systemie ERTMS/ETCS2 – cz. I.* [Aspects of GSM-R communication in the ERTMS / ETCS2 system – part I]. Publ. Infrastruktura Transportu, 2010, nr 3 (in Polish).

[13] Markowski, R.: *Aspekty łączności GSM-R w systemie ERTMS/ETCS2 – cz. II.* [Aspects of GSM-R communication in the ERTMS / ETCS2 system – part II.] Publ. Infrastruktura Transportu, 2010, nr 3 (in Polish).

[14] *Network Rail. Guidance on GSM-R Cell Planning Consultation.* 12–2007, Association of Train Operating Companies.

[15] Noldus, R.: *CAMEL: Intelligent Networks for the GSM, GPRS and UMTS Network.* Chichester: Wiley-Blackwell, 2006.

[16] Pawlik, M.: *Polski Narodowy Plan Wdrażania Europejskiego Systemu Zarządzania Ruchem Kolejowym ERTMS* [Polish National Implementation Plan of the European Railway Traffic Management System ERTMS]. Publ. Technika Transportu Szynowego, 2007, nr 1 (in Polish).

[17] Pawlik, M.: *Wprowadzenie do ERTMS – europejskiego systemu zarządzania ruchem kolejowym* [Introduction to ERTMS – European Rail Traffic Management System]. Publ. Transport i Komunikacja, 2010, nr 2 (in Polish).

[18] PKP Polskie Linie Kolejowe, S.A: *Wstępne planowanie radiowe GSM-R dla linii kolejowych objętych Narodowym Planem Wdrażania ERTMS w Polsce* [PKP Polskie Linie Kolejowe, S.A: Pre-planned GSM-R radio planning for railway lines covered by the National ERTMS Implementation Plan in Poland]. Warsaw, 2012 (in Polish).

[19] Praca CNTK "Projekt Narodowego Planu Wdrażania Europejskiego Systemu Zarządzania Ruchem Kolejowym w Polsce," Etap III "Wstępnego studium wykonalności wdrożenia systemu ERTMS w skali sieci PKP PLK S.A.," Główny referent: dr inż. Marek Pawlik, Zespół: mgr inż. Andrzej Toruń, dr inż. Andrzej Białoń, mgr inż. Witold Olpiński, mgr inż. Marek Ucieszyński, mgr inż. Piotr Chyliński, dr inż. Jacek Kukulski, mgr inż. Alfred Szymański, mgr inż. Paweł Gradowski, mgr inż. Iwona Wacławiak, mgr inż. Beata Piwowar, mgr inż. Bogusław Bartosik, inż. Rafał Iwański, dr inż. Janusz Poliński, mgr inż. Robert Kruk, mgr inż [CNTK's work "Project of the National Implementation Plan for the European Rail Traffic Management System in Poland," Stage III "Preliminary feasibility study of ERTMS system implementation in the PKP PLK S.A. network scale," Main clerk: Ph.D. Marek Pawlik, Team: M.Sc. Eng. Andrzej Toruń, Andrzej

Białoń, M.Sc. Eng. Witold Olpiński, MSc. Marek Ucieszyński, M.Sc. Eng. Piotr Chyliński, Ph.D. Jacek Kukulski, M.Sc. Eng. Alfred Szymański, M.Sc. Eng. Paweł Gradowski, M.Sc. Eng. Iwona Wacławiak, MSc. Beata Piwowar, MSc. Bogusław Bartosik, Eng. Rafał Iwański, Ph.D. Janusz Poliński, M.Sc. Eng. Robert Kruk, M.Sc. Eng.]. Warsaw: Zdzisław Wiśniewski, 2007 (in Polish).

[20] Siergiejczyk, M.: *Wybrane zagadnienia systemów sterowania ruchem i łączności dla Kolei Dużych Prędkości w Polsce* [Selected problems of traffic control and communication systems for the High-Speed Rail in Poland]. Publ. Logistyka 3/2012. Poznań, 2012 (in Polish).

[21] Siergiejczyk, M., and Gago, S.: *Wybrane problemy niezawodności i bezpieczeństwa transmisji informacji w systemie GSM-R* [Selected problems of reliability and security of information transmission in the GSM-R system]. Publ. Problemy Kolejnictwa Vol 58 Issue 162 Warsaw, 2014. ISSN 0552–2145 str. 111–124 (in Polish).

[22] Siergiejczyk, M., Palik, M., and Gago, S.: *Safety of the New Control Command European System.* "*Safety and Reliability: Methodology and Applications* – Proceedings of the European Safety and Reliability Conference ESREL 2014" Edited byNowakowski, T., Młyńczak, M., Jodejko-Pietruczuk, A., and Werbińska – Wojciechowska S. Publ. CRC Press/Balkema, 2015. pages. 635–642; ISBN 978-1-138-02681-0.

[23] Siergiejczyk, M.: *Aspekty środowiskowe planowania i wdrażania sieci GSM-R w Polsce.* Rozdział w monografii "Projektowanie, budowa i utrzymanie infrastruktury w transporcie szynowym" [Environmental aspects of planning and implementing GSM-R networks in Poland. Chapter in the monograph "Designing, construction and maintenance of infrastructure in rail transport"]. Publ. Wydawnictwo Naukowe ITE-PIB Radom 2014, ISBN 978-83-7789-269-5 str. 228–245 (in Polish).

[24] Siergiejczyk, M.: *Design of Radio Digital Mobile Communications in Conditions of the Polish Railway: Applied Mechanics and Materials.* ISSN: 1662–7482, Vol. 817/2016. Trans Tech Publications, Switzerland.

[25] UIC Project EIRENE, *Functional Requirements Specification – version 7.1., GSM-R Functional Group*, FRS v 7.1, 17 May 2006 Reference: PSA167D005

[26] UIC Project EIRENE, *System Requirements Specification, GSM-R Operators Group*, SRS v 15.1, 1 June 2010

[27] Urbanek, A.: *Komunikacja kolejowa GSM-R* [Rail communication GSM-R]. Networld, 2005, nr 1 (in Polish).

[28] Supplementing the Feasibility Study in the area of the GSM-R digital radio communication system, technological communication and IT systems related to the operation of the planned railway line of the Pomeranian Metropolitan Railway. Research work under the direction of M. Siergiejczyk, Warsaw, 2011 (in Polish)

[29] Wesołowski, K.: *Systemy radiokomunikacji ruchomej* [Mobile radio communication systems]. Warsaw: Publ.Wydawnictwa Komunikacji i Łączności, 2006 (in Polish).

[30] Winter, P.: *International Union of Railways, Compendium on ERTMS.* Hamburg: Eurail Press, 2009.

[31] Winter, P.: *Global Perspectives for ERTMS, ETCS and GSM-R UIC.* Paris: Editions Techniques Ferroviaires, 2007.

[32] Włodkowska, J.: *Pierwsze wdrożenia systemu ERTMS/ETCS w Polsce* [First implementations of the ERTMS / ETCS system in Poland]. Publ. Transport i Komunikacja, 2010, nr 2. (in Polish).

[33] www.erlang.com/calculator/erlb/

[34] *Wybór wymagań na GSM-R dla PKP z EIRENE FRS 7.0 i FRS 6.0.* Materiały PKP Polskie Linie Kolejowe S.A. Warszawa, lipiec 2009. [Selection of requirements for GSM-R for PKP with EIRENE FRS 7.0 and FRS 6.0. Materials of PKP Polskie Linie Kolejowe S.A.] Warsaw, July 2009. (in Polish)

[35] Zarządzenie nr 23: *Prezesa Urzędu Komunikacji Elektronicznej z dnia 7 czerwca 2006 r. w sprawie planu zagospodarowania częstotliwości dla zakresów 876–915 MHz oraz 921–960 MHz* [Ordinance no. 23. of the President of the Office of Electronic Communications dated 7 June 2006 on the frequency management plan for the bands 876 – 915 MHz and 921 – 960 MHz.] (in Polish).

Chapter 14

Significance of rail track diagnostics to high-speed rails

Henryk Bałuch

14.1. Introduction

Diagnostics of rail track is the specialty oriented on the detection of risks related to construction and operation of rail track structure. Such threats often originate from railroad design. In practical aspect, diagnostics is the area of correlating measurements and observations. The origins of diagnostics as an isolated theoretical specialty can be traced back to the monograph [3] defining relationships between tools, technologies, and scientific methods, and – as Dolby calls it [8] – conceptual framework.

Hazard is the key category of safety theory [10]. There are at least a few hazard concepts known. In the broader sense, hazard is the factor that under the effect of external circumstances can cause risk. Common definition says that hazard is the source of risk, and risk is the potential for the occurrence of an undesirable effect, at a certain time and under certain circumstances [21].

Assuming such definition, rail track hazard is the state which under the effect of external circumstances may lead to undesirable event. This definition is broader than the definition in the (EC) Commission Regulation,[1] where hazard is only the state that can lead to accident. A broader conceptual range of hazard, also assumed in other disciplines, determines the field of application of rail track diagnostics, covering all phases of surface lifecycle – programming, design, construction, and operation.

When analyzing diagnostics of high-speed rails, we cannot avoid specific features of a given railway network, as this would render the analysis incomplete. We have to stress, therefore, that this section refers to diagnostics of the railway network in Poland, where high-speed rails are only being introduced, and where there is a lot to be done to make railroads ready for high speeds. The major task in this respect is reaching high speeds not only in lines where the speed of 200 km/h has already been introduced. Rail track diagnostics as the technique of quality verification shall be decisive in this respect.

14.2. Hazards in rail track and areas of diagnostics application

In a broad sense, hazards inherent in rail tracks can be divided into two categories:

1 Leading to derailment (catastrophic);
2 Leading to other damage, e.g. due to design errors, poor quality of workmanship or selecting inadequate remedial measures following a mistaken determination of the cause of derailment.

---

1 No. 352/2009 of 24 April 2009.

This division is actually not so sharp, as only some cracking in rail leads to derailment. Forms of hazards in rail tracks assumed by the author are shown in Figure 14.1.

A broader approach to rail track hazards expands the area of diagnostics application. Not only does it cover operational tasks, such as investigating rail track for necessary repairs and checking that the railroad condition is suitable for maximum operating speeds and axle loads, but also operations referring to upgrade of railroads. In this area, diagnostics should provide answers when possibilities of remedial actions are exhausted and offer grounds for the assessment of upgrade works quality, deciding on further expenditure on railroad maintenance (Figure 14.2).

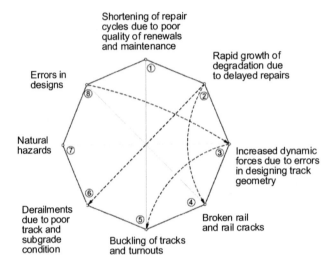

Figure 14.1 Most common rail track hazards and related consequences

Source: Author's own estimation

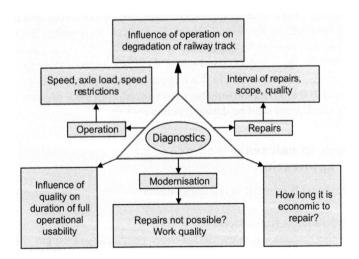

Figure 14.2 Rail track diagnostics areas of application

Source: Author's own estimation

14.3. The role of rail track diagnostics in preparation of railway lines for introduction of high speeds

High-speed rail track maintenance demands on maintaining low permitted deviations. At the operating speed of 200 km/h, horizontal irregularities of the track should not exceed 4 mm, vertical irregularities 3 mm, and twist 1‰. Meeting such requirements demands a very high quality of upgrade works. The notion covers accuracy obtained following the completion of works, with deviations approximately half as high as those during operation, and effectiveness, which means the long-term ability to maintain obligatory tolerances.

In Poland, the objective assessment of rail track quality is the synthetic track condition index J, postulated in the study [6] and adopted throughout the network as the basic parameter of track geometric condition. The index is based on standard deviations of four geometric values. It does not therefore depend on operating speed or on permitted deviations. It is expressed by the equation:

$$J = \frac{S_z + S_y + S_w + 0,5S_e}{3,5} \tag{14.1}$$

where: S_z, S_y, S_w, S_e – standard deviations: vertical irregularities, horizontal irregularities, track twist, and gauge, respectively.

Quality of works on high-speed railroads, where operating, is very high. Two examples should suffice. Figure 14.3 represents the synthetic view of track condition index J, which was used on the Lisbon-Oporto line in Portugal, with a maximum speed of passenger operation at 220 km/h [7]. This line, length 337 km, is also used for the operation of freight trains, with an operating speed of 80 km/h and axle load of 225 kN [1]. Concrete sleepers of 600 mm, Vossloh rail fasteners, and spacers with rigidity 450 kN/mm were used there. The purpose of the test, in addition to the determination of J index role, was the determination of track changes over time. Condition of the analyzed track section, represented by J index ≈ 0.5 mm, changed during the period from approximately day 1200 to approximately day 2650;

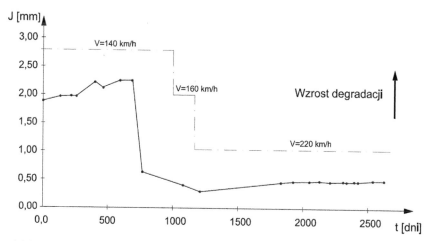

Figure 14.3 Synthetic track condition index J on high-speed rail line in Portugal
Based on: [7]

that is, during 3.97 years between 0.4 mm to 0.6 mm, which also means very high efficiency of works, equal to 0.05 mm/year. Such a level of work quality significantly extends the interval of repairs [4, 5, 6, 20].

Another example is the assessment of tamping quality observed on Swedish railroads (Figure 14.4). On the horizontal axis there are located standard deviations of vertical irregularities in track before tamping (σ_b), and on the vertical axis – standard deviations after tamping (σ_a). Tamping of the track with an original standard deviation 2.0 mm should be considered as good when deviation after tamping is in the range of 0.5 mm and 1.3 mm .

[It is worth contrasting these two examples with values obtained on railroads in Poland. Figure 14.5 represents synthetic indices J on one of the upgraded sections. Only on one hectometer value of the index J during guarantee acceptance was = 0.7 mm. In the majority of hectometers, the value of J index exceeded 1.0.

Quality of works on some upgraded sections of PLK S.A. railroads is low. In extreme cases, attempts were observed of the handing over of tracks in the condition warranting their prompt deformations. Complete absence of drainage, in the form of blocked ditches (Figure 14.6), and too small depth of ballast (Figure 14.7) would cause out-of-tolerance deformations on high-speed railroads. Poor subgrade and insufficient tamping would cause deformations as shown in Figure 14.8. Using a ballast inadequate for railroad, that is a ballast containing marl, would cause slaking of the ballast, resulting in the need for costly overhaul (Figure 14.9).

Extreme cases presented indicate that without integrating diagnostics in quality control of upgrade works, at all phases, maintenance of rail track on lines with operating speed 200 km/h or more would be very costly or would not ensure a peaceful journey.

Obtaining the quality of rail track as in Figures 14.3 and 14.4 also requires another approach to the selection of contractors, whose number will increase in the course of expected increase in volume of upgrade works and ongoing exchange of rail track. They should have experience, not only assessed by the number of projects completed but most of all by quality

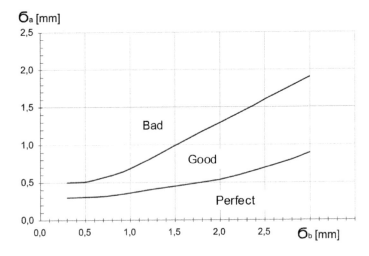

Figure 14.4 Quality of track tamping by Swedish railways

Based on: [2]

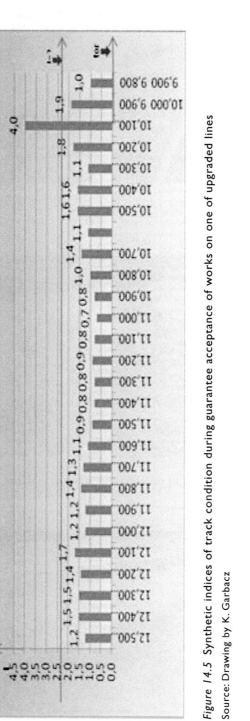

Figure 14.5 Synthetic indices of track condition during guarantee acceptance of works on one of upgraded lines

Source: Drawing by K. Garbacz

Figure 14.6 Track after the upgrade; no securing of cross-cut slopes and blocked ditches

Source: Photo from G. Gawroński

Figure 14.7 Too small thickness of ballast under sleepers, absence of drainage

Source: Photo from G. Gawroński

Figure 14.8 Track irregularities in the new railroad after 6 months of operation

Source: Photo from IK archives

Figure 14.9 Slaking ballast, two years after the upgrade
Source: Photo from RI archives

of their work, owned machinery and equipment, and experienced personnel, which all contribute to a high quality of works. Present practice proves that the above conditions are not always met.

14.4. Diagnostic techniques

High operating speeds of trains require adequate diagnostic techniques and in-depth interpretation of survey results. Fundamental features of HSR rail track diagnostics are:

1   increased frequency of driving survey vehicles on railroads;
2   necessity to survey track irregularities with long and short waves;
3   high accuracy of surveys.

In the UK, railroads with operating speeds of 115 mph and 125 mph (185 km/h and 201 km/h) are inspected with a survey vehicle every 28 days. In each railroad section of 1/8 mile (201 m), standard deviations of vertical and horizontal irregularities are calculated. Exceeding a standard deviation of 4.7 mm requires track overhaul or reduction in operating speed [16]. Track surveys carried out using passenger cars integrated with passenger trains are more and more common, as they ensure a high frequency of surveys [19].

Geometrical surveys carried out using motor cars and survey cars offer many benefits, such as high efficiency, usually hundreds of kilometer per day, automatic processing of survey data, surveying of track under load, instantaneous exchange of data from dangerous sections to executive units, and collection of great quantities of data. Track surveys carried out using hand equipment on conventional railroads are also of great importance to rail track diagnostics.

Survey cars and motors cars performing surveys based on the fixed chords representation of results are characterized by transition functions. A transition function expresses the output signal transform to input signal transform at zero initial conditions. The transition function of the measurement system, with chord $c = a + b$, where $a + b$, at the wave length different than L, is described by the equation:

$$F = \frac{\left[\begin{array}{l} a^2 + b^2 + (a+b)^2 - 2b(a+b)\cos(\lambda a) - 2a(a+b)\cos(\lambda b) + \\ + 2ab\left(\cos(\lambda a) \times \cos(\lambda b) - \sin(\lambda a) \times \sin(\lambda b)\right) \end{array}\right]^{\frac{1}{2}}}{a+b} \tag{14.2}$$

where: $\lambda = \dfrac{2\pi}{L}$

Transition function described by such equation is shown in Figure 14.10. It also shows transition function at $c = a + b$, where a $= b$.

The form of this function suggests that the image of blown-up track irregularities at symmetrically divided chord is approximately 7 m to approximately 20 m, and at the chord $4 + 6 = 10$ m – from approximately 6 m to approximately 23 m. At low frequencies of car specific vibrations, in the range 0.5 Hz and 3.0 Hz, such representation does not give the view of longer waves, and does not provide an opportunity to assess possible resonance phenomena (Table 14.1). The table says that good representation according to transient functions shown in Figure 14.10 demonstrates irregularities with waves in grey fields. It is worth mentioning that analyzing the natural frequency of vibrations in vehicles is not simple, and e.g. when inspecting bridgeworks, attempts were made to convert results obtained at low speeds and on more easily accessible rolling stock to much faster trains. Inspection of the bridgework at the speed of 40 km/h using the locomotive ET-44 and referring the conclusions to Pendolino trains, may be a good, if controversial, example [11].

Modern survey trains are fitted with inertia survey systems for detecting vertical and horizontal irregularities and with an optical track gauge survey system. This also makes it possible to include irregularities with long waves [12]. The inertia system generates 3D images, thanks to gyro lasers combined with GPS. Some survey vehicles, such as the two-segment unit MERMEC Group, also perform surveys of the traction system and ballast pile at the survey speed of 160 km/h. Two-segment diagnostic train ERA from the Samara plant measures 120 values characterizing infrastructure [17], although it requires a separate locomotive. Presently manufactured survey vehicles are usually self-powered.

Table 14.1 Wavelengths vs. speed and specific frequency rate of cars

Speed [km/h]	Frequency (Hz)				
	0.5	1.0	2.0	3.0	4.0
100	55.6	27.8	13.9	9.3	6.9
150	83.3	41.7	20.8	13.9	10.4
200	111.1	55.6	27.8	18.5	13.9
250	138.9	69.4	34.7	23.1	17.4
300	166.7	83.3	41.7	27.8	20.8
350	194.4	97.2	48.6	32.4	24.3

Source: Author's own estimation

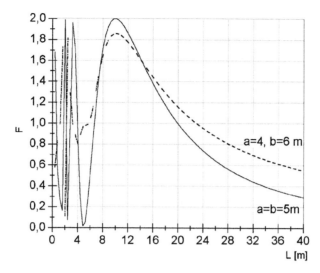

Figure 14.10 Transition function of survey vehicles with 10 m chord

Source: Author's own estimation

There are devices known, such as Railvision, that record rail track damages at the speed of 350 km/h and have a sampling rate in the range of 100–200 mm. This is not, however, the top accuracy; according to the study [14], the combination of inertia and optical system can provide a survey accuracy in the range of 0.1–0.5 mm at the speed of 430 km/h. In addition to recording irregularities in rolling surfaces of rails, modern survey cars offer the possibility of measuring many other geometric parameters, such as optical recording of 15 rail head cross-sections, with accuracy 0.25 mm, and the profile of wavy wear of rails, with waves 30–500 mm and accuracy 0.1 mm, at the speed of 25 km/h to 250 km/h [9].

There are measuring devices installed in survey vehicles or passenger cars that record track irregularities and rail track damages at 150 km/h in combination with GPS data. Measurement of accelerations defining a peaceful journey is currently the standard in presently used survey vehicles. New rail track diagnostic methods are sought on an ongoing basis, such as diagnostics of rail fixtures by analyzing vibration frequency caused by hammer impact [15].[2] Also, works are in progress on faster and more accurate methods of interpreting survey data, e.g. utilizing artificial neural networks.

A separate area of diagnostics application is one in combination with controls of rail machines. Results of track irregularity surveys are transferred remotely to tamper controllers. Satellite transfer technique has been used in this application for years now. Such systems ensure high accuracy of location of tamped and slid tracks in the absolute values, which are in reference to the national coordinates grid. Survey motor cars used in Poland so far do not meet all requirements required to maintain HSR rail tracks. Planned changes in this respect are completely justified.

---

2 In the 1970s, similar research on the determination of logarithmic decrement of attenuation, as the index of subgrade compaction, was carried out by Kazimierz Towpik of COBiRTK (presently the Railway Research Institute).

14.5. Computer-assisted assessment of rail track survey and observation results

Information technology (IT) is the foundation of all modern survey systems. In addition to large survey data processing systems, sending data in real time to databases, and automatic generation of exceeding alert conditions, rail track diagnostics also utilize smaller assistance systems dedicated to narrow problems. Polish railways have been using such tools as rail track repair hierarchy planning (SOHRON), determination of permitted operating speeds and axle loads (UNIP), assessment of rail track structure (SOKON), etc., all created by the Railway Institute.

A new track inspection system, involving both on-foot inspections and vehicle inspections, shall be introduced on PLK S.A. in 2016. It is characterized by the reduced frequency of on-foot inspections, which are dangerous on HSR lines, and stresses the correct interpretation of survey results and rail track observation. The assumed principle is shorter interval of inspections, both on foot and in vehicles, when track condition is worse. Evaluation of inspection frequency is based on calculations for each kilometer of track and results of observations. Calculations consider results of geometric surveys, rail wear, age of sleepers, curve radiuses, train speeds, and intensity of operations. The result of calculations and observation is the classification of each track section to one of four inspection frequency classes.

14.6. Summary

The condition-based maintenance strategy of operating rail tracks is presently in force in the majority of countries and requires a good level of diagnostics and projections concerning imminent changes [19]. Tasks of diagnostics in Polish railways are much harder at present than earlier, for two reasons:

1   The need to adapt survey techniques and equipment to high operating speeds, especially introducing new survey motor car, providing some minor equipment, and modifying some systems assisting decision-making process.
2   Sending the most-experienced diagnosticians to carry out thorough acceptance of upgrade works and rail track replacement works, in order to improve the quality and reduce the lifecycle costs of rail tracks. This scope also includes regulation of some formal problems [13].

The management of PLK S.A. makes all efforts to update the professional knowledge of their diagnostic inspectors. Training courses, covering tens of hours of curriculum, were completed by approximately 200 persons. Nevertheless, there is a real need for post-graduate courses for higher engineering personnel, which are currently missing. Also, some research problems related to diagnostics await their resolution.

Bibliography

[1]   Androde, A.R., and Teixeira, F.: Uncertainty in Rail-Track Geometry Degradation: Lisbon-Oporto Line Case Study. *Journal of Transportation Engineering ASCE*, March 2011.
[2]   Arosteh-khoug, I., Larsson Kraik, P.O., Nissen, A., and Kumar, U.: *Track Geometry Degradation in Swedish Heavy Haul Railroad*. Lulea Teknisko Universitet. 2011
[3]   Bałuch, H.: *Diagnostyka nawierzchni kolejowej* [Diagnostics of railway track]. Warszawa: WKŁ, 1978.

[4] Bałuch, H.: Jakość robót nawierzchniowych i metody jej oceny. [Quality of railway track works and methods of its assessment] *Problemy Kolejnictwa [Railway Reports]*, no. 128, 1998.

[5] Bałuch, H.: *Wpływ jakości robót na trwałość nawierzchni kolejowej. [The impact of the quality of works on the durability of the railway track]*. Międzynarodowa Konferencja Naukowo-Techniczna "Technologie modernizacji i utrzymania linii kolejowych – 50 lat doświadczeń." International Scientific and Technical Conference "Technologies for modernization and maintenance of railway lines – 50 years of experience," Wrocław 2004-06 – 3-4.

[6] Bałuch, M.: Estymacja nierówności toru kolejowego. [Estimation of irregularities in the railway track]. *Archiwum Inżynierii Lądowej, Archives of Civil Engineering*, Nos. 3-4, 1989

[7] Berawi, A.R.B., Delgado, R., Calcada, R., and Vale, C.: Evaluating Track Geometrical Quality Through Different Methodologies. *Journal of Technology*, 2010, pp. 38–47.

[8] Dolby, R.G.A.: *Niepewność wiedzy – Obraz nauki w końcu XX wieku. [Uncertainty of knowledge – A picture of science at the end of the 20th century]*. Wydawnictwo Amber, Warszawa, 1998. Amber Publisher

[9] Eglseer, F.: Recording Track and Catenary Geometry at 250 km/h. *Railway Gazette International*, 1995, 8, pp. 507–509.

[10] Fehler, W.: *Zagrożenie – kluczowa kategoria teorii bezpieczeństwa. Współczesne postrzeganie bezpieczeństwa. [Threat – a key category of the security theory. Contemporary perception of security]*. Materiały Konferencji zorganizowanej przez Wyższą Szkołę Administracji w Bielsku-Białej, 23.11.2006. Materials from the Conference organized by the College of Administration in Bielsko-Biała

11] Gołębiowski, K., and Zielińska, M: Modelowanie oddziaływań dynamicznych pociągu typu Pendolino na konstrukcję zabytkowych mostów kolejowych.[Modeling dynamic interactions of a Pendolino train for the construction of historic railway bridges]. *Przegląd Budowlany*, 2015, no. 1, pp. 27-32

[12] Ise, K.: Present Condition and Future Prospects of the Track Condition Monitoring. *Japanese Railway Engineering*, 2012, 176.

[13] Kampczyk, A., and Bort, T.: Odbiory torowych robót nawierzchniowych.[Acceptance of railway track works]. *Technika Transportu Szynowego*, Rail Transport Technique, 2014, no. 1–2, pp. 38–44.

[14] Larionov, D: *Inertial and Optical Technologies for Rail Track Diagnostics*. International Conference on Mechanical Engineering, Automation and Control Systems (MACS), 2014.

[15] Oregui, M., Li, Z., and Dollevoet, R.: Identification of Characteristic Frequencies of Damaged Railway Track Using Field Hammer Test Measurements. *Mechanical System and Signal Processing*, March 2015, 54–55, 224–242.

[16] Prescott, D., and Andrews, J.: A Track Ballast Maintenance and Inspection Model for a Rail Network. *Journal of Risk and Reliability*, 2013, 227, 3, pp. 251–266.

[17] *Russian Railway Infrastructure Monitoring Using State-of-the – Art Multifunctional Diagnostic Train. Transport i Komunikacja* web, 21. 10, 2014.

[18] Stencel, G.: Metody pomiarów nawierzchni kolejowej wykorzystywane przy ocenie jej trwałości. [Methods of measuring the railway track used to assess its durability]. *Problemy Kolejnictwa [Railway Reports]*, no. 165, 2014.

[19] Schenkendorf, R., Groos, J.C., and Johannes, L.: *Strengthening the Rail Mode of Transport by Condition Based Preventive Maintenance*. IFAC – International Federation of Automatic Control hosting by Elsevier Ltd, PapersOnline 48–21, 2015.

[20] Wenty, R.: Plasser & Theurer Machines and Technologies Applied for Track Maintenance on High-Speed Railway Lines: A Selection. *Rail Engineering International*, 2007, 1, pp. 9–12.

[21] Zieliński, K.R.: *Analiza zagrożeń i ocena ryzyka ich wystąpienia [Threat analysis and risk assessment]*. Conference Ekomilitaris, Zakopane, 2012.

Chapter 15

Requirements relating to designed and modernized engineering structures at high-speed rail lines

Juliusz Cieśla and Łukasz Topczewski

15.1. Introduction

PKP Polskie Linie Kolejowe S.A. (PKP PLK S.A.) has been established under the Act taken by Polskie Koleje Państwowe S.A. (PKP S.A.). The basic task of PKP Polskie Linie Kolejowe S.A. is to manage railway lines. The railway infrastructure administered by the company includes the track superstructure (tracks, interchanges) and roadbed, level crossings, railway engineering structures (bridges, overpasses, underpasses, culverts, tunnels, footbridges, retaining walls), buildings and facilities, overhead lines, and railway automation devices.

Railway engineering structures are one of the most important elements of railway infrastructure. Their technical efficiency, determined by the operating parameters (speed, axle load limit and load per 1 meter, construction gauge), i.e. values that characterize the operation side of the given object, determines the performance of the railway lines on which they are located.

At the end of 2015 on railway lines managed by PKP Polskie Linie Kolejowe S.A., there was a total of 27,727 railway engineering structures, with a total length of 775,819.8 meters, including:

- 3561 bridges – with a total operational length of 137,071.1 meters;
- 3191 overpasses – with a total operational length of 137,071.1 meters;
- 427 underpasses – with a total operational length of 13,506.0 meters;
- 19,555 culverts – with a total operational length of 379,762.2 meters;
- 26 line tunnels – with a total operational length of 22,195.2 meters;
- 166 pedestrian bridges – with a total operational length of 15,895.9 meters;
- 801 retaining walls – with a total operational length of 119,464.7 meters.

The technical conditions currently applicable to PKP Polskie Linie Kolejowe S.A. apply to planned and already operating railway engineering facilities, located on the standard-gauge public railway lines, where maximum permissible speed of trains does not exceed 160 km/h.

From the technical point of view, the most important problems of the railway infrastructure managed by PKP PLK are [1]:

- aging of all infrastructure elements as a result of a drastic reduction in repairs after 1990;
- a high percentage of track laid on wooden sleepers, many of which exceeded their nominal service life (17–18 years for softwood);
- track layouts at stations not adapted to modern needs;

- platforms that do not provide travelers with comfort when getting on and off the train;
- poor condition of engineering objects, including speed limits;
- the low level of automation of traffic control devices;
- lack of train control systems that allow operation of trains with speeds exceeding 160 km/h;
- insufficient number of multi-level intersections with roads;
- a small number of crossings equipped with active protection devices (only approximately 20%).

The scope of infrastructure investments carried out from the first half of the 1990s was far from satisfying the needs.

Those investments have not yet shown results in the form of a significant improvement of parameters over entire transport lines and a reduction of journey times within the railways network. The main reason is the lack of funds sufficient for the proper maintenance of the infrastructure and its modernization.

The diagnosis of the current state of rail transport in Poland and the SWOT analysis performed within [1] found that the most important factor that inhibits the development is the degradation of the infrastructure.

Low top speeds on a large part of the railway network are the symptoms of that degradation [1]. In many cases, these speeds are much lower than the construction speeds that were once applicable on given sections. In addition, there is a large number of speed limits. Both reduced maximum speeds and limits mean that journey times in a number of services are now significantly extended compared to the shortest times achieved for each service in the past.

To counter this state, three levels of investment activities, differing in material scope, the level of costs, and the implementation period are provided within the Master Plan [1]:

- development of a new, high–standard railway infrastructure (including in particular the high-speed lines);
- modernization of the existing railway infrastructure, with particular emphasis on lines belonging to the Trans-European Transport Network (TEN-T);
- investments that restore the normal parameters of the railway infrastructure on the lines considered as relevant in this Master Plan (restoring investments).

It is assumed that the target maximum speeds on individual railway lines will result from the tasks of these lines within the railway network. These tasks have been defined in commercial importance categories of individual services (and constituent lines).

Lines that are the busiest and link Warsaw with the largest urban agglomerations in Poland, by 2030, will be serviced by high-speed trains running on lines adapted for the speed of at least 250–300 km/h and 200 km/h in case of sections of modernized existing lines.

Figure 15.1 shows the map of planned and carried out major tasks within the scope of construction and modernization of railway lines. Construction of the Y line, originally planned for 2008–2020, is likely to be implemented in subsequent years.

As a result of the development strategy of the railway network adopted by the government, which aims at increasing the speeds of conventional lines up to 200 km/h for conventional rolling stock and up to 250 km/h for tilting rolling stock, it has become necessary to develop technical standards – detailed technical conditions for the modernization and construction of rail lines [8], adapted to these requirements. This task has been commissioned to a consortium that is coordinated by Instytut Kolejnictwa (Railway Institute). The section relating to engineering structures has been carried out by a special team at the Road and Bridge Research Institute in Warsaw.

According to the data provided in the Master Plan [1], one of the reasons of inability of increasing the speeds on a given line is the poor condition of engineering structures. Therefore it seems perfectly sensible that there is an urgent need for developing new standards, which would relate to the construction and modernization of engineering structures and take into account technological progress and new legal conditions, including the implementation of structural Eurocodes and by [6, 7].

It should be noted that, following the new standards, the maintenance and technical standards would be created – the detailed technical conditions for the modernization or construction of railway lines of Vmax ≤ 350 km/h.

15.2. Outline of the new standards assumptions

The scope of standards [8] covers the principles of designing new engineering structures and modernizing structures already present on high-speed lines, i.e. those where trains travel at speeds above 160 km/h.

By developing the new standards, the Road and Bridge Research Institute team adopted, as one of the priorities, the introduction of general provisions with the introduction of the designer responsibility principle and the principle of a certain freedom in the adoption of design solutions.

From the calculations side, new requirements for the limit values of load-carrying capacity have been introduced. Structure dynamic and aerodynamic impacts and the fatigue load capacity were taken into account. In the scope of serviceability limits, the acceptable displacement of the structure the following have been considered: acceptable vibrations and accelerations of the track superstructure at the facilities and construction gauge, resulting from the speed ranges of railway lines on which the structures will be located. Within the scope of these requirements, the requirements contained in the technical specifications for interoperability (TSI) for infrastructure and tunnels [6, 7], as well as the requirements of structural Eurocodes (Eurocode 0–9), have been taken into account.

In the case of tunnels, new requirements for their development and fire safety have been introduced and standardized in accordance with EU directives.

The requirements relating to the foundation of engineering structures have been modified, taking into account the increased requirements resulting from the increased dynamic impacts.

Taking into account the dynamic impacts, designing of pedestrian bridges on the lines with speeds above 160 km/h is not provided, and the existing pedestrian bridges over the modernized lines will be removed.

Increased requirements in relation to durability of the engineering structures along with the provision of the rules of ensuring the durability have been introduced.

15.3. Requirements relating to engineering structures at high-speed rail lines

As mentioned in the previous section, the standards introduce many new provisions resulting from the new standards and interoperability specifications. This was caused by the need to standardize the use of Eurocodes and the provisions of the TSI.

Directive 96/48/EC on the interoperability of the trans-European high-speed rail lines [2] was issued in 1996. The provisions of the directive have obliged the European Commission to introduce detailed technical specifications relating to interoperability (TSI) for various subsystems. In relation to the infrastructure, those technical requirements were introduced by the European Commission through the Decision 2002/732 [3].

This article focuses on engineering structures already present on high-speed lines, i.e. those where trains travel at speeds above 160 km/h. Those issues primarily cover the computational consideration of the effect of speed on the safety and serviceability of structures, including testing the structures at the limit conditions and taking into account the aerodynamic impact caused by the passing trains.

15.3.1. Computational consideration of the impact of speed on safety and serviceability of structures

One of the most important issues related to the design of structures on high-speed lines is computational consideration of the effect of speed on the safety and serviceability of structures. This influence is taken into account mainly within the dynamic calculations. PN EN 1991–2 standard [4] provides the rules on considering the dynamic effects for train speeds of up to 350 km/h. In case of a local maximum speed exceeding 200 km/h, the standard requires performing dynamic analysis. The standard provides an algorithm specifying the cases in which it is necessary to perform dynamic calculations also at speeds below 200 km/h. Dynamic analysis should be performed taking into account the characteristic values of the load in relation to real trains.

When selecting the real trains, take into account any possible configuration permitted to travel on the structure with speeds exceeding 200 km/h.

On international lines where interoperational European high-speed criteria are used, the dynamic analysis should be performed with the use of the HSLM load models [4].

HSLM load models consist of two separate universal HSLM-A and HSLM-B trains with cars of variable length and variable wheelbase, a variable number of cars, and a concentration of forces, imitating the pressures of axes. Both trains present the effects of dynamic load of tilting, conventional, and regular passenger high-speed trains.

Dynamic impact of the real train is represented by a series of movable concentrated forces. The calculations allow skipping the effects of synergy of vehicle and structure weights.

In accordance with PN EN 1991–2 [4], in dynamic calculations in every real train and HSLM load model, a range of speeds up to the maximum computational speed should be taken into account. It is recommended to use the maximum design speed as 1.2 of a local maximum linear velocity.

15.3.2. Testing the structure at the limit conditions

Engineering structures on the high-speed rail as on other lines should be tested under limit load-carrying capacity and serviceability conditions. However, there are some additional elements related to high speeds of trains. These are, above all, high acceleration values and significant aerodynamic impacts occurring here.

Checking the maximum peak of span acceleration should be treated as traffic safety prerequisite checked at the serviceability limit condition to prevent instability of the track. If a dynamic analysis is needed, then the results of this analysis should be compared with the results of static analysis multiplied by the dynamic factor Φ.

The maximum permitted computational peak values of bridge span acceleration calculated along the track line should not exceed the values recommended in PN EN 1990, A.2 [5].

The maximum permissible acceleration values for the span, due to track stability, set along each track should not exceed 3.5 m/s² for ballasted track, and in the case of direct bridges 5 m/s².

In order to ensure traveler comfort, the maximum values of vertical acceleration of a span, recommended in PN EN 1990 A.2 [5] should not exceed 1.0 m/s² at a very good level, 1.3 m/s² at a good level, and 2 m/s² at a sufficient level.

In order to limit the vertical acceleration of the vehicle to values specified in the standard, limit the maximum allowable values of vertical deflection δ along the axis of the track at the railway bridges, which are a function of:

- span length L [m],
- train speed V [km/h],
- the number of spans, and
- static scheme of the bridge (free-ends beam, continuous beam).

Vertical deflections δ should be determined based on the Load Model 71 multiplied by the dynamic factor Φ for the value $\alpha = 1$, in accordance with EN 1991–2 [1].

Alternatively, the vertical acceleration b_v can also be determined by a detailed dynamic analysis, taking into account the interaction of the vehicle with a bridge.

In the case of bridges with two or more tracks, only one track should be loaded. The maximum allowable deflection of the railway span is shown in Figure A 2.3 of the PN EN 1990 A.2 [5].

The limit values L/δ shown in Figure A 2.3 of the PN EN 1990 A.2 [5] relate to $b_v = 1$ m/s², which can be regarded as a "very good" level of comfort in the vehicle at speed V [km/h].

For other comfort levels and the associated maximum permissible vertical accelerations of the b'_v value, L/δ shown in this figure can be divided by the permissible b_v [m/s²] values.

L/δ values shown in Figure A 2.3 of the PN EN 1990 A.2 [5] relate to a number of free-ends beams with three or more spans. For a single span, or two free-ends beams or two continuous beams, L/δ values should be multiplied by 0.7.

For three or more continuous beams, L/δ values should be multiplied by 0.9.

L/δ values provided in Figure A 2.3 of the PN EN 1990 A.2 [5] are valid for span lengths of up to 120 m. Longer spans require a special analysis. The above requirements do not apply to the comfort level of temporary bridges.

In the case of performing a detailed dynamic analysis aiming at checking the passenger comfort level, PN EN 1990, A.2 [5] requires to take into account the following elements:

- the range of the speed of the vehicle, up to a certain maximum speed;
- characteristic load of real trains defined in the individual project documentation in accordance with EN 1991–2,6.4.6.1.1 [4];
- dynamic interaction of the masses between the cars of the real train and the structure;
- damping and stiffness characteristics of the car suspension;
- a number of cars sufficient to cause maximum load effects on the longest span;
- a sufficient number of spans in the multi-span structure that is necessary to cause resonance effects in the car suspension.

When taking into account the requirements relating to irregularity of the track in the dynamic analysis of the interaction between the vehicle and the bridge, when checking the passenger comfort, the conditions may be specified in the individual project documentation.

15.3.3. Aerodynamic impacts caused by passing trains

When designing structures adjacent to railroad tracks, the aerodynamic impacts of passing trains has to be taken into account. When the rolling stock passes, it subjects every structure located near the track to a traveling wave of alternating pressure and suction (Figure 6.22 to Figure 6.25 of the code EN 1991–2 [4]).

The magnitude of this impact depends mainly on:

*   squared train speed;
*   aerodynamic shape of the train;
*   structure shape;
*   location of the structure, the distance between the vehicle, and the structure in particular.

When checking in limit load-carrying capacity, serviceability, and fatigue conditions, those impacts may be approximated by equivalent loads on the front and rear of the train. Recommended characteristic values of equivalent loads are provided in respective figures; they are dependent on the speed of the train.

The maximum design speed V [km/h] should be assumed as the local maximum line speed.

It is recommended to multiply the equivalent loads for aerodynamic impacts by an increasing dynamic factor of 2.0, at the beginning and the end of structures adjacent to the tracks, over a 5 m distance, measured in parallel from the tracks, from the beginning and the end of structure.

In case of dynamically sensitive structures, the above increasing dynamic factor may not be sufficient and it may be needed to calculate it through a special analysis. Such analysis should take into account the dynamic characteristics of structures including the support conditions, speed of the adjacent rail traffic, and accompanying aerodynamic impacts, and a dynamic response of the structures that takes into account the speed of deflection wave developed in the structure. In case of dynamically sensitive structures, it may be necessary to use an increasing dynamic factor for the middle parts of the structure between the support points on the bearings.

Aerodynamic impacts in case of simple vertical surfaces are shown in Figure 6.22 of the code EN 1991–2 [4].

Characteristic values of pressures relate to trains with poor aerodynamics and may be reduced through:

* $k_1 = 0.85$ coefficient in case of trains with a smooth lateral surface;
* $k_1 = 0.6$ coefficient in case of rolling stock with a streamlined shape, e.g. ETR, ICE, TGV, Eurostar, and similar.

If a small part of the wall with the height of < 1.00 m and a length of < 2.50 m is considered, e.g. sound screen element, then q_{1k} impacts should be increased by a $k_2 = 1.3$ coefficient.

In case of simple horizontal surfaces located over the track, e.g. protective ceiling constructions, the characteristic values of impacts $\pm q_{2k}$ are provided in Figure 6.23 of the code EN 1991–2 [4].

The loaded width of the examined structural element extends up to 10 m on both sides of the track axis.

In the case of trains passing in opposite directions, their impacts should be added to each other. Only the load of trains on two lines should be taken into account.

The q_{2k} impacts may be reduced by k_1, coefficient, just like for vertical surfaces.

Impacts occurring at the edge bands of the wide structures that are perpendicular to the track can be multiplied by a 0.75 coefficient for the belt with a width of 1.50 m.

In case of simple horizontal surfaces adjacent to the track, e.g. platform shelter without the vertical wall, the characteristic values of impacts, $\pm q_{3k}$, are provided in Figure 6.24 of the code EN 1991–2 [4] and are applied regardless of the aerodynamics of the train.

In every position along the designed structure, the q_{3k} value should be determined as a function of the a_g distance from the nearest track. These impacts should be added if the tracks are located on both sides of the construction element concerned.

If the h_a distance exceeds 3.8 m, then the q_{3k} impact can be reduced by a k_3 coefficient:

$$k_3 = \frac{(7,5-h_g)}{3,7} \quad \text{for } 3.8\text{m} < h_g < 7.5\text{m} \tag{15.1}$$

$$k_3 = 0 \qquad \text{for hg} \geq 7.5\text{m} \tag{15.2}$$

where:
h_g is a distance from the top surface of the rail to the bottom of the structure.

In the case of multi-surface structures with vertical, horizontal, or inclined surfaces that are located along the track, e.g. sectional acoustic screens, platform shelter with vertical walls, etc., the characteristic values of impacts, $\pm q_{4k}$, shown in Figure 6.25 of the code EN 1991–2 [4] should be applied perpendicularly to the considered surfaces. It is recommended to take these impacts from Figure 6.22 of the code EN 1991–2 [4] diagrams, assuming the track distance as the lesser of the following values:

a'$_g$ = 0.6 min a$_g$ + 0.4 max a$_g$ or 6 m wherein the minimum distance a_g and the maximum distance a$_g$ are shown in 6.25 of the code EN 1991–2 [4].
If max a$_g$ > 6 m, then adopting the value of maximum a$_g$ = 6 m is recommended.
It is recommended to use k_1 and k_2 coefficients defined above.

In case of a surface containing the construction gauge of the track over a limited length of 20 m, where there is a horizontal surface above the tracks and at least one vertical wall, e.g. scaffolding or temporary structures, all impacts should be applied up to the entire vertical surfaces, regardless of the aerodynamic shape of the train, adopting the pressure value of $\pm k_4 q_{1k}$, wherein q_{1k} should be determined in accordance with Figure 6.22 of the code EN 1991–2 [4] and tk_4 = 2.

In the case of horizontal surfaces the pressure value of $\pm k_5 q_{2k}$ should be assumed, wherein q_{2k} should be determined in accordance with Figure 6.23 of the code EN 1991–2 [4] for only one track, and k_5 = 2.5 if one track is fenced or k_5 = 3.5, if both tracks are fenced.

Designing tunnels is another key issue that takes into account the aspects of aerodynamic effects of passing trains. According to the provisions of the new standards, tunnels should be designed so that the maximum pressure change defined as the difference between the extreme value of the positive and negative pressure does not exceed 10 kPa. This applies to the maximum speed of passage of a train for a given design of the tunnel. This requirement should apply to both interoperable trains, as well as any other train licensed for operation in this tunnel.

The design of the tunnel should not limit the train speeds, the structural conditions of the laying of railway tracks and drainage, or the suspension of electric traction.

15.4. Summary

As a result of the development strategy of the railway network adopted by the government, which aims at increasing the speeds of conventional lines up to 200 km/h for conventional rolling stock and up to 250 km/h for tilting rolling stock, it has become necessary to develop technical standards – detailed technical conditions for the modernization and construction of rail lines [8], adapted to these requirements. This task has been commissioned to a consortium that is coordinated by Instytut Kolejnictwa (Railway Institute). The section relating to engineering structures has been carried out by a special team at the Road and Bridge Research Institute in Warsaw. It is assumed that the target maximum speeds on individual railway lines will result from the tasks of these lines within the railway network. These tasks have been defined in commercial importance categories of individual services (and constituent lines).

Computational issues computing cover primarily the consideration of the effect of speed on the safety and serviceability of structures, including testing the structures at the limit conditions and taking into account the aerodynamic impact caused by the passing trains.

Technical Standards [8] have been introduced for application by a decision of the Board of PKP PLK S.A. and are a binding legal act on all lines belonging to PKP PLK S.A.

Bibliography

[1] Master Plan for railway transport in Poland until 2030, Ministry of Infrastructure, Warsaw, August 2008.
[2] Directive 96/48/EC on the interoperability of the trans-European high-speed rail lines dated 23.07.1996.
[3] Decision 2002/732/EC of the European Commission dated 30.05.2002.
[4] PN EN 1991–2 Actions on structures – Part 2: Movable loads on bridges.
[5] PN EN 1990, A.2 – Basis of structural design, Annex A2.
[6] Commission Decision dated 20 December 2007 concerning a technical specification for interoperability relating to the "Infrastructure" subsystem of the trans-European highspeed rail system along with the Annex – Technical Specification for Interoperability relating to the subsystem "Railway traffic," the "Infrastructure" subsystem.
[7] Commission Decision dated 20 December 2007 concerning a technical specification for interoperability relating to "Safety in Railway Tunnels" along with the Annex – TSI "Safety in Railway Tunnels."
[8] Technical standards – detailed technical conditions for the modernization or construction of railway lines for speeds v ≤ 200 km/h (for conventional rolling stock) / 250 km/h (for tilting rolling stock), PKP PLK S.A., Warsaw, 2009.

Chapter 16

Diagnostics of structural health of rapid rail transportation

Andrzej Chudzikiewicz

16.1. Introduction

Over the last half-century, we have observed a considerable development related to the design of vehicles, which is forced by striving after an increase in ride speed and comfort. At strict requirements concerning safety, the development successively forced undertaking research aimed at the development of new methods of diagnosing rail vehicles, which are capable of meeting requirements set by the market of passenger and freight transportation. Increased ride speeds, reaching up to 300–350 km/h, resulted in the fact that railway collisions, which had happened, started to result in considerable casualties. Besides, fatigue and wear processes affecting elements and subassemblies of vehicles, which have so far not always been taken into consideration in the related analyses, have started to be considered a significant source of phenomena influencing the durability of a structure, its dynamic behavior, ride comfort, and the resultant safety of travelling.

Technical capabilities, which appeared in the late 1990s, enabling the development of a ride speed over 500 km/h, at the same time striving to ensure a high safety level as well as high ride comfort, made both the designers and the manufacturers, as well as the carriers, undertake works related to passive and active safety. Problems related to passive safety of rail vehicles had been not developed very dynamically until the beginning of the 1980s, since serious railway collisions with casualties took place very seldom. This could be credited to a high active safety level of this branch of land transport. Nevertheless, serious railway collisions unfortunately had happened (e.g. in 1998 – Eschede, Germany; in 1995 – India), which intensified works on improvement of passive and active safety of rail vehicles.

A motivation for elaborating recommendations concerning passive safety were European projects TRAINCOL (1991–1995) and SAFETRAIN (1997–2001) [20]. Making use of statistical data related to rail crashes, event scenarios were defined and analysis of collision strength of rail vehicles was performed. Within these projects, rightness of a concept was proved. The concept assumed that a basis for analyses and scenarios of collisions should be management of the collision energy. The effects of the works realized within the projects were not only some design solutions, but also some methods of validation of theoretical models and techniques of prediction in practical tests.

The first document in the European Community defining requirements, which must be met by passenger railway rolling stock in the case of a collision, is EN 15227 standard (PN-EN 15227 standard in Poland) [32], which went into effect in 2008. Before, designers had only some guidelines in a form of UIC 566 Card [25] and EN 12663 standard [31] regarding strength of the body under conditions related to standard operation. Key issues became to ensure a survival zone, which is not deformable during collision, to protect against the climbing phenomenon and to limit the delay experienced by the crew and the passengers. In order to check if a vehicle meets this demand, there has been introduced a requirement of carrying out crash tests according to scenarios described in the standard. However, tests carried out on the track under real conditions result in high costs. In order to limit the number of real tests, which ensures reduction of the related costs, simulation studies have been proposed, which enable an assessment of the vehicle structure under extreme conditions that occur in reality. This makes it possible to improve the structure already while working on the prototype, in cases when the results are not satisfying. Simulations can be realized that make use of professional software packages based on the finite element method (FEM) [11].

With regard to the active safety, the problem boils down to the assessment of dynamic behavior of the vehicle under operating conditions. Generally, the assessment amounts to answering the question of whether some safety criteria are met. The criteria are safety with regard to derailment, magnitudes of forces interacting between the vehicle and the track, and evaluation of the level of motion stability. Practically, limit values of safety indicators are determined in UIC 518 Card, describing also the way they should be measured [40]. The card contains a set of rules concerning performance of tests and analyses of their results within the tests related to approval and certifications of rail vehicles (conventional vehicles, special vehicles, and structures based on modern technologies) from the point of view of motion dynamics, as related to safety, especially:

- derailment;
- evaluation of track forces occurring between train unit and track;
- analysis of dynamic stability of vehicle motion;
- evaluation of ride quality;
- analysis of structure gauging.

As proven by numerous events and practice related to current operation of rail vehicles, homologation tests aimed at approval are not sufficient while taking into consideration the safety and the foreseen running time of vehicle, estimated over a few tens of years. Therefore, along with increasing the riding speed, methods of diagnosing structural health of the vehicle and the track are starting to be improved and implemented, in order to additionally monitor in real time the structural health of such complicated mechanical system at high-speeds while running [1], [6–8], [15], [17], [24], [26], [27], [34].

16.2. Introduction to diagnostics of mechanical systems

A notion of diagnostics had been borrowed from medical science and started to be referred to in technical sciences only at the end of the 1960s. First works, which introduced a notion of technical diagnostics and laid foundations of development of the new discipline of science, defined technical diagnostics as a field of knowledge dealing only with problems related to identification of state of an object in a given moment of time [1]. However, first Polish works concerning problems of diagnosing health of technical objects appeared at the beginning of the 1970s [2]. As the related works have advanced, the definition of diagnostics has been evolving. According to [2], the following definition of diagnostics can be formulated: a field of knowledge concerning identification of health of technical systems in a current, past, and future moment. The definition does not include any information on kind of investigations, measurement methods, or applied devices – all the ways leading to determination of the system health are acceptable. Neither does the definition define a moment of time when the investigations should be carried out. Thus, it should be presumed that such investigations should be carried out for the needs of designing, manufacturing, and operating technical

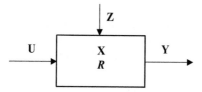

Figure 16.1 Technical object as a system

systems/objects. A classical notion of a system is as follows: set of elements or objects, which are integrated or connected into a whole by certain relationships or dependencies. A visualization of a system, commonly appearing in the relevant literature and shown in Figure 16.1, consists of the object itself and sets of signals: set U of input signals (also called excitations), set Y of output signals (called diagnostic parameters in diagnostics), set X of signals (called state [health] parameters), and set Z of signals called disturbances. Relationships or dependencies between elements or objects of the system create its particular structure R.

A common understanding of the meaning of the word *health* describes it in a satisfactory way. Nevertheless, in the relevant literature, e.g. [6], a *state (health) of a system (or an object)* is meant as the numerically smallest set of coordinates (parameters), which are sufficient for predicting behavior of that system in future (or at present). Health of technical objects is conditioned by factors related to design, technology, or factors resulting from the process of operating certain object. Therefore, a situation is possible, when objects having identical design and manufactured using the same technology, despite the same time of operation, will considerably differ in technical health due to extremely different operating conditions. These conditions include the following factors [6]:

- operational, e.g. conditions of lubrication, loading;
- external, e.g. air temperature, air humidity;
- related to human engineering, e.g. skills of the operator.

An equation of state, which can be found in the relevant literature most frequently, can be expressed as follows:

$$\frac{dS}{dt} = F(S(t_0), \Omega(t), t) \tag{16.1}$$

where:
$S(t)$ – object state in moment of time t,
$S(t_0)$ – object state in moment of time t_0,
$\Omega(t)$ – set of factors influencing the object state for a given moment of time t.

As aforementioned, the technical health (state) of the object $S(t)$ is a set X of values of parameters of state $x_1(t)$, $x_2(t)$, . . . , $x_m(t)$ defined for a given moment of time t. The parameter is meant as a qualitative measure characterizing the state of the system, and value of the parameter is its quantitative measure. So, the state of an object can be represented in a form of an ordered series of numeric values of the state parameters $x_i(t)$ (i−1, 2, . . . , m), which are also called variables or coordinates of the state, and can be regarded as the following vector:

$$X(t) = \left[x_1(t), x_2(t), ..., x_m(t)\right]^T \tag{16.2}$$

Generally, the state variables $x_i(t)$ can have an arbitrary form, i.e. they can be numbers, functions, vectors, etc.

A complete description of an object state consists of a set of characteristics (features, parameters, symptoms) revealing all the aspects of the object's existence and functioning. The following uniform terminology, related to the aforementioned notions, is accepted [6]:

- *feature of the state*, connected with a property of the object; a physical quantity having its own measure, standard, and reference level, which unequivocally describe value of the component of the state vector of the object at a given moment of time t_0;
- *diagnostic parameter*, always connected with an observable description of the object diagnosed by means of diagnostic signals (processes), determining directly values of the state features of the object;
- *diagnostic symptom*, damage-oriented measure of the diagnostic signal, reconstructing certain type of damage (component of the signal vector).

Thus, technical health of an object can be evaluated based on the measured signals and successively calculated values of the diagnostic parameters, contained in the observable output processes, provided the relationships between the state features and the diagnostic parameters are known. Knowing the structure R of an object makes it possible to determine the character of these dependencies, based on the capabilities that result from modeling a process and diagnostic models of technical objects.

16.2.1. Signal and its application in diagnostics of an object

A course of any physical quantity, which is an information carrier, is called a *signal*. The components of the signal, which contain information, are called signal parameters. A source of signals are operating technical objects, which generate signals due to physico-chemical processes, e.g. energy transfer, energy conversion, vibroacoustic vibration, etc. An exemplary course of a signal representing a change of lateral displacement of the mass center of a wheelset of two-axial bogie, depending on the structural health of the wheel profiles, while riding along the Central Rail Line (CMK) at the speed of 120 km/h, is presented in Figure 16.2.

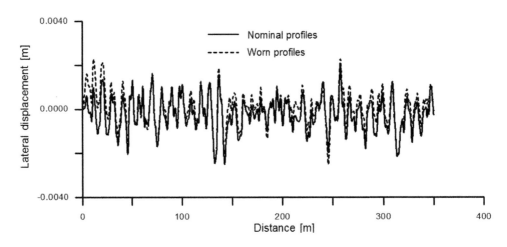

Figure 16.2 Trajectory of the second wheelset at new and worn-out profiles

Diagnostic signal is a signal, which is an output variable (Figure 16.1), whose parameters must meet requirements related to sensitivity, uniqueness, stability, and informative character with respect to a change of the object state [5].

Sensitivity K of the signal is a ratio of increment $\Delta y_n(t)$ of the diagnostic parameter to increment of the state parameter $\Delta x_m(t)$, i.e.:

$$K = \max\left(\frac{\Delta y_n}{\Delta x_m}\right) \tag{16.3}$$

At the same time, a better one is the diagnostic parameter, whose ratio in Equation (16.3) takes on a higher value.

Uniqueness means that only one value of the diagnostic parameter should be associated with each value of the state parameter.

Stability means that the value of a diagnostic parameter should be insignificantly variable under steady conditions of diagnosing an object, e.g. temperature, load.

Informative character means that in order to determine the state of the diagnosed object we should select the diagnostic parameters that provide the largest amount of information on the state of this object.

The problem of selecting the state parameters and the diagnostic parameters is a very important issue because of the costs of object diagnosing.

Diagnostic signal can be described by only one value of the parameter, e.g. temperature of a factor; in such case it is called a *one-dimensional signal*, whereas when we deal with a description in a form of a set of values of many parameters, *a multi-dimensional diagnostic signal* is referred to.

An important notion is an ability state w^1 and an inability state w^0 of a technical object. These states can be also defined by means of diagnostic parameters. A technical object is an ability state w^1, if values of all the parameters of diagnostic signals stay within the permissible ranges. However, if a value of even one of the diagnostic parameters (diagnostic symptoms) is out of the permissible range, then the diagnosed technical object does not satisfy certain requirements, so it is in an inability state w^0.

Input signals are excitations, which act upon the diagnosed technical object. The diagnosing process can proceed under standard operating conditions, and then the input signals are the real excitations, or under laboratory conditions, and then the excitations are some preset signals. The excitations are interactions of diagnostic devices with the studied object. As in the case of a diagnostic signal, the input signals (excitations) can be represented in a form of an ordered series of numeric values of the signal parameters $u_j(t)$ (j = 1, 2, ... , L) and dealt with as with an input vector (of excitations):

$$U(t) = \left[u_1(t), u_2(t), ..., u_l(t)\right]^T \tag{16.4}$$

Disturbing signals are not measured directly. Their existence can be stated by an analysis of the output signals; their sources are usually neighboring objects and the environment. *Disturbances* can be measureable and controllable, measureable and not controllable, or not measureable and not controllable, whereas signals being a representation of disturbances are of a probabilistic character. Analogously to other cases of the signals discussed above, disturbances can be represented in a form of an ordered series of numeric values of disturbance parameters $z_l(t)$ (l = 1, 2, ... , J) and dealt with as with a disturbance vector:

$$Z(t) = \left[z_1(t), z_2(t), ..., z_j(t)\right]^T \tag{16.5}$$

The technical object presented in Figure 16.1 has been characterized by defining sets, which determine: inputs (excitations) U, outputs Y, disturbances Z, state X and relationship R determining the structure of the object. Diagnosis of the state of a given technical object is usually a complex process; it is assumed that the process progresses under definite, i.e. constant conditions, forcing operation of the object. So, it is assumed that the diagnosing process, i.e. process of determining a form of the state vector takes place at fixed excitations $U = const$. This is a fundamental assumption, which makes it possible to formulate the essence of technical diagnostics. In Figure 16.3, a technical object is presented as a subject of diagnosing, whose input is now set X determining a form of state S, and the output is set Y of parameters of the output signal.

Under real conditions, parameters determining the health (state) of a technical object are usually not available for a direct measurement, thus we measure some parameters of diagnostic signals that are directly available for us, which are parameters of the output vector, from a formal point of view.

So, the problem of technical diagnostics can be formulated as follows: *a problem of direct measuring and determining values of parameters of the health (state) of a technical object based on parameters of the measured diagnostic signals.*

The problem can be expressed in a form of two relationships, which are fundamental for the diagnostic process [6]:

$$Y = H(X) \tag{16.6}$$

$$X = G(Y) \tag{16.7}$$

where:
H – operator transforming set X into set Y,
G – operator transforming set Y into set X.

Equation (16.6) corresponds to an existing relationship, which takes place under standard operating conditions of a given technical object. State parameters contained in set X in a given moment of time are transformable into elements of set Y (containing parameters of diagnostic signals). This transformation is realized by operator H. However, the problem of diagnostics of the state of a technical object that has been formulated before refers to determining the state of an object on the basis of parameters of the vector of diagnostic signals, what is expressed by Equation (16.7). However, a question arises: when is it possible?

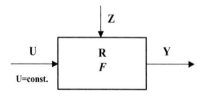

Figure 16.3 Technical object as a subject of diagnosing

If operator H is a vectorial function, then Equation (16.6) can be expressed in a form of a system of equations:

$$f_1(x_1(t), x_2(t), ..., x_m(t), y_1(t), y_2(t), ..., y_n(t)) \qquad (16.8)$$

where:
$f_1, f_2, ..., f_j$ – functions, with regard to which it is assumed that they are definite and differentiable within a certain (n+m)-dimension space.

Making use of an implicit function theorem [6], conditions for existence of a solution of system (16.8) can be specified, i.e. conditions for determining $x_i(t)$:

$$x_1(t) = x_i(y_1, y_2, ..., y_n) \qquad (16.9)$$

that is determining the health (state) of a technical object on the basis of the parameters of the diagnostic signal.

So, operator G exists then, and a relationship defined by Equation (16.7) is possible.

It is obvious that a necessary condition for obtaining a solution in a form of Equation (16.9) is the fact that the number of independent parameters of the diagnostic signal is no lower than the number of independent parameters of the state, i.e. the following condition must be satisfied: $n \geq m$.

A result of practical application of Equations (16.6) and (16.7) in a process of diagnosing the state is the fact of facing problems due to the fact that assumptions of the implicit function theorem are not always satisfied, as well as due to occurrence of disturbances of the diagnostic signals during the experiment.

16.2.2. Diagnostic model of the object

Modeling is meant as a process consisting in creation of a certain *composition* (description, diagram, standard, formalism, structure) resembling the studied physical object, in order to obtain characteristics, which we are interested in, of the studied object by means of the created *composition*, called a model. Many various definitions of a model can be found in the relevant literature. One of them is [28]: *Model is such a system that can be thought of or can be materially realized, which is capable of replacing the studied subject, while reconstructing or reproducing it, in such a way that the study provides a new knowledge about the subject.*

On the other hand, the following definition of a diagnostic model of an object can be found in the literature related to the problems of diagnostics of technical objects [6]:

Generally, a diagnostic model of an object is a relationship (dependence) between parameters of diagnostic signals and its states.

As based on the above definition of the diagnostic model, the following kinds of diagnostic models of technical objects can be distinguished:

1. Model of a type:

    parameters of state signals ⇨ parameters of diagnostic signals what can be expressed in a form of a certain transformation,

$$Y = F_1(X) \tag{16.10}$$

where: Y – set of parameters of diagnostic signals,
X – set of parameters of state signals,
F_1 – certain transformation.

2. Model of a type:

parameters of diagnostic signals ⇨ parameters of state signals
what can be expressed in a form of a certain transformation,

$$X = F_2(Y) \tag{16.11}$$

where: Y – set of parameters of diagnostic signals,
X – set of state parameters,
F_2 – certain transformation.

3. Model of a type:

measure of operation ⇨ parameters of diagnostic signals
what can be expressed in a form of a certain transformation,

$$Y = G_1(L) \tag{16.12}$$

where: Y – set of parameters of diagnostic signals,
L – set of measures of operation,
G_1 – certain transformation (it can be linear).

A measure of operation may be e.g. time of operation, mileage (in km), number of cycles, etc.

4. Model of a type:

parameters of diagnostic signals ⇨ measure of operation
what can be expressed in a form of a certain transformation,

$$L = G_2(Y) \tag{16.13}$$

where: Y – set of parameters of diagnostic signals,
L – set of measures of operation,
G_2 – certain transformation.

5. Model of a type:

states ⇨ parameters of diagnostic signals
what can be expressed in a form of a certain transformation,

$$Y = K(W) \tag{16.14}$$

where: Y – set of parameters of diagnostic signals,
W – set of states,
K – certain transformation.

6. Model of a type:

parameters of state signals ⇨ states
what can be expressed in a form of a certain transformation,

$$W = H_1(X) \tag{16.15}$$

where: X – set of parameters of state signals,
W – set of states,
H_1 – certain transformation.

Models (16.10–16.15) are of a *"black box"* type and describe relationships between input and output signals with no regard to the internal structure of the analyzed models of the objects. The signals can be, for example, slowly variable or even constant parameters (temperature, pressure of the working medium, percentage fraction of inclusions or impurities), or quickly variable signals determining dynamic behavior of the diagnosed object. In the case of vibroacoustic signals, use of these models in a diagnosis process is connected with application of methods employing one of many analyses of vibroacoustic signals. The models can be also used both in *on-line* as well as *off-line* methods [6].

A model, in which there are taken into account relationships between these signals (or their parameters), states of the object model, and parameters of the model such as mass, elasticity, damping, and excitation is called a structural model. A formal notation of such model can be represented in the following form:

7   Model of a structural type:

excitation + structure $\Rightarrow$ parameters of diagnostic signals
what can be expressed in a form of a certain transformation,

$$y(v) = F_3(m, k, c, p(v)) \tag{16.16}$$

where: m – mass,
k – rigidity,
c – damping,
$p(v)$ – excitation,
$y(v)$ – parameters of diagnostic signals.

In the case of an object of a type of rail vehicle-track system, a structural model can be mostly used in *off-line* diagnosing methods, because of long times of simulation of the mathematical models that constitute the model of this object.

16.2.3. Methods of diagnosing technical objects

Methods of testing technical objects suffer from errors resulting from indication errors of the measurement instruments, errors of method, as well as randomness and lack of repeatability. The situation is similar in the case of diagnostic methods. Generally, diagnostic testing consists in recording a signal generated by the object, its analysis, which consists in separating in it characteristic features, and comparing them with features of standard signals corresponding to other states. A process of diagnostic testing can be presented in a form of a series of actions, as presented in Figure 16.4.

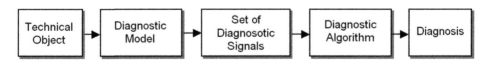

Figure 16.4 Process of diagnostic tests of a technical object [6]

Classic repartitions of diagnostics methods, as can be found in the related literature [30], were created according to the following criteria:

a. subjectivity of diagnostic tests (organoleptic and instrumented methods);
b. kind of the diagnostic model;
c. degree of automation of the tests.

A basic repartition, which results from phases of object existence, concerns methods used at the following stages:

a. evaluating (heuristic diagnostics);
b. designing (design diagnostics);
c. manufacturing (control diagnostics);
d. operating (operating diagnostics).

Dependent on the methodologies applied in the process of state diagnosing, the used methods can be repartitioned to:

a. not instrumented;
b. instrumented.

Not instrumented methods, also called organoleptic methods, have been known for a long time and are used by persons having a rich experience and an industrial practice. However, nowadays they are slowly disappearing due to the complexity of modern machines and devices.

In many cases, it is not possible to directly determine the health of a certain element in the tested machine, e.g. health of a bearing race. In this case, assessment is realized indirectly, e.g. by measuring bearing vibration. So, the diagnostic methods can be repartitioned into:

c. indirect;
d. direct.

However, in the practice related to diagnostic processes, direct methods are applied more and more often, since they do not require additional application of processing of the diagnostic signal or the application of processing procedures, which sometimes are quite complex.

Taking into account application of the state of the diagnostic model in the diagnosis process, the methods can be repartitioned into:

a. determined;
b. undetermined.

The repartition in this case can be continued, taking into account classification used in modeling theory, which results in yet another repartition of the models.

Still another repartition, resulting from accessibility of the tested object operating under real conditions, is as follows:

c. on-line methods;
d. off-line methods.

Application of on-line methods can be referred to whenever the process of recognizing the state of a tested technical object under operation is carried out in real time. On the other hand, wherever the state of a system, which we are interested in, will be tested and recognized directly, off-line methods will be used. Taking an example of a mechanical system of rail vehicle – track [11], in the case of monitoring and diagnosing at the same time the health of the vehicle or the track, while riding, methods used in this process should be included among on-line methods, whereas all the tests aimed at the determination of the health of that system or its elements, realized using models or test rigs, should be numbered among tests carried out by means of off-line methods. Examples of diagnosing the health of an axle-box bearing (bearing of the running gear) of a rail vehicle using an on-line and off-line method is shown in Figure 16.5a and Figure 16.5b.

In the case of high-speed trains, their current diagnostics, realized while operated, takes place at Service Centers. An example of such center can be a Service Center for ED250 trains built in Warsaw, in the region of Olszynka Grochowska (Pendolino) (see Figure 16.6). The area of the center is over 12,000 m² and is divided into a storehouse/service part and a zone of office premises. The workshop hall is equipped with three maintenance tracks and two repair tracks and is capable of servicing seven-car trains. The facility contains also a storehouse with spare parts and an automatic washing stand for trains. Within this investment, there were built almost 2 km of tracks with the traction. In order to ensure more effective realizations related to maintenance and diagnostics, the center makes use of a TrainTracer diagnostic technology offered by Alstom. The TrainTracer system informs the center about the status of a train in real time, making it thus possible to quickly detect and remove any abnormalities. The system also enables

Figure 16.5 Exemplary diagnostics of monitoring a bearing of the running gear of a rail vehicle (a) on-line method, actual measurement at the vehicle and (b) off-line method, tests under laboratory conditions

Figure 16.6 Service Center for Pendolino trains in Warsaw; (a) storage hall, (b) subgrade lathe

a current determination of the train position using a GPS system. Another tool applied is a RAIL-SYS platform offered by Alstom; it is software that makes it possible to manage the process of maintenance and repairs by the way of monitoring wear of some components and planning their replacements and repairs. One of the crucial diagnostic devices for wheels of the wheelsets is a subgrade lathe installed at the center, which consists of two lathes that can be operated independently, yet simultaneously, on two bogies. The lathe is controlled from a switching station. It is the only one device of this type in Poland and one of the few all over Europe [43].

In this chapter, we will deal with a problem of monitoring the structural health of a rail vehicle and a track during their current operation, using an on-line method.

16.3. Monitoring the rail vehicle – track system

16.3.1 Review of the works

Once a rail vehicle has been recognized as an object of diagnosing and monitoring, it should be noted that while dynamically testing behavior of a running rail vehicle or analyzing its technical health resulting from the operating conditions, one must always take into account the fact that the rail vehicle itself is a part of a mechanical system [7], [8], which consists of a track, a contact region between the wheel and the rail, and a vehicle. There appear some interactions between these three elements during the operation; the interactions are characterized by an exchange of signals, such as forces, accelerations, velocities, and displacements. Considering the vehicle only, with no regard to interrelations with the other two elements of the system, is appropriate only in some cases, such as e.g. quasistatic analyses, whereas generally, in the case of dynamic analyses, such an approach should not be assumed. External interactions (e.g. atmospheric conditions), that have been regarded in this system only recently due to their influence on the operating health of particular elements of the system or their influence on dynamic interactions (e.g. wind, aerodynamic interactions inside of tunnels while riding at a high speed), are an element, which should be included in that system as a missing link, while monitoring a railway rolling stock for rapid transportation.

In Poland, works of a fundamental character or R&D works in the aforementioned fields (diagnostics, monitoring) have been carried out for many years with respect to the object of rail vehicle – track. Problems related to diagnosis of track health is dealt with by many centers, and results of these works can be found in numerous publications or monographs [1], [3], [15], [26], [27], [38]. As far as the vehicle and its subsystems are concerned, the problem of diagnosis is referred to in such works as [4], [29], [30], whereas the rail vehicle – track system is considered from the above point of view in works [7], [23], [36].

16.3.2. Accepting assumptions

As aforementioned in Section 16.2, one of the methods of health monitoring is a diagnosis according to an on-line method. In the case of a rail vehicle, the idea of diagnosing structural health according to an on-line method is shown in Figure 16.7. Measurement of diagnostic signals takes place in real time, during a ride under conditions typical for a standard operation. Each of the monitored vehicles (car, member of a train unit) is equipped with systems recording the diagnostic signals installed at selected points. The systems are interconnected by means of a bus bar and create a net, which transmits signals from each vehicle to a data acquisition unit (DAU) – an on-board computer, where the signals are recoded and processed in order to obtain information on the health of particular elements and systems in the vehicles.

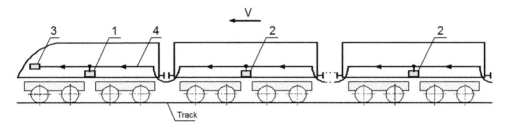

Figure 16.7 Diagnosis of structural health of a rail vehicle using an on-line method; 1, 2 – local data acquisition units, 3 – central data acquisition unit, 4 – bus bar

A DAU sends wirelessly the recoded signals to the system server for monitoring purposes. In the case when an emergency situation occurs, e.g. the permissible level of a measured parameter is exceeded, the mechanic who operates the train, having appropriate information on the situation, can make a decision on a further ride or in a critical situation, the monitoring system automatically activates the emergency braking system, thus terminating the ride in this case.

Besides, the system monitoring the structural health of each car is capable of recording signals collected from a given car, which enables a diagnosis of its structural health during a cyclic inspection or damage repairs.

Basing on the concepts of monitoring the state of a rail vehicle, as presented in [7], [12], [14], [17] within the Program [36], some works were undertaken aimed at elaborating a concept, building a prototype, and implementing a system for monitoring states in the rail vehicle – track system. One accepted, the following are assumptions concerning the system that has been built:

- monitoring both the vehicle and the track will be realized at the moving vehicle;
- a vibroacoustic acceleration signal will be used to evaluate the state in the monitoring process;
- evaluation of the monitored phenomena will be qualitative and will be based on monitoring cases when the permissible levels of vibration are exceeded;
- monitoring of the track structural health will consist in evaluation of the health of the track geometry and of the track substructure. The monitoring process will be based on measurements and an analysis of the acceleration signals measured at the bearing casings of the wheelsets;
- monitoring of the vehicle will be aimed at monitoring the health of the elements of the compliant systems of the I- and II-grade of springing as well as the temperature of the bearings in the axle-boxes of the wheelsets;
- taking into account the related costs, the monitoring system should be characterized by simplicity and accessibility.

It was also assumed that the elaborated concept of the system should be easily adaptable for the needs of a monitoring system of light rail vehicles, such as a tramway or a rail bus [21].

16.3.3. Procedure of monitoring structural health of the vehicle

The testing procedure of a rail vehicle with respect to ride safety, dynamic properties of the running gear, and interactions with the track is based on the requirements specified in EN 14363 standard [35] and UIC 518 Card [40]. Taking into consideration a rolling stock designed for high-speed transportation, requirements contained in TSI [18] should also be regarded. Therefore, the elaborated procedure of monitoring the structural health of a rail vehicle pays regard to the requirements specified in these documents concerning critical values of accelerations measured at the bogie and at the vehicle body.

Vertical and lateral accelerations measured at the bogie make it possible to evaluate safety of running in a simplified way. Besides, they enable monitoring of the vehicle behavior while riding. On the other hand, vertical and lateral accelerations measured at the vehicle body are used for evaluating ride quality of the vehicle.

The UIC 518 Card determines the maximal values of quantities, whose measure is a basis for an official certification of railway vehicles equipped in bogies:

- $\ddot{y}^{+}$ – lateral acceleration of the bogie frame (measurement at the end of the frame side-member, above the wheelset, at one side);
- $\ddot{y}^{*}$ – lateral acceleration of the body (measurement on the floor above the bogie center);
- $\ddot{z}^{*}$ – vertical acceleration of the body (measurement on the floor above the bogie center).

These values have been accepted in the procedure as the normative critical values.

The accepted concept of the algorithm for evaluation of the health of the I- and II-grade suspension based on measurements of acceleration associated with vibration at the bogie frame and at the vehicle body assumes a comparison of the data obtained by the way of acquisition with a standard of accelerations associated with vibration. The role of the standard can be played by values obtained for a fully efficient suspension, under the same operating conditions or the aforementioned limiting values specified in the relevant standards.

Damages to the suspension elements will result in changes of its behavior and in generation of signals, for which the computed measures will differ from values of the accepted measures of the standard signal. In such a case, the diagnostic symptom will be a difference between these values for the standard signal and the experimental signal. If the difference exceeds a certain predefined level (detection threshold of the damage), it will signalize the occurrence of an abnormality state in the suspension system.

The presented method enables monitoring of I (bogie frame) and II (body) grades of suspension. Implementation of the presented idea of the algorithm assumes measurement of accelerations in three directions (X, Y, Z) at two points, i.e. bogie frame and vehicle body. The algorithm is presented in Figure 16.8 in the form of a diagram.

Three levels of diagnostics are foreseen with regard to the elaborated procedure, based on this algorithm; the diagnostics consists in measuring acceleration at specified points of vehicle elements, such as wheelsets, bogie frame and body:

Level I   –   Comparison with normative values [18], [35], [40]; these values are independent on the vehicle type.

Level II   –   Comparison with values obtained by means of an analysis of standard courses, which are dependent on the vehicle type; in practice, they will be obtained while carrying out an official certification for a given vehicle type.

Level III   –   Comparison with values obtained from courses recorded during the monitored process of the same train set, while passing over the same section of the track. Comparison of the results will be realized for cars of the same type.

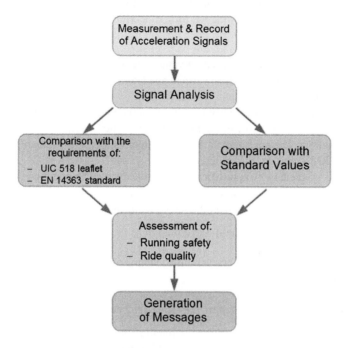

Figure 16.8 Algorithm for assessment of the suspension system of a rail vehicle using measurement of accelerations related to vibration of the bogie frame and the vehicle body

Comparison of the recorded signals at level II and III is realized using the following statistic measures [13]:

1. Pearson correlation coefficient;
2. concordance correlation coefficient;
3. intraclass correlation coefficient;
4. peak-to-peak value;
5. standard deviations;
6. kurtosis;
7. RMS;
8. quartile range ;
9. skewness.

An important issue is to determine the levels of a critical value for the above measures. Within the works realized in project [36], simulation studies were performed, testing behavior of the above measures in the case of simulating damages of elements of I- and II-grade of vehicle springing, using the existing mathematical models of vehicles of a type of a passenger car or EMU [13], [14].

In order to obtain critical values of these measures, an active experiment realized on an experimental track can be used. Besides, it might turn out that not all the above measures must be used in the final form of the algorithm.

While striving for an improvement of the algorithm, in the considered case it has been developed and supplemented, defining the form of a critical alarm and warnings by introducing the following options:

1. Critical Alarm – exceeding normative values specified in UIC518 Card and EN 14363 standard; the measured quantities: vertical and lateral accelerations; the analyzed statistical parameters: root-mean-square values (RMS).
2. Warning 1 – exceeding by approximately 40% the values recorded as standard values for a given type of the vehicle; the measured quantities: vertical, lateral, and longitudinal accelerations; the analyzed statistical parameters: peak-to-peak amplitude, standard deviation, RMS, and quartile range.
3. Warning 2 – exceeding by approximately 20% the average among the values recorded for other monitored cars of the same type from the same train set, or among cars of the same type from another train set, yet over the same section of the track and at the same ride speed.

(20% – an initial value; it should be verified once the experiment is completed).

The measured quantities: vertical accelerations, lateral accelerations, longitudinal and angular accelerations.

16.3.3.1. Temperature monitoring of the wheelset bearing

Referring to the technology of measuring temperature of bearing casings that has been developed so far, wireless sensors installed within the infrastructure of tracks are the majority. In the developed system, it was proposed to measure the temperature by means of sensors located on the bearing casing of the wheelset. The bearing casing of a rail vehicle, according to the relevant standard, should be designed in such a way that the maximal temperature difference between the loaded region of the bearing and the outer region, measured using a method specified in Annex 6 of standard [41], does not exceed a predefined value. In the considered case, it was proposed to signal a temperature difference between the casing and the outer region, which exceeds 30 °C as well as a temperature difference between the casings of different bearing of a given bogie having value of 30 °C, 40 °C, and over 50 °C. The proposal has been based on the following temperature distribution related to a stationary case (i.e. when the heat source acts infinitely long), shown in Figure 16.9.

As results from Figure 16.9 present a stationary temperature distribution along the wheelset, it is necessary to monitor the temperature of each bearing casing separately.

16.3.3.2. Monitoring of the structural health of the track

A quantity denoted by W_t has been proposed as a numeric parameter for evaluation of the structural health of rails at given ride velocity; it is similar to the one applied while evaluating the ride quality, which, in the case of a track description that employs spectral densities, is expressed as follows:

$$W_t = c_t \left[\int_0^\infty S_\alpha(\omega)\omega^b d\omega \right]^c \qquad (16.17)$$

where:
c_t – constant factor,
$S_a(\omega)$ – spectral density of accelerations related to a point moving over irregularities of the track with the velocity v

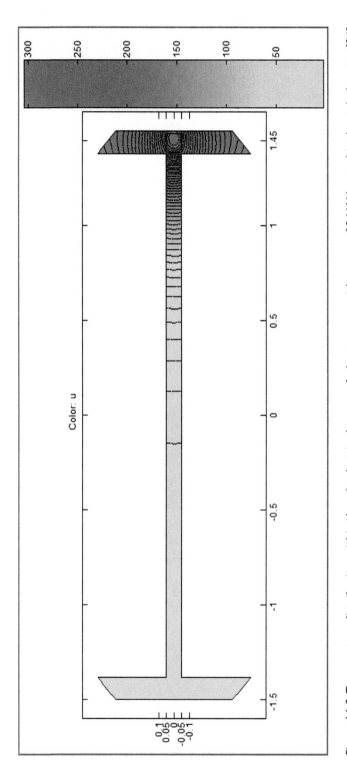

Figure 16.9 Temperature distribution within the wheelset in the case of a heat source with power of 2 kW located in the right bearing [36]

Values of coefficients c_t, b, and c were determined using a method of successive trial in such a way that in the case of the tested tracks the evaluation indicator W_t takes on values in specific ranges, e.g. from 0 to 3, or 0 to 5. The proposed expression for parameter W_t may be considered a generalization of the formula, characterized by an indicator of ride quality, which regards kinetic energy of a material point attributed to its mass and a derivative of the acceleration of this point with respect to time, called jerk.

16.4. Computer simulation of the monitoring system

Verification of the accepted assumptions was carried out using methods based on computer simulation, applying for this purpose a mathematic model of the vehicle and the real input kinematic functions in a form of geometrical irregularities of the track. The following assumptions needed to be verified:

- the accepted points of installing the sensors;
- the accepted statistical measures of the acceleration signals.

The aim of the verification was to determine a minimal number of sensors and accept for evaluating the structural health of the vehicle in the process of monitoring only those measures, which are characterized by a "sensitivity" to a change of the structural health of the vehicle.

For the purposes of simulation studies, one accepted a discrete model of the vehicle of a type of a passenger car [12], described by a set of nonlinear II-order ordinary differential equations. Model of this type is a low-frequency model and can describe in a satisfactory way vibration of the system up to 30 Hz. Structure of the mathematical model [7] is the following:

$$M\ddot{\overline{q}} + C\dot{\overline{q}} + K\overline{q} = F(t,\overline{q},\dot{\overline{q}}) \tag{16.18}$$

$$F(t,\overline{q},\dot{\overline{q}}) = f_w(t) + f_c(\overline{q},\ddot{\overline{q}})$$

where:
M, C, and K – matrices of masses, moments of inertia, damping and rigidity, respectively,
q – vector of generalized coordinates,
F – vector of generalized forces,
f_w – vector of exciting forces,
f_c – vector of forces occurring in the contact region between the wheel and the rail.

The accepted program of studies assumed performing a simulation for two variants: undamaged and damaged car, at variable parameters such as:

- ride speed;
- maintenance condition of the track, upon which acceptance of the maximal ride speed was dependent.

In the case of the damaged car, the following were accepted:

- loss of elastic or damping properties of the elastic elements of the I-grade system in horizontal and vertical direction, between the first wheelset and the frame of the first bogie;

- loss of elastic or damping properties of the elastic elements of the II-grade system in horizontal and vertical direction, between the first bogie and the body.

The loss was simulated by a change of a value of coefficients assigned to elastic and damping elements of these systems.

Structural health of the track, in accordance to the existing relevant regulations [40], was characterized by parameters describing the track as one of the following three categories: QN1, QN2, or QN3. Such an approach made it possible to simulate rides with speeds in the range of 60 km/h to 160 km/h. The simulation studies were carried out according to the elaborated program and procedure. The obtained results made it possible to propose an arrangement of the sensors over the wheelsets, the bogies, and the body, which is illustrated in Figure 16.10.

The obtained simulation results, having a form of acceleration values at the selected points of the vehicles (for preset conditions of the ride), made it possible to calculate values of measures of statistical parameters characterizing a given diagnostic situation. Exemplary results, related to the case of a damage of a spring element of the I-grade of springing, while the vehicle rides at the speed of 160 km/h, are listed in Table 16.1.

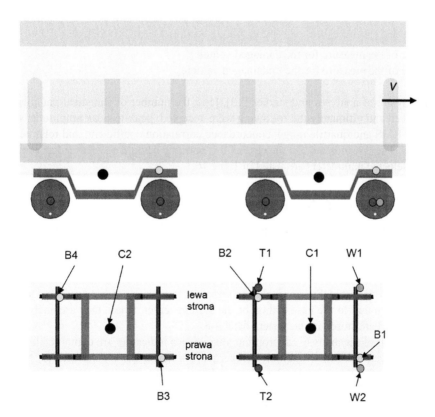

Figure 16.10 Arrangement of the sensors over the wheelsets, bogies, and body T – temperature sensors, B, C, W – accelerometers

Table 16.1 Damage of a spring element in the I-grade of springing kz

Statistical Parameter	$k_z = 0.5\,k_{znom}$		$k_z = 0.5\,k_{znom}$	
	$\ddot{z}^+$	$\ddot{y}^+$	$\ddot{z}^+$	$\ddot{y}^+$
Pearson correlation coefficient	0.71	0.75	0.16	0.57
Concordance correlation coefficient	0.66	0.75	0.12	0.57
Relative value of peak-to-peak (%)	30.10	13.18	29.00	12.18
Kurtosis of the "standard signal"	15.88	26.00	15.88	26.00
Kurtosis of the "damaged signal"	33.40	48.81	0.86	32.01
Relative change of RMS value (%)	48.34	6.38	127.47	3.36
Relative change of quartile range value (%)	53.23	88.19	10.50	31.33
Skewness of the "standard signal"	1.47	−0.17	1.47	−0.17
Skewness of the "damaged signal"	3.66	0.06	−0.02	0.80

In order to assess usability of a given statistical parameter in a process of diagnostic inference, a measure of a relative usability assessment M_W has been defined as:

$$M_w = \left(1 - \frac{D_U}{D_N}\right) \cdot 100\% \tag{16.19}$$

where:
D_U – value of the measure for the damaged vehicle,
D_N – value of the measure for the undamaged vehicle.

As a result of analyses and studies [13], [14], the number of statistical parameters had been reduced and ultimately the following were accepted: peak-to-peak amplitude, standard deviation, RMS and quartile range, concordance correlation coefficient, and relative change of kurtosis. The last indicator, in the case of this warning, should be regarded as supplementary. The selected parameters are listed in Table 16.1.

16.5. Structure of the monitoring system

Functional structure of the system is presented in Figure 16.11. Signals generated by the sensors installed on the vehicle are transferred via a cable to a local data acquisition unit, also installed on the vehicle, where they are sorted and appropriately recorded in a local database. Then, the signals are sent (also via a cable) to the central data acquisition unit, where they are preliminary analyzed. Successively, the signals are wirelessly transferred from the central data acquisition unit to the base station or the system server, where they are analyzed more thoroughly and are stored in the system database.

As a result of the analysis, appropriate diagnostics indicators are computed (amplitude, mean square value, kurtosis coefficient, interquartile range, quality indicator of the track), which characterize the structural health, and then qualitative information on the structural health is generated. The information can be sent in real time to appropriate staff supervising the traffic of the vehicle and the staff responsible for the technical and operational health of the vehicle and the track. Motion of the vehicle is described by geographic coordinates and is visible on an electronic map of Poland, therefore information on the vehicle and the health

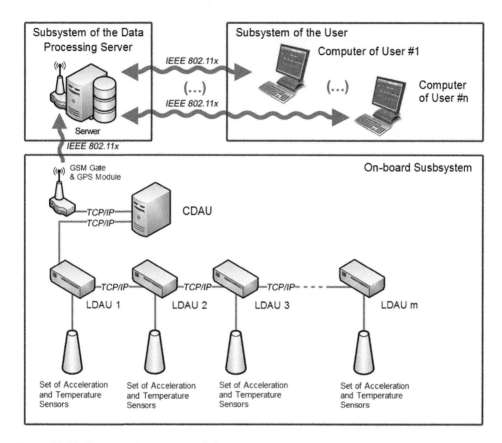

Figure 16.11 Functional structure of the system

of the track can be univocally identified with a spot in the field. Therefore, operator of the vehicle can receive information of behavior of the vehicle over appropriate route.

The information that is collected in the database of the system server is also used for evaluation of the technical health of springy and damping elements of the I- and II-grade of springing of the vehicle as well as for evaluation of the health of the track, regarding a longer period of time. On the basis of this information, it is possible to make decisions on the necessary repairs, overhauls, or replacements of vehicle elements.

Communication between the central data acquisition unit and the subsystem of the server for data processing is realized wirelessly, by means of a GSM telecommunication system.

Algorithms managing the correct operation of the system and algorithms managing the procedure of monitoring and diagnosing the structural health of the vehicle and the track are a part of the system, which is as important as the functional structure of the system.

A necessary condition for monitoring the structural health is a correct operation of the monitoring system itself. The most crucial part of the system can be regarded as the electric harness developed on the vehicle, and especially electric connections between the sensors and the data acquisition units, as well as an internal system of the data acquisition units and communication modules. The system server is also a subject of monitoring its correct operation.

The following are under control:

- all the input and output circuits (e.g. control of the permissible maximal and minimal values);
- communication (loss of communication);
- volume of a free space of the hard disk/non-volatile memory;
- volume of free RAM;
- CPU load.

In the case of any problems in any of these scopes, basic information necessary for identification and location of an inefficiency is generated and saved. At the next successful attempt at communicating, these events will be sent to the server as a priority. An enhanced variant may include sending automatic information in a form of a SMS or an e-mail from the server to the person in charge, in the case of important alarms.

In the case of a failure of the communication module located between the data acquisition unit and the server, it is impossible to directly determine a cause of such state. In the designed system, there are two ways of checking operation of the communication:

1. Even when all the subsystems operate in a correct way, information on this fact is cyclically sent to the server. Lack of such information within a certain time interval results in the generation by the server an event – alarm indicating a communication error.
2. It is possible to send a signal from the server to a certain data acquisition unit (*ping*) provided the communication operates in a correct way; then the data acquisition unit generates a return signal of correct operation.

If the communication is undisturbed, the data packages are recorded in a non-volatile internal memory of the data acquisition unit. At the moment when storage occupancy of the memory space exceeds a certain level, e.g. 70%, the program erases a certain number of data packages, starting with the oldest. Prior to erasing, the data within the package are preliminarily evaluated by the program with respect to e.g. maximal RMS values (or other indicators) of the recorded signals. In the case when the RMS value of the recorded signal exceeds a certain level (for a given ride speed), the package is classified as ready to be sent, and at the same time information about this event is also generated and saved in a separate file. Transmission of the data packages is realized using a GSM system.

16.6. Prototype of the monitoring system

The system for monitoring the structural health and diagnosing of rail vehicles and tracks consists of the following subsystems:

- on-board subsystem;
- user subsystem;
- subsystem of the data processing server.

The on-board subsystem has a modular structure. It consists of a module of the central data acquisition unit and a certain number of modules of local data acquisition units; this number results from the number of units belonging to the rail vehicle.

The on-board subsystem consists of the following components:

- central data acquisition unit;
- local data acquisition units (a subsystem predefined for the needs of the system);
- acceleration sensors;
- temperature sensors;
- software.

The central data acquisition unit (CDAU) is an industrial computer used in railway engineering, compatible with the EN 50155 Standard.

The central data acquisition unit is located at one of the terminal units of the rail vehicle. Each rail vehicle has one central data acquisition unit. Its role is to collect signals from the local data acquisition units, perform analyses based on these signals, and transfer data packages and control packages to the subsystem of the data processing server. Wire communication between the central data acquisition unit and the local data acquisition units is realized by means of Ethernet. A hierarchical structure of the system for monitoring and diagnosing of the vehicle – track system is shown in Figure 16.12.

The following functions are connected with the above levels of the structure of the system monitoring the suspension system of a rail vehicle:

Sensors (input transducers):

- provide information on the recorded physical quantity.

Data acquisition units:

- acquire signals from the transducers;
- condition the signals – generally, an output signal from a sensor must be conditioned prior to its further analysis and presentation by a computer or other device. Exemplary operations of signal conditioning are amplification, suppression, and filtering.

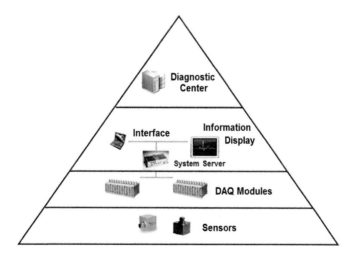

Figure 16.12 The hierarchical structure of the system

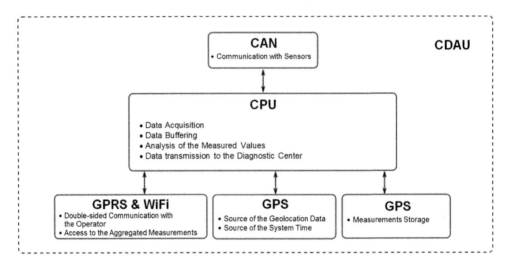

Figure 16.13 Schematic presentation of the tasks realized by the DAU

The functions listed above are realized by means of software, which is a part of the data acquisition unit – DAU (Figure 16.13).

Software of the presented system was developed in C language, compiled by means of a compiler package and programming tools based on GCC line. The programming environment was the Eclipse CDT software. The software was prepared for working under the Nut/OS operation system, designed for built-in systems.

Processing and communication unit carries out the following tasks:

- loading data from the acquisition units;
- validation of the measuring circuits;
- data processing;
- control of situations when the alarm thresholds are exceeded;
- recording in the database;
- reporting emergency situations to the operators;
- transferring diagnostic parameters to the diagnostic center in the case of messages, such as alarm or warning.

Diagnostic center realizes the following tasks:

- collecting diagnostic data concerning the monitored structural health of the rail vehicle;
- a complex diagnostics of the structural health of the rail vehicle, i.e. detection, location, and determination of intensity of damages.

Modules and sensors belonging to the subsystem are connected by means of a commonly applied high-capacity CAN 2.0B bus.

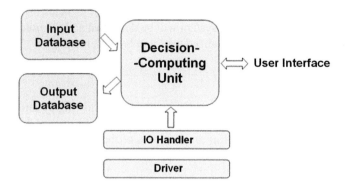

Figure 16.14 Schematic presentation of the tasks realized by the DAU. Driver: element of the control system; IO Handler: element of the User Interface

The following functional blocks can be distinguished in the software structure of the system, presented in Figure 16.14:

- block of data loading form the acquisition system;
- decision-computing block;
- input database;
- output database;
- user interface.

Integration of hardware elements with the installed created software yielded a whole – a prototype of the system.

16.6.1. Testing of the system

Experimental study in the case of each prototype plays an important role in the process of creating a new object or its elements. In some cases, it is the only possibility of learning the studied phenomenon or some dynamic properties of a given object. In the case of a mechanic system such as rail vehicle – track, experimental study is used for supplementing the information on the vehicle and the track that has already been gained, which makes it possible to improve the existing models and research tools. In the case of a monitoring system, this kind of study was used for verification of the accepted assumptions as well as for a preliminary check if the system operates in a correct way at the stage of building its prototype. A first generation of the system prototype was subjected to tests. The prototype consisted of sensors connected to the data bus, wiring, data acquisition and processing unit with its internal software, module of communication with the server, as well as the system server with necessary software.

The studies were carried out in two stages. The first stage was a study realized under laboratory conditions, aimed at evaluating if the software and hardware components of the system operate in a correct way. In order to do that, there were employed some standard test rigs designed for testing simple dynamic systems. The second stage was a study realized at an experimental track in Żmigród, during which it was possible to evaluate functional and substantial quality of the system built for this purpose within vehicles of the measurement train. Later in the text, the developed system will be described and the process of its testing along with exemplary results of the realized tests will be presented.

16.6.1.1. Laboratory tests

The aim of the testing was to check co-acting within the following system: sensors – data acquisition unit (DAU) – server. The testing included also verification of the software of the DAU and the server, as well as performing analyses of the recorded signals at the level of the DAU and the server. It should be emphasized that testing under laboratory conditions is aimed only at checking functionality of the monitoring system within a specific scope.

The testing process was divided into the following stages:

- testing of the system hardware;
- testing of the system software.

Testing of the system with regard to the hardware was aimed at checking operational functionality of the designed solution, connections between sensors and particular elements of the system, the possibility of a fast exchange of one of the elements of the system, etc.

However, testing related to the software was aimed at checking the essential correctness of particular programs, and especially correctness of analyses, computed indicators, generation of alarms, and correctness of the diagnostic inference process. Besides, within this part of testing, one carried out an assessment of the user interface installed on the system server and on the data acquisition units.

Diagram of the test rig for testing the prototype of the monitoring system is presented in Figure 16.15. A vibroacoustic signal was generated in one of the test rigs belonging to a laboratory of the Group of Modeling and Studying Dynamics of Mechanical Systems at the Faculty of Transport, Warsaw University of Technology. Then, by means of a wire link, the signals were transmitted to a local data acquisition unit (LDAU – Figure 16.16), where an initial processing and analysis of the signal is realized. The information that had been processed this way, was sent then to a central data acquisition unit (CDAU) – optionally, and successively to a system server by means of a wireless link.

The process of testing was carried out according to a program that had been prepared beforehand. Experimental measurements were performed at a test rig for studying free and forced vibration, as well as a test rig for diagnosis of failures of rolling bearings and rotating shafts.

Accelerations in the frequency range of approximately 1–20 Hz were studied during free or forced vibration and were recorded by the first test rig. On the other hand, vibration with frequencies reaching up to few hundreds of Hz was generated and recorded by a test rig for diagnosing shaft damages. Using these two test rigs made it possible to test the monitoring system over a wider range of frequencies and in the presence of failures. An exemplary view of the test rig for testing the system is shown in Figure 16.17.

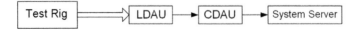

Figure 16.15 Diagram of the test rig for testing the prototype of the system

Figure 16.16 Data acquisition unit

Figure 16.17 View of the test rig for testing the acquisition system for acceleration signals (a test rig was employed in laboratory classes within a didactic subject of machine diagnostics)

Table 16.2 Channel#12, sampling frequency of 1500 Hz

Channel #12 (VIB)

Sampling Frequency 1500 Hz

Measurement	f_w [Hz]	RMSr [g]	RMSj [g]	PVr [g]	PVj [g]	PPr [g]	PPj [g]
#1	20	0.05	0.07	0.30	0.37	0.56	–
#2	60	1.17	1.42	4.29	4.93	7.75	–
Measurement	f_w [Hz]	Kr [–]	Kj [–]	Er [ ]	Ej [ ]	Tr [°C]	Tj [°C]
#1	20	3.64	4.02	0.05	0.03	0.27	0.26
#2	60	2.99	3.09	1.16	1.03	1.19	1.20

Indicator	*f = 20 Hz*		*f = 60 Hz*	
	Δ	[%]	Δ	[%]
RMS	0.02	33.33	0.25	19.31
PV	0.07	20.90	0.64	13.88
K	0.38	9.92	0.10	3.29
T	0.01	3.77	0.01	0.84

Table 16.3 Analysis of the signals recorded by means of DAU and NI acquisition module

	Analysis in DAU, control package	Analysis of data package from DAU by means of monit_ik program	Analysis of data file from the NI acquisition module by means of monit_ik program
Kurtosis	3.40	3.31	3.32
RMS	0.67	0.59	0.55
Zero-Peak	8.70	–	–

The test studies were carried out according to a program that had been prepared beforehand. Exemplary results in the case of channel#12 are presented in Table 16.2.

The obtained results were compared with the results of analysis obtained by means of commercial software packages for analysis of random signals. An exemplary comparison is shown in Table 16.3.

Faults of functioning of the system revealed during the testing have been eliminated.

16.6.1.2. Tests at the experimental track

Test studies of the system prototype were carried out at the experimental track in Żmigród (Figure 16.18). The basic technical parameters of the track are as follows.

1. Track of the experimental loop: length of 7725 m, jointed track, UIC 60 rails.

   • Sleepers: made of hardwood, softwood, and concrete, fasteners of K and S type, number of ties: 1733 pcs/km.

- Straight sections: maximal length of a straight section of 1314 m, other straight sections have lengths of 27 m, 54.5 m, 535 m.
- Curves and lengths of sections covering curves:

  - $R = 600$ m, 1568 m, transition curve 130 m,
  - $R = 700$ m, 249 m, transition curve 101.2 m,
  - $R = 800$ m, 33.5 m, transition curve 80 m,
  - $R = 900$ m, 3182 m, transition curve 120 m.

Permissible speed over the curves:

  - $R = 600$ m, $v_{max} = 110$ km/h[1)], superelevation h = 150 mm, a = 0.58 m/s^2,
  - $R = 700$ m, $v_{max} = 110$ km/h, superelevation h = 115 mm, a = 0.58 m/s^2,
  - $R = 800$ m, $v_{max} = 110$ km/h, superelevation h = 90 mm, a = 0.58 m/s^2,
  - $R = 900$ m, $v_{max} = 120$ km/h[2)], superelevation h = 100 mm, a = 0.58 m/s^2,

  [1)] in fact, it is possible to develop $v_{max} = 130$ km/h; appropriate locomotive is then required;
  [2)] in fact, it is possible to develop $v_{max} = 140$ km/h; appropriate locomotive is then required.

2. Length of tracks of the workshop station (designations of the tracks as in Figure 16.18):

- No. 1 – 1065 m,
- No. 2 – 973 m,
- No. 3 – 470 m,
- No. 4 – 200 m.

3. Permissible axial thrusts from 225 kN to 250 kN.
4. S-curve track [42].

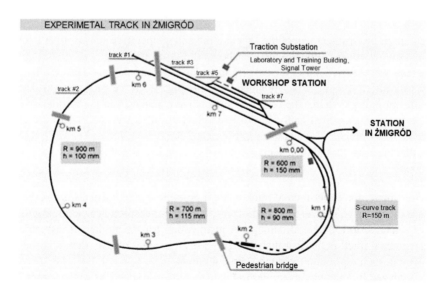

Figure 16.18 Schematic of the IK experimental track in Żmigród

Source: www.ikolej.pl/zaklady-laboratoria-i-osrodki/osrodek-eksploatacji-toru-doswiadczalnego

The tests were carried out using a passenger and a freight car, and consisted of two parts. After each installation of the sensors, the first part consisted in checking functionality of the system configurator as well as algorithms and procedures related to monitoring of the structural health of the object. The second part consisted in performing series of measurement rides, recording the signals by means of DAU and transmitting them to the server, realizing initial analyses. In parallel, the same signals were recorded synchronously by means of a measurement system used by the Railway Institute (IK). The IK system is certified and satisfies UIC requirements. In this case, it was considered a reference system. Set of the measurement train is presented in Figure 16.19.

Aim of the testing was to check correctness of the accepted concept and realization of the system as far as its hardware and software are concerned, as well as functionality of the system, taking into consideration the tasks to be realized in the future. Because of the aim of the study, the number of sensors was increased as compared to the number obtained on the basis of simulation tests.

An assumption was accepted that the number and locations of the measurement sensors must ensure monitoring of members of the vehicle suspension within a certain range of structural health of the track and the temperature of the bearing nodes of the wheelset axles. Ultimately, the number of measurement signals (channels) should be as low as possible. Since monitoring structural health of rail vehicles refers to a vehicle with an approval to exploitation, it is necessary to perform an on-line evaluation of the quantities, which are a basis for making a decision on vehicle's approval to exploitation. Set of these quantities in a simplified measurement method is defined by EN 14363/UIC518 standard [40,42]. The quantities are the aforementioned four acceleration measurement points, which were called UIC518 points:

- $\ddot{y}^+$ – lateral acceleration of the bogie frame (measurement at the end of the frame side-member, above the wheelset, at one side);
- $\ddot{y}^*$ – lateral acceleration of the body (measurement on the floor above the bogie center);
- $\ddot{z}^*$ – vertical acceleration of the body (measurement on the floor above the bogie center);
- $\ddot{y}$ – lateral acceleration of the wheelset (measurement at the axle-box in axis of the wheelset, at one side).

Diagram of the arrangement of the acceleration and temperature sensors over a four-axle car for test studies corresponded to the one shown in Figure 16.10.

The tests were carried out in the case of an undamaged and damaged car, realizing rides at various speeds. The damage was realized by removing one of the springs and a damper of the I-grade springing system. The recorded signals were analyzed in the frequency domain and subjected to a statistical analysis. An exemplary result corresponding to the cases of undamaged and damaged car are shown in Figure 16.20.

The obtained results and their successive analysis made it possible to draw the following conclusions:

- The testing on the experimental track made it possible to make a decision on the choice of a representative measurement point, in which the measured acceleration will be the most sensitive to a potential damage. A justification for the studies of test cars on the experimental track is the fact that they provide an amount of data that is sufficient for indicating such representative measurement points.
- The analyzed fragmentary examples indicate that the ride speed, as well as other factors, have a significant influence on detection of vehicle damage by the way of using statistical parameters of RMS, PV, and kurtosis, as well as power spectral densities of accelerations – determined for the signals recorded at selected spots on this vehicle.

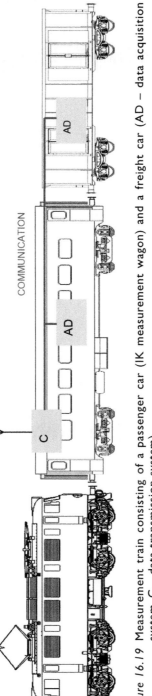

Figure 16.19 Measurement train consisting of a passenger car (IK measurement wagon) and a freight car (AD – data acquisition system, C – data transmission system)

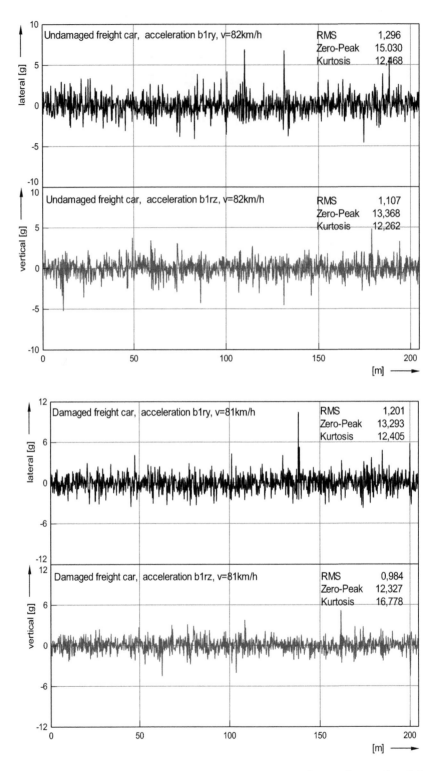

Figure 16.20 Lateral and vertical accelerations at point b1r of an undamaged and damaged freight car, speeds of 82 km/h and 81 km/h; statistic parameters of the signal

- Condition of technical maintenance of the track is also very significant while detecting failures. As results from the experiences of the team realizing the project, it is also possible that the excitations connected with the track mask or amplify the failure symptoms.
- The tests made it possible to specify RMS (or current RMS) as relatively the best diagnostic indicator. The other statistical parameters can be regarded supplementary. Spectral analyses are more complicated characteristics of the measured signals and can be performed at the level of the operator station, e.g. in the cases reported by the warning systems.

16.7. Tests of the system under operating conditions

The system for monitoring a rail vehicle was installed on an electric articulated train ED74 manufactured by PESA Bydgoszcz Company Ltd. ED74 is a four-unit electric articulated train, whose units are supported on five bogies (Figure 16.21).

During the supervised operation, the following were verified:

- quality of operation of elements (procedures and devices) of the system;
- location and number of measurement points;
- protection against damages of the installed elements of the system;
- quality and reliability of transmission of the measurement signals;
- correctness of operation of various configurations of the system;
- operational parameters of the monitoring system (length and geometry of the measurement sections of the track, range of riding speed, values of threshold accelerations, indicator of track structural health, and temperature of the axle-boxes).

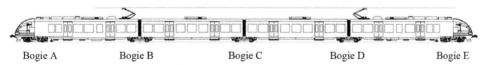

| Bogie A | Bogie B | Bogie C | Bogie D | Bogie E |

Figure 16.21 ED74 articulated train

The following were installed on the vehicle:

- central data acquisition unit – in one of the frontal units, cubicle in the driver's cab;
- five local data acquisition units (one for each bogie);
- temperature sensor at all the axle-boxes;
- acceleration sensors (at one frame of the driving truck E; for vertical direction only).

Altogether, there are approximately 50 measurement channels (related to vibration and temperature) of the monitoring system for one articulated train.

After completion of these works:

- the software was modified in order to connect a GPS module via UDP;
- the software was modified in order to connect five local data acquisition units (LDAU);
- an initial launch of the system on an ED-74 unit was realized.

In Figure 16.22, an example of the sensors installed on the frame of the trailer bogie is presented.

In Figures 16.23 and 16.24, elements of the system installed on the vehicle are shown: module of the central data acquisition unit and sensors along with the wiring installed on the bogie.

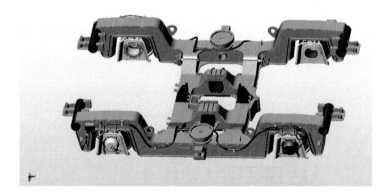

Figure 16.22 Trailer bogie – visualization of the wiring arrangement

Figure 16.23 Central data acquisition unit

Figure 16.24 Fragment of the body with sensors and wiring

During operational rides with an installed prototype of the system for monitoring the rail vehicle – track system, the following were tested: functionality of positioning the vehicle and correctness of functioning of the system as far as monitoring of the structural health of the vehicle and the track are concerned. In Figure 16.25, the geographical location of the vehicle is presented within the period of being operated, i.e. October 2011–June 2012.

Functionality of positioning the vehicle is a correct visualization of its geographical location on the map, as well as a possibility of tracking values of selected operational parameters and indicators of the structural health of the vehicle. The operational parameters are speed of the vehicle, direction of the ride, number of a bogie (number of a local data acquisition unit LDAU assigned to the bogie), and name and number of the vehicle (along with the number of the central data acquisition unit CDAU assigned to the vehicle). Indicators based on statistical measures of acceleration and temperature signals measured at appropriate points of the vehicle were accepted as the basic indicators of the structural health of the vehicle and the track. The measures are the following: root-mean-square value RMS, peak value PV, kurtosis K, and track quality TQ. Permissible values of RMS and PV indicators were determined with regard to the regulations contained in [37] and in standards [36, 40]. The permissible values of TQ and K indicators as well as additional significant values of RMS and PV indicators were accepted on the basis of statistical characteristics of the acceleration signals for typical and extreme operating conditions of a given vehicle over its whole route. These conditions are dependent, among other things, on operational speed, track quality, and structural health of the vehicle.

In order to determine additional values of the indicators, i.e. typical and extreme values characteristic for the studied vehicle – track system, average arithmetic values and standard deviations were calculated on the basis of all the data packages recorded over the whole route. The related results are presented in Table 16.4.

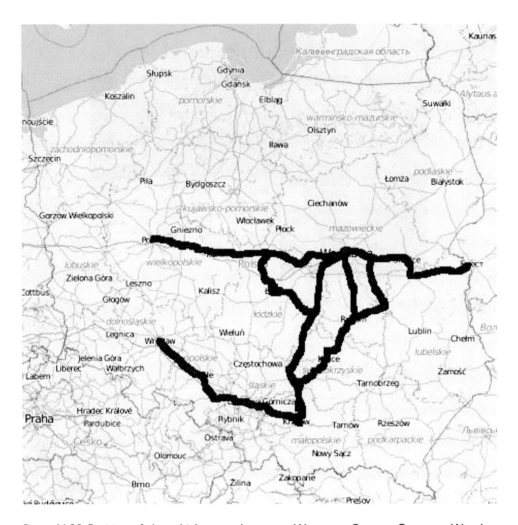

Figure 16.25 Position of the vehicle over the routes Warsaw – Cracow, Cracow – Wroclaw, Warsaw – Poznan, Warsaw – Terespol within the period October 2011–June 2012

Table 16.4 Average values and standard deviations of particular indicators of the structural health of the vehicle for ED74 vehicle – track system over the route Cracow – Warsaw – Poznan – Warsaw – Terespol (on the basis of acceleration signals measured at bogie D)

	Mean Square Value [m/s²]		Amplitude [m/s²]		Kurtosis Factor		Track Quality (Average)
Acceleration Direction	y	z	y	z	y	z	z
Average Value	0.7	1.1	4.0	5.6	17.3	9.6	1.3
Standard Deviation	0.6	0.7	4.8	4.9	30.8	14.8	0.5

Particular ranges were determined on the basis of a probability of occurrence of average/typical as well as extreme features for a process, whose probability distribution is near-normal. The way of determining is described in Table 16.5.

In the case when the extreme values, which result from a natural limitation of the variability of the process, exceed the normative values or are close to them, then the normative values are to be accepted as the limits of the range. The calculated values for indicators characterizing the signals recorded at the vehicle and values of the track indicator are listed in Tables 16.6, 16.7, and 16.8.

Table 16.5 Way of determining typical and extreme values of a given indicator under different operating conditions; V_{av} – average value, s – standard deviation

	Indicator W		
	Typical/Correct Condition	Deteriorated Conditions/Warning	Adverse Conditions/Alarm
Statistical Range	$W < V_{av} + s$	$V_{av} + s < W < V_{av} + 2·s$	$W > V_{av} + 2·s$ or a normative value
Percentage of observations	≈ 68%	≈ 28%	≈ 4%

Table 16.6 Calculated ranges for the PV peak value indicator

	PV Peak Value for acceleration signals measured at the bogie [m/s²]		
Acceleration Direction	Typical/Correct Condition	Deteriorated Conditions/Warning	Adverse Conditions/Alarm
y	PV < 8.5	8.5 ≤ PV ≤ 10.9	PV > 10.9
z	PV < 15	15 ≤ PV ≤ 20	PV > 20

Table 16.7 Calculated ranges for the RMS indicator

	RMS for acceleration signals measured at the bogie [m/s2]		
Acceleration Direction	Typical/Correct Condition	Deteriorated Conditions/Warning	Adverse Conditions/Alarm
y	rms < 2.3	2.3 ≤ rms ≤ 5	rms > 5
z	rms < 2.5	2.5 ≤ rms ≤ 7	rms > 7

Table 16.8 Calculated ranges for the track quality indicator TQ

Track Quality		
Typical/Correct Condition	Deteriorated Conditions/Warning	Adverse Conditions/Alarm
TQ < 1.8	1.8 ≤ TQ ≤ 2.3	TQ > 2.3

As a result of studies and analyses, the following definition of the indicator determining the track quality [29, 36], having the form of Equation (16.3), has been accepted,

$$jt(v) = c_{t5} \left\{ \frac{v^4}{v_e^4} \lim_{T \to \infty} \left[\frac{1}{T} \int_0^T a^2(t)dt \right] \right\}^p \tag{16.20}$$

where:
a – acceleration of the axle-box in vertical direction; the following values were accepted: exponent $p = 0.225$, reference speed $v_e = 120$ km/h, coefficient $C_{t5} = 2.4$.

The following picture (Figure 16.26) presents an example of geographical location of a vehicle over a selected route with a corresponding level of particular indicators, according to the pattern described in Table 16.5. In order to prepare the maps, cartographical data of the Openstreetmap project and a program designed for displaying these data were used. The source code of the program is available under an open GPL license, which made it possible to introduce small modifications in the program in order to visualize the data in a better way.

On the basis of evaluation of the collected data, one did not state a distinct dependency between the number of cases when the permissible values had been exceeded, magnitude of the exceeding and ride speed. The dependencies are presented in Figures 16.27 and 16.28. Lack of a distinct statistical dependency may occur when a rail vehicle rides with speeds adjusted to the infrastructure capabilities in most of the cases.

To each vehicle position marked on the map shown in Figure 16.26, there is assigned a data package containing the recorded signals and operational parameters, e.g. direction of the ride (geographical azimuth expressed in degrees) and values of particular indicators, which are expressed in percent of the permissible value on the map. These data are available at any moment, once a chosen position on the map has been indicated. A more accurate identification of the spots, where e.g. the permissible values were exceeded, can be realized using a railway POS database containing a detailed data concerning the railway infrastructure. Besides, in order to perform an initial identification, satellite photographs, available via the Internet, can be used; such an example is shown in Figure 16.29.

An effective way of positioning a rail vehicle has been presented; it consists in visualizing a geographical location on the map, as well as in a possibility of tracking at these spots values of operational parameters and indicators of the structural health of the vehicle. Such analysis can be useful especially when there is a relation between values of appropriate indicators of the structural health of the vehicle over a given route section and the quality of the infrastructure, i.e. a geographical location. It can be also helpful, while interpreting results and correlating them with characteristic sections of the route, which are located e.g. within a region of a rail station or a railway crossing. Such spots can be used for calibrating the system. Lack of appropriate calibration as well as an evident and considerable variability of the operating conditions may make the inference concerning the structural health of the vehicle difficult, especially with respect to its suspension. However, an attempt of an approximate inference with regard to the structural health of the vehicle suspension can be also carried out using a comparative method or by the way of studying a statistical trend. The comparison may be carried out for particular bogies of the vehicle running over the same section of the track or by compiling a map with values of the indicators of the track structural health and

Figure 16.26 Geographical position and speed of the vehicle against the corresponding RMS value and peak value PV (as based on the acceleration signal measured at the bogie frame); on the top: data related to November 2011; on the bottom: data related to January 2012

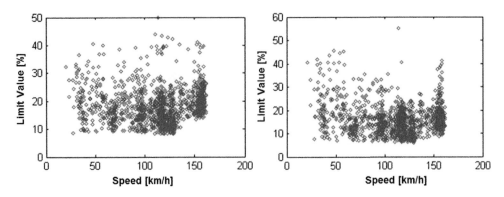

Figure 16.27 Acceleration of a corner of the bogie frame, value of RMS indicator for measurement data recorded within a period of one month; on the left: vertical direction; on the right: lateral direction

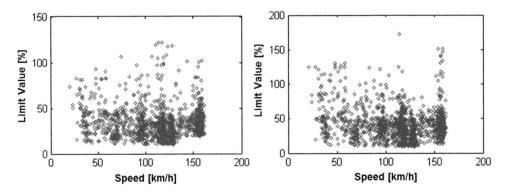

Figure 16.28 Acceleration of a corner of the bogie frame, value of PV index for measurement data recorded within a period of one month; on the left: vertical direction; on the right: lateral direction.

Figure 16.29 Satellite photograph [29] of a track section, over which the permissible PV peak value was exceeded at the speed of 154 km/h; a region of a railway crossing has been marked at the photograph

a map with values of indicators of the structural health of the vehicle. Inference based on trends requires long periods of collecting data, just as the comparative method can be practically implemented only at the level of the server, to which the data are sent. Another way of inference can be based on an assumption that a rail vehicle will be operated in most of the cases in a manner adjusted to the capabilities of the infrastructure. Then, operation of the vehicle – track system can be considered a process, whose extreme features are subjected to natural statistical limitations. So, this process can be controlled by the permissible values of appropriate indicators. In such a case, a systematic exceeding of the permissible values could indicate a deterioration of the structural health of the vehicle as well as the track.

16.8. Summary

A monitoring system for a rail vehicle and its track has been developed in two versions: for standard and light rail vehicles. Monitoring is realized in on-line mode, i.e. in real time during a ride of a vehicle with the system installed. Diagnostic signals, recorded by the sensors installed on the vehicle, are initially processed, and then wirelessly transmitted to the system server, where they are analyzed and stored in databases. The created software enables management of the signals as well as a full identification of events, which took place over the route, while riding under operating conditions. Identification of events makes it possible to answer the following questions: where a certain event took place (with an accepted accuracy), what the structural heath of the track is at this spot, and how the vehicle behaved along this route (its dynamics, ride comfort, or other parameters influencing the safety of the passengers) prior to the ensuing event. The system is also capable of sending a warning signal to the operator of the train or a manager of the railway traffic.

Functionality of the system is not limited by the vehicle speed, thus the system can be used for monitoring the structural health of rolling stock and track infrastructure designed for rapid rail transportation.

As far as the presented solution of monitoring the structural health of a rail vehicle and its track is concerned, a patent has been applied for at the Polish Patent Office, entitled Monitoring System for Structural Health of a Rail Vehicle and its Track.

Bibliography

[1] Bałuch, H.: *Diagnostics of Track Structures*. Warsaw: Transport and Communication Publishers WKiŁ, 1978 (in Polish).

[2] Będkowski, L.: *Elements of General Theory Technical Diagnostics*. Warsaw: Appendix to WAT Bulletin, 1981, 3, 343 (in Polish).

[3] Bogacz, R.: On Dynamics and Stability of Continuous Systems Subjected to Distributed Moving Load. Warsaw: Ing. Archiv, 1983, pp. 57–69.

[4] Bogacz, R., Chudzikiewicz, A., and Frischmuth, K.: Friction, Wear and Heat, in: L. Bobrowski and A. Tylikowski (eds.): *Simulation in R&D*, Warsaw, 2005, pp. 67–76 (in Polish).

[5] Bogacz, R., and Frischmuth, K.: *Simulation of Wheelset in Curved Track*. Warsaw: Machine Dynamics Problems, 2002, 25, 3/4 (in Polish).

[6] Chudzikiewicz, A.: *Elements of Diagnostics of Rail Vehicles*. Radom and Warsaw: ITEe PiB, 2002 (in Polish).

[7] Chudzikiewicz, A.: *A System of Studying Dynamic Phenomena in the Track – Rail Vehicle Mechanical System*. Warsaw: Publishing House of the Warsaw University of Technology, 1988 (in Polish).

[8] Chudzikiewicz, A., Droździel, J., Kisilowski, J., and Żochowski, A.: *Modeling and Analysis of Dynamics of the Track – Rail Vehicle Mechanical System*. Warsaw: PWN, 1982 (in Polish).

[9] Chudzikiewicz, A., Droździel, J., and Sowiński, B.: *Practical Solution of Rail Vehicle and Track Dynamics Monitoring System: Key Engineering Materials*. Structural Health Monitoring II Book Series, 518, 2012, pp. 271–280 (in Polish).

[10] Chudzikiewicz, A., and Firlik, B.: Light Rail Vehicle Dynamics from a Running Safety Perspective. Warsaw: *Archives of Transport*, 2009, 3–4 (in Polish).

[11] Chudzikiewicz, A., and Melnik, R.: Simulation Evaluation of Collisions in the Case of a Rail Vehicle. Poznań: *Rail Vehicles* (Pojazdy Szynowe), 2010, 3, pp. 1–5 (in Polish).

[12] Chudzikiewicz, A., Radkowski, S., and Uhl, T.: *Problems of Diagnosing Rail Vehicles*. Warsaw: II International Congress of Technical Diagnostics, September, 2000 (in Polish).

[13] Chudzikiewicz, A., and Sowiński, B.: *Problems of Choosing Statistical Parameters in the Process of Monitoring the System of Railway Vehicle*. Budapest: BOGIE '10 – The 8th International Conference on Railway Bogies and Running Gears, Budapest, Hungary, 13–16 September, 2010, Proceedings CD, pp. 1–14 (in Polish).

[14] Chudzikiewicz, A., and Sowiński, B.: *Selection of Diagnostic Parameters in the Process of Monitoring the Rail Vehicle's Conditions*. IX International Conference on Degradation of Technical Systems, Liptovský Mikuláš, Slovakia, 7–10.04.2010, Proceedings CD, pp. 1–10 (in Polish).

[15] Czyczuła, W., and Topik, K.: *Problems of Modeling and Identification of Models of a Contact Track*. Railway Reports (Problemy Kolejnictwa), No. 128, CNTK, Warsaw, 1998, Wroclaw University of Technology (in Polish).

[16] Droździel, J., and Sowiński, B.: *Simulation of Influence of the Rail Vehicle and Track in Presence of Anomalies*. XV Scientific and Technical Conference Rail Vehicles 2002, Vol. 1, Publishing House of the Wroclaw University of Technology, Wroclaw, 2002.

[17] Droździel, J., and Sowiński, B.: *Simulation of Railway Track Deterioration Influenced by Ballast Stiffness and Dry Friction*. WIT PRESS. Computers in Railway XI, 2008, pp. 693–702.

[18] Directive 96/48 of the European Parliament and of the Council on the Interoperability of High-Speed Rail System. TSI – High-Speed Rolling Stock, 1996.

[19] Elkins, J.A., and Carter, A.: Testing and Analysis Techniques for Safety Assessment of Rail Vehicles: The State-of-the Art. *Vehicle System Dynamics*, 1993, pp. 185–208.

[20] Erskine, A.: *Structural Crashworthiness Literature Search*, www.rssb.co.uk/pdf/reports /research. pdf.

[21] Firlik, B., Czechyra, B., and Chudzikiewicz, A.: *Condition Monitoring System for Light Rail Vehicle and Track: Key Engineering Materials*. Structural Health Monitoring II Book Series, 518, 2012, pp. 66–75.

[22] Gronowicz, J., and Kasprzak, B.: *Diesel Locomotives*. Warsaw: Transport and Communication Publishers (WKiŁ), 1989 (in Polish).

[23] Bednarz, J., Barszcz, T., and Uhl, T.: *An Example Diagnostic System Based on OMA(X) Method and NARX Models for Rotating Machinery*. Proceedings of the Fourth European Workshop Structural Health Monitoring, 2008. DEStech Publication, Inc., Lancaster, Pennsylvania.

[24] Kardas-Cinal, E.: *Studying Ride Comfort as Dependent on Parameters Characterizing the Rail Vehicle – Track System*. International Conference 21st Century Transport (Transport XXI wieku), 2004, Proceedings, (in Polish).

[25] UIC 566 Card; Loads of bodies of passenger cars and added parts (in Polish).

[26] Koc, W., and Grulkowski, S.: *Influence If Longitudinal Forces in Rails on the Course of Geometric Adjustments of the Track*. Conference Design, Building and Maintenance of the Infrastructure in the Rail Transport (Projektowanie, budowa i utrzymanie infrastruktury w transporcie szynowym). PKP PLK Company Ltd., Zakopane, 9–11.04.2008, Proceedings, (in Polish).

[27] Krużyński, M.: *Selected Problems of Designing and Building the Track Substructure*. Conference Design, Building and Maintenance of the Infrastructure in the Rail Transport (Projektowanie,

budowa i utrzymanie infrastruktury w transporcie szynowym). PKP PLK Company Ltd., Zako-pane, 9–11.04.2008, Proceedings, (in Polish).

[28] Uhl, T.: *Computer Assisted Identification of Mechanical Structures*, Warsaw: WNT, 1998 (in Polish).

[29] Melnik, R., and Kostrzewski, M.: *Rail Vehicle's Suspension Monitoring System – Analysis of Results Obtained From Tests of the Prototype. Key Engineering Materials.* Structural Health Monitoring II Book Series, 518, pp. 281–288, 2012.

[30] *Methodology of Diagnosing a Rail Vehicle by On-line and Off-line Methods.* KBN grant No. T12C 046 15, supervised by Chudzikiewicz A., Warsaw, 1998–2001 (in Polish).

[31] PN-EN 12663 Standard – Requirements related to design and strength of the rail bodies (in Polish).

[32] PN-EN 15227 Standard – Collision requirements for the bodies of rail vehicles (in Polish).

[33] Opala, M.: *Analysis of Experimental Data in the Context of Safety Against Derailment at a Rail-way Vehicle: Using the Energy Method. Key Engineering Materials.* Structural Health Monitoring II Book Series, 518, 2012, pp. 16–23.

[34] Opala, M.: *Simulation Method of Evaluating Traffic Safety of Freight Wagons of an Increased Axle Load.* Doctoral dissertation, Publishing House of the Warsaw University of Technology, Warsaw, 2006 (in Polish).

[35] PN-EN 14363:2007 Railway applications – Testing for the acceptance of running characteristics of railway vehicles – Testing of running behaviors and stationary tests (in Polish).

[36] "MONIT" Project: Structural Health Monitoring of Structures and Evaluation of Their Lifespan – Innovative Economics Operational Program – Action 1.1: Support for scientific research in build-ing a knowledge-based economics, Warsaw, 2008–2013, (in Polish).

[37] Romaniszyn, Z., and Wolfram, T.: *Modern Railway Rolling Stock.* Warsaw: Published by the Railway Scientific and Technical Centre, 1997 (in Polish).

[38] Szcześniak, W.: *Statics, Dynamics and Stability of Track Structure and Substructure.* Warsaw: Scientific Works of the Warsaw University of Technology, Building Engineering Series, No. 129, Publishing House of the Warsaw University of Technology, 1995 (in Polish).

[39] Tomaszewski, F.: *Possibilities of Applying Diagnostics in the Process of Rail Vehicles Mainte-nance.* Poznań: Rail Vehicles (Pojazdy Szynowe), 4, 1999, pp. 52–60 (in Polish).

[40] UIC Code 518 – Testing and approval of railway vehicles from the point of view of their dynamic behavior – Safety – Track fatigue – Ride quality.

[41] PN-EN-12082 Standard – Axle-boxes.

[42] UIC Code 530 – Wagons – Sécurité de circulation.

[43] Service Center for Pendolino, http://inforail.pl/centrum-serwisowe-dla-pendolino-_more_47781. html.

Chapter 17

High-speed rail versus environmental protection

Krzysztof Polak

17.1. Introduction

The environmental situation in Poland has been improving in spite of the fact that it still significantly differs from the state of other European Union countries. This necessitates taking further action and incurring more expense in order to improve the status of particular environmental elements.

Rail transport is considered to be one of the most environmentally friendly modes of transport. Pollutants' emission by rail transport means is substantially lower than for road transport, which is responsible for 80% of pollution [1].

Moreover, rail transport generates smaller (than for other means of transport) external costs connected, inter alia, with accidents or pollution of the environment, which amount only to 1.6% of general external costs of all transport modes [1].

Considering this, it is very important to develop rail transport, including high-speed rail, so that it can compete with other means of transport. It must be remembered, however, that this development needs to include the principle of sustainability, which consists of social and economic development in which the process of political, economic, and social activities' integration takes place together with maintaining natural environment balance and permanence of basic natural processes in order to guarantee basic needs of specific communities or citizens of both current and future generations [2].

All issues relating to pursuing railway investment and its further operation must be in accordance with the requirements of environment protection. Recognizing and taking into account essential principles of environmental protection and related processes will significantly improve the pre-design process as well as high-speed rail construction and operation. Such a big project entails a great challenge to minimize various factors in the environment, not only such obvious and easily measurable ones as pollution emission or noise, but also those which are difficult to specify (landscape or heritage). It is known that combining all the factors to a common denominator is difficult, almost impossible, to achieve. However, defining crucial environment components and their sensitivity, taking into consideration the assessment of nature in areas surrounding railway lines, will allow identifying major threats that could negatively influence environmental status and show possible solutions or defenses against their negative impact.

17.2. Law versus environment

The high-speed rail project will be connected with the implementation of administrative procedures, necessary to take further steps in the investment process (leading to launching construction works). One of the most significant and time-consuming elements of this process will be to obtain a decision on environmental conditions.

The Polish Act of Law of 3 October 2008 on the Provision of Information on the Environment and its Protection, Community's Participation in Environmental Impact Assessments (Art. 71) [3] defines requirements on how to obtain a decision on environmental conditions for planned:

1. projects that could have a significant impact on the environment – Group 1;
2. projects that could potentially significantly influence the environment – Group 2.

Types of investment plans comprising the above-mentioned projects were outlined in the Regulation of the Council of Ministers of 9 November 2010 on projects that may significantly impact the environment [4]. According to this regulation, railway investments are included in both groups of projects that could have a significant impact on the environment. The projects that significantly affect the environment (Group 1) include "railway lines constituting the trans-European rail system, as defined in the provisions of the Act of Law of 28 March 2003 on railway transport (Journal of Laws of 2015, item 1297, with further amendments)" [4]. However, railway lines and transshipment facilities in intermodal transport, other than that mentioned above, as well as bridges, viaducts, tunnels on railway lines, and sidings with at least one railway track of over 1 km length, are included in projects that may potentially affect the environment (Group 2).

The projects included in Group 1 comprise obligatory conducting an environmental impact assessment (EIA report) and environmental conditions. In order to precisely define the scope of the EIA report, the investor is required to present the Project Information Card (PIC) to the competent authority, on the basis of which they stipulate which report elements should provide additional details.

For projects that were qualified to projects that could potentially greatly influence the environment (Group 2), conducting an environmental impact assessment report is optional. The investor is obliged to present the Project Information Card, and based on that the competent authority states whether an environmental impact assessment report is required. When carrying out environmental impact assessment is not necessary, the competent body issues a decision on environmental conditions. However, when an environmental impact assessment report is required, the competent body is obliged to define the scope of the report in which they identify which elements should be analyzed more broadly.

The Project Information Card, pursuant to the Act of Law of 3 November 2008 on the provision of information on the environment and its protection, public participation in environmental protection, and environmental impact assessments [3], contains basic information relating to a planned investment project, including data on kind, scale, and siting of the enterprise; area of the real estate property and its vegetation coverage; possible enterprise variants; forecast quantity of water, fuels, resources, and energy used; solutions applied to protect the environment; possible transboundary environmental impact; forms of nature protection; and ecological corridors.

An EIA report is an elaboration which covers a much broader range of information than the Project Information Card. The environmental documentation for the EIA report should include, inter alia:

• the description of the planned project, in particular the characteristics of the entire enterprise and conditions of use of the site during construction and operation or use; forecast kinds and amounts of emissions, waste including that resulting from planned functioning of the enterprise, information on biological diversity, exploitation of natural resources, soil, water, and ground surface including information on energy demand and its consumption, risk of major construction accidents or natural disasters while taking into consideration substances used and technologies applied, including the risk connected with the change of climate;

- description of natural elements of the environment falling within the scope of the forecast impact of the planned enterprise on the environment, including forms of environmental protection as well as hydromorphological, physiochemical, biological, and chemical characteristics of waters;
- results of inventory of nature and wildlife;
- an inventory and description of protected monuments and sites in the neighborhood or within the range of the planned project's impact;
- description of landscape;
- information on connections with other projects, in particular impact aggregation of ongoing, accomplished, or planned projects;
- description of options analyzed, such as:
- defining predicted environmental impact of these analyzed options;
- description of forecast activities aiming at avoiding, protecting, limiting, and natural compensation of negative impact on the environment;
- analysis of potential social conflicts with regard to the planned project.

Consequently, an environmental impact assessment report requires significantly far more time, finance, and subject matter efforts. The main reason for this is the necessity to elaborate the natural inventory, which must include all phenological dates – the full inventory should take a minimum of one year. Also taking into account the time to conduct other EIA report elements (including acoustic issues) and administrative procedures, the mean time to receive a decision on environmental conditions amounts to approximately 2–3 years, depending on the project's complexity. Otherwise, issuing a decision on environmental conditions without preparing an EIA report takes about 6–12 months.

According to the Regulation of the Council of Ministers of 9 November 2010 on projects which may significantly affect the environment [4], high-speed rail building may be included as projects that may significantly affect the environment (Group 2), because currently this kind of railways does not constitute a trans-European rail system in the meaning of the Act of Law of 28 March 2003 on railway transport. Acquiring a decision on environmental conditions will be associated with prior elaboration of the Project Information Card, basing on which a competent authority will state the need or its lack of conducting the environmental impact assessment report. Associated infrastructures, including access roads, power supply lines, and GSM-R antennas connected with railway infrastructure will also be classified as enterprises that could always potentially affect the environment, in spite of the fact that they may be a separate project listed in the Regulation of the Council of Ministers of 9 November 2010 on projects which may significantly affect the environment [4]. Apart from the activities themselves to obtain all necessary permits, such a long-term and time-consuming project as high-speed rail (HSR) building will have to face the problem of continuously changing law provisions. A constant evolution of law requirements may result in the necessity of a permanent update of already existing developed issues, good practices, or decisions made/issued.

17.3. Environmental status in Poland

The environmental situation in Poland presented below was outlined based on data pooled from the Chief Inspectorate of Environmental Protection [5].

Poland is perceived as one of the most developed countries for which industry, transport, and urban engineering pose major threats for the environment. Their degree and kind vary spatially; however, their main pressures are observed on large agglomeration areas and in their neighborhoods. Due to its geographical position, Poland is characterized by its natural and landscape wealth. The existence of rare plant and animal species requires Poland to undertake measures aimed at the protection and improvement of the natural environment.

There has been great progress in environmental protection improvement over the last 20 years; nevertheless, it is still necessary to take actions to reduce pressures on the environment. The main undertakings should be focused on making the Polish economy greener and increasing ecological effectiveness, as it is forecast that

> economic balance of enterprises' performance will be more and more dependent on the necessity to implement eco-innovation as well as resources and energy saving. Not only will the reduction in the consumption of raw materials result in the decrease in reduced costs for the economy in the future but also in decrease of pressures on the environment.
>
> [5] (p. 186)

A vast territory, Poland is covered by areas of precious natural value that are mostly subject to statutory legal protection (forms of nature conservation). Forms of nature conservation are a crucial element of pursing natural environment protection goals, as each form plays a different role in the natural environment system and serves a different purpose. One of the most widespread forms of nature conservation is the network of Nature 2000 (*Natura 2000*), which was created to protect the most endangered species and natural habitats in Europe. This network covers 20% of the country. Unfortunately, the state of most endangered species and nature habitats has been assessed as unsatisfactory. The main threats affecting the impact on quantitative and qualitative condition of species and habitats include, inter alia, the development of road, tourist, industrial, and energy (wind power stations) infrastructures.

Ambient air quality in Poland, in spite of many activities undertaken, still needs to be improved. An essential problem relies on an excessive concentration of fine particulate matter PM2.5 and too high tropospheric ozone concentration (in summertime) as well as too high concentration of excessive concentration of suspended particulate matter PM10 and benzo(a) pyrene (in wintertime). Investment works relating to state-of-the-art transport infrastructure, including the construction of high-speed rail connections, should have a great influence on the improvement of air quality in Poland.

A significant majority of Poland's soils (over 90%) is used for agriculture and forestry. Small changes of land use have been noticed recently, i.e. increase of land area designated for construction and urbanization. Furthermore, over 96% of arable soils can be classified as high-quality soils (minimal excess of heavy metal concentration). It should also be added that the purity of soils used for agricultural purposes in Poland is very good.

The quality of surface flowing water and lakes is still insufficient, in spite of the building of industrial and urban wastewater treatment plants as well as innovations in existing technologies. According to the report on environmental status in Poland [5], in 2010–2012 only 30% of natural (rivers) surface water bodies (SWB) reached at least good ecological potential. Similar results referred to artificial SWB, where only 30.7% reached good or very good ecological potential. As far as natural lakes SWB are concerned, as many as 65% of them did not reach an expected ecological potential, whereas 44.5% artificial lakes SWB reached at least good ecological potential. In the case of bodies of groundwater (BG), 90% of BG quantitative and chemical status was characterized as good.

Water resources in Poland per capita are among the lowest in the European Union. Therefore it is extremely important to manage them in a reasonable manner. Consequently, it is vital to achieve and maintain a good status of surface and underground waters.

Thus, the environmental situation in Poland should be assessed as good, as the quality of particular elements of the environment is constantly improving. The above-mentioned data show that air pollution still poses the greatest problem.

17.4. High-speed rail versus environment

Facing such an investment as high-speed rail building will be connected with designating new areas for rail infrastructure. A new railway route can contribute to a barrier effect emerging for both people (difficult access to fields, places of work, etc.) and the environment (separating precious natural areas). A new investment would also introduce new elements to the landscape and would also be an additional source of pollution emissions, e.g. noise.

Dangers for the environment resulting from building high-speed rail may appear both in the construction stage and in the operational one. The main threats include:

* occupying new areas designated for railway infrastructure and associated infrastructure (access routes, other buildings);
* changes in the topography of the site and the landscape, affecting
* precious natural areas and protected areas;
* protected plant and animal species as well as their habitat;
* noise and vibration emission;
* emission of harmful pollutants to the air;
* influence on the climate;
* affecting surface and underground waters;
* emission of electromagnetic fields;
* influence on monuments and historical sites;
* generation of waste;
* cumulative impacts;
* transboundary impacts;
* serious breakdowns and accidents.

17.4.1. Site occupancy for railway infrastructure construction

In accordance to technical standards for high-speed rail construction, it is necessary to separate railway infrastructure from other spaces. A railway line must be fenced and protected against human and animal access. Moreover, all junctions with roads and pedestrian crossings must be made collision-free. The necessity of such a solution while building a new railway infrastructure will cause the occupancy of new areas and may even lead to the appearance of social barriers (destroying local cohesion) and/or nature barriers (animal migration).

Occupying new areas for building HSR and associated infrastructure (inter alia, electric power lines, roads) will cause a change in the way the areas adjacent to railway investment are managed and used. This may consequently decrease the use of space adjacent to the railway line.

Furthermore, in the course of carrying out the investment, it will be necessary to occupy additional space for the construction backup facilities, materials, and equipment bases or access routes. The impact of construction works will be short and temporary (except for the permanent occupation of a strip of land).

It is essential to take into consideration precious natural areas (protected areas, ecological corridors), development important from the viewpoint of local communities, housing development, including, while preparing the investment, the site selection. HSR line needs to be designed so as to first avoid such terrains which are precious as regards environmental protection. If it is not possible, losses connected with HSR building need to be minimized as much as possible. The inability to reduce losses to a minimum would entail the necessity of introducing compensations (e.g. planting new trees and bushes in place of the lost ones).

Such an approach will largely allow minimizing negative impacts related to the new infrastructure.

17.4.2. Changes in topography, including landscape

Changes in the landscape due to the investment carried out will mainly result from direct interference in landscape. Activities directly affecting landscape value will be connected with adding or removing certain components from the investment vicinity. HSR building will be related to emerging new elements in the environment such as the railway line itself, overhead contact line, engineering facilities, or associated infrastructure. The deterioration of visual values may result from cutting down trees and bushes, changes to the topography of the site (levelling the ground), or regulation of watercourses.

The HSR lines' impact on landscape will be greater than that of traditional railways. Designing the path of an HSR line is quite limited due to technical standards which, inter alia, forbid building pedestrian and level crossings on the level of a railway line or requirements regarding curve radii. This will contribute to the disturbance of the landscape harmony near the newly built HSR line and will interfere with scenic views [6].

It must be remembered, however, that a railway line is not as dominant in the landscape as other infrastructure buildings are. A railway line in many cases complements the landscape when it does not destroy its order, and even the installation of necessary fences (galvanized steel fencing net with increasing upward mesh) should not cause serious disturbances.

Acoustic screens will be one of the few elements that are completely new and strongly interfere with landscape values. However, they will mostly be mounted in an urban landscape, close to housing or residential areas. Using appropriate materials and rational land use through planting ornamental shrubs, trees, and flowers will reduce acoustic screens' negative impact on the environment and will improve the comfort of life of residents exposed to transport noise pollution.

17.4.3. Impact on valuable natural areas, including protected areas

In order to protect nature resources, the Nature Conservation Act of 16 April 2004 [7] defines forms of nature conservation subject to legal protection, which include:

1. national parks;
2. nature reserves;

3. landscape parks;
4. protected landscape areas;
5. Nature 2000 areas;
6. nature monuments;
7. documentation sites;
8. ecological grounds;
9. nature and landscape complexes;
10. protection of plant, animal, and fungi species.

One of the measures of conservation introduced by the above-mentioned law is the ban to carry out any activities or investments that would significantly affect protected areas, their conservation goals, or integrity with other forms of conservation. These bans do not refer to activities of overriding public interest, including social or economic requirements where there are no alternative solutions. Taking into account the fact that HSR line will be included to the overriding public interest, during setting the path of the railway line, collision situations with protected areas will be limited to a minimum and all of the above conditions will be complied with.

In the whole territory of Poland, ecological corridors (migration routes of plants and animals) have been established to ensure the ecological communication and cohesion of protected areas, including Nature 2000 network and other precious nature areas, as well as enable free migration of plants and animals throughout Poland and Europe.

The line investment construction in the intersected areas is connected with the possibility of natural links fragmentation. This is a long-term phenomenon. The range and scope of limiting the spread of species as well as decreasing the population in divided ecosystems will depend on the location of the investment. The attempt to assess the impact of possible fragmentation is difficult to evaluate; possible effects may appear in time. In addition, an issue that makes the above estimation difficult is the possibility of appearance of impacts cumulated with other investments (e.g. roads) or urban development (bring residential development closer to a railway line).

However, it is possible to limit an adverse scope though detailed search of HSR locations that will avoid precious natural areas as well as protected species of animals and plants.

Due to this approach, a negative impact of building a railway line will be significantly minimized.

17.4.4. Impact on animal and plant species and habitats

As was mentioned in the previous section, building an HSR line may be connected with interference and fragmentation of natural habitats and consequently can have a negative impact on protected animal and plants species. It may also lead to changes in the composition and numbers of animals and plants.

It must be taken into consideration that in case of natural habitats and protected plant species that are located on the lines of crossing the projects, they will be occupied by the investment. Further impacts on natural habitats and vegetation will be connected with the necessity of cutting down trees and shrubs, which results from the collision of designed railway and auxiliary infrastructure and safety reasons (15 m strip from the outermost axle of the track).

Moreover, the fragmentation of natural habitat may cause the decreasing of volume, isolation of patches, and lowering the habitat quality. Possible changes in hydrological conditions (resulting from deep trenches for the need to build foundations of engineering objects) may also affect the quality and status of habitat.

As far as the impact on protected animal species is concerned, the following threats are expected to appear:

- direct damage to habitats (reproduction, bird nesting places, feeding grounds, hideouts);
- deterioration of habitats connected with worsening of acoustic climate, change in hydrological conditions, or change in the land use;
- barrier effect – deterioration of ecological corridors' functionality.

It is inevitable to avoid completely the above-mentioned impacts while building new infrastructure. However, conducting design and construction works due to the binding law, in particular in connection with protection of the environment, and using the best information about the environment (inter alia, natural inventory) as well as good practices (like designing and building animal crossings), should significantly limit the impacts mentioned.

17.4.5. Vibration and noise emissions

17.4.5.1. Noise

Transport noise is currently one of the main factors of environmental pollution. However, railway noise in Poland has been decreasing lately due to conducted repairs and renovations, exchange of rolling stock into quieter ones, and also closures of unprofitable sections of the line [6].

Train movement on a railway line is a source of a few kinds of noise:

- rolling – on the contact area of the track and wheels, dominating in conventional rail;
- aerodynamic due to irregular flow of air during the train movement, dominating in HSR;
- noise as an effect of train breaking and powering.

The technical condition of track and rolling stock has a big influence on the level of generating noise.

It is worth paying attention to the constantly repeated myth that rolling stock driving faster than 250 km/h significantly increases energy consumption and impact on the environment (noise). In fact, this problem appeared at the turn of the millennium. However, there are now train sets based on modern technologies, in which the speed was increased to 350 km/h maintaining the same energy intensity and acoustic level as the rolling stock moving at 250 km/h [8].

Building HSR lines is a source of noise in the environment both in the construction stage (short-term and momentary impact) and in operation (long-term and permanent impact). Newly built railway lines, both conventional and HSR, will also be a source of noise in areas where they have not appeared before. Moreover, HSR will only be a passenger line, which will contribute to reducing automobile traffic. However, this phenomenon will not affect significantly the improvement of the acoustic climate near roads, because trucks produce the greatest noise. Considering this, it has to be stated that the acoustic climate in the neighborhood of a railway line will be worsened.

The negative impact of noise on people and animals may be ensured or limited through an appropriate investment location (outside protected by law areas) and through anti-noise protection measures (e.g. dampers, acoustic screens, sanding rails, new rolling stock).

The most frequently applied methods to limit railway noise on the road of propagation from the source to the receiver:

- acoustic screens;
- dampers;
- earth embankments;
- strips of vegetation;
- locating the tracks in a trench;
- organizational activities, e.g. grinding rails, mounting lubricators.

17.4.5.1.1. NOISE LEVEL LIMITS

Noise level limits in the environment caused by specific groups of noise sources, (excluding noise caused by aircraft taking off, landing, and flying), as well as electric power lines, and various kinds of land designated for protection, are included in the Annex to the

Table 17.1 Sound level limits [dB]

No.	Kind of terrain	Sound level limits in [dB]			
		Roads or railway lines		Remaining objects, facilities, and source of noise activity	
		L_{Aeq} D time interval reference equals 16 hours	L_{Aeq} N time interval reference equals 8 hours	L_{Aeq} D time interval reference equals least favourable consecutive 8 hours	L_{Aeq} N time interval reference equals least favorable night hours
1.	a) Protection zone "A" sanatoriums b) Hospital premises outside towns	50	45	45	40
2.	a) Development residential housing areas b) Development areas connected with permanent or temporary stay of children[1] c) Premises of public nursing homes d) Hospital premises in towns	61	56	50	40
3.	a) Residential multi-family areas b) Farm building areas c) Recreation areas d) Housing and services areas	65	56	55	45
4.	Land in the urban zone of cities of over 100 thousand inhabitants[2]	68	60	55	45

[1] In case of not using the terrains according to their function, sound level limit is not binding at night.
[2] Urban zone of cities over 100 thousand inhabitants – this is an area of compact civil administrative, commercial, and service objects. When a city has more than 100 thousand residents district, it is possible to determine a city zone if it is characterized by compact development with the concentration of administrative, commercial, and service objects.

Minister's of the Environment Regulation of 14 June 2007 on the noise level limits in the environment [9]. It defines noise level limits depending on its source, designation of the land, and time of day.

Pursuant to article 114 and 115 of the Act of Law on protection of the environment, the classification of specific categories mentioned in Article 113 of the Law should take place while preparing local spatial development plan. If there is not a plan like this, actual land use of the terrain is decisive.

17.4.5.2. Vibrations

Protection of people and objects against unfavorable impact of vibrations was defined in executive regulations to the Act of Construction Law [10]. According to Para. 326, item 1 of the Minister's of Infrastructure Regulation on technical conditions that buildings and their locations should meet [11],

> the level of noise and vibrations penetrating houses, multi-family dwelling and public buildings, excluding buildings for which it is required to meet protection against noise, cannot exceed limit values defined in Polish Standards regarding protection against noise in building and assessment of vibration impact on people in buildings.

The standards mentioned in the above-mentioned regulation include:

- PN-85/B02170 "Assessment of the harmfulness of vibrations passed by the ground to buildings";
- PN-88/B-02171 "Assessment of the impact of vibrations on people in buildings."

The standard of assessment of the impact of vibrations on people in buildings defines limit values of acceleration or vibrations in buildings of various purposes. The other standard defines methods of assessment of vibration impacts on construction and shows the effective value of vibration speed limit due to the possibility of different devices working. The estimation of vibration impact should be based on measurement results in bands divided in thirds or by using appropriate filters [6].

The basic way to maintain vibrations on a low level, similarly to noise emission, is to ensure a good state of tracks and rolling stock, and through sanding rails and exchanging deformed train wheels. The key element to minimize vibrations is an appropriate location of particular railway infrastructure elements (turnouts, switches), i.e. far away from protected areas.

Highly effective for eliminating vibrations from a railway line are padding systems, e.g. insulated rubber mats and sleeper pads. Sometimes the only solution for endangered buildings is to ensure a flexible support for foundations.

17.4.6. Emission of pollutants into ambient air

In recent years, there has been a change of global emission, which in turn brought about change of areas where improper air quality is found. The exceedance of air pollution in a smaller degree also appears near business entities close to city centers and towns with high traffic intensity [12].

The protection of ambient air is designed to ensure its best quality, in particular:

- maintaining levels of substances below the limit levels or at least on the same levels;
- decreasing the level of substances in the air to at least some limit when they are not met;
- decreasing and maintaining the level of substances in the air below the targeted levels.

The necessity to maintain levels of substance limits from the area of planned enterprise and its realization and exploitation refers to the following pollutants: sulphur dioxide, oxides and nitrogen dioxide, carbon monoxide, benzene, aliphatic and aromatic hydrocarbons, lead, and particle matters PM10 and PM2.5.

The emission of pollutants to the air is hard to define due to lack of information concerning the construction stages, kind of building machines, and specialized equipment. It should be assumed that building machines (vehicles) used at this stage will have current technical check-ups, whereas equipment used to realize the investment should meet the requirements regarding emission parameters in accordance with provisions of the Minister's of Economy and Labour Regulation, which concern detailed requirements for diesel engines regarding limiting emission of gaseous pollution and particulate matters by these engines [13].

The HSR line being built will be fully electrified; consequently, from the point of view of emission, the planned investment will not be a source of pollution emission. Emissions connected with railway operation and management will remain in power generation (power plants and CHP plants).

However, situations (not having the influence on the total volume of pollution) in which diesel engines will run on that line are not excluded. But it will be sporadic and connected with maneuverability and exploitation: shunting, snow clearing, emergency situations, etc.

HSR line functioning will also cause the decrease of dispersed emissions of other means of transport (mainly air and road transportation) for the sake of commercial power that is subject to restrictive emission measures. Taking this into consideration, it could be stated that HSR lines construction will have a positive influence on the quality of ambient air in Poland.

17.4.7. Impact on climate

Rail transport, including HSR, will ensure a high quality of passenger transport, at much smaller greenhouse gas emissions, in comparison to aviation and road transport.

It is predicted that the implementation and operation of HSR will be connected with greenhouse emission that has the biggest influence on climate changes.

During the construction works, short-term inconveniences may appear that are connected to greenhouse gas emission, mainly carbon dioxide. These disturbances will result from the process of fuel combustion in vehicle engines and machines used during the building period, mainly heavy building machines (bulldozers, loading vehicles, and truck transport). It has to be emphasized that these activities will have a temporary and transient character (with a ceasing of impact when the work is completed), as well as a relatively short duration.

Due to the HSR line construction, the decreasing of terrains that ensure the sequestration of carbon dioxide has to be taken into account. The loss of green areas may have a slight impact on the general balance of greenhouse gases. The final transformation of green areas will largely depend on design solutions, including locations.

The HSR line will also be an indirect source of greenhouse gas emission resulting from using electric power produced in power plants.

Shortening the journey time (in comparison with existing railway connections) and increasing safety and comfort during travelling will have an impact on the resignation of individual and collective passenger road transport. Taking over some passengers from car transport will have an influence on greenhouse gas emission from fuel combustion in motor vehicles.

In light of the foregoing, it is stated that the construction of HSR will have a positive impact on limiting climate changes from a long-term perspective.

17.4.8. Impact on surface and ground waters

The basic aim for surface (surface water bodies) and ground (groundwater bodies) waters is to reach and/or maintain the state approximate to the natural one. The most important ways to do this are to limit substances that pollute surface and ground waters and to restore (or maintain) a minimal good biological and qualitative state of waters.

The key element of HSR line assessment on the entire impact on particular water bodies is to define factors that could affect the ecological water state and then carry out an analysis if hydromorphological, biological, and physicochemical elements of waters could be changed. Such an approach will allow a full assessment of impact on particular elements of water conditions and will show if the environmental aim resulting from Water Framework Directive [14] established for given homogenous part of waters will be reached.

The provisions of the Water Framework Directive defining environmental aims for water bodies and conditions to use applications for deviating from these aims were transferred to Polish Water Law [15].

The analysis and review conducted by railway infrastructure managers of national investments showed that none of them influenced essentially the condition of surface and ground waters. Nevertheless, in order to define if the construction and later operation of HSR line will prevent reaching environmental aims, it is necessary to conduct a detailed assessment of water bodies [16].

Railway areas, pursuant to the Regulation of the Minister of the Environment concerning conditions that should be met to release sewage into the water or onto a land and substances particularly hazardous for water environment [17], do not constitute a source of sewage in the form of pollutant from snow and rain waters. It was also confirmed in research conducted by PKP Polskie Linie Kolejowe (PKP Polish Railway Lines), a railway infrastructure manager in Poland, regarding physicochemical parameters, i.e. total suspension and oil derivatives.

The tests included 100 water samples, characterized by various operational conditions: developed and non-developed properties, railway stations, railway section. None of the tested samples showed exceeding oil substances (maximum permitted content, i.e. 15 mg/l). For 79 samples, the petroleum hydrocarbons content was defined below the limit of quantification content, i.e. 0.1 mg/l. In the remaining samples, the petroleum hydrocarbons content amounted to 0.1–0.77 mg/l.

In the case of total suspended solids, in 94% of samples, no permissible value exceedances, which amounts to over 100 mg/l, were recorded. For 21% of trials, the content of suspension was below the lower limit of quantification (< 2 mg/l). Exceedances were identified in only six rainwater trials and occurred on track sections (along electrified double track line) and on railway stations

In the light of the foregoing, it can be stated that HSR after its construction will not be a source of pollution into waters, because waters discharged from railway tracks area will not be sewage but unpolluted rain or snowmelt waters. Nevertheless during the course of construction there could be a temporary impact on that water environment in connection with improper organization of a building site or ground works.

Rainwater drained to reconstructed trenches will have to meet the requirements of law. The Water Law Act [15] designates to water facilities the range of the amount of water discharged and methods to secure the banks and bottoms of these recipients in a way that would guarantee safe discharge of rain and snowmelt waters without harmful effect on adjacent lands.

17.4.9. Electromagnetic field emission

The main source of electromagnetic fields in the case of railway lines is the European Railway Traffic Management System (ERTMS). This system enables, among others, raising safety levels of railway traffic and increasing the liquidity and capacity of a railway line. The European Railway Traffic Management System consists of two systems:

- the European Train Control System – ETCS;
- the Global System for Mobile Communications-Railways – GSM-R.

The ETCS system, through sending information to the train driver's cabin from devices installed on the track, ensures a high safety of train driving. The most essential elements of this system are Eurobalises (installed in the track axis) and electronic controls (sending information between transmission devices and control command and signaling installations). This system is not a source of electromagnetic fields.

The GSM-R system is a digital radio communication designated both to provide voice communication and to ensure digital data transmission for the ETCS system. The main GSM-R equipment are the GSM-R base stations, which are a source of electromagnetic fields (antennas) [18].

The most important element in assessing electromagnetic field impact on the environment is establishing whether the main axis of radiation of particular GSM-R antennas is located in accessible space for a population.

In compliance with railway infrastructure manager's requirements imposed on contractors to cover the ERTMS/GSM-R system, the location and technical parameters of radio-telecommunications facilities should be selected to minimize the necessity to acquire the decision on environmental conditions, and so that the investment would not have a negative impact on objects of protection of Nature 2000 protected area, cohesion and system integrity of Nature 2000 and other protected areas [19].

The table shows data presenting values of equivalent isotropically radiated power and distances of locations accessible for the population from the electric center of the antenna. These values determine if the project could be included to projects that have always a significant impact on the environment (the need to carry out an assessment of the impact on the environment) or to projects that could potentially impact the environment significantly.

Moreover, pursuant to provisions of the Description of the Subject-Matter of the Contract [19], the location and parameters of base stations should guarantee that the limits of electromagnetic fields levels in the environment defined in the Regulation of the Minister of the Environment of 30 October 2003 on permissible levels of electromagnetic fields levels in the environment [20] will not be breached.

Table 17.2 Qualification of radio-communication, radio-navigation, radio-location (except radio lines) installations for projects that could potentially influence the environment significantly, pursuant to the Regulation of the Council of Ministers of 9 November 2010 on projects that could impact the environment significantly

Projects that could impact the environment significantly, requiring conducting proceedings in respect of environmental impact assessment			Projects not requiring conducting proceedings in respect of environmental impact assessment
Equivalent isotropically radiated power EIRP [W]	Projects for which EIA report is obligatory	Projects for which EIA report is optional	
	Distance from accessible for the population places from the electrical center of the antenna along the main axis of radiation of this antenna [m]	Distance from accessible for the population places from the electrical center of the antenna along the main axis of radiation of this antenna [m]	Distance from accessible for the population places from the electrical center of the antenna along the main axis of radiation of this antenna [m]
1	2	3	4
≥ 15 and < 100	–	≤ 5	> 5
≥ 100 and < 500	–	≤ 20	> 20
≥ 500 and <1000	–	≤ 40	> 40
≥ 1000 and < 2000	–	≤ 70	> 70
≥ 2000 and < 5000	≤ 100	> 100 and ≤150	> 150

Source: PKP Polskie Linie Kolejowe

It is also a standard that the railway infrastructure manager in the design stage requires that electromagnetic fields impact should be maximally reduced. Due to this, it can be stated that in most cases the ERTMS/GSM-R system on HSR line will not affect negatively the environment, including people's health and life.

A similar situation is likely to occur while solving the problem of power supply for HSR lines that will be based on 2×25 kV alternate current – AC (currently railway lines are supplied with 3 kV direct current – DC). This solution is characterized by a higher energy efficiency than a DC system – the ratio is 87% to 74% [21], but also a higher impact of electromagnetic fields. Nevertheless, the above-mentioned impact will largely depend, as it happens in case of ETCS/GSM-R systems, on proper location of power supply infrastructure with regard to areas accessible to people. Due to optimal solutions still in the design stage, it will be possible to a large extent to minimize the impact of electromagnetic fields.

17.4.10. Impact on historical sites and monuments

The execution of the investment, which HSR construction may be connected with, might have a risk impact on historical sites and monuments. A monument is described as a real property or movable property, their parts or sets that are a human creation or associated with human activity and revealing evidence of bygone era or event, whose maintenance is in the public interest due to its historical, artistic, or scientific values.

At the line of infrastructure investments, the most endangered will be immovable elements such as cemeteries, valuable town planning units, buildings, parks, or archaeological sites.

Protection of historical monuments in Poland is defined in legal regulations regarding monuments [21], by implementing the forms of protection consisting in:

- entering it into the monument register;
- awarding it historical monument status;
- establishing a culture park;
- setting up protection by listing it in the local development plan, deciding on the terms and conditions of the development, and setting up the location and permit for a public purpose investment, a railway line, the permit for a road works investment, and in regard to airports open to public use.

The most restrictive form of historical monuments protection is to enter it into the monument register, which puts that monument under the supervision of the voivodship conservator of monuments. Any works that could lead to damage of the monument or modification of its appearance require granting a permit by the voivodship conservator of monuments. The demolition of the monument is possible only in the case of loss of historical, scientific, or artistic value, either due to the protection of human health and life, important interests of the state, or in order to avoid damage to the national economy.

In the case of historical monuments, there are no special duties and demands concerning them, because these objects/facilities had to enter the monuments register earlier or should be protected as a culture park. Therefore, duties and requirements to such objects/facilities result from conditions referring to the two above-mentioned forms of protection [22].

The next form of monuments' legal protection is to create a culture park, on the premises of which could be set up bans and limitations regarding conducting activity there.

The last form of monuments' legal protections is setting up protection by listing it in the local development plan.

The most important way of limiting negative impact on monuments will be to take into consideration the location of object/facilities of historical status while demarking of the routing of an HRS line. It is assumed that those monuments located in the range of works, in the direct neighborhood or 50 m distance from the railway line, are most exposed to impact from the HSR line.

17.4.11. Generation of waste

Building high-speed rail and its later operation will be connected with generating waste. Pursuing the project of HSR line will generate waste that will not appear when the construction is abandoned. In principle, the stage of building will be largely connected with generating waste from the construction sites and repairs and demolition of buildings and road infrastructure (including excavated soil from contaminated sites).

In the course of a railway line operation, generating waste will be mostly connected with carrying out the works of railway and technical infrastructure maintenance, as well as clean-up work on the premises of internal transport routes (maneuver areas, car parks, and access roads), platforms, pedestrian tunnels, and footbridges generating municipal waste including separately collected fractions.

All activities regarding waste economy emerging in the construction and operation stage, including ways to minimize the impact on the environment, must comply with the provisions of the Act of Law on waste [23].

The basic principle of sustainable waste economy is to undertake actions aimed at limiting the volume of generated waste and using the waste as a substrate of another process in the investment area.

Conducting a responsible waste economy compliant with the law regulations, taking into account limiting the volume of waste, will reduce the impact on the environment to a negligible level and will be limited to a short-term impact on particular sections of works. This impact will be connected with occupancy of the ground surface of the investment premises where the waste is going to be temporarily stored.

17.4.12. Aggregated impact

The accumulation phenomenon consists in showing significant kinds of impact in all stages of the investment stemming from HSR lines together with defining their range and conducting identification of other sources of impact of similar character which are being or have been carried out, located in the area where the projects are planned to be executed and in the area of the project's impact or whose impacts are located in the area of the planned project's impact – in the range that their impacts can lead to aggregate impacts with the planned project. Taking into account the impact mentioned above, there should be analyzed phenomena that could have the biggest influence on the environment such as terrain transformations, areas' urbanization, infrastructure solutions, changes in climate and wind conditions, changes in water conditions, and emergency situations. In case of a railway line, the most frequent impact aggregated with other projects is noise coming from road infrastructure.

17.4.13. Transboundary impact

Transboundary impact should be understood as any impact not having a global character in the area subjected to jurisdiction of The Party, caused by planned activity whose physical reason is entirely or partly located on the area undergoing jurisdiction of the Other Party [24]. Assuming this interpretation, in order to conclude the transboundary impact, there should occur indications showing that a given investment may affect the areas of neighboring country. Such impacts include interference in protected areas, ecosystems, pollution emission, and migration barriers.

Due to limited knowledge concerning the location of the investment (in particular as regards the possibility of extending the HSR line to the state borders), it is difficult to define explicitly a potential character and scale of impact that may appear beyond the territory of Poland. It seems that currently there are no indications to find a significant negative impact of transboundary character. All possible identified threats will rather have a local character that will not affect particular environment components in a physical way.

17.4.14. Serious accidents and breakdowns

In compliance with the regulations of the Environmental Protection Law [2], a serious accident or a breakdown are understood as an event, in particular emission, fire, or explosion that occurred in the course of an industrial process, storage, or transport where appears one or more hazardous substances leading to the immediate threat of life or health of people or the environment or appearance of such a threat with delay.

Taking into account the fact that high-speed rail will not be used for transporting hazard-ous materials but only for passenger transport, the threat of serious accidents and breakdowns will not appear.

17.5. Conclusions

The main aim of high-speed rail in Poland is to create an alternative and competitive mode of transport in comparison to road and air transport, which is currently faster and more friendly than rail transport. A measurable effect of undertaking HSR construction will be the growth in the share of rail passenger transport and, consequently, reduction of pollutant emission in the entire passenger transport system.

Building a new railway line means the necessity to occupy new lands for investment. This will be connected with new railway and associated infrastructure (engineering objects and facilities, access and technological routes) in the landscape, as well as may separate important areas important for communities or valuable natural areas (e.g. migration corri-dors, protected areas). Moreover, the HSR line will become a new source of emissions into the environment, including, inter alia, noise.

As it was mentioned in previous sections of this chapter, identified impacts should be reduced to the limits pursuant to the law regulations and also accepted by society. The impacts generated by the new investment could be limited in the course of operation by applying solu-tions protecting the environment: acoustic screens, animal bridges. However, the biggest influ-ence on the final impact volume will have an appropriate demarking of the HSR line route.

Additionally, it has to be taken into account that together with the appearance of a new investment, there may break out social conflicts connected with other impacts or simply resulting from interference into an existing harmony of life.

However, it must be taken into consideration that also some positive impact will result from the HSR construction. The biggest benefit will be to create a public purpose investment significantly improving the quality of travel between the biggest Polish cities. Due to all the resources necessary to ensure proper functioning of the HSR line, many jobs in the railway sector itself and in other industries (e.g. in collective road transport, in order to provide a bet-ter transport in places neighboring the HSR line) will be created. Furthermore, the HSR line will contribute to taking over a numbers of passengers from individual and collective road transport, which will bring measurable advantages, inter alia, reducing road traffic (improv-ing safety) and greenhouse gas emissions to the environment.

Bibliography

[1] Tomaszewski, F., and Wojciechowska, E.: *Transport kolejowy a ochrona środowiska [Rail Trans-port – Environmen Protection]*, Biblioteka Cyfrowa Politechniki Krakowskiej, Kraków, 2011.
[2] The Polish Act of Law of 21 April 2001 on the Environmental Protection Law [Ustawa Prawo ochrony środowiska z dnia 21 kwietnia 2001]. *Journal of Laws*, 2016, item 672.
[3] The Polish Act of Law of 3 October 2008 on the provision of information on the environment and its protection, community's participation in environmental impact assessments [Ustawa o udostępnianiu informacji o środowisku i jego ochronie, udziale społeczeństwa w ochronie środowiska oraz o ocenach oddziaływania na środowisko z dnia 3 października 2008 r.]. *Journal of Laws*, 2016, item 353.
[4] The Regulation of the Council of Ministers of 9 November 2010 on projects which may sig-nificantly affect the environment [Rozporządzenie Rady Ministrów w sprawie przedsięwzięć

mogących znacząco oddziaływać na środowisko z dnia 9 listopada 2010 r]. *Journal of Laws*, 2016, item 71.

[5] Chief Inspectorate of Environmental Protection. Report 2014 [Główny Inspektorat Ochrony Środowiska, Stan środowiska w Polsce. Raport 2014], Warszawa, 2014.

[6] Ministry of Infrastructure, Forecast of High-Speed Rail Construction and Service Impact on the Environment in Poland [Prognoza oddziaływania na środowisko Programu budowy i uruchomienia przewozów kolejami dużych prędkości w Polsce], Warszawa, 2008.

[7] The Nature Conservation Act of 16 April 2004 [Ustawa o ochronie przyrody z dnia 16 kwietnia 2004 r.]. *Journal of Laws*, 2016, item 2134.

[8] Raczyński, J.: Rządowy program budowy linii dużych prędkości w Polsce [Government High-Speed Rail Construction Programme], *"Technika Transportu Szynowego"* 2008, 9.

[9] The Minister's of the Environment Regulation of 14 June 2007 on the noise level limits in the environment [Rozporządzenie Ministra Środowiska z dnia 14 czerwca 2007 r. w sprawie dopuszczalnych poziomów hałasu w środowisku]. *Journal of Laws*, 2014, item 112.

[10] The Act of Construction Law of 7 July 1994 [Ustawa Prawo budowlane z dnia 7 lipca 1994 r.]. *Journal of Laws*, 2016, item 290.

[11] The Minister's of Infrastructure Regulation of 12 April 2002 on technical conditions that buildings and their locations should meet [Rozporządzenie Ministra Infrastruktury w sprawie warunków technicznych, jakim powinny odpowiadać budynki i ich usytuowanie z dnia 12 kwietnia 2002 r]. Journal of Laws, 2015, item 1422.

[12] Hajto, M., Kacprzyk, K., Kamieniecka, J., Karaczun, Z., Kassenberg, A., Kędra, A., Rąkowski, G., Wójcik, B., and others., Rzeszot, U.: Prognoza oddziaływania na środowisko projektu Programu Operacyjnego Rozwój Polski Wschodniej na lata 2007–2013 [The Forecast of the Impact on the Environment in the Operational Draft Programme of Eastern Poland's Development for 2007–2013], published by the Ministry of Regional Development, Warszawa, 2007.

[13] The Minister's of Economy and Labour Regulation of 30 April 2014 concerning detailed requirements for diesel engines regarding limiting emission of gaseous pollution and particulate matters by these engines [Rozporządzenie Ministra Gospodarki w sprawie szczegółowych wymagań dla silników spalinowych w zakresie ograniczenia emisji zanieczyszczeń gazowych i cząstek stałych przez te silniki z dnia 30 kwietnia 2014 r.]. *Journal of Laws*, 2014, item 588,

[14] Directive 2000/60/EC of the European Parliament and of the Council setting out framework for Community action in the field of water policy of 23 October 2000 [Dyrektywa 2000/60/WE Parlamentu Europejskiego i Rady ustanawiająca ramy wspólnotowego działania w dziedzinie polityki wodnej z dnia 23 października 2000 r.]. *Journal of Laws*, L 327, 22.12.2000, p. 1.

[15] The Water Law of 18 July 2001 [Ustawa Prawo wodne z dnia 18 lipca 2001 r.]. *Journal of Laws*, 2015, item 469.

[16] Hobot, A.: Ekspertyza dotycząca sposobu realizacji zaleceń Dyrektywy 2000/60/WE Parlamentu Europejskiego i Rady z dnia 23 października 2000 r., ustanawiającej ramy wspólnotowego działania w dziedzinie polityki wodnej w projektach kolejowych PKP Polskie Linie Kolejowe S.A. planowanych do realizacji w latach 2014–2020 [Expert work concerning implementation of recommendations of Directive 2000/60/EC of the European Parliament and of the Council setting out framework for Community action in the field of water policy in PKP Polskie Linie Kolejowe S.A. railway projects planned for 2014–2020], Gliwice, 2015.

[17] Regulation of 18 November 2014 of the Minister of the Environment concerning conditions that should be met to release sewage into the water or onto a land and substances particularly hazardous for water environment [Rozporządzenie Ministra Środowiska w sprawie warunków, jakie należy spełnić przy wprowadzaniu ścieków do wód lub do ziemi, oraz w sprawie substancji szczególnie szkodliwych dla środowiska wodnego z dnia 18 listopada 2014 r.]. *Journal of Laws*, 2014, item 1800.

[19] Dyduch, J., and Cholewa, A.: Kierunki działań po katastrofie kolejowej pod Szczekocinami [Direction of Actions After a Railway Disaster at Szczekociny], *"Technika Transportu Szynowego"* 2013, nos. 2–3.

[20] PKP Polskie Linie Kolejowe, S.A.: Opis Przedmiotu Zamówienia dla zadań: na wykonanie projektów, pozyskanie decyzji administracyjnych oraz wykonanie planowania radiowego dla podsystemów radiowych systemu ERTMS/GSM-R realizowanych w ramach projektu POIiŚ 7.1–36.2 „Budowa infrastruktury systemu GSM-R na liniach kolejowych zgodnych z harmonogramem NPW ERTMS, FAZA I – PRACE PRZYGOTOWAWCZE" [Description of the Subject of the Contract for tasks to carry out projects, acquire administrative decisions and performing radio planning for *ERTMS/GSM-R* radio sub-systems conducted *within POIiŚ 7.1–36.2 project „Building the GSM-R system infrastructure on railway lines in compliance with NPW ERTMS schedule, PHASE 1 – PREPARATORY WORK*," https://zamowienia.plk-sa.pl.

[21] The Minister's of the Environment Regulation of 30 October 2003 on permissible levels of electromagnetic fields levels in the environment and measures to check meetings these levels [Rozporządzenie Ministra Środowiska z dnia 30 października 2003 r. w sprawie dopuszczalnych poziomów pól elektromagnetycznych w środowisku oraz sposobów sprawdzania dotrzymania tych poziomów]. *Journal of Laws*, 2003, 192, item 1883.

[22] PKP Polskie Linie Kolejowe S.A. Biuro Linii Dużych Prędkości [Office for high-speed lines], Program budowy linii dużych prędkości w Polsce. Uwarunkowania społeczne i ekonomiczne [Programme of High-speed Lines Construction in Poland. Social and Economic Conditions], Warszawa, 2010,

[23] The Act of Law of 23 July 2003 on the protection and care of monuments [Ustawa o ochronie zabytków i opiece nad zabytkami z dnia 23 lipca 2003 r.]. *Journal of Laws*, 2014, item 1446.

[24] Ministry of Infrastructure and Development: *Prognoza oddziaływania na środowisko dla Dokumentu Implementacyjnego do Strategii Rozwoju Transportu do 2020 r. z perspektywą do 2030 r* [The forecast of impact on the environment for Implementation Document to Transport Development Strategy to 2020 with the perspective to 2030], Vol. I, Warszawa, 2014.

[25] Act of Law on waste of 14 December 2012 [Ustawa o odpadach z dnia 14 grudnia 2012 r.]. *Journal of Laws*, 2016, item 1987.

[26] The Espoo Convention on Environmental Impact Assessment in a Transboundary Context of 25 February 1991 [Konwencja o ocenach oddziaływania na środowisko w kontekście transgranicznym, sporządzona w Espoo z dnia 25 lutego 1991 r.]. *Journal of Laws*, 1999, 96, item 1110.

Chapter 18

Personnel education for the high-speed rail needs

Wojciech Wawrzyński

18.1. Introduction

According to [12], the high-speed rails (HSR) constitute a subsystem of the railway passenger transport characterized by a much higher commercial speed of the trains in comparison with the remaining types of railway transport. A problem of adaptation of railways to the increased speed in the national literature was noticed a long time ago. Monograph [7] gives evidence of it.

In Poland, the appearance of the HSR concerns as for now a certain perspective; however, an initiated increased speed of the Pendolino type of trains already in operation and a perspicuous trend of the Polish rolling stock producers aiming to become a player in this field with their own products create a real demand for engineers educated in the Transport and HSR specializations field of study.

The education of transport engineers in the field of high-speed rails should embrace five years anticipation in the perspective of their employment in the studied profession, whereas obtaining a Master of Science extends this period to seven years.

A development in the increase of admissible speed of the railway traffic is conditioned by multiple processes which occur in parallel to each other. By this is intended an improvement of the technical railway lines infrastructure – including railway superstructures construction, rail vehicles construction, and in particular the process of replacing the conventional trains, together with the traction units with the train sets, the improvement of railroad infrastructure concerning the signaling systems and assuring the safety of railway traffic, as well as changes in the traffic organization. Those issues should be taken into consideration during the educational process.

18.2. Formal determinants of education

Education regarding the specialization in high-speed trains should be conducted during studies of the first degree of a general academic profile of the Transport field of study.

Studies in the field of Transport include learning about transport systems, management systems, and means of transport in the aspect of their construction, use, and safe and effective exploitation. Technologies of transportation are the subject of inquiries. Special attention is paid to the signaling and traffic management systems in this field. As a result of the education, students gain abilities regarding modelling, construction, modernization, and rules of exploitation of transport systems.

In particular, graduates of the studies of the first degree in the field of Transport have knowledge of and abilities in forecasting and planning the development of transport networks, together with the ability of assessment of demand for transport services, shaping the infrastructure and planning of the transport networks, selection of transport means and technologies of transport accordingly to the demanded tasks, automatization of processes in the transport systems, management and control of the traffic processes, projecting and applying of the telecommunication and telematics systems in transport, construction and application of intelligent transport systems, examination and evaluation of a reciprocal interaction of technical means of transport with the infrastructure and the environment, designing of a functional technical transport background, and examination and evaluation of infallibility and safety of the transport systems. According to the rules of the National Qualifications Framework for higher education [9], the effects of education were defined for technical science. They embrace 11 effects in the field of knowledge, 16 in the field of abilities (including six in the field of general abilities that are not connected with the engineering field, six basic engineering abilities, and four abilities related directly to solving engineering issues), and seven in the field of social competences.

The effects of education in the field of knowledge regard mainly the basic subjects for an engineer, such as mathematics or physics, and the key issues regarding the studied field, development trends in the science fields and scientific disciplines that are crucial for the studied field, knowledge of engineer techniques in selection of methodology of solving separate issues together with selection of tools and materials. Furthermore, they regard an essential knowledge of management, including quality management.

The effects of education in the field of abilities include the means of obtaining information from different data sources and ways of their use, the abilities of communication in the professional environments, the abilities of preparing speeches and their presentation, as well as having the habits of self-education.

The engineering abilities are concentrated around planning and conducting experiments together with measurements and computer simulations as well as a correct interpretation of results. Ability of a system view on the problems under resolution is important.

Abilities related directly to the resolution of engineering tasks include critical analysis of existing technical solutions, identification and specification of engineering tasks, and evaluation of usefulness of routine methods and tools which serve to solve practical tasks.

The effects of education in the field of social competences constitute a sphere of a general approach to perceive the reality by a technician. In particular, it relates to the fact of a constant need of knowledge supplementation (a life-long education), an understanding of out-of-tech aspects of engineering activity, the ability to co-act in a group, the defining of priorities of engineering activity, and the presentation of opinions in a commonly comprehensive way.

The field of study of Transport is directly related to a scientific discipline of Transport as well as to other scientific disciplines that pertain to different fields of education, such as mathematics, informatics, economy, management science, mechanics, construction and exploitation of machines, telecommunications, electronics, electrotechnics, energetics, automatics and robotics, construction, geodesy and cartography, materials science, environmental protection, and ecology.

The effects of education for the Transport field of study have to cover the above-mentioned effects of education for the technical science in all of the three areas, i.e. in the field of knowledge, abilities, and social competences. A graphic illustration of this coverage is a matrix of conformity of the educational effects (Figure 18.1).

Presented in Figures 18.1 and Table 18.1 effects of education in the field of education of the Transport were marked with the "Tr1A" symbol (numeral "1" meaning the studies of the first degree and the letter "A" a general academic profile). Subsequently, a letter informing about the area of the educational effect is added, i.e.:

W – knowledge, U – abilities, K – social competences. After those symbols, a double-digit number is added that represents a subsequent number of an educational effect in a defined area of education.

The effects of education for the Transport field of study are elaborated by a Commission (or a Team) for the National Qualifications Framework operating by a Faculty Board which conducts studies in the field of Transport.

The effects of education for the Transport field of study are beyond the sphere of generality and orient knowledge, abilities, and competences towards areas closer to the needs of transport. It concerns e.g. transport infrastructure, technical means of transport, transport systems, movement engineering, technical exploitation, telematics of transport, transport economics, safety, ecology, and others.

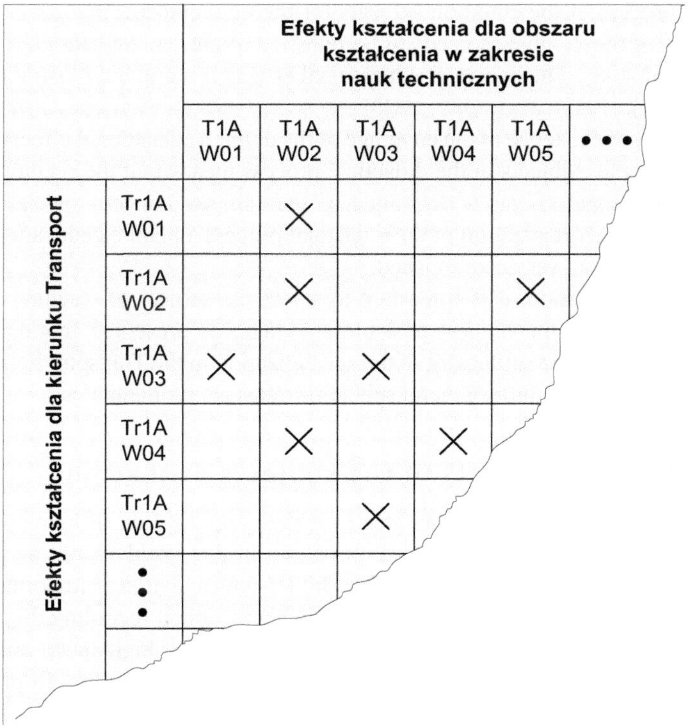

Figure 18.1 Fragment of compatibility matrix of the educational effects for technical science with the effects of education for the Transport field of study; X – compatibility

Table 18.1 Fragment of a matrix of the educational effects, which illustrates a coverage of the defined effects of education in the Transport field of studies with exemplary subjects; X – coverage existence

Subjects	Effects of education in the field of Transport				
	TrIA_W01	TrIA_W02	TrIA_W03	TrIA_W04	· · · · · ·
Railway vehicles and traction		X			...
European Train Control System				X	...
Railway road infrastructure			X		...
....	...	...	...	...	...

Within abilities e.g. there is accepted a fluency in describing a system of exploitation of a technical object, taking into consideration its maintenance and readiness together with an analysis and a synthesis of its infallibility structure.

Reaching effects of education in the field of Transport is possible thanks to realization of a didactic process of the common subjects for the field and specific subjects for the specialization. Any of those subjects needs to be reflected in chosen educational effects for the Transport field of study. The matrix of educational effects is a graphic illustration of this reflection (Table 18.1).

An ascertainment of existence of the coverage of the defined educational effects throughout realization of separate subjects in the didactical process means that the accomplishment of the expected educational effects was granted. If the subjects concern specialization, it means that throughout their realization in the didactical process a planned educational effect related to this specialization was accomplished.

For the HSR specialization, it is important to select the subject of teaching and their contents in a way enabling the highest possible coverage, as presented fragmentarily in Table 18.1. It guarantees a deep settlement of the specialization in the Transport field of study and links this specialization with the general sphere of transport problematics.

18.3. The outline of contents of the subjects of teaching

As far as the contents of teaching are concerned for the specialization of high-speed rails, the legal conditions of HSR construction in Europe and in the world should be presented in the part concerning general knowledge. Within the European area, it is important to take into consideration the applicable directives of the European Council regarding this problem. An economical aspect is also important, justifying construction of high-speed railws, their proper use, cost analysis of exploitation, and balance of advantages resulting from their use for particular regions [5].

Within the subject *Infrastructure of Railway Road*, there should be taken into consideration the problems of shaping of HSR lines course and the prerequisites of their determining, shaping of geometrical lines if HSR, radiuses of curvature, designing of track cant and its surplus, transition spirals, and shaping track systems [1], [6]. The analysis of mechanisms of impact of vehicles on railway superstructure and the environment of the HSR lines is important. The problems of construction of railway superstructures (rails, fastenings, sleepers, ballast), unconventional superstructures, HSR lines substructures, and other structures are worth taking into account. A separate problem regards a vibroacoustic impact and environmental protection on the HSR lines [13].

What is especially important in the subject *Traffic Management System in the High-speed Trains* is to present traffic management with use of combined signaling as well as knowledge about European Railway Traffic Management System (ERTMS) and the European Train Control System (ETCS), ETCS levels, and the basic systems' functions, as well as balises – their functions and configurations, radio block systems (RBC), the concept of reference position, and permission to drive based on the rule of block section. Knowledge of onboard and track-side ECTS devices [3], [14] is also important.

In the subject *Rolling Stock for the High-Speed Railways*, what shall be taken into account are the specific characteristics that differentiate this type of means of transport from conventional solutions. Presentation of the driver's cab equipment and techniques of its use during realization of procedures and available functions of traffic management is crucial. Apportionment of propulsion alongside the train and the brakes – their construction and functioning, realization of automatic braking – constitute important components of knowledge about the HSR rolling stock. Knowledge about computer system of train management [8] is also necessary.

In the subject *Railway Telecommunication for HSR*, especially important is a presentation of the railway's digital network, GSM-R. Learning the architecture of this network, its functionality, types of services, transmission safety, as well as functions of separate devices of the digital radio data transmission [4], [11] is important.

In the subject *Exploitation and Maintenance of HSR*, special attention shall be given to the technical background of rolling stock maintenance, realization of reviews and their types, repairs, and organization, and the system of ensuring the quality of rolling stock maintenance. In issues related to railway road, the superstructure maintenance, diagnostics of track state, technology and organization of the process of the HSR line maintenance [2] is essential.

The importance of the organization of personnel education, which anticipates in time an extension of HSR even on separate sections, may be shown from the British example.

In Great Britain, there is an extension of HSR planned for a so-called High-speed 2 (HS2) section from London to Birmingham (224 km). Those plans are anticipated by localizing in this town a railway academy called *High-speed Rail College* [17], which will be preparing personnel for service of this railway section. It was stated that the number of engineering staff educated until then in the British academies is not sufficient for so developed an HSR network.

The new academy will be conducted by employers of the engineering and railway sectors, and will fulfill the standards in the teaching field based on a modern technology and use of objects of technical HSR background. This educational institution will ensure the specialized professional trainings for the next generation of engineers working on the HS2.

A certain base for the educational concept of the *High-speed Rail College* was a conference on high-speed rails [15], which had taken place at the University of Birmingham on the occasion of the 50th anniversary of the creation of the first high-speed line – Shinkansen in Japan. It is interesting to trace some of the issues undertaken during this conference. Among them were infrastructure, railway superstructure, construction and stability of land bridges, modelling and measurements of critical effects arising during the critical speed of train, esthetics and functionality of high-speed rail stations, remote monitoring of the railway substructure state, modelling of the ground vibrations, power supply, aerodynamical noise from high-speed trains, factors influencing the capacity limitations, and dynamic safety monitoring.

As may be seen, the problems troubling a great majority of modern scientists working in the HSR field are coherent with the proposed earlier outline of program content of the specialization, which confirms that the choice of issues presented there was right.

A need to popularize knowledge about HSR enforces actions of an academy such as organization of distance learning. There is a good example in one of the American universities, which organized a teaching course within distance learning regarding *Introduction to the High-Speed Railways* [16]. The education is divided into modules. The module *Concepts, Definitions and General Construction of the HSR System* includes a general introduction to the specifics of these railways. A special emphasis is put in this module on the difference between HSR and conventional solutions. There are discussed the subsystems and the component parts of these subsystems. The module *History, Advantages and Vision of Development of High-Speed Railways* includes an introduction describing the examples of construction of these railways as well as a vision of development in this field.

It embraces also the national plans regarding modernization of existing transport corridors as well as plans of construction of the new ones. HSR module titled *Demand, Levels of Services and Participation in the Market* enables a formal perspective to be formulated on the evaluation of functioning and optimization of the railway performance parameters.

Organization of the distance learning within HSR confirms the existence of demand for popularization of knowledge about this kind of railway passenger transport. It is a positive proof of the increase of technical consciousness of societies in the field of the modern transport.

18.4. Summary

This chapter presented the formal conditions of education on the specialization of high-speed rails within the Transport field of study at the academic level. Moreover, an outline of educational content was presented, realized in the particular subjects of this specialization.

The educational process has its rights. First, it requires time; therefore, the preparatory actions of the new specialization should be undertaken with a proper prediction enabling a possibility of calm further realization of the intentions.

Second, it requires a properly prepared teaching staff as well as a laboratory background of the academies at its disposal, and an active role should be played by producers and exploiters acting in the field of high-speed rails.

Third, program contents, especially of the specialized subjects, have to be corrected continuously because of the changing reality of the transport systems.

The chosen examples of the foreign centers presented confirm the importance of education in the specialization of high-speed rails and its impact on the development of the transport networks using this type of railway transport.

Bibliography

[1] Bałuch, H., and Bałuch, M.: *Układy geometryczne toru i ich deformacje*. Warszawa: Kolejowa Oficyna Wydawnicza, 2010. [Bałuch, H., and Bałuch, M.: *Track Layouts Geometry and Their Deformations*. Warszawa: Kolejowa Oficyna Wydawnicza, 2010].

[2] Bogdaniuk, B., and Towpik, K.: *Budowa, modernizacja i naprawy dróg kolejowych*. Warszawa: Kolejowa Oficyna Wydawnicza, 2010. [Towpik, K.: *Construction, Modernization and Repairs of Railway Roads*. Warszawa: Kolejowa Oficyna Wydawnicza, 2010].

[3] Dąbrowa-Bajon, M.: *Podstawy sterowania ruchem kolejowym*. Warszawa: Oficyna Wydawnicza Politechniki Warszawskiej, 2014. [*Basics of the Railway Traffic Management*. Warszawa: Oficyna Wydawnicza Politechniki Warszawskiej, 2014].

[4] Gago, S.: *Niektóre problemy praktyczne występujące w układach sterowania i telekomunikacji KDP*. Logistyka 3/2012, ISSN 1231–5478, str. 583–592. [*Several Practical Problems Occurring in the Control an Telecommunications Systems of HSR*. Logistyka 3/2012, ISSN 1231–5478, str. 583–592].

[5] Gorlewski, B.: *Kolej dużych prędkości: Uwarunkowania ekonomiczne*. Warszawa: Oficyna Wydawnicza Szkoła Główna Handlowa w Warszawie, 2012. [*High-Speed Raiwlays: Economical Conditions*. Warszawa: Oficyna Wydawnicza Szkoła Główna Handlowa w Warszawie, 2012].

[6] Massel, A.: *Projektowanie linii i stacji kolejowych*. Warszawa: Kolejowa Oficyna Wydawnicza, 2010. [*Design of Lines and Railway Stations*. Warszawa: Kolejowa Oficyna Wydawnicza, 2010].

[7] *Przystosowanie kolei do zwiększonych szybkości i dużych przewozów*. Praca zbiorowa. Wydawnictwo Komunikacji i Łączności, Warszawa, 1969. [Adjustment of railways to the increased speeds and big traffic. Collective work. Wydawnictwo Komunikacji i Łączności, Warszawa, 1969]

[8] Rojek, A.: *Tabor i trakcja kolejowa*. Warszawa: Kolejowa Oficyna Wydawnicza, 2010. [Rojek, A.: *Rollind Stock and Railway Traction*. Warszawa: Kolejowa Oficyna Wydawnicza, 2010].

[9] Rozporządzenie Ministra Nauki i Szkolnictwa Wyższego z dnia 2 listopada 2011 roku w sprawie Krajowych Ram Kwalifikacji dla Szkolnictwa Wyższego. [Regulation of the Minister of Science and Higher Education of the 2 November 2011 regarding the National Qualifications Framework for the higher education].

[10] Sanczewicz, S.: *Nawierzchnia kolejowa*. Warszawa: Kolejowa Oficyna Wydawnicza, 2010. [Sanczewicz, S.: *Railway Superstructure*. Warszawa: Kolejowa Oficyna Wydawnicza, 2010].

[11] Siergiejczyk, M.: *Wybrane zagadnienia systemów sterowania ruchem i łączności dla Kolei Dużych Prędkości w Polsce*. Logistyka 3/2012, ISSN 1231–5478, str.1991–2022. [Siergiejczyk, M.: *Chosen Aspects of Railway Signaling and Communication Systems for the High-Speed Trains in Poland*. Logistyka 3/2012, ISSN 1231–5478, str.1991–2022].

[12] Słownik pojęć strategii rozwoju transportu do 2020 roku. Ministerstwo Transportu, Budownictwa i Gospodarki Morskiej, Warszawa, 2014. [The dictionary of notions of the transport development strategy until 2020 roku. Ministerstwo Transportu, Budownictwa i Gospodarki Morskiej, Warszawa, 2014].

[13] Towpik, K.: *Koleje dużych prędkości. Infrastruktura drogi kolejowej*. Warszawa: Oficyna Wydawnicza Politechniki Warszawskiej, 2012. [Towpik, K.: *High-Speed Railways: Railway Road Infrastructure*. Warszawa: Oficyna Wydawnicza Politechniki Warszawskiej, 2012].

[14] Żurkowski, A., and Pawlik, M.: *Ruch i przewozy kolejowe: Sterowanie ruchem*. Warszawa: Kolejowa Oficyna Wydawnicza, 2010. [Żurkowski, A., and Pawlik, M.: *Railway Traffic and Transport: Movement Management*. Warszawa: Kolejowa Oficyna Wydawnicza, 2010].

[15] www.birmingham.ac.uk/research/activity/railway/events/high-speed-rail-conference.aspx – Page Regarding the Conference for the 50th Anniversary of Construction of the Shinkansen Railways, March 2015.

[16] www.rail-learning.mtu.edu/ – Page Regarding Distance Learning in the Subject of HSR, March 2015.

[17] www.tunneltalk.com/Education-and-Training-15Oct2014-UK-High-Speed-Rail-College-to-be-built.php – Page Regarding Construction of High-Speed Rail College, March 2015.

Chapter 19

From backwardness to modernity

High-speed rail – the strategic element
of national program for development of
rail transport – strategic and political
considerations

Marek Bartosik and Sławomir Wiak

19.1. Problem overview

High-speed rail (HSR) is a means of public transit offering service quality not available to other means of land transport. Safety and reliability of HSR, exceeding all other means of transport, allow for the creation of reliable and timely intercity links. From the viewpoint of national development strategy and policy, in transport and economy as well as in foreign policy, HSR cannot be construed as competition to the network of highways and expressways. Such an approach, only understood in lobbyist promoting of road transport, is not acceptable in political circles responsible for the fate of the nation and its position in a continental and global perspective [1], [3].

One of the gravest errors in Poland's rail transport policy is breaching the principle of sustainable development of road and railway transport, due to prolonged domination of interests of fuel-car-road sector, having overwhelming effect of adopted economic and political solutions in transport sector (sec. 19.2.2.2; Figures 19.17–19.19) [1], [2].

Prior experience of EU-15 countries demonstrates that good functioning of national economies is not possible without efficient interregional railway transport (RLT). The European Union is presently the place where national railways are becoming more integrated over time, consistently converted into elements of the Trans-European RLT System [1], [3].

Rail transport is one of the weaknesses of Poland's economy. The time it will take for Polish railways to catch up is estimated at 30 years. This is a strong barrier in the development of the society and economy on both national and regional levels.

Ambitious development projects for individual cities and regions undertaken by local government are very often competitive but inefficient without strong communication links with other cities and regions of the country (sec. 19.2.2.1).

Upgrade or revitalization works on railway lines allow an increase in the operation speed of trains sufficient for average-class intercity connections. In practice, only large metropolitan areas, such as Warszawa, Gdańsk, Wrocław, or Kraków, receive preferential treatment. Condition precedent for qualitative change in the current situation of Poland's RLT is restoration and acceleration of all works on Poland's HSR, in particular on the so-called Y line, complementing the upgraded CMK line. In combination with commuter railways (CR), this shall improve nationwide integration of rail transport, especially in terms of commuter rail [1], [45].

Demand for high-quality rail transport in Poland is sufficiently high (to justify investments). As per governmental projections concerning construction and launching of HSR, by the year 2030 the volume of railway transport can increase nearly five-fold [2].

EU plans for the establishment of a uniform European Rail Transport Area also strongly stimulate the development of Poland's HSR. A modern and efficient railway network is indispensable for the country's development and drawing benefits from its location as a transit area. Abandoning the construction and implementation of modern rail transport will cause marginalization of rail transport in Poland and marginalization of Poland in Europe.

19.2. Condition of Poland's railway transport[1] (RLT)

19.2.1. Poland's RLT and HSR from a continental perspective

19.2.1.1. Poland's RLT vs TEN-T core network and rail traction systems

Building a highly competitive Trans-European RLT system means for carriers from individual member states the need to improve both the quality and competitiveness of their transport services. For over 20 years,[2] the process has been determined by consecutive decisions by the EU with regard to the network (TEN-T)[3] concept, revised to a great extent following EU expansion in 2004 (UE 25) and 2007 (UE 27). A greater than two-fold increase (from 14 to 30) was observed then in the number of transport-related projects [4].

Consecutive general revision of TEN-T policy (2009) increased the EU stress on definition of the core network covering major intra-EU links and cross-border EU links, while keeping the principle of maintenance and improvement of the existing networks of roads, railways, inland waterways, sea ports, and airports through their upgrade. In plans, the core network was supposed to substitute 30 earlier priority projects. Strategic objectives of that policy are territorial, economic and social cohesion, and improvement in global competitiveness of EU, strengthening of internal market, sustainable development of regions, improved mobility of people and goods, the well-being and security of citizens, and environmental protection (climate, reduction in CO_2, contamination, etc.). Actions taken have to conform to the assumptions of the "Europa 2020" strategy, developed following the failure of the Lisbon Strategy. The principle of EU transport policy continuity is still in force [1].

The multilateral debate on the form of TEN-T has been going on for 6 years now. Among others, the concept of a two-tier integrated network was adopted, comprising the core network (on which Community efforts should be focused, in particular cross-border sections, missing links, multimodal links, and major bottlenecks), to be established by 2030, and the comprehensive (complementary) network to be established by 2050. The core network shall

---

1 Although the author focuses on RLT issues, some issues can impact all rail transport RT, including urban rail transport.

2 The Treaty of Maastricht (1992): implementing Trans-European Network of Transport (TEN-T), Telecommunication (e-TEN) and Energy (TEN-E) as strategic factors contributing to establishment of single market and to improve EU's economic and social cohesion; priority areas and projects, with implementation schedules; and financial instruments supporting their implementation.

3 TEN-T – EU program referring to land, waterway, and air transport, comprising nine networks: 1 – road network, 2 – rail network (conventional and high-speed), 3 – inland waterway network, 4 – sea port network, 5 – airport network, 6 – intermodal transport network, 7 – waterway management and information network, 8 – air traffic management network (covering uniform European airspace and SESAR program – the future European air traffic management system), 9 – positioning and navigation network (Galileo). Program implementation is the responsibility of the TEN-T Executive Agency. This publication analyzes mostly selected issues related to network 2, also referring to networks 1 and 6.

ensure connections between major European hubs, while the comprehensive network shall establish communication links of each EU region with the core network. The expression of new TEN-T policy is an approximately three-fold (up to EUR 26 billion) increase in EU expenditure on transport in 2014–2020, with financing focus on the core network. Its development, implemented by means of nine established transport corridors, shall stimulate construction and development of the comprehensive network [5], [4], [7].

There are two core network corridors routed through the territory of Poland: the Baltic – Adriatic corridor and the North Sea – Baltic corridor (Figures 19.1a and b) [4], [10].

Financial preferences of the EU for RLT under the TEN-T project are estimated at approximately 25%, among others based on the appropriation of TEN-T funds for 2013, amounting to EUR 280 million, out of which EUR 70 million was appropriated to RLT (second place by amount of outlays),[4] in particular to projects under European Rail Transport Management System (ERTMS) [6]. Poland approved those proposals, after a series of negotiations, as all key communication links – road, railway, and air links – have been considered the components of TEN-T [5], [7], [8]. Premises for Poland's transport policy resulting from the above are clear.

Poland may use the resources from TEN-T funds since 2004; in 2004–2010, Poland received financing in the amount EUR 74.01 million for 27 projects, for a total value EUR 157.31 million. Nearly all projects were studies supporting preparation of documents necessary for obtaining permits and starting projects. There were two implementations. In 2011, Poland received the consent for co-financing four more projects (three implementations, one study), with a total value of EUR 53.51 million. Of that amount, the EU contribution is EUR 19.66 million [5].

Out of nine projects planned in Poland as per new EU guidelines for 2014–2020, one refers to ports and eight to railway lines (including the upgrade of three domestic and six cross-border railway lines; ERTMS/ETCS for E20/CE20; on-board ETCS for locomotives "Husarz" and their line tests). This means a focus on RLT expansion. TEN-T contribution covers only a small part of costs. In the transport area, European funds can be used to co-finance:

- maximum 50% of eligible costs for studies;
- 20% eligible costs for implementation projects related to, among other, rail transport; this limit can be increased up to 30% for projects eliminating bottlenecks, and up to 40% for cross-border projects [5].

Maximization of the use of UE funds available in consecutive editions of competitions requires ensuring domestic co-financing. However, due to the accumulation of costly implementation projects focused on railway infrastructure, domestic financing may be inhibited by financial and execution limitations. This makes us analyze thoroughly the principles of the current transport and economic policy, as well as the principles of foreign policy to the extent of RLT, concerning implementation of solutions allowing cooperation, on equality and partnership basis, between all EU member states involved in the integration of national railways and, consequently, their conversion into components of the Trans-European Rail Transport System.

---

4 EUR 80 million shall be appropriated to "maritime highways" projects (item 1, 28.6%)

Another limitation inhibiting this process is a very high difference in engineering level of railway transport in new member states, especially in Poland, as compared with former EU15 member states, adding to border limitations and differences in traction power supply. In order to compete successfully within the European market, Polish railways have to increase gradually their engineering capacity and economic efficiency of services, both in passenger transport and in freight services, overcoming technical obstacles resulting from differences in railway traction power supply systems, as shown in Figure 19.1b. Polish trains operating on foreign lines in passenger and freight traffic should be adapted to use voltage used in foreign traction systems, and therefore a number of locomotives should be two-, three-, or multi-system type.

Corridors of TEN-T core network (see Figure 19.1a) are routed through territories with various rail traction systems (see Figure 19.1b). Those systems are DC1 – direct current, 3000 V (Belgium, Croatia, Czech Republic, Poland, Slovakia, Slovenia, Italy, and former USSR); AC1 – alternating current, 15 kV/16.7 Hz (Austria, Germany, Norway, Sweden, and Switzerland); DC2 – direct current, 1.5 kV (France, the Netherlands, part of the United Kingdom); AC2 – alternating current, 25 kV/50 Hz or 60 Hz (34 countries; AC2 system is used in the largest area in the world).

Detailed characteristics of individual traction systems are described exhaustively in [11], and related problems are outside the scope of this study.

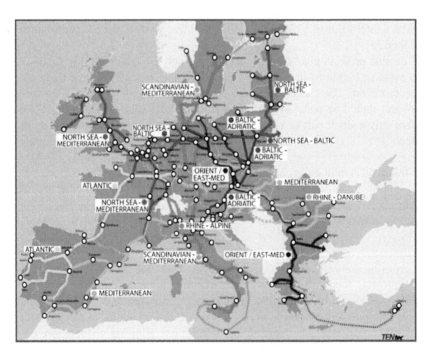

Figure 19.1 (a) Nine railway corridors of TEN-T core network [4]; (b) rail traction systems in Europe [10]. Own additional descriptions (Figure 19.1a) and supplements (Figure 19.1b). Figure 19.1b shows additional two corridors of the TEN-T core network routed through the territory of Poland: North Sea – Baltic and Baltic – Adriatic, and two adjacent corridors: Scandinavian – Mediterranean and Rhine – Danube (routes simplified).

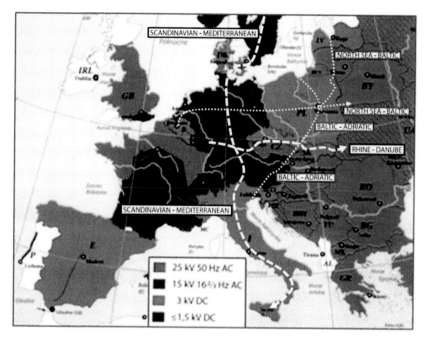

Figure 19.1 (Continued)

Traction systems DC1 and AC1 (Figure 19.1b) have long common borders, and rail transport from DC to DC (e.g. from Poland to Brussels or Rome) is through the area of AC systems. The Polish railway system is not directly adjacent to DC2 systems.

From the viewpoint of transcontinental rail transport, in both the north–south and west–east direction, Poland's geographical location in Europe is extremely beneficial, providing great opportunity to take over the share of transit freight, as long as we are able to take it.

19.2.1.2. Transit role of Poland in TEN-T network

The analysis of Figure 19.1 gives the general premises for reorienting the current transport policy, as well as economic and foreign policies of Poland. Even three years ago the situation seemed relatively predictable. Developments in Ukraine, with their effect on Poland and the world, significantly reduced such predictability. By optimistic assumption, that current situation is going to clear and would probably be stable within 2–3 years; any prior concepts and studies related to establishment to railway link between Europe and Asia shall be unsuspended, updated, and implemented. Poland has to be ready to take that challenge. Otherwise we may lose a lot.

While keeping in mind the above misgiving, analyses mentioned below, referred to some actions taken on railway transport markets east of Poland, are valid for the situation of 3 years ago. There is no new data, and the possibility of update at present is none. However, the strategic problems of Poland remain valid.

The global economic hub is gradually moving to the East and to Southeast Asia. The trend is getting stronger and stronger.

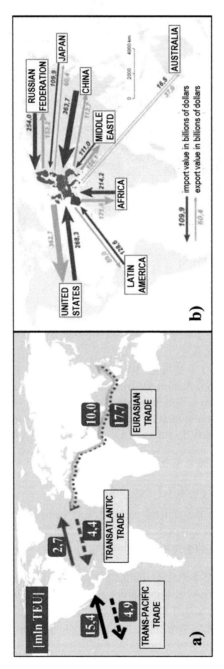

Figure 19.2 (a) Major maritime routes of global trade and container traffic in 2007 [million TEU⁵] [12]. (b) European Union trade balance with major partners in 2008 [13].

Source: Own listing and additional description

---

5 Volume units (for ports and ships): TEU – (20-foot equivalent unit) – the volumetric equivalent of 20-foot container; in addition: FEU – (40-foot equivalent unit) – the volumetric equivalent of 40-foot container; 1 FEU = 2 TEU.

Sea transport caters to over 80% of global trade in goods. This is one of the pillars of international trade and the whole globalization process. This is shown in Figure 19.2.

Data concerning the volume of marine trade are expressed in tons or ton-miles, and they are not easily compared with currency-based statistics. This is why estimation of global marine trade in a currency aspect is difficult. In addition, available data fail to include interferences caused by the global economic crisis. A general view of the situation can be based on data from [12], [13], and [14].

In 2007, the global fleet carried goods weighing over 8 billion tons, completing the transport work of over 32 thousand billion ton-miles. The United Nation's Conference on Trade and Development (UNCTAD) estimated the revenues of commercial freightliners at USD 380 billion in the global economy (5% value of global trade). Annual growth trend (since 1970) in this respect is estimated at 4.8%.

Data in Figure 19.2a indicate that the Euro-Asian trade makes up approximately half (50.3%) of global trade. Considering the rising trend, it can be concluded that this means an estimated annual revenue of freightliners on the order of USD 200 billion. Those are financial resources more than sufficient to bring attention to the highly attractive concept of using the intermodal transport, combining marine and rail freight, between the Baltic, the North Sea, and the Far East.

In the case that some of that trade is taken over rail freight, the corresponding share of above revenues shall go to individual operators involved in the process. This is not only a particularly important aspect of rail transport development in Europe and Asia, and the interest of individual carriers, but also a great strategic challenge to the Trans-European Transport Network being implemented, which also demands avoiding any potential conflicts of national and sectoral interests.

Specific geographic considerations characterizing this globally unique alternative transport system are shown in Figure 19.3.

There are a number of detour options available to Euro-Asian (or even Euro-Russian) transit companies, directly stemming from the situation shown in Figure 19.1b and described in sec. 19.2.1.4.

As in the case of the Baltic Sea area (Figure 19.3), B to A road via rail corridor Baltic – Adriatic (i.e. RB + BAA) is approximately 3000 km shorter than the Gibraltar route, trains are usually faster than ships and use much less energy, and the expansion of rail corridors RB and BAA seemed well grounded and was supported by EU as a competing solution that allows material reduction in time of transport, fuel consumption, and CO_2 emissions. Condition precedent for RLT competitive advantage is the existence of electrified rail corridors with the possibility of capacity and speed, along the north–south axis, from Tallin to Venice. Unfortunately, in the very center of this route there is Poland, with a condition of rail transport as described in sec. 19.2.2.1 and 19.2.2.2. Our neighbors are fully aware of that. We are presently a bottleneck of international rail freight (see sec. 19.2.2. D).

EU trade with Asian countries is going to grow. Many goods can be transported by rail via Russia, other goods can be transported by sea via the Suez Canal and the Mediterranean Sea. European ports on the Mediterranean Sea are in good position. As compared with the Gibraltar route, their distance from Asia is smaller than the distance of the North Sea by over 2000 km and the Baltic Sea by over 3000 km. This allows significant reduction in time of transport, fuel consumption, and CO_2 emission, conditional to the existence of usable and fast rail corridors connecting northern Europe with the Adriatic Sea (referred to as N–S corridors). Similar transcontinental corridors east–west (referred to as W–E corridors) stimulate

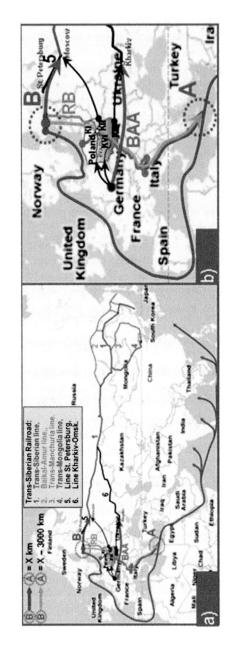

Figure 19.3 Transport routes from the Baltic sea area: (a) listing of rail transport routes from the Baltic to China via trans Asia networks 1–6 and marine transport (via Gibraltar and Suez), as well as rail – marine combined transport via Baltic – Adriatic corridor (see Figure 19.3) covering projects RB – Rail Baltica and BAA – Baltic Adriatic Axis (sec. 19.2.1.3.2, Figure 19.8), where KI–VI are corridors of TEN-T (see Figure 19.1). B, A – calculated starting point and end point of the route; X – distance from B to A via Gibraltar; (b) blown up fragment of Figure 19.3a showing transit opportunities for Poland. Other symbols in the text.

development of mass land transport of materials and goods between Europe and Asia. Components of such a system have been created and developed by EU right from the initial phase of TEN-T development, first as selected priority projects, then evolving into strategic transport concepts, in particular in Germany and Nordic countries, with critical importance to Poland.

The principles of continuity and cohesion in EU transport policy support the validity of previous projects and activities, as well as their evolutionary and adaptive potential, due to subsequent development concepts and decisions in the TEN-T area. This stems from UE legislation[6] [15].

19.2.1.3. UE projects determine the future of Poland's RLT in TEN-T network

19.2.1.3.1. EU PROJECTS COMPETING WITH POLAND'S RLT AND POLAND'S ECONOMY

It is in the strategic interest of Poland to make use of the transit location described in sec. 19.2.1.2 to ensure economic benefits, in the form of possibly high and sustainable revenues from taking over international transit operations in any direction possible. To that end, Polish pro-development operations in RLT have to consider the complementary or competitive nature of major strategic projects in RLT, created in our neighborhood. Under competing efforts with regard to future transport markets, multiple initiatives are initiated in the area of rail transport, which in the future might have both beneficial and adverse consequences to the Polish economy. Major competing projects are presented in this section, whereas major beneficial projects are presented in sec. 19.2.1.3.2. First and foremost, projects competing against potential Polish projects are two major projects located in two adjacent countries, at the cross-section of two core TEN-T corridors: Scandinavian – Mediterranean and Rhine – Danube, as shown in Figure 19.1a, and in a simplified way in Figure 19.1b. Detailed specifications of those corridors, omitted here, are included in [15]. Three projects, known as SCANDRIA, EWTC II, and more recently WGL RUSA, are the most important ones.[7]

The Berlin Declaration, adopted in 2007, confirmed starting of the initiative to establish attractive and competitive transport infrastructure connecting all Nordic countries with the Adriatic Sea (via German territory). Complementary projects, i.e. SONORA (south–north axis), TRANSITECTS (Transalpine Transport Architects), COINCO – North (Corridor of Innovation and Cooperation), have been implemented since 2008. In 2012, multilateral cooperation agreements were signed with regard to transnational projects SCANDRIA, East–West Transport Corridor II, and TransBaltic.

SCANDRIA ® PROJECT: Scandinavian-Adriatic Corridor for Growth and Innovation, [16], [17] [18], [19], [20].

The scope and nature of the SCANDRIA project is shown in Figures 19.4 and 19.5. Thus far the share of rail freight in transport between Nordic countries and Germany is much smaller

---

6 [38] Article 30. *Intermediate regulations*. This Regulation does not affect continuing or changing, including complete or partial cancellation of relevant projects including their closing, or financial aid granted by the Commission subject to regulations . . .

7 WGL RUSA – own acronym. German name is *Breitspur-Anbindung des Twin-City Raumes Wien/Bratislava*.

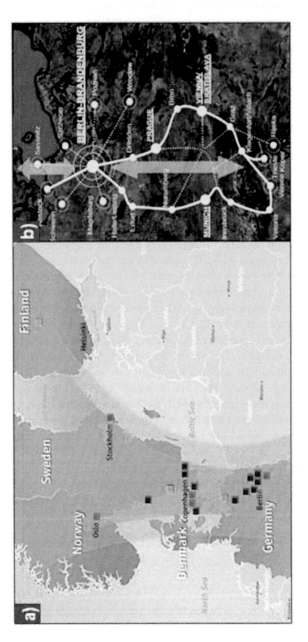

Figure 19.4 SCANDRIA – scope and key partners of the project, with Berlin – Brandenburg as the strategic hub: (a) Area North, Germany and Nordic countries [18]; (b) Area South, with major transport corridors to ports on the Adriatic Sea [19]

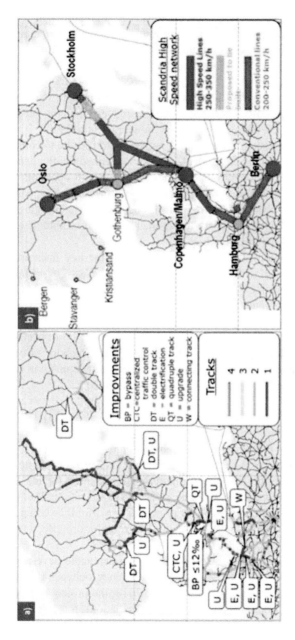

Figure 19.5 SCANDRIA – major rail corridors: (a) proposed modifications in core network infrastructure; (b) high-speed rail lines [19]

than the share of either road or marine transport. One of the goals, therefore, is increasing the attractiveness of rail services to forwarders, by means of improving the efficiency of railway system and reducing costs. Between Hamburg and Copenhagen, the maximum length of the train was increased from 750 m to 835 m. Further increase in length shall be obtained by extending the route of such trains to Sweden, mostly Malmö and Hallsberg, and necessary construction of new or upgrade of the existing tracks, in order to reach the axle load of 25 tons and an increase in freight operations up to 120 km/h. Infrastructure and rolling stock will suit the variety of goods, including intermodal transport, with suitable lines already existing in Finland, Netherlands, Germany, Norway, Sweden, and United Kingdom, including also the fleet of intermodal wagons. Bottlenecks (single-track sections or too-steep sections, etc.) within some lines of SCANDRIA corridor shall be eliminated, and the main corridor lines in Germany will be electrified. Above projects are shown in Figure 19.5a.

In order to solve the issue of international passenger traffic and to compete successfully with road and air transport, the SCANDRIA corridor is planned to offer centralized information system, coordination of operations and ticket sales for all operators (also online), an increase in the number of direct links, etc. There are projections for construction of HSR within the SCANDRIA corridor, with transit times between large cities less than 3–4 hours, which is critical when competing with air transport. There are no common standards or strategies as yet for HSR in all Nordic countries.

For passengers travelling the SCANDRIA corridor, the key lines are Southern Main Line, connecting Stockholm with Malmö/Copenhagen, and West Coast Main Line, connecting Oslo/Göteborg with Malmö/Copenhagen. Both lines shall be linked to Germany via Fehmarn Belt with permanent road and railway bridge and tunnel, around the year 2022. Above projects are shown in Figure 19.5b.

The Swedish government proposed in 2012 the construction of a high-speed rail line between Södertälje and Linköping, named "Ostlänken," and Göteborg – Boras (partly operational now, 200–250 km/h). This is the first stage of HSR development in Sweden, relieving the capacity of main lines for freight trains and regional trains.

EWTC II PROJECT: East–West Transport Corridor [21], [22], [23], [24], [25].

Scope and nature of EWTC II Project are shown in Figure 19.6.

Rapid increase in Europe–Asia trade, referred to in sec. 19.2.1.2, creates conditions for a sustainable system of transport from the southern Baltic region, via Belarus, Russia, and Central Asia countries to China, and via Ukraine to the Black Sea region. Rail transport through both continents reduces the duration of cargo delivery two-fold, as compared with marine transport, and is more environmentally friendly, safer, and more effective in long haul, as compared with road transport.

EWTC II is the project of transport corridor for sustainable development of business interests and improvement of East-West transport efficiency, between Baltic Sea Region and countries of the Far East, especially China. Since 2010, the project has been implemented by EWTCA (East–West Transportation Corridor Association). Association members are 12 countries (Belgium [EIA], Belarus, China, Denmark, France, Kazakhstan, Lithuania, Mongolia, Germany, Russia, Sweden, and Ukraine. Among the major stakeholders from transport and logistics sector are national railway operators; associations of road carriers, marine carriers, intermodal carriers (including European Intermodal Transport Association, EIA); a variety of regions, municipalities, and ports; high schools and research institutes, etc. EWTCA

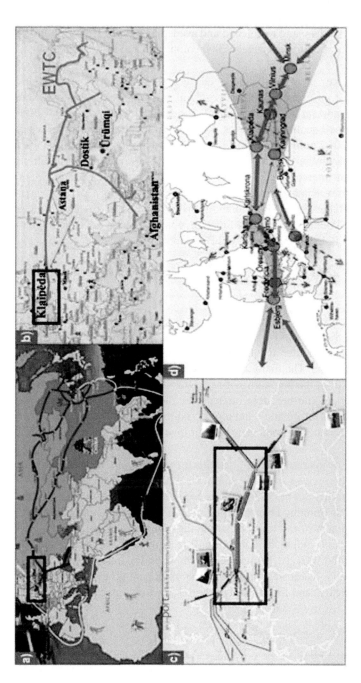

Figure 19.6 EWTC II – project scope in global and regional aspect: (a) EWTC in global context – comparative listing of rail transport routes (via Trans Asian trunk routes) and maritime transport routes (via Gibraltar and Suez) from the Baltic Sea Region to China; (b) Asian EWTC routes; (c) intermodal terminal hub in Karlshamn (Sweden), diagram of EWTC corridor towards the east, with major logistic hubs; (d) southern part of BSR[8] with detailed diagram of links between EWTCA logistic hubs. Rectangles ▭ in Figures 19.6a, b, and c symbolize the BSR area as per Figure 19.6d.

8 BSR – Baltic Sea Region. EWTCA – East–West Transport Corridor Association.

generates added value on maintaining and expanding trade between Europe and Russia, Belarus, Ukraine, Kazakhstan, China, and other countries of the Far East. EWTCA will be intensifying cooperation with government bodies, carriers, and logistic companies; operators of intermodal transport, senders, and recipients; and academic and research institutions. Presently, EWTC groups consist of approximately 2000 companies and institutions from 12 countries (listed above).

Cooperation covers five key areas: political, social, technological, administrative, and business, which is critical for ensuring reliable flow or rail freight. Specific actions were taken in order to expand transport services to include intermodal rail freight, eliminate obstacles and limitations and improve quality of services. Research of the new transport services market and analyses of feasibility of implementing new technical solutions in terminals were completed; international work groups contribute to ensure coordination of initial conditions to launch a few operations of intercontinental container trains (Sun, Viking, Mercury, and Saule), etc. as well as rail ferries. Klaipeda in Lithuania is quickly becoming a strategic intermodal hub of the southern Baltic region. Poland and Polish business and research entities are not among the members. Poland attended the founding conference but never acceded to EWTCA. EWTC II bypasses Polish rail network and Polish ports.

WGL RUSA PROJECT: Broad-gauge Line Russia – Ukraine – Slovakia – Austria [26], [27], [28].

The WGL RUSA project is based on the concept of extending the broad-gauge line from Russia through Ukraine and Slovakia to twin cities Vienna and Bratislava, the joint capital of CENTROPE macro-region,[9] and later also further west to handle the intercontinental rail transport between Europe and Asia. The first train is scheduled for 2025.

Considerations referred to in the beginning of sec. 19.2.1.2, related to the situation in Ukraine, prevent projecting any further developments in the not-so-long-ago-ambitious WGL RUSA project. Absence of information and global and local political uncertainty concerning solutions for Ukraine issues are premises enough to state that in years to come the project shall remain frozen.

However, strong groups of interest who inspired this project have not disappeared. The concept of WGL RUSA can be revived when the conflict is over.

In the near future, this project can be considered as pro-development, allowing recovery of losses in the East, as well as bringing economic profits to all involved countries and entities interested in the development of rail transport along the core network corridor Rhine – Danube of TEN-T. Concepts and studies shall then be unfrozen, updated, and implemented.

The scope and nature of the WGL RUSA project are shown in Figure 19.7.

Location of the CENTROPE macro-region at the corner of four countries (Austria, Czech Republic, Slovakia, and Hungary) and four languages corresponds with the development of

---

9 CENTROPE is a joint initiative of Austria – Vienna, Lower Austria and Burgenlandu, Czech Republic – South Moravia, Slovakia – Regions of Bratislava and Trnava, Hungary – counties Gyor-Moson-Sopron and Vas, including communities: Bratislava, Brno, Eisenstadt, Győr, Sopron, St. Pölten, Szombathely, and Trnava. The grounds for establishment of the macro-region was Kittsee Declaration from 2003. EU support for CENTROPE expansion takes the form of INTERREG III Project and CENTROPE CAPACITY Project (presently the main project co-financed under the EU program for Central Europe) [55].

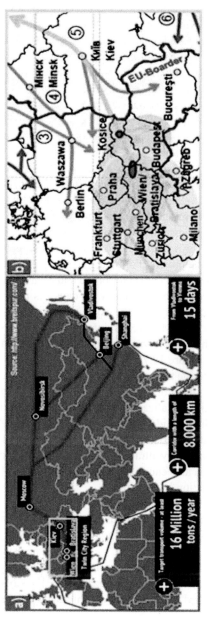

Figure 19.7 WGL RUSA concept: (a) scope and impact of the project in global scale – comparative listing of rail transport routes (via Trans Asia core network) and marine transport (via Gibraltar and Suez) from Central Europe of CIS and China[10]; (b) WGL RUSA in macro-regional scale (within the core corridor Rhine – Danube, with broad -gauge logistics terminal near Vienna-Bratislava, and distribution of rail transport throughout the CENTROPE and westwards (yellow lines). Rectangle ☐ in Figure 19.7a represents the area as per Figure 19.7b.

10 For animation acc. to Fig. 14.7.a and detailed information see the website www.breitspur.com/ Breitspur Planungsgesellschaft mbH started the procurement procedure for a "Feasibility Study". 31st of August 2013.

CENTROPE contacts with a number of other partner regions and cities. CENTROPE regions have been working together for 10 years towards establishing solid foundations for sustainable development and innovative economy in this area. The population of this area is over 6.5 million. The concept of WGL RUSE as broad -gauge railway connection between Asia and Europe is an ideal solution for stimulation of economic development and integration of CENTROPE, especially for Austria. Without improving their links with the East, Austria may never be able to benefit from developmental opportunities. The benefits are obvious. This line and development of related logistical hubs will bring permanent and accelerated economic growth to CENTROPE area. Capitals Bratislava and Vienna are only 60 km apart; Brno and Győr, as well as other cities with transnational significance, are engines of the macro-region economy. They complement each other. They have efficient and export-oriented industry, a consistent network, and centers offering specialized services. They have polycentric structure allowing effective cross-border cooperation. They have both potential and political will to implement that project.

Russian railways express their need for broad-gauge line that could be used for rail freight to southern Europe. Transport of goods between Europe and Asia is nearly completely marine transport. Its present value is approximately USD 600 billion. CENTROPE has good access to sea ports in northern and southern Europe. Thanks to WGL RUSA, rail freight can take up to 15% of freight over from marine transport [26]. In 2009, Austrian, Russian, Ukrainian, and Slovak railways founded a joint venture company called Breitspur Planungsgesellschaft to implement the project. Following completion of the implementation phase, the duration of rail freight operation between Europe and Asia (especially China) was reduced from approximately 30 days to approximately 17 days. As per earlier projections, line capacity is 43 trains, 1 km long, per day, and the capacity of new rail terminal in Vienna is 20–30 million tons of cargo a year [26]. More recent information [58] specifies daily capacity at approximately 66 trains, each way, carrying mostly containers from the Far East and fuel (via Trans Asia core network – Figure 19.7a).

Eliminating the monopoly of marine transport on this route is possible, but it will require better coordination between railway operators, significant investments, removal of numerous bureaucratic barriers, and most of all normalization of the present geopolitical situation.

Previously, Russian sources[11] did not hide that the broad -gauge line to Vienna originated also from the need to bypass transit countries disloyal to Moscow (the route bypassing the territory of Belarus and Poland seemed very attractive from the viewpoint of Russian foreign policy). Political changes in Slovakia also cause trouble to the project [61]. With the current situation in Ukraine, no new information is available.

Presently, the most eastward intermodal terminal of the LHS broad-gauge line is Sławków in Poland. This link has a status of regional line and is not subject to European or Russian development plans. In addition, LHS does not meet EU criteria, as the company PKP LHS is both the administrator and the operator.[12] Ever since the establishment of PKP LHS on1 December 2001, there have been numerous concepts as to the target ownership and organizational structure of the company and its privatization [62]. As of present, 10 trains operate on a single-track, not electrified LHS line, five of which carry iron ore for Mittal (line

---

11　Russian newspaper *Kommiersant* (in [65]).
12　PKP Linia Hutnicza Szerokotorowa (PKP LHS) – rail operator and infrastructure manager of railway line no. 65 (LHS = BGML - Broad-Gauge Metallurgy Line), are owned by the state (99.99%) and PKP S.A. (0.01%), and are the companies of PKP Group holding.

capacity 11 pairs of trains per day). PKP LHS rolling stock has been systematically modernized since 2005. For more than a dozen years now, the discussion concerns the use of the line for a projected Europe–Asia link.

In the near perspective, Sławków Terminal and 397 km long LHS may lose their advantage to the planned route to Vienna via Slovakia. This resulted in intervention in the form of a protest submitted by local Silesian authorities [58] and a parliamentary[13] question [60]. The vague answer (authorized by the president of the Council of Ministers) to this question, in addition to selected facts, included also two fundamental statements: "Poland has been systematically developing the concept for development of Euroterminal in Sławków (. . .) and it should be expected that in the period of next 2 years necessary investments shall be completed so as to handle transit from the East"; and "one should realize that Euroterminal in Sławków will not be the only land logistic hub handling transport operations CIS/Far East – Central Europe, and that it has to work with similar enterprises, including the logistic hub in Vienna." The majority owner of LHS did not seem perturbed.

Effects of the project can have impact not only on Euroterminal Sławków, but also terminals at the border with Ukraine and Belarus, and the whole region of Upper Silesia, reducing their competitive advantage and moving logistic hubs outside of Poland. Cheap Russian coal will be sold right outside our borders. We may lose the markets of Czech Republic (25% of total exports), Slovakia, and Austria; even our export to Germany (50% of total exports) may be at risk. A postulated countermeasure is to increase LHS capacity from present 10 trains per day to approximately 450 trains per day. Feasibility studies for the second track of LHS line [30] had already been started. broad-gauge line LHS should be a part of European transit corridors. But on the end of this line there is no powerful CENTROPE.

WGL RUSA is obvious competition to Polish interests. Its implementation can result in creating a new market for rail transport from Asia to Central Europe. The Russian-Austrian-Slovak Project, highly competitive also to marine transport via Suez Canal, will put the position of Poland, Belarus, and Baltic States at risk as regards transport of containers along the W–E axis [58], [59], [60], [61].

19.2.1.3.2. EU PROJECTS BENEFICIAL TO POLAND'S RLT AND POLAND'S ECONOMY

First and foremost, projects with beneficial effect to Poland's economy are major projects located in the cross-section of two core TEN-T corridors: North Sea – Baltic and Baltic – Adriatic, shown in Figure 19.1a, and, in simplified form, also in Figure 19.1b. Detailed specifications of those corridors, omitted here, are included in [15].

In the N–S axis, two strategic concepts require special attention: Rail Baltica (RB) and Baltic – Adriatic Axis (BAA). In the W–E axis, restoration of the Y line project, complementing the existing CMK line, has strategic importance, forming the base for a future HSR network. Railway corridors related to three concepts are shown in Figure 19.8.

RB PROJECT: Rail Baltica [31], [32].

The concept of Rail Baltica first appeared in 1994 in the common political document "Visions and Strategies for the Baltic Sea 2010" as a projected interoperable railway

---

13 Parliamentary Question No. 9226 by Member of Parliament Jerzy Wenderlich (2007).

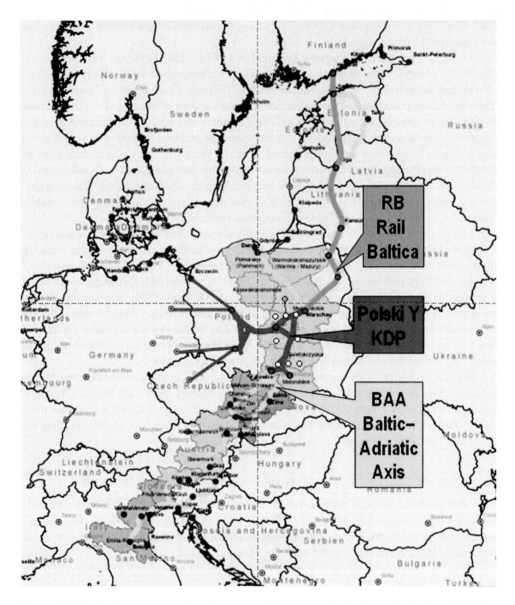

Figure 19.8 Indicative routes of railway corridors per strategic projects beneficial to Poland's economy: RB – Rail Baltica; BAA – Baltic – Adriatic Axis; HSR – Polish "Y line" and CMK – high-speed rail

corridor within the N–S axis, linking Baltic States with Poland and the rest of the EU rail network. It has key importance from the viewpoint of development of rail transport in the Baltic Sea region. Figure 19.1 presents the RB Project as priority project TEN-T PP27. It is complemented by the Road Baltica project. Appearance of the Y line project (see below) provided new possibilities for Rail Baltica HSR connection via Łódź and Wrocław to Dresden and Praha.

BAA PROJECT: Baltic – Adriatic Axis [33], [34]

On 12 October 2006, Poland, Czech Republic, Slovakia, Austria, and Italy signed a letter of intent with regard to the intermodal Baltic – Adriatic Corridor, linking regions of Baltic Sea and Adriatic Sea and their corresponding ports (Gdańsk, Gdynia and Triest, Venice). The corridor is supposed to be established in the next revision of TEN-T, by way of extending the rail corridor (TEN-T PP23) and road corridor (TEN-T PP25) (see Figure 19.1), linking the following cities: Gdańsk – Warszawa – Katowice – Ostrava – Brno – Vienna (insec. branching Katowice – Zilina – Bratislava – Vienna) – Graz – Klagenfurt – Villach – Udine – Triest – Venice to Bologna.

On 6 October 2009, in Brussels, an Alliance of 14 Regions was signed, which is the declaration concerning development of TEN-T PP23 rail corridor. Extending the rail section of this corridor, PP23 TEN-T, to Italy is very important for economic development of Central Europe. The BAA concept matches the objectives of TEN-T policy, meets its criteria, and is adapted to the future core network [35]. Among the17 regions, signatories[14] of the Alliance, there are only two regions interested in this extremely attractive program: Pomorskie and Śląskie.

Other regions were not interested in the project. The BAA concept also covers the following projects: AB Landbridge [36] and its extension SoNorA [37], aimed at the development of transport infrastructure in the N–S axis and improvement of land links between the southern Baltic region and the Adriatic Sea.

Y LINE PROJECT: Polish High-Speed Rail (HSR).

Increase in HSR operation volumes is much higher than for other means of transport. Thanks to an ever-expanding network of HSR operations, the share of railways in the transport market increases, becoming a more and more important element of international and national transport systems, affecting economic growth. This is why the plans of starting the high-speed rail system in Poland have been shaping for years now [2], [38]. HSR is the subject of multiple publications and studies, and they were characterized exhaustively in section III[15]of this monograph. Here we present only those elements of the problem that are necessary for complete understanding of the situation analyzed.

Construction of the Y line , connecting Warszawa, Łódź, Wrocław, and Poznań, had been approved by the government on 19 December 2008 [43]. It was assumed then that the investment should be completed by 2020, at the latest. Following completion of the feasibility study for the high-speed Y line, three variants of the route were prepared for consulting [40].

Even before selecting the variant, the Ministry of Transport, quoting high construction costs (approximately PLN 20 billion [15]), postponed implementation of the project until

---

14 Number of regions-signatories: Austria: Carinthia, Styria, Vienna; Czech Republic: Moravian-Silesian, Olomouc R., South Moravia R., Zlín R.; Italy: Emilia-Romagna, Friuli-Venezia Giulia, Veneto; Poland: Pomorskie, Śląskie; Slovak Republic: Bratislava, Nitra R., Trencin R., Trnava R., Zilina R.

15 Raczyński J.: The development of the concept of high-speed rail in Poland. Studies, elaborations, concepts. The state of preparatory work.

2020 [42]. It has to be stated clearly that this decision was non-substantive and a grave strategic error. In addition, it was contrary to a resolution of the Government of Poland [43].

The Y line Project is the only one to consider comprehensively linking the large metropolitan areas of Warszawa, Łódź, Wrocław, and Poznań, ensuring also relieving the section Łódź – Warszawa, as well as providing an attractive communication link for Warszawa, Wrocław, and Poznań (due to assigning the E20 route to rail freight).

In addition, it would allow future extensions of the Y line to neighboring countries, allowing integration with the European high-speed network. The Y line was introduced to the updated TEN-T.

The proposed route of Y line (variant III) [40] is shown in Figure 19.9. Later political disputes about the project are more proof of inconsistency of Poland's rail transport policy with EU rail transport policy.

The project, developed for many years, was even included in the EU Master Plan and new plans of TEN-T [5], [15], [35], and was effectively blocked by the Minister of Transport not even two weeks after he took his official position, *contrary to valid decisions of the government*, resulting in the number of negative emotions and adverse political and economic effects.

The above-mentioned concept of HSR in Poland is extended [38]. New additions and plans and projects appear that involve the use of a 25 kV/50 Hz system and improved use of the existing 3 kV DC system, to create target consistent system of HSR and near-HSR systems, ensuring significant improvement in access to major metropolitan areas in Poland and better international links. Subject to analyses are presently:

- by 2020 two HSR lines: Warszawa – Łódź – Poznań/Wrocław (i.e. Y) and Warszawa – Katowice/Kraków (i.e. CMK);
- beyond 2020 four new HSR lines: Gdańsk – Poznań/Łódź/Warszawa, Poznań – Szczecin, Warszawa – Lublin – Rzeszów, Warszawa – Białystok; in addition international links: Poznań – Berlin, Wrocław – Praga, Katowice – Ostrava, and, in the future, to Kiev, Minsk, Moscow, and Baltic States [38].

As compared with the original concept of the Polish Y line, this means a many-fold increase in the length of rail network 25 kV/50 Hz.

The rail transport development process, clearly outlined in reference documents, including the HSR Project and complementary projects, affecting the population of 10–15 million, received unanimous support of all research and engineering circles, expressed formally in materials of the 14th KTP [5]. Such a program would have beneficial effect to the majority of Poland's territory, due to the assumed interoperability of high-speed trains with conventional trains, translating, among others, into being within direct or indirect reach of HSR not only for Warszawa, Łódź, Kalisz, Poznań, and Wrocław, but also for Kraków, Katowice, Rzeszów, Opole, Jelenia Góra, Zielona Góra, Gorzów Wielkopolski, Szczecin, Gdańsk – Sopot – Gdynia, Olsztyn, Białystok, and Lublin, as well as minor towns situated near larger agglomerations (assuming links to HSR hubs and intensifying the development of commuter rail systems).

Postponing the construction of the Y line and the whole HSR network indefinitely poses a threat to Poland's transport policy, and as well as its economic policy, and might affect further formation of TEN-T, with adverse effects to Poland and its position in EU.

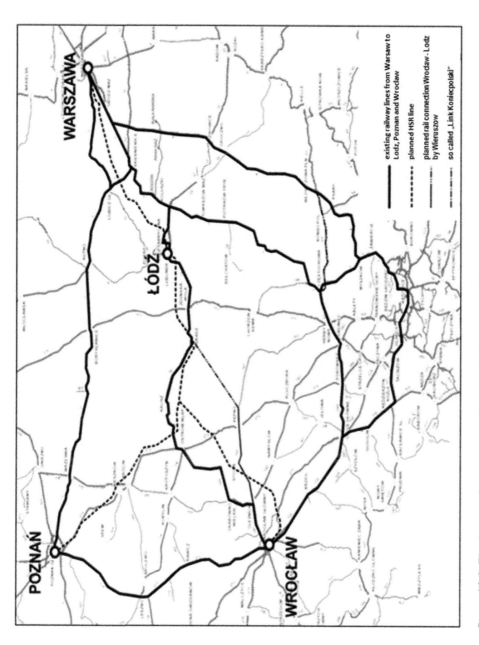

Figure 19.9 The layout of currently used railway links between Warszawa, Łódź, Wrocław, and Poznań, including variant 3 of Y line HSR [39], [40].

Construction of an HSR line will not only launch Polish railway into the 21st century, but will also increase the capacity of the conventional rail network, for both passenger and freight operations. This will have particular significance to the W–E axis, due to ever-increasing volume of EU trade with countries of Eastern Europe and Asia, referred to in sec. 19.2.1.2. Rail corridors within the N–S axis and W–E axis via the territory of Poland (Figure 19.1, see also Figure 19.10), should take over a possibly large share of passenger and freight traffic, especially when our geographical location favors doing so. Other aspects of this matter were described in sec. 19.2.2.

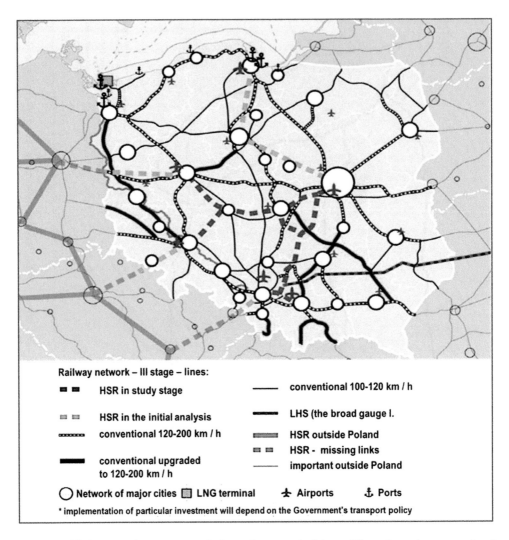

Railway network – III stage – lines:

■ ■	HSR in study stage	———	conventional 100-120 km / h
▦ ▦	HSR in the initial analysis	▬▬▬	LHS (the broad gauge I.
·······	conventional 120-200 km / h	▦▦▦	HSR outside Poland
		▫ ▫	HSR - missing links
▬▬	conventional upgraded to 120-200 km / h	———	important outside Poland

○ Network of major cities   ■ LNG terminal   ⚓ Airports   ⚓ Ports

* implementation of particular investment will depend on the Government's transport policy

Figure 19.10 Projected expansion of the rail network (phase III) against the network of airports and sea ports. The date of completion of the phase is not determined in [47]. Also the possibility of updating CLDP 2030, in part or in whole , is considered, subject to developments in social and economic considerations and land development. Earlier phases I and II are omitted. Description on the Figure is simplified [47], [48].

19.2.1.4. List of transport projects in adjacent countries

Intense operations of our neighbors in the area of rail infrastructure projects located within priority transport N–S and W–E corridors keep undermining Poland's transit position in international intermodal transport, including in particular rail transport. This is demonstrated in Figure 19.11, where all the above-described transport projects, roughly in one scale, are overlaid on the map of Central Europe.

In existing EU studies related to the Baltic – Adriatic transport system, many earlier concepts and projects (omitted in this study) provided fertile ground for the development of strategic solutions in Figure 19.11, competing with Polish systems. All around Poland, competing infrastructure is being developed, allowing a bypassing our country. Development of this infrastructure is faster than in Poland.

The RB concept is among the oldest of projects (soon it will have 20 years) for interoperable rail corridor within the N–S axis, connecting Baltic States with Poland and the rest of the EU rail network. Poland still remains a bottleneck. The BAA concept, more recent than RB, consisting in using an intermodal transport link between the Baltic Sea

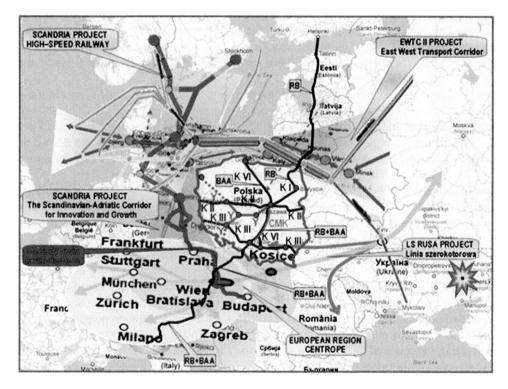

Figure 19.11 List of projects in Poland and in adjacent countries: RB, BAA, and Y – HSR, described in sec. 19.2.1.3.2, shown in Figure 19.8 and Figure 19.9, against competing projects SCANDRIA, EWTC II, and WGL RUSA described in sec. 19.2.1.3.1 and shown in Figures 19.4 to 19.7, with additional elements shown in Figure 19.10. Legend as in Figure 19.3 to 19.10; additional descriptions in the body of the text. Conflict zone.

region and Far East, marine and rail, had good starting conditions and could be very profitable for ports and railways. This solution was and is competitive to only marine transport for geographical, energy, and economic reasons (see sec. 19.2.1.2, Figure 19.3). Establishment of the Baltic – Adriatic core corridor routed through the territory of Poland improves conditions for BAA, but in practice it shall only provide access to freight from the eastern part of the Baltic Sea via RB (without freight via St. Petersburg or Klaipeda (EWTC II) to Moscow and Trans Asian core network), or from the territory of Poland. Without implementing HSR and complementary projects, it is hard to imagine any competition in the N–S direction to the core Scandinavian – Mediterranean corridor and projects in the SCANDRIA group.

Poland's lack of interest in EWTC II was an irreversible political mistake, which has to be noted especially in the context of its significance due to political blockage of the EGL RUSA project because of conflict in Ukraine. The conflict zone, shown in Figure 19.11, where military actions are observed, is relatively far from Russian-Ukraine rail infrastructure (linking Kiev with Trans-Siberian line), which shall probably remain intact.

Perspectives for both sides seem to be good (no damage is reported so far). As specified in sec. 19.2.1.2, if the situation is going to clear and stabilize within 2–3 years, the matter of intercontinental transport by broad-gauge lines between Europe and Asia may by unlocked. This business is too big to be shrugged off by EU15 member states. Poland should use the intermediate period of blocking the WGL RUSE project for maximum development of LHS and Euroterminal Sławków, improvement of competitive advantage, and creating the capacity to compete effectively for Europe – Asia transit, in the N–S axis and in particular in the W–E axis. Otherwise we may lose a lot.

Strategic interest of Poland demands that corridors North Sea – Baltic Sea and Baltic – Adriatic (Figure 19.1b) are key components of EU transport strategy, integrated by HSR with other corridors of the TEN-T network.

Momentum, strength, and effectiveness of competing projects make RB and BAA concepts look life ineffective half-measures, their future unpredictable, which is very dangerous to Poland's interests. There is no alternative concept for Poland – we have to use all possibilities of integrating and coordinating not only RB and BAA, but also with other corridors of the TEN-T network, via HSR. This is a very complex and difficult task, but nevertheless a necessary and practicable task. Coordination of works shall require establishment of a dedicated organization system.

Comprehensive and coordinated implementation of projects on the national level has a series of obvious organizational, engineering, economic, and political benefits, allowing prioritization of goals and implementation phases, optimization of works and use of resources, achieving complementarity of projects and synergy effect of undertaken efforts, as well as increased political impact in the EU. This will also open new integration potential towards the W–E axis, connecting Western and Eastern Europe. The condition precedent is a prompt unfreezing and acceleration of all works on Poland's HSR system, especially the Y line, binding all above projects together.

Complementarity of all railway projects, both new and upgraded, has to be analyzed from this viewpoint.

All that requires a serious revaluation of Poland's economic, transport, and foreign policies, especially related to the East.

19.2.2. Technical, economic and political aspects of RLT condition

19.2.2.1. Diagnosis of the present condition of Poland's RLT

Non-governmental assessment of the condition of rail transport in Poland, included in materials of the 24th Congress of Polish Engineers (KTP) in 2011 [5] and other sources, e.g. [66] and Annual Report 2013 by PLK PKP, in the context of sec. 19.2.2.2 and 19.2.3, indicate that the present condition of railway transport is the consequence of years of underfinancing. Railway infrastructure is worn out and outdated. Nearly half of it (47%) received a good score (that is, meets the assumed utility parameters), and more than half (53%) received a bad score, including 27% satisfactory (reduced parameters, track sections for replacement), 26% unsatisfactory (significantly reduced parameters – speed and axle loads, pavement for replacement).

After the 1991 upgrade, works in Poland were never sufficient. During last 25 years, the railway infrastructure was reduced by over 25%. Bottlenecks are commonplace. All that is a strong barrier for development of RLT. The current state of affairs is illustrated by the data below and by Figure 19.12.

A. Rail infrastructure expenditure RLE vs. road infrastructure expenditure RDE (detailed data in sec. 19.2.2):

- 1998: RLE/RDE = 1321/2269 million PLN (58%);
- 2011: RLE/RDE = 3700/40,253 million PLN (9%).

B. Reduction in infrastructure (differences in line length in 2013: GUS **19,328** km, *PLK 18,533 km*):

- total length of railway network in 2013/1990: *18,533*/**19,328**/26,228 [km] (*70%*/**74%**), – including *[GUS 2013]*: 8699 km – double track (45.0%), 11,868 km – electrified (61.4%);

  - plans for further dismantling of approximately 1–2 thousand km of lines (public data[16]; discrepancies observed[17]; (~5–10%);

- density of rail network in years 2013/1990: 62/84 [km/1000 km^2]; (74%).

C. Insufficient upgrade works:

- 1991/2011, **average annual**: needs ~1550 km/r, overhauls ~428 km/r (28%).

D. Average commercial speed of cargo trains: RP/UE15: ~20/~50 km/h (40%).
E. Average cost of transport in PL/EU15: ~5.5/~2.5 €/km (220%).
F. % Share of road freight RD versus railway freight RL *[GUS 2014]*:

- 2005/2013:

  - RDT à 75.9%/84.0% (**increase by 8.1%**),
  - RLT à 18.9%/12.6% (**decrease by 6,3%**).

---

16 http://forsal.pl/galerie/689406,duze-zdjecie,1,linie_kolejowe_do_likwidacji_pkp_likwiduja_tory
17 Acc. to [49] PKP PLK SA plans included 1049 km of lines as of 31 December 2011.

G. Bottlenecks (Figure 19.12):

- reduced operation speed to ensure traffic safety;
- threat of single-track operation;
- threat of line closures.

Bottlenecks in Poland are a limitation for international transit through Poland within TEN-T, including transcontinental transit Europe – Asia. For example, currently only approximately 10% of Russian transit through Belarus uses Poland's rail infrastructure.

As it stems from Figure 19.12 those processes intensify, limiting the possibility for reduction in transit time and resulting improvement of railway competitive advantage. Progressing aggravation of rolling stock and infrastructure is the reason for decreasing share of railways in transport. Little competitive advantage of Polish railways is the major technical and economic barrier for implementation of sustainable development of transport principle. For the Polish economy, this means negative feedback, with a counter-progress effect.

Interesting conclusions can be drawn from comparison of development trends in major categories of land transport in Poland and EU [10], see Figures 19.13, 19.14, and 19.15 (data according to a extensive report prepared for the European Commission [68]).

Figure 19.13 indicates that before the year 2000, the total length of railway lines (both conventional and high-speed) decreased by approximately 5%; that is, some lines were closed, and then the matter stabilized. HSR (isolated for illustrative purposes) expanded significantly during that time. HSR-related investments accumulated in EU15 member states.

Figure 19.14 indicates that during the decade 1995–2005, the road network in Poland expanded (+16%), and the rail network shrunk (−19%) (Figure 19.14a). Only road passenger transport was expanding (Figure 19.14b). In 2005, the share of private car transport increased by 78% compared to 1995, which was due to significant increase in car ownership (in 1995 – 195, and in 2005 – 323 cars per one thousand persons), linked to ongoing increase in the standard of living.

Similar situation was with freight (Figure 19.14c), where the share of road freight increased by 120%. Pursuant to data in Figure 19.14d, in the decade 1995–2005, the total share of road transport in passenger transport increased from 82% to 91%, whereas the share of rail freight decreased from 15% to 7%.

Commuter rail usage decreased by approximately one-third. The share of road freight increased from 38% to 60%, whereas the share of rail freight decreased from 51% to 27%, which is nearly twofold. Consequences are far-reaching and serious. Diversification of accessibility to multi-modal capacity for Poland's regions as compared with average EU MDTE is large and increasing.[18] Accessibility for western and northwestern regions is in striking contrast with lack thereof in eastern and northern regions of Poland.

Consequently, the integrity of Polish economic and social areas is slow, and the duration of rail journey prevents integration of labor markets between any pair of metropolitan areas.

Metropolitan areas do not have any convenient rail links due to inefficiency of operators and infrastructural limitations. The absence of perspectives regarding the development of HSR only makes matters worse.

---

18 Example MDTE values in 2004 in following regions: Warszawa (120–140%); Katowice (100–120%); Eastern and Central Pomorze (40–60%); part of Mazurskie (20–40%); other (60–80%).

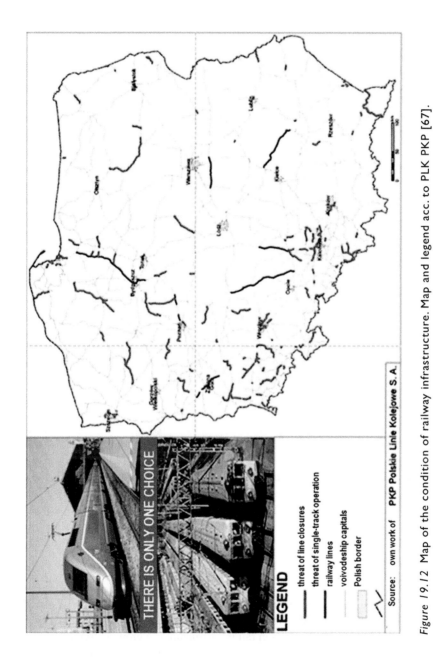

Figure 19.12 Map of the condition of railway infrastructure. Map and legend acc. to PLK PKP [67].

Source: Own listing

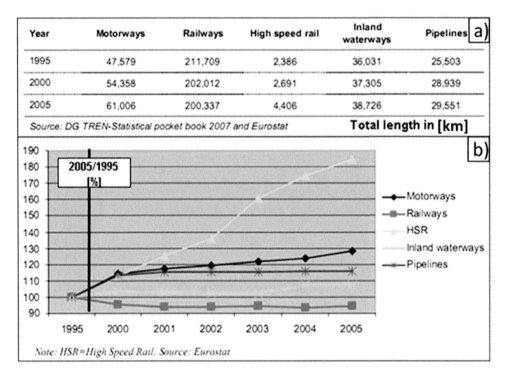

Year	Motorways	Railways	High speed rail	Inland waterways	Pipelines	a)
1995	47,579	211,709	2,386	36,031	25,503	
2000	54,358	202,012	2,691	37,305	28,939	
2005	61,006	200,337	4,406	38,726	29,551	

Source: DG TREN-Statistical pocket book 2007 and Eurostat   **Total length in [km]**

Note: HSR=High Speed Rail. Source: Eurostat

Figure 19.13 Characteristics of development trends of major categories of land transport in EU25 in the decade 1995–2005: (a) changes in total lengths of individual components of transport infrastructure [km]; (b) percent indices for data according to Figure 19.13a.

Source: [10], own listing and additional descriptions.

The personnel gap RLT is widening – currently estimated at approximately 1500 engineers, there are practically no specialists and no engineering and production base for new HSR technologies. Higher and medium-level education facilities for RT sector purposes are insufficient.

19.2.2.2. Financing Poland's RLT as the expression of Poland's transport policy contrasted with global and community processes

The share of rail investments in total land transport investments increased for the whole OECD from 17% to 23% in 1995–2001. The trend depends on developments in situation, especially in Western countries, where investments in rail infrastructure is a growing trend. Countries in Eastern and Central Europe invest more in roads. Details are shown in Figure 19.15.

In Western Europe, the share of investments in railway infrastructure increased (from approximately 20% in 1975) to 30% in 1995, and then to 40% in 2011, as the share of total investments in land modes of transport. In countries of Central and Eastern Europe, it was

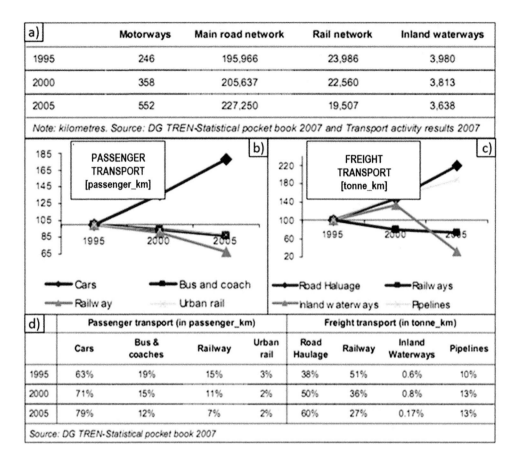

a)	Motorways	Main road network	Rail network	Inland waterways
1995	246	195,966	23,986	3,980
2000	358	205,637	22,560	3,813
2005	552	227,250	19,507	3,638

Note: kilometres. Source: DG TREN-Statistical pocket book 2007 and Transport activity results 2007

d)	Passenger transport (in passenger_km)				Freight transport (in tonne_km)			
	Cars	Bus & coaches	Railway	Urban rail	Road Haulage	Railway	Inland Waterways	Pipelines
1995	63%	19%	15%	3%	38%	51%	0.6%	10%
2000	71%	15%	11%	2%	50%	36%	0.8%	13%
2005	79%	12%	7%	2%	60%	27%	0.17%	13%

Source: DG TREN-Statistical pocket book 2007

Figure 19.14 General characteristics of development trends of major categories of land transport in Poland in the decade 1995–2005 [5], [12]: (a) changes in total lengths of individual components of transport infrastructure [km]; (b) changes in passenger transport structure; (c) changes in cargo transport structure; (d) percent indices for share of major modes of land transport, passenger and freight

Source: [12], own listing and additional descriptions

the opposite. The share of investments in road infrastructure increased, from 66% in 1995 to 84% in 2005, and the share of investments in rail infrastructure decreased from 24% in 1995 to 13% in 2011.

Transport policy of the Russian Federation is of great importance to the Europe – Asia transit. In the similar period of 1995 to 2011, the share of investments in road transport, as share in land transport infrastructure decreased, respectively, from 60% in 1995 to over 45% in 2011, and the share of investments in rail infrastructure increased from 37% to over 53%, respectively.

Comparison of investments in road and rail infrastructure in Poland is shown in Table 19.1.

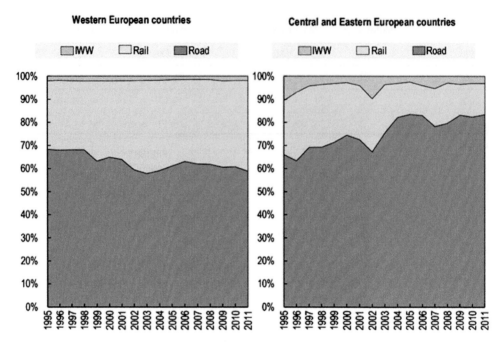

Source: International Transport Forum at the OECD.

Figure 19.15 Shares [%] of infrastructural investments in Western Europe Countries (*WECs*[19]) and Central and Eastern Europe Countries (*CEECs*[20]) in land transport (Road), railway transport (Rail) and inland waterways transport. OECD foreign exchange in EUR, as per present prices and exchange rates.

In Poland, in 1995 to 1999, the percent share of investments in rail infrastructure $RLI_\%$ = (RLI / Σ_I) × 100 [%] in total investments in land transport Σ_I = RLI + RDI were in the order 28–61%, and after the year 2000 they decreased many-fold, reaching approximately 10–16%. Similarly, percent share of railway infrastructure maintenance expenditures $UK_\%$ = (UK / Σ_U) × 100 [%] in total expenditure on maintenance of transport infrastructure $\Sigma_{U=}$ RLM + RDM were in the order of (40 to 81)%, in years 1995–1999, and after the year 2000 they decreased many-fold to approximately 2–8%.

In total road and rail transport investments Σ_T = Σ_I + Σ_U percent share of total rail infrastructure investments $RLIM_\%$ = {(RLI + RLM) / Σ_T}] × 100 [%] were 47.4% in 1995 and 9.6% in 2011, respectively, and approximately 13.7% in 1995–2011. Compared with similarly estimated total expenditure $RDM_\%$ = 86.3% on road infrastructure, this means increasing, average annual six-fold underfinancing of Poland's railway network as compared with road network.

19 As per OECD nomenclature: *WECs*: Austria, Belgium, Denmark, Finland, France, Germany, Greece, Iceland, Ireland, Italy, Luxembourg, Netherlands, Norway, Portugal, Spain, Sweden, Switzerland, and the United Kingdom.
20 *CEECs*: Albania, Bulgaria, Croatia, Czech Republic, Estonia, FYROM, Hungary, Latvia, Lithuania, Montenegro, Poland, Romania, Serbia, Slovakia, and Slovenia.

Table 19.1 Poland: Investments in road infrastructure RDI and rail infrastructure RLI, including maintenance – RDM and RLM, respectively, in years 1995–2011.

Investments in road and rail infrastructure [EUR million]

Year	1995	1996	1997	1998	1999	2000	2001	2002	2003	2004	2005	2006	2007	2008	2009	2010	2011	Total 95-11
Rail **RLI**	248	282	307	337	237	198	113	108	194	220	236	353	647	904	650	690	925	6649
Road **RDI**	638	180	227	299	297	1019	1094	1035	1010	1237	1875	2605	3443	4508	5340	6510	8319	39636
Σ_i = RLI + RDI	886	462	534	636	534	1217	1207	1143	1204	1457	2111	2958	4090	5412	5990	7200	9244	46285
$RLI_\%$ [%]	28.0	61.0	57.5	53.0	44.4	16.3	9.4	9.5	16.1	15.1	11.2	11.9	15.8	16.7	10.9	9.6	10.0	14.4

Infrastructure maintenance expenditure [EUR million]

Year	1995	1996	1997	1998	1999	2000	2001	2002	2003	2004	2005	2006	2007	2008	2009	2010	2011	Total 95-11
Rail **RLM**	585	406	249	118	83	59	45	39	68	77	82	67	100	36	157	213	239	2623
Road **RDM**	287	98	134	111	136	449	666	791	721	1055	1263	1670	1515	2006	2341	2636	2678	18557
Σ_u = RLM + RDM	872	504	383	229	219	508	711	830	789	1132	1345	1737	1615	2042	2498	2849	2917	21180
$RLM_\%$ [%]	67.1	80.6	65.0	51.5	37.9	11.6	6.3	4.7	8.6	6.8	6.1	3.9	6.2	1.76	6.3	7.5	8.2	12.4

Data according to [51] in EUR million, OECD foreign exchange in EUR, as per current prices and exchange rates. Own listing and calculations of indicators.

Rail transport is among the sectors of the economy that require active support of the state for its development and operations. Countries that want and are able to do so effectively and in the long term usually have sustainable and effective transport systems.

A modern state should have the policy and systemic mechanisms in place to ensure sustainable development of transport systems. Market mechanisms are not always sufficient to meet social needs and ensure national interests in this respect. In Poland, the extent of state participation in the development of RLT is fluctuating. This is due to the state of public finances, assumptions of social and economic policies (including transport policy), and EU requirements. Total transport investments in Poland in 2010 are estimated at 4.1% GDP [51]. Comparative listing of public aid for RLT in Poland and selected EU member states is shown in Figure 19.16.

The information and situation described in sec. 19.2.2.1 is complemented by Figure 19.17, illustrating the structure of public expenditure on major modes of transport.

As shown in Figure 19.17, expenditure on road infrastructure was nearly six times bigger than expenditure on rail infrastructure (82/13 = 6.31; see sec. 19.2.2.1 A). That was our original condition at the date of accession to European Union, and that is the main reason of increasing marginalization of rail transport in Poland, which has to be stopped as soon as possible.

The Supreme Audit Office (NIK) report confirms that the trend of financial discrimination of rail transport, as compared with the road transport, is solid [49].

Expenditure on railway lines was much smaller than expenditure on national roads, with similar lengths of both networks[21] (Figure 19.18).

In 2004–2011, total expenditure on national roads administered by the General Director of Public Roads and Motorways was PLN 108.4 billion, and total expenditure on railway lines administered by PLK SA – PLN 18.3 billion. In that period, expenditure on national roads were nearly six-fold bigger than expenditure on railway lines. The European Commission recommends that expenditure on road and rail is divided 60:40 [55]. In Poland, this ratio in the analyzed period was 86:14. This had adverse effects to competitive advantage of rail transport versus road transport and to implementation of sustainable development of transport policy [49].

For a variety of reasons, if only due to the energy situation in Poland, especially as regards petroleum-based fuels, we should be particularly partial to development of transport towards the model dominating in the EU [55], creating advantageous conditions for expansion of rail transport, including high-speed rails and electric urban transit. It's quite the opposite. This is demonstrated in Figure 19.19, where known diagram of PKP PLK for years 1998 to 2009 was complemented with more recent data and relevant conversion factors.

During the last 15 years, expenditure on road transport RDT has been increasing annually five to six times more than expenditure on railway transport RLT. The "motorway action" due to EURO 2012 shall certainly increase those discrepancies even further. In the analyzed period, the expenditure ratio RDE/RLE increased from 1.72 to 10.88. As it shows, all above-mentioned sources convey the same message.

---

21 The length of national roads network as of late 2011 was 18,801 km (GUS Statistical Yearbook).

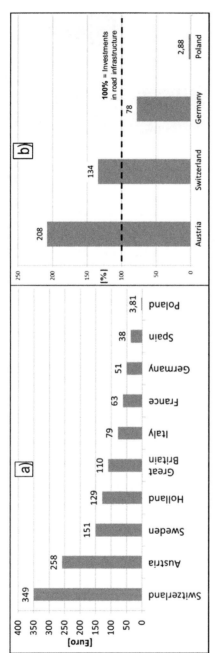

Figure 19.16 Public aid expenditure on rail infrastructure in 2012 in Poland and selected UE member states: (a) per 1 citizen [EUR]; (b) compared with road infrastructure expenditure [%]

Source: [54]; own listing and additional descriptions

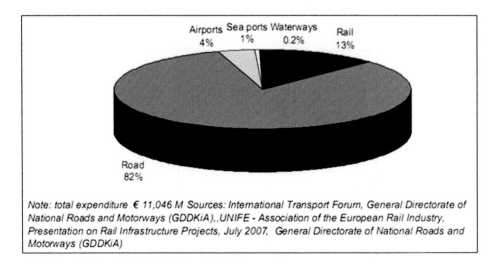

Note: total expenditure € 11,046 M Sources: International Transport Forum, General Directorate of National Roads and Motorways (GDDKiA),,UNIFE - Association of the European Rail Industry, Presentation on Rail Infrastructure Projects, July 2007, General Directorate of National Roads and Motorways (GDDKiA)

Figure 19.17 Poland's public investments in transport infrastructure for various modes of transport in years 2000–2006 (total expenditure) [1], [10]

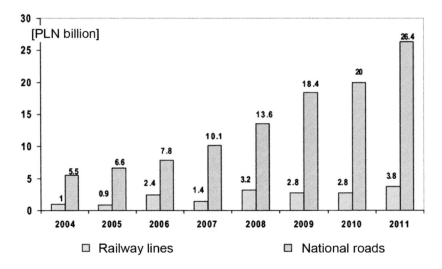

Figure 19.18 Comparison of expenditure on railway lines and national roads (in PLN billion) [49]

19.2.3. Internal and external assessment of Poland's RLT condition

The information in sec. 19.2.1 and 19.2.2 can be concluded as follows: presently, Poland continues its transport policy without considering current trends in the East and in the West. Expenditure on railway is dramatically small compared to expenditure on roads. There is only one conclusion.

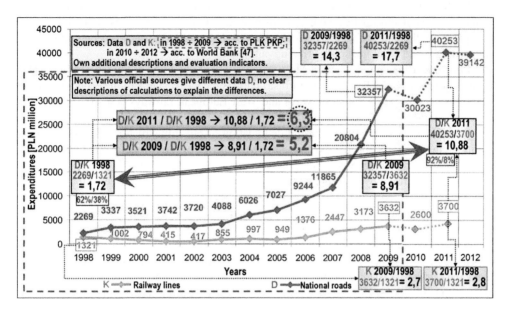

Figure 19.19 Comparison of expenditure on road and rail infrastructure in Poland in years 1998 to 2012

Sources: PLK PKP; The World Bank [68], own estimation of indicators

For years now Poland's transport policy has been dominated by interests in fuel-car-road sector, which led to the breach of equality in financing and development of road and rail transport.

This is contrary to development principles of European Union transport strategy and policy. The White Book of Transport 2011 [55], being the continuation of prior EU documents related to transport, defines transport objectives for next 10 years, and beyond, resulting in the need to implement general changes in transport. Particular attention is drawn to the need for deep reconstruction of the transport system, promoting independence from petroleum, creating modern infrastructure and multimodal mobility, assisted by smart management and ICT systems. Two problems, clearly defined in the White Book, have to be considered, which are particularly important to the subject of this study:

1. By the year 2030, 30% of road transport of goods for distances exceeding 300 km should be switched to other means of transport, such as railways or inland waterways, and by 2050 that share should be over 50%. Achieving that objective should require suitable infrastructure.

2. Completion of Trans-European high-speed rail by 2050; three-fold growth of the existing network of HSR by 2030, and maintenance of dense rail network in all member states. By the year 2050, the majority of passenger traffic should be by rail.

Implementing structural changes necessary to compete effectively with other modes of transport, and to take over the larger share of medium- and long-haul freight, is a challenge. This shall require material investments to increase the capacity of railway network [55].

Strategic interest of Poland in the EU requires that our operations are consistent with EU economic and transport policy, which means re-evaluation of the priorities in Poland's transport policy, in the manner ensuring sustainable development of transport systems and proper support of Poland's rail transport, utilizing public aid.

Long underfinancing is the root cause of present poor condition of Poland's railway. This is the major reason why Poland's rail infrastructure is the weakest link of the economy. Thirty years' distance between Polish railways and West European railways is a strong barrier in development of society and economy on both national and regional level.

Such situation results in increasing degradation of rail infrastructure and rolling stock, reduction in transport capacity and competitive advantage of rail transport, and consequently pushing Polish carriers from rail transport markets.

This is confirmed by the NIK, whose reports are sought after sources of information for authorities and society alike.

In the information [49] on results of the audit of infrastructural expenditures by PKP PLK in 2010 to 2012, and the performance assessment of government administration and rail infrastructure management with regard to ensuring sufficient resources for financing rail expenditure, with general negative performance assessment of rail infrastructure management and critical assessment of the minister responsible for transport, provides a few conclusions with significance to transport policy (own selection and simplification of text):

- progressing technical degradation of railway lines was and still is among the major reasons for decreasing competitive advantage of rail transport in the market;
- the fundamental reason for unsatisfactory technical condition of rail infrastructure is years of underfinancing;
- the EC recommended ratio of expenditure on rail and road infrastructure, i.e. 60% on roads and 40% on railways, was not observed; in 2004 to 2011 this ratio was 86:14;
- this had adverse effects to competitive advantage of rail transport versus road transport and to implementation of sustainable development of transport policy;
- financing of rail investments was not provided in time and in the amount allowing full absorption of EU funds;
- lack of national contribution may result in abandoning certain projects, or in failure to complete such projects in prescribed time or scope;
- negative assessment of organizational and institutional preparation for using EU funds;
- years of stagnation in construction and upgrade of railway infrastructure reduced competitive advantage of railway contractors;
- in the period subject to audit there were two governmental programs defining infrastructural investments of PLK S.A.:

  - Program of establishment and launching of High-Speed Rail operations in Poland[22];
  - Multiannual Railway Investment Project by 2013 with perspective to 2015 (WPIK);

- tasks defined in Multiannual Railway Investment Project by 2013 with perspective to 2015 are at the risk of not completion, which shall result in the risk of losing some EU financing and the need for financing already commenced projects entirely from own funds.
- As regards the HSR Program, the position of NIK is not consistent with law.

---

22 Adopted by Resolution No. 276/2008 by the Council of Ministers on 19th of December 2008

NIK considers the decision by minister responsible for transport, taken on 27 January 2012 on halting the implementation of "Program of establishment and launching of High-Speed Rail operations in Poland," adopted by the Resolution by the Council of Ministers as justified.

No earlier resolution by the Council of Ministers on halting the HSR Program is known.

Ministers are obligated by CM resolutions and they have neither authority nor competence to change decisions by government.

Justification of NIK support for minister's breaking the law is unheard of.

In the consequence the Interdepartmental High-Speed Rail Team,[23] and later HSR Center of PLK SA, were closed. The status of frozen HSR Program and its further implementation require clear governmental declarations.

External assessment of Poland's transport policy is also highly critical. Report by the World Bank [68] is perfectly clear about that. Some theses of the report are presented below:

- The government should not delay implementing recommendations proposed by EU's Sustainable Transport Initiative;
- Actions should be taken, and soon, that shall force material changes in transport policy;
- More uniform distribution of investments among transport modes could accelerate its sustainable development;
- The method of dividing resources among individual transport modes or improvement in the railway competitive advantage can significantly improve the financial balance of the whole sector;
- Such changes shall require institutional reforms that should be implemented now in order to ensure long-term sustainable development and efficiency of the whole sector;
- Delaying decision (among others on HSR – author's note) indefinitely may lead to significant increase in costs and prevent Poland from completing both national and community goals.

Also, the European Commission strongly criticized Poland's transport policy [50], [57]. Some theses of the report are presented below:

- Threats to the stability of transport system are mostly due to insufficient expenditure on maintenance of rail infrastructure, preferential treatment of road transport, unsatisfactory functioning of public transit, low integration of various modes of transport, and insufficient use of Smart Transport Systems (STS);
- The rail sector is underinvested. During the last 25 years, underinvested rail transport was significantly degraded, with lines being closed and rail traffic reduced;
- Very high tack access fees, as compared with other European countries, halt the expansion of this mode of transport versus the road transport;
- Inflexible and unsure government financing of rail projects. Insufficient administrative and organizational skills of institutions managing rail infrastructure;
- Consequently, rail transport lost its competitive advantage over road transport;

---

23 Called and closed by Prime Minister – respectively: Res. no. 15 of 28/2/2008 and Res. no. 36 of 27 April 2012.

- Power consumption in the transport sector increased by 6% in 2005–2011, resulting in increased dependence of Poland on petroleum products, which is the highest in the EU (3.5% PKB);
- Bottlenecks and inefficient transport network, power grids, and ICT networks keep hindering the growth potential in Poland.

Also in this case all above-mentioned sources are unanimous.

19.3. National program for development of RLT and construction of HSR

19.3.1. Complementary elements of Poland's HSR condition diagnosis

The view of Poland's transport reality resulting from vast documentation presented in sec. 19.2 offers premises to the conclusion that Polish railway transport requires not only a development program, which may prove insufficient, but immediate rescue operations.

For many years now the major inhibitors of development are instability of Poland's transport policy, breaking the continuity of decision-making process, incorrect structure of financing various areas of transport, and poor engineering condition of rail transport.

Consecutive governments keep avoiding strong involvement in railway matters, counting on the possibility of indefinitely delaying costly or unpopular decisions that could remedy some of the railway issues. Without determination in Parliament and government actions, the obsolete rail transport shall decelerate development of the country and expose Poland to ridicule at the EU level. We will lose revenues and jobs. We will lose the battle for Europe – Asia rail transit. Delaying the HSR project widens the gap between Polish railways and EU railways by decades, which shall eliminate Poland from effective use of TEN-T corridors.

There is no lack of strategic vision or concept in the railway sector. There are various long-term strategic documents developed regularly, with objectives and tasks of the state in the area of transport system development, including rail, especially railway, transport. They are, however, consistently disregarded and not implemented, remaining in the sphere of declarations, and policy and decision-makers are not brought to political responsibility due to relatively short election cycle, determining discontinuity of transport policy.

In practice, there is no political will to create systemic components for organized, sustainable, and long-term development of transport, such as a stable system for sustainable financing of (all the modes of) transport; completion of structural transformations in the sector; pro-development considerations and stimulation for cooperation of science and business, leading to increase in innovative advantage of rail transport system; an efficient managing mechanism providing administrative and organizational capacity for proper implementation of rail transport investments and utilization of appropriated financial resources; stable optimization system of decision-making processes and maintaining the lobbying, to the extent permitted by laws; solid political tool for consistent cooperation in transport, economic, foreign, scientific, and educational policy, etc. Hence there is an absence of modern, long-term, and ongoing transport policy, making use of good experiences of other countries, advanced analytical methods and models of transport and logistic systems.

In this context, it is worth noting the highly critical assessment of Poland's transport policy by the Ministry of Infrastructure (2010).

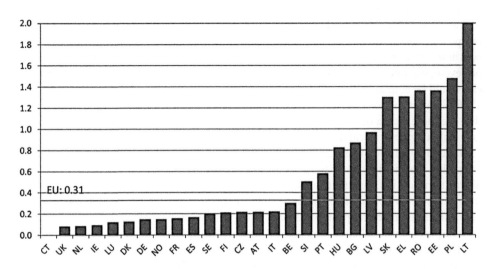

Figure 19.20 Fatalities in EU: Fatalities per million train-kilometers (2009–2011) [54]

"The technical condition of rail infrastructure in Poland has to be generally assessed as poor and declining." Small public resources appropriated in 1991 to 2006 for upgrade, overhaul, and maintenance led to such degradation that in many cases they cannot be made up for. The result is abandoning rail transport, in many areas of Poland, resulting in limited accessibility to transport, which is among the worst in EU to begin with [69].

Persistent underfinancing of railways led, among others, to dramatic reduction in expenditure on substitution and overhaul projects following 1990, poor technical condition, aging and degradation of engineering objects, rolling stock, and the whole infrastructure, making it insufficient for modern needs, reduced traffic safety (low degree of traffic control automation, absence of operation safety systems for speeds exceeding 160 km/h, insufficient number of two-level crossings with roads, etc.)

This results in very low safety of Poland's rail transport (Figure 19.20). Our fatality rate is five times the average rate in the EU.

Such barriers have a serious impact on social and economic development, both at the national and regional levels. Condition of RLT limits mobility of workforce and possibilities of integration of labor markets of Poland's metropolitan areas. It also hurts jobs related to railway transport RLT, especially in the mind of the youth. Vocational education is ruined. Population of academic staff is dwindling,[24] and there are no HSR professionals. Possibilities for ongoing education to improve skills of the industrial workforce and adaptation of the education structure to the needs of the market place are getting smaller and smaller.[25]

24 Pursuant to estimations, the number of professors and assistant professors who occupy themselves with rail transport problems is approximately 50, the number of Ph.D.s and M.Sc.s over 250, and 200 engineers (mostly employees of departmental institutes). Personnel, especially top personnel, is getting older and older; there are no designers, process engineers, or professionals in rolling stock, infrastructure, and control systems areas.

25 In 1980 and 2000, when rail freight and passenger traffic in Poland decreased dramatically, as expenditure on upgrades and development of infrastructure were reduced, so was the production of rail industry, bringing the closure of many industrial plants. Departments and faculties of high schools dealing with rail transport were gradually closed. All railway technical schools were closed. The process continues.

Rail transport industry offers approximately 100 thousand jobs. When railway operations are reduced, with no modernization effort, both production and services catering to rail transport sector shall be quickly reduced. Future restoration of specialized personnel and fixed assets shall be costly and time-consuming, and a great portion of EU funds earmarked for developing rail infrastructure and rolling stock may be lost due to the absence of people who would be able to implement rail transport projects.

Railway laboratories of rail transport research centers are obsolete and underinvested. They are more than a dozen years behind. Research works on the development of rail transport, including HSR research, is next to impossible in such an environment.

Discontinued and inconsistent transport policy is not compliant with research and education policy. This is the reason for years of marginalization of R&D work related to rail and railway transport in financing from sources managed by consecutive Ministers of Education. The phenomenon is amplified by the system of research project evaluation, dominated by more numerous and better positioned lobby groups from other scientific areas. Consequently, rail transport is mishandled in the National Research Program.

The present situation gives grounds for stating the current condition of RLT as perpetual underdevelopment, caused by adverse effects of errors in Poland's transport policy accumulating for years now.

The present condition of rail transport, which is the weakest link of Poland's economy, has a very serious impact on social and economic development, both at national and regional levels.

The Polish market, a part of international commercial exchange, can be consistently marginalized, which in addition to adverse economic effects can also have strong negative impact on Poland's position within the EU.

In order to reverse negative trends in rail transport, Poland has to take a huge and long-lasting effort, both politically and economically, with the strategic goal of reaching after 2030 the average European level (as compared with UE15 member states) in rail transport. This process should start as soon as possible.

19.3.2. Orientation of Poland's transport policy

If our goal is to overcome those 30 years of RLT backwardness as compared with UE15, we have to root for two-speeds Europe, only ours needs to be higher.

When financing of rail transport and road transport is equilibrated, and the transport policy is fully consistent with the EU's, only then the progress of underdevelopment can be stopped.

Seeing the special role played by RLT as one of the main factors conditioning the development of sustainable growth of the country and the regions, bear in mind the bigger picture – development of Poland's whole rail transport network, not only HSR and other elements of Polish Trans-European Network TEN-T, but also metropolitan railways combined with HSR and multimodal urban and interregional transport, appreciating the civilizational role of society's access to transport and its contribution to social integration and development, especially economic development.

Such orientation of Poland's transport policy is required to respect the principle of continuity of such policy and political will to finance and implement it for next 20 years or more.

Development of rail transport, especially high-speed rail transport, is the fundamental requisite of the Polish economy's jump into the future. The matter is exceeding the competences and capacities of the RLT sector, and requires open-minded political care about the development strategy in this area. Development of consistent RLT network, including

HSR, plus implementation of the Multiannual Program (sec. 19.3.3) will have far-reaching consequences; bringing Polish society the expected and desirable system and utility benefits, among others, the following:

- system benefits: increased mobility of the society, integration of labor markets, reduction of unemployment, improved accessibility of cities and regions, better integration of the country, accelerated and sustainable development of regions, securing the strategic role of Poland in transcontinental system of rail transport (in particular along W–E axis) and Poland's position within EU;
- utility benefits: high quality and efficiency of railways, reduced transit times, increase in transport capacity, creating the portfolio of competitive rail services, improvement in rail traffic safety, energy savings, and reduction in environmental pollution, road congestion, traffic accidents, and internal costs of the transport sector.

Only the Parliament and the government can ensure consistency of work and policy of continuation in the area of rail transport, elimination of lobbying in transport policy and restoration of sustainable development of rail and road transport policy with healthy financing ratios.

Achievement of the above requires a significant breakthrough in politicians' and society's thinking. It requires agreement across the political scene and understanding of the fact that transport has fundamental significance to economies of Europe and Poland. Its development should be considered as the national priority, determining the pace of our development, without political aspects.

We also need society's support and understanding of such development strategy, defined as national development program for rail transport (RT), especially railway transport (RLT), including HSR – with maximum contribution of Polish educational, scientific, engineering, and industrial circles related to RT.

Without creating a strong, uniform transport system, including railway transport system, to replace the existing patchworks of rail links, Europe will not by capable of rapid development and citizens shall not experience wealth. Poland, being a EU member state, has to be an equal participant in this process, possibly competitive in its own rights.

Polish science and industry should be more and more involved in such national program.

Poland has no other option for development. Managing such a critical situation requires exceptional political will and wisdom, responsibility, courage, effectiveness in decision-making process, and efficient acting on the part of all relevant state bodies, including in particular the Parliament and the government, for at least the coming decade.

It is worth noting here that the financial framework for 2014–2020 provides for an increase in EU financing of the railway sector by 85% – up to EUR 10.2 billion. Additional EUR 4.5 billion shall be invested in sustainable urban transit systems [50]. This may assist decision-making processes and quickly reverse negative trends in our passenger transport and freight, provided we can use it.

19.3.3. Multiannual program (MAP) "By Railway to the 21st Century"

With underdeveloped railway transport (RLT), the regress of its intellectual resources – both in education and R&D, necessary for quick restoration of pro-development environment, requires taking special actions in order to restore pro-development environment for RLT. Implementation of the national program for development of rail transport RT, especially railway transport RLT, including for HSR, requires timely training of specialists for RLT, including for HSR.

Organized efforts have been undertaken in RT sector, providing support for scientific, engineering, and educational needs of RT, in order to establish a country-level system for comprehensive merit support for such enterprise. National scientific and research bodies and business entities participating now in development of the project Multiannual Program MAP titled *"By Railway to the 21st Century – Scientific, Engineering and Educational Support System for Development of Rail Transport and Integrated Systems of Regional Transport."* The Rail Transport Consortium (RTC) called for joint implementation of the MAP, which is in the organization phase. RTC groups practically all research, development and education bodies (R&D&E) related in their core business with RT, including twelve public high schools, eight research institutes, four academies and federations, as well as GRUPA PKP. A group of business entities, especially plants producing for rail transport, are now in the process of creating a business consortium. The consortium is a structure open to all national and international entities interested in collaboration for development of Poland's RT.

The strategic objective of RTC for 2015 to 2030 and beyond is education of personnel and mobilization of research and financial resources, both national and European, for the purpose of comprehensive, interdisciplinary development of modern rail transport, and in particular creating broad potential, on a European level, for development of modern research, technologies, and implementations of new RT products, as well as non-engineering aspects of RT – from economics, to safety, to ecology, and in addition also a comprehensive upgrade of research and education facilities and formation of specialists for RT sector. Far-reaching effects of the MAP should be visible in years 2022–2030.

Implementation of MAP should involve, in particular, organization and education of scientific and didactic personnel, restoration and (re)construction of specialized research, and didactic, material, and engineering facilities; development of engineering concepts, interdisciplinary research projects, and advanced RT technologies; upgrade and development of Polish RLT system, including HSR, as the component of TEN-T, in order to improve the competitive advantage of RLT; development of commuter rail systems (CRS), smart transport systems (STS), and intermodal transport systems (ITS), integrated with HSR; development of intermodal freight; rationalization of energy consumption in RT and development of energy-saving power supply technologies; reduction in environmental impact of RT; timely formation of specialists for upgraded conventional railways and innovative technologies for RT, including HSR; maximization of contribution by Polish science and industry in RT development program, including HSR; tightening of international cooperation and participation in EU programs referring to RT; and safety of transport in the context of global crisis related to petroleum-based fuels.

From the international perspective, the essence of the matter is ensuring economic benefits, in the form of possibly high and sustainable revenues from taking over of international transit operations. Strategic geographic location of Poland, in the center of Europe, allows Poland to play a key role in international rail transport, both in the W–E and N–S axis, and collecting significant economic benefits in the process. This is our big, national interest. To that end, Polish pro-development operations in RLT have to consider complementary or competitive nature of major strategic projects in RLT, including HSR, created in our neighborhood and show why the HSR is indispensable for Poland's growth. Launching of HSR will release the capacity of existing lines in TEN-T network for freight. Abandoning the HSR project means marginalization of the whole rail transport, resulting in marginalization of Poland's position in Europe. Poland can no longer be a European bottleneck.

From a national perspective, the essence of the matter is civilizational importance of RLT, including HSR. Availability of transport determines sustainable growth of regions and the country, as well as integration processes, including economic integration processes. Anticipated effects of MAP implemented by RTC entail:

- effects to the country: sustainable development of regions, improved social and economic cohesion and integration of labor markets, expansion of new industry sectors servicing the RT sector, including HSR, reduction of unemployment, development of interdisciplinary research for the benefit of the economy, innovative concepts and technologies promoting efficient consumption of electric energy, rationalization of energy consumption, improvement of national energy security, improvement of transport safety, improvement in environmental protection;
- effects to RT sector: development of new RT areas in Poland and of international cooperation, integration of RT circles around the common objective, participation in European framework programs and utilization of EU resources, development of advanced technologies for RT, development of interdisciplinary research for RT, coordinated development of energy saving technologies for RT, propagation of cartridge-based power supply systems in RT, recuperation of electric energy in RT.

Due to above considerations, establishment by the Government of Multiannual Program titled *By Railway into 21st Century*, and execution of the program by relevant scientific, R&D, and economic entities, shall be one of the tools for implementation of the new strategy and policy in Poland. In 2015, the MAP project will be submitted for consulting and approval to relevant ministers.

Undertaking, on a national level, of substantial and organizational works on MAP and RTC, initiated by TS sector, is the expression of support by non-governmental organizations within the RT sector to government actions with regard to the new transport strategy and policy

19.3.4. Final remarks

It is government's duty to relieve rail transport from its current crisis, with the profit for both Poland and European Union, which contributes to make than happen.

Only the government can, and should, prepare the national development program for rail transport (RT), especially railway transport (RLT), including HSR – with maximum contribution of Polish educational, scientific, engineering, and industrial circles related to RT sector, and then establish such plan in suitable legal form. There is no doubt that all above circles related to RT shall support the government in doing so. The process of self-organization of such circles, that already began, is the proof enough of that.

The government should also support efforts aiming at effective establishment of Multiannual Program *"By Railway to the 21st Century"* by the Council of Ministers, and undertaking its implementation by relevant scientific, R&D, and business entities. Then, and only then, quick achievement of expected results, as below, shall be possible:

- improvement of the present condition of rail transport;
- obtaining by 2030 the average European level in rail transport, including HSR;
- safety of transport in the light of global crisis in petroleum-based fuels [70];

- timely formation of specialized personnel for new technologies in RT, including HRS, and in particular:

  - organization and training of scientific and didactic personnel;
  - establishing dedicated scientific and didactic facilities;
  - establishing dedicated material and engineering facilities;

- maximization of Polish science contribution to RT development;
- maximization of Polish industry contribution to RT development.

Creating the high-speed rail system in Poland, thanks to application of state-of-the-art technology and maintaining strictest environmental standards, should make HSR the symbol of modern passenger railway in Poland, allowing successful improvement of rail transport image and making it attractive among the public, which is extremely important for supply of new persons to work for RT, RLT, and HSR and for the acceptance of RT development program by the public.

Finally, three detailed strategic problems solving which significantly facilitate implementation of proposed development projects.

- The new transport strategy and policy have to be implemented in close coordination with energy policy. Increase in the share of rail transport in passenger and freight operations and preferences for electric traction shall require significant changes in power supply systems, both substations DC 3 kV – to increase the speed on main railway lines, and substations AC 2×25 kV/50 Hz – projected for new applications in RT (including HSR). In both cases increase in the power rating shall be required, by way of complete overhaul of the power supply system for DC 3 kV and introducing completely new solutions for AC 2×25 kV/50 Hz. Expansion of MV and HV power supply and distribution grids and traction substations, development of mass manufacture of electric devices and appliances, and equipment for RT, studies, research, and implementations of energy saving solutions, container control systems and traction drives, allowing energy recuperation, etc.
- Important political instruments for creation of pro-development conditions for RT are, among others, following instruments, postulated for years and objected to by Minster of Finance although proven everywhere else in the world:

  - economic stimulators aiming at increase in BERD (business expenditure on R&D), in the form of the system of tax relief for business entities, e.g. CIT (Corporate Income Tax) relief;
  - the principle of appropriating funds for R&D (e.g. 5%) in investment expenditure;
  - increase in the level of public expenditures on R&D in the RT area.

An intelligent and permanent political umbrella will be required to protect the development of RT, especially RLT and HSR.

It is worth mentioning that proposed enterprises are fully consistent with assumptions of current state transport policy, which provides for:[26]

Halting degradation and gradual improvement of the condition of railway infrastructure in on the network level. This requires simultaneous implementation of modernization

26 "State Transport Policy 2006–2025"

projects within core railway corridors and "restoration" projects, allowing elimination of speed restrictions and restoring normal operating speeds and other engineering parameters on other lines; this can be achieved by providing state infrastructure managers with resources, in the form of central subsidies, necessary to provide own contribution when applying for co-financing from EU funds, as well as for financing of projects outside such EU co-financing plans.

Bibliography

[1] Anuszczyk, J., Bartosik, M., and Wiak, S: *Strategiczne problemy rozwojowe transportu kolejowego w Polsce 2012.* (*Strategic development problems of rail transport in Poland 2012*). XV Ogólnopolska Konferencja Naukowa Trakcji Elektrycznej SEMTRAK 2012. Kraków, 2012, s. 187–202.

[2] Raczyński, J.: *Rządowy program budowy linii dużych prędkości w Polsce.* (*Government program for building high-speed rails in Poland*). Technika Transportu Szynowego TTS, nr 9, 2008, s. 17–26.

[3] Anuszczyk, J., Bartosik, M., and Wiak, S.: *Linia Y kolei dużych prędkości jako rozwinięcie korytarza wzrostu Rail Baltica.* (*The Y line of high-speed rails as a development of the Rail Baltica growth corridor*). XV Ogólnopolska Konferencja Naukowa Trakcji Elektrycznej SEMTRAK 2012. Kraków, 2012, s. 177–186.

[4] European Commission. *Trans-European transport network: TEN-T priority axes and projects 2005.* Luxembourg: Office for Official Publications of the European Communities, 2005. ISBN 92-894-9837-4, http://ec.europa.eu/transport/infrastructure/maps/doc/ten-t_pp_axes_projects_2005.pdf.

[5] Adamiec, J.: *Transeuropejskie sieci transportowe (TEN-T).* (*Trans-European transport networks TEN-T*). *Studia BAS*, 2012, nr 4(32), s. 63–78, http://orka.sejm.gov.pl/WydBAS.nsf/0/9B2B8A8 0F3495587C1257ADB003C1B28/$file/Strony%20odStudia_BAS_32-5.pdf; www.bas.sejm.gov. pl.

[6] Ministerstwo Inwestycji i Rozwoju: *FunduszTENT. Konkursy w roku2013.* (*TEN-T fund. Competitions in 2013*). www.mir.gov.pl/aktualnosci/Transport/Strony/FunduszTENT_Konkursy_w_roku2013_5122013.aspx

[7] Ministerstwo Transportu: *Europa zbuduje połączoną sieć transportową.* (*Europe will build a connected transport network*). http://logistyka.wnp.pl/europa-zbuduje-polaczona-siec-transportowa,167171_1_0_0.html.

[8] Ministerstwo Inwestycji i Rozwoju: *Korytarze_transportowe_ważne_dla_rozwoju.* (*Transport corridors important for development*). www.mir.gov.pl/aktualnosci/Transport/Strony/Korytarze_transportowe_wazne_dla_rozwoju .aspx.

[9] European Commission: *Trans-European Transport Network. TEN-T core network corridors.* http://ec.europa.eu/transport/themes/infrastructure/ten-t-guidelines/corridors/doc/ten-t-corridor-map-2013.pdf

[10] Anuszczyk, J., Bartosik, M., and Wiak, S.: *Strategia edukacyjnego i naukowego wsparcia rozwoju KDP i ŁKA.* (*Strategy for educational and scientific support for the development of HSR and LAR*). Politechnika Łódzka, CTS CETRANS. Uwarunkowania zewnętrzne, założenia merytoryczne założenia organizacyjne. Studium projektowe. Łódź, 2010.

[11] Szeląg, A., and Mierzejewski, L.: *Systemy zasilania linii kolejowych dużych prędkości jazdy.* (*Power systems for high-speed rail lines*).Technika Transportu Szynowego tts, nr 5–6, 2005, s. 80–90.

[12] Hildebrandt, A.: *Międzynarodowy handel morski.* (*International maritime trade*). Instytut Badań nad Gospodarką Rynkową. Gospodarka morska nr 2/2009, 41, http://ppg.ibngr.pl/gospodarka-morska/miedzynarodowy-handel-morski.

[13] Wydawnictwa Edukacyjne WIKING 2005-2008: *Handel zagraniczny w Europie*. (*Foreign trade in Europe*). www.wiking.edu.pl/article.php?id=297

[14] Wydawnictwa Edukacyjne WIKING: Czołowi eksporterzy i importerzy na świecie w 2008 roku. (*Leading exporters and importers in the world in 2008*). www.wiking.edu.pl/definiction.php?id=296&width=700&height=540.

[15] Rozporządzenie Parlamentu Europejskiego i Rady (Regulation of the European Parliament and of the Council) nr 1316/2013 z dnia 11 grudnia 2013 r. Dziennik Urzędowy Unii Europejskiej (Official Journal of the European Union) 20.12.2013, L 348/129, http://eurlex.europa.eu /LexUriServ/ LexUriServ.do?uri=OJ:L:2013:348:0129:0171:PL:PDF.

[16] The Scandria project. *The Scandinavian-Adriatic Corridor For Growth And Innovation*. https:// www.scandria-corridor.eu/index.php/en/projects/scandria

[17] Baltic sea Region Programme 2002-2013. www.transbaltic.eu/wp-content/uploads/2009/12/ TransBaltic-Project-Data-Form.pdf

[18] Baltic Sea Region Project #026: *Scandinavian Adriatic Corridor for Growth and Innovation*. www.scandriaproject.eu/templates/File/dl-results/Wp%203.2/3.21-4_Scandria_Railway_ Corridor_Performance_120907-Report.pdf.

[19] Homann, Jespersen: *SCANDRIA Green Corridor*. Roskilde University, 2011. www.transbaltic.eu/ wp-content/uploads/2011/03/Scandria-Per-Homann-Jespersen.pdf.

[20] Scandria®2Act. *Sustainable and Multimodal Transport Actions in the Scandinavian-Adriatic Corridor*. https://www.scandria-corridor.eu/index.php/en/projects/scandria2-act.

[21] Šakalys, A., President of the EWTC Association: *The East–West Transport Corridor. Association – an innovative tool for cooperation along EWTC*. Proc. Of The East West Transport Corridor II Final Conference in Vilnius, Lithuania June 7–June 8 2012, www.ewtc2.eu /ewtc/project-news/ final-conference-documentation.aspx.

[22] East–West Transport Corridor Association. http://www.ewtcassociation.net/

[23] Naudužas, V., Ambassador for Energy and Transport Policy: *East West Transport Corridor II. Opportunities Greater Than Difficulties*. www.ewtc2.eu/ewtc/project-news/final-conference-documentation.aspx.

[24] Olsson, M.: *Port of Karlshamn – For Tomorrow's Business*, www.ewtc2.eu/ewtc/project-news/ final-conference-documentation.aspx.

[25] Zurba, S.: Coordinator of the EWTC II WP: *Business Opportunities in Railway Transport*. EWTC II Newsletter. Baltic Transport Journal, 2011, 5, p. 42.

[26] Stefańska, A.: *Mniejszy tranzyt przez Polskę?* (*Smaller transit through Poland?*) Rzeczpospolita. Ekonomia 24, 04.06.2012. www.ekonomia24.pl/artykul/706254,886571-Polska-wykluczona-z-tranzytu-koleja-.html.

[27] Kapczyńska, K.: *Rosjanie budują kolejowy objazd Polski* (*The Russians are building rail detour of Poland*). Puls Biznesu, 14.09.2012, http://logistyka.pb.pl/2666682,17901,rosjanie-buduja-kolejowy-objazd-polski.

[28] Kummer, S., Fürst, E., Stranner, G., and Ploberger, R.: *Die Breitspur-Anbindung des Twin-City Raumes Wien/Bratislava: Was geschieht, wenn nichts geschieht!?*(*The Breitspur connection of the Twin-City area Vienna / Bratislava: What happens if nothing happens?*) Schwarzbuch Institut für Transportwirtschaft und Logistik. Wirtschaftsuniversität Wien, 2009, www.wu.ac.at/itl/forschung / forschungsberichte/081017_breitspur_schwarzbuch_final.pdf.

[29] *CENTROPE, multinational region in Central Europe*. http://en.wikipedia.org/wiki/Centrope

[30] *Linia szerokotorowa do Berlina?* (*Broad-gauge to Berlin?*) onet BIZNES, 13.01.2011. http:// biznes.onet.pl/linia-szerokotorowa-doberlina,18493,4106623,3045147,92,news-detal?utm_ source=google&utm_ medium=cpc&utm_campaign=allonet1_ informacjesem_dsa _01.

[31] European Commission: *Draft regulation on guidelines for the development of the Trans-European Transport Network – maps of the comprehensive and the core networks*. www.consilium.europa. eu/ueDocs/cms_Data/docs/pressData/en/trans/129080.pdf.

[32] *Rail Baltica*. Final Report. Executive summary. AECOM Transportation, May 2011.

[33] *The Baltic Adriatic Axis. Element of the future European TEN-T Core Network.* bmvit. https://www.tsk.sk/buxus/docs//europskaunia/Broschure_BAA_100910_RAUMUMWELT_fin.pdf

[34] European Commission: *Rail. The Baltic – Adriatic Corridor.* Mobility and Transport. www.baltic-adriatic.eu/en/baltic-adriatic-axis/history; www.baltic-adriatic.eu/en/infopool.

[35] European Commission: *Transport Infrastructures – TEN-T. Revision of TEN-T Guidelines,* http://ec.europa.eu/transport /themes/infrastructure/revision-t_en.htm.

[36] *Adriatic – Baltic Landbridge.* Final Report, 2007.

[37] Gather M. (ed), Luttmerding A. (ed.): *SoNorA. South North Axis.* Proceedings of the 2nd SoNorA University Think Tank Conference. University of Applied Sciences Erfurt, Germany, 2009. ISSN 1868–8411.
https://www.fh-erfurt.de/fhe/fileadmin/Material/Institut/Verkehr_Raum/Publikationen/Think_Tank_4_Proceedings_v4.pdf

[38] PKP PLK S.A. Centrum Kolei Dużych Prędkości, Instytut Kolejnictwa: *Kierunki rozwoju kolei dużych prędkości w Polsce* (*Directions of high-speed rail development in Poland*). Warszawa, 2011. (Studium analityczne).

[39] PKP PLK S.A., *Mapa linii kolejowych w Polsce* (Map of railway lines in Poland). Warszawa, 2011.

[40] IDOM: *Prezentacja przebiegu trasowania KDP* (*Presentation of the HSR route*). Łódź, 22.09.2011.

[41] *Bez pieniędzy z UE na polskie KDP* (*Without EU money for Polish HSR*). Rynek Kolejowy, 20.10.2011. http://www.rynek-kolejowy.pl/wiadomosci/bez-pieniedzy-z-ue-na-polskie-kdp-45626.html

[42] *Nowak: Nie przełożyliśmy "Ygreka" o 20 lat* (*Nowak: We did not postpone "Ygrek" for 20 years*). Rynek Kolejowy, 08.02.2012. http://www.rynek-kolejowy.pl/wiadomosci/nowak-nie-przelozylismy-ygreka-o-20-lat-43122.html

[43] UCHWAŁA Nr 276/2008 Rady Ministrów z dnia 19 grudnia 2008 r., w sprawie przyjęcia strategii ponadregionalnej – "Program budowy i uruchomienia przewozów kolejami dużych prędkości w Polsce" (*Program for construction and launch of high-speed rail transport in Poland*).

[44] Stępień, R.: *Węzły multimodalne na przykładzie węzła łodzkiego* (*Multimodal nodes on the example of the Łódź node*). Materiały konferencji – "Development of multimodal hubs in Europe – Rail Baltica Growth Corridor." Łódź, 31.05j–1.06.2012.

[45] Wróbel, I., Calvet, B., Kruk, R., and Wiśniewska, K.: *Zasadność utworzenia systemu kolei dużych prędkości w Polsce* (*The legitimacy of establishing a high-speed rail system in Poland*).Warszawa: Instytut Kolejnictwa, czerwiec 2011, nr 4473/11.

[46] PKP PLK S.A. *Studium Wykonalności dla budowy linii kolejowej dużych prędkości "Warszawa – Łódź – Poznań/Wrocław"* (*Feasibility study for the construction of a high-speed railway line "Warszawa – Łódź – Poznań / Wrocław"*). Raport nr 16, Rewizja 2.1. Dokument końcowy. Warszawa, lipiec 2013 r.

[47] *KPZK2030, Koncepcja Przestrzennego Zagospodarowania Kraju 2030* (*The concept of the Spatial Development of the Country 2030*). Uchwała nr 239 Rady Ministrów z dnia 13 grudnia 2011 roku. Monitor Polski, poz. 252, Warszawa, dnia 27 kwietnia 2012 r, www.mir.gov.pl/rozwoj_regionalny/polityka_przestrzenna/kpzk/strony/koncepcja_przestrzennego_zagospodarowania_kraju.aspx

[48] *KPZK2030 – jw. streszczenie* (*– as above, summary*). www.mir.gov.pl/rozwoj_regionalny/Polityka_przestrzenna/KPZK /Documents/Streszczenie_KPZK2030_PL_small.pdf.

[49] Najwyższa Izba Kontroli: *Inwestycje Infrastrukturalne PKP PLK S.A. Informacja o wynikach kontroli* (*Infrastructural Investments PKP PLK S.A. Information about the results of the inspection*), nr ewid. 2/2013/P/12/078/KIN; KIN-4101–04/2012. Warszawa, 2013. www.nik.gov.pl/plik/id,4754,vp,6175.pdf.

[50] Komisja, Europejska: *Sprawozdanie krajowe – Polska, 2015 r.* (*Country Report Poland 2015*). Dokument roboczy służb komisji SWD(2015) 40 final. Bruksela, 26.2.2015 r.

[51] *Spending on Transport Infrastructure 1995–2011: Trends, Policies, Data.* International Transport Forum. OECD/ITF 2013, www.internationaltransportforum.org.

[52] Rosik, P., Komornicki, T., Kowalczyk, K., Szejgiec, B.: *Inwestycje i działania konieczne do podjęcia przez Polskę w celu wdrożenia korytarza sieci bazowej TEN-T Morze Północne – Bałtyk na terytorium Polski – w ujęciu krajowym i wojewódzkim, w średnim oraz długim horyzoncie czasowym (do i po 2020 r.) (Investments and actions necessary to be taken by Poland to implement the TEN-T core network corridor, North Sea – Baltic Sea, on the territory of Poland – in the national and voivodship terms, in the medium and long term, to and after 2020).* Polska Akademia Nauk, Instytut Geografii i Przestrzennego Zagospodarowania im. S. Leszczyckiego. Raport Końcowy, Warszawa, 27.11.2014.

[53] Korolewska, M.: *Wydatki publiczne na infrastrukturę transportu lądowego w Polsce. (Public expenditure on land transport infrastructure in Poland).* Studia BAS, 2012, nr 4(32), s. 79–124, www.bas.sejm.gov.pl.

[54] Sitarz M. (red.) z zespołem (raca zbiorowa): *Strategia rozwoju przemysłu i transportu szynowego w oparciu o innowacje oraz badania naukowe w Polsce do 2030 roku, Pol-Tech-Kol-2030, cz. I – III (Strategy for the development of industry and rail transport based on innovations and research in Poland until 2030, Pol-Tech-Kol-2030, part I – III).* Dokument Polskiej Platformy Technologicznej Transportu Szynowego, Warszawa, maj 2014 r.

[55] PL. Komisja Europejska: *BIAŁA KSIĘGA. Plan utworzenia jednolitego europejskiego obszaru transportu – dążenie do osiągnięcia konkurencyjnego i zasobooszczędnego systemu transportu (WHITE BOOK. Roadmap to a Single European Transport Area – Towards a competitive and resource-efficient transport system).* SEK(2011) 359, 358, 391 (wersje ostateczne). Bruksela, 28.3.2011, KOM(2011) 144, wersja ostateczna. www.senat.gov.pl/download/gfx/senat/pl/defaultopisy/296/3/1/026a.pdf

[56] European Commission. Mobility and Transport. Trans-European Transport Network TENtec. http://ec.europa.eu/transport/infrastructure/tentec/tentec-portal/site/en/maps.html.

[57] European Commission. SWD(2015) 40 final, Commission Staff Working Document: *Country Report Poland 2015.* {COM(2015) 85 final}. Brussels, 26.2.2015. https://ec.europa.eu/info/sites/info/files/file_import/cr2015_poland_en_0.pdf

[58] BUSINESS INSIDER POLSKA: *Apel samorządu woj. śląskiego ws. rozwoju linii kolejowej LHS (Appeal of the self-government province Silesia on the development of the railway line BGML)* PAP, 10.032014. https://archiwum.businessinsider.com.pl/kraj/apel-samorzadu-woj-slaskiego-ws-rozwoju-linii-kolejowej-lhs/6mvkr

[59] *Polska chce szerokiego toru do Wiednia? (Poland wants broad-gauge railway to Vienna?).* Kurier Kolejowy, 16.04.2013. www.kurierkolejowy.eu/aktualnosci/13577/Polska-chce-szerokiego-toru-do-Wiednia.html.

[60] Odpowiedź podsekretarza stanu w Ministerstwie Transportu – z upoważnienia prezesa Rady Ministrów – na interpelację nr 9226 w sprawie budowy linii kolejowej z Rosji do Wiednia z wyłączeniem Polski (Response of the Undersecretary of State in the Ministry of Transport – under the authority of the Prime Minister – to interpellation No. 9226 regarding the construction of a railway line from Russia to Vienna, excluding Poland). http://orka2.sejm.gov.pl/IZ5.nsf/main/2D6DD757.

[61] *Kolej szerokotorowa do Wiednia (Broad-gauge railway to Vienna).* www.porteuropa.eu/slowacja/gospodarka/5041-szerokie-tory-do-wiednia.

[62] *PKP Linia Hutnicza Szerokotorowa (PKP Broad Gauge Metallurgy Line).* http://pl.wikipedia.org/wiki/PKP_Linia_Hutnicza_Szerokotorowa.

[63] Ministerstwo Transportu, Budownictwa i Gospodarki Morskiej. *Strategia Rozwoju Transportu do, 2020 roku (z perspektywą do 2030 roku) (Transport Development Strategy up to 2020 (with a prospect until 2030).* Warszawa, 22.01.2013 r.

[64] Ministerstwo Transportu, Budownictwa i Gospodarki Morskiej. *Dokument Implementacyjny Do Strategii Rozwoju Transportu Do 2020 r. (z perspektywą do 2030 r.) (Implementation Document*

for the Transport Development Strategy Until 2020, with a prospect until 2030). Warszawa, sierpień 2014 r.

[65] Marcysiak, A., Pieniak-Lendzion, K., Lendzion M., and Drygiel T.: *Rozwój infrastruktury transportu kolejowego w Polsce w ramach II Paneuropejskiego Korytarza Transportowego* (*Development of railway transport infrastructure in Poland as part of the 2nd Pan-European Transport Corridor*). Uniwersytet Przyrodniczo-Humanistyczny W Siedlcach, Zeszyty Naukowe nr 97 Seria: Administracja i Zarządzanie, 2013.

[66] Jerzy Kwaśnikowski, J., Gramza, G., and Medwid, M.: *Transport kolejowy a system logistyczny Polski* (*Railway transport and Poland's logistics system*). Prace Naukowe Politechniki Warszawskiej, z. 76, Transport, 2010.

[67] PKP Polskie Linie Kolejowe S.A. *Mapa: Planowane modernizacje dróg kolejowych.* (*Map: Planned modernization of railways*). www.pkp.pl/node/344, pobr. 26.05.2012.

[68] The World Bank. Report Nr. 59715-PL. *Polska. Dokument dotyczący polityki transportowej. W kierunku zrównoważonego rozwoju transportu lądowego* (*Poland. Transport policy document. Towards the sustainable development of land transport*). Luty, 2011.

[69] Ministerstwo Infrastruktury. *Program działań dla rozwoju transportu kolejowego do roku 2015* (*Program of activities for the development of rail transport until 2015*). Warszawa, lipiec–sierpień 2010.

[70] Bartosik M.: *Apetyt energetyczny cywilizacji a szanse jej przetrwania* (*Energy appetite of civilization and the chances of its survival*). Artykuł w monografii PAN „Czy kryzys światowych zasobów?" Publisher: PAN, Komitet Prognoz "Polska 2000 Plus," Warszawa, 2014, ISBN 979-83-7151-591-0, s. 47–87.

Chapter 20

The effectiveness and financing of the high-speed rail structure

Tadeusz Dyr and Karolina Ziółkowska

20.1. Introduction

Specific features of the transport infrastructure decide about exceptional character of infrastructure investments, including the high-speed rails. A long planning and lifetimes, technical and economic indivisibility, high capital intensity and stock intensity, spatial and functional immobility, presence of externalities, and the significant participation of public means in financing have direct influence on the scope of assessment of its effectiveness. Consequently, the public distinguish infrastructure investments from those carried out in other sectors of the economy [4]. These special features are reflected mainly in:

- the need to invest relatively big financial means;
- a long investment, structure, and operation preparatory phase;
- majority of modernization investments, perceived as activities with a smaller risk than the structure of new facilities;
- temporal changeability of operating costs, dependent on the structure and direction of executed transports, that are the factors on which the infrastructure manager has only a limited impact;
- a significant impact on the external environment, in particular to the natural environment.

Factors listed above have a significant influence on the methodologies for assessing the investment projects. Exchanged factors have a significant influence on the methodology of the assessment of investment projects. The studies concerning usefulness of creating the high-speed rail system in Poland which have been carried out for years confirm that. These works include both evaluation of variants consisting of the structure of new lines as well as modernization of the existing infrastructure in a scope that will enable conducting the high-speed rail (above 200 km/h).

Considering the premises given, a conception of the methodological assumptions of the effectiveness evaluation of the investment concerning the new high-speed rail structure are presented in this chapter. Also, conclusions of chosen studies including the authors' ones, containing the evaluation of such investments' effectiveness, are presented and characterized.

20.2. Methodology of assessing the efficiency of the high-speed rail structure

An important issue in the assessment process of the effectiveness of transport infrastructure investments, including the structure of new high-speed rail, is to determine its scope and structure. The analysis scope refers to its types. The structure is related to establishing detailed actions undertaken as part of each analysis type.

The specific features of facilities and infrastructure devices in the transport, as well as requirements of entities co-financing investments from public means (e.g. from the European Union funds), determine the need of the investment effectiveness evaluation in this sector both from the investor's point of view and from the socioeconomic dimension. The conducted analysis should answer two key questions:

- whether the project *is worth co-financing*;
- whether the project *needs co-financing* [7].

Answers to these two key questions are given by the evaluation of the economic and financial effectiveness. Its measures are:

- economic net present value (ENPV), economic rate of return (ERR), and B/C ratio in economic assessment;
- financial net present value (FNPV) and financial rate of return (FRR) in financial assessment.

The primary purpose of a financial analysis is to examine whether the project is financially effective and, on that basis, make a statement about whether it requires funding from public funds and to determine the funding level. Investments that demonstrate the financial effectiveness do not need such support. It is possible to obtain funds for their implementation from the capital market. Financially ineffective undertakings can be funded partially from the European Union's funds or from other public funds. However, their economic effectiveness is a condition. The flowchart in Figure 20.1 illustrates the financial assessment of infrastructure projects.

The public character of transport infrastructure investments and generating essential external effects decide on priority importance of the economic assessment. It enables the identification of undertakings which are socially valuable and should be carried out, because they bring external benefit, contributing to the region's development where they are located. The algorithm of the economic analysis is presented in Figure 20.2.

Achieving essential objectives of the financial assessment of train infrastructure investments (Figure 20.3), led from the investor's point of view (infrastructure manager's), requires:

- drawing the forecast of the cash flows' size and schedule;
- calculating efficiency indicators;
- establishing the size of funding the project from public funds;
- examining the financial sustainability.

The economic analysis appraises the project's contribution to the economic welfare of the region or country. It is made on behalf of the whole of society instead of just the owners of the infrastructure, as in the financial analysis. The key concept is the use of accounting shadow prices, based on the social opportunity cost, instead of observed distorted prices [8]. The structure of the economic assessment is presented in Figure 20.4.

Time horizon is the maximum number of years for which forecasts are provided. Forecasts regarding the future of the project should be formulated for a period appropriate to its economically useful life and long enough to encompass its likely mid- to long-term impact

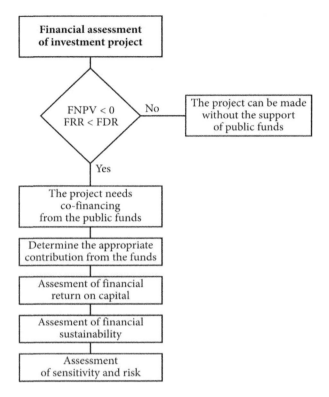

Figure 20.1 The algorithm of the financial assessment for the investment project in railway transport

Source: Own study

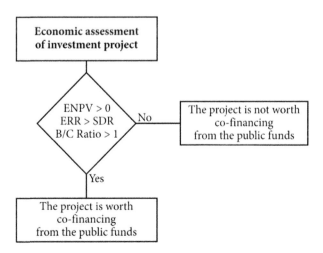

Figure 20.2 The algorithm of the economic assessment for the investment project in railway transport

Source: Own study

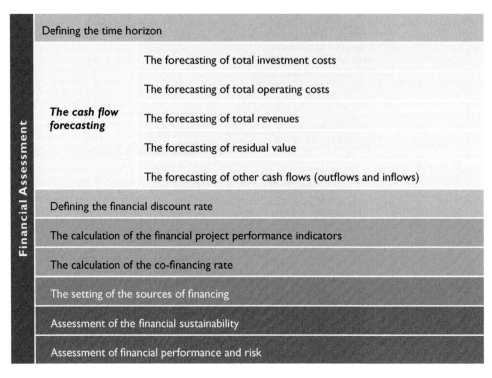

Figure 20.3 The structure of the financial assessment for the investment project in railway transport

Source: Own study

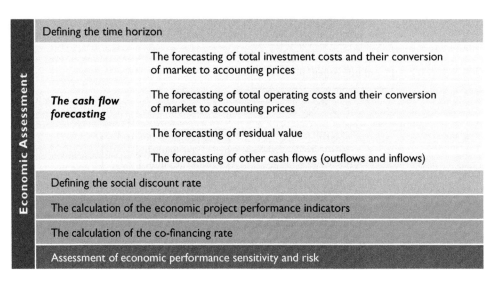

Figure 20.4 The structure of the economic assessment for the investment project in railway transport

Source: Own study

[8]. Reference time horizon recommended for railways investment projects is 30 years. This time horizon was used in the feasibility studies of building new high-speed rail lines in Poland and modernization of existing lines.

The discount rate used in the financial analysis should reflect the opportunity cost of capital to the investor. This can be thought of as the foregone return on the best alternative project. In Programming Period 2007–2013, the Commission recommended that a 5% financial discount rate in real terms be used as an indicative benchmark for public investment projects co-financed by the funds [7]. In the feasibility studies of building new high-speed rail lines in Poland was used the discount rate recommended by the Commission, i.e. 5%.

In the economic analysis, the Commission recommends using a social discount rate. Based on long-term economic growth and pure time-preference rates, the Commission proposed in Programming Period 2007–2013 the following indicative benchmarks for the social discount rate:

- 5.5% for the Cohesion countries;
- 3.5% for the other countries [7].

The Ministry of Regional Development, in agreeing with the Committee, recommended using the social discount rate for the transport sector projects in amount of 5%. Such a rate value was applied in feasibility studies of the high-speed rail structure. In some studies, an additional rate in the amount of 3.5% was used.

Current infrastructure investments in the rail transport in Poland have concentrated on modernization and revitalization of existing lines. Since half of the first decade of the 21st century, studies concerning the structure of new line adapted to the speed of at least 300 km/h have begun. In 2006 in the Railway Research Institute in Warsaw (Instytut Kolejnictwa), a pre-feasibility study of the structure of Warszawa – Łódź – Poznań/Wrocław line was developed [23]. In the following years, this concept has been an object of detailed research in the technical, organizational, institutional, and economic scope. In 2011–2012, feasibility studies of the line's structure [20], and in 2015 a pre-feasibility study of its lengthening from Poznań to Berlin and from Wrocław to Prague, were developed [24]. Using results of the analyses included in these documents, methodological assumptions and an evaluation of economic and financial effectiveness of Warszawa – Łódź-Poznań/Wrocław line's structure were presented.

The presented methodological assumptions constitute a general framework for completion of the investment projects' feasibility studies in the rail transport. The adoption of detailed assumptions concerning cash flow forecasting is a crucial issue. These issues are presented in the following sections of the chapter.

20.3. Financial effectiveness of the high-speed rail structure in Poland

Experience in the implementation of cost-benefit analysis shows that the cash flows with the greatest impact on financial effectiveness of the structure and modernization of railway lines investments are investment expenditure and operating cost (of railway lines maintenance and use). The operating income has a smaller impact. For the infrastructure manager they are income from track access charges incurred by operators. Charges for the minimum access to the infrastructure should be established after the cost that is directly incurred as the result of performing train transportation. In such a situation, the considerable part of running costs of the line maintenance is covered from the budget grant, not included in the project's cost effectiveness calculation. In such a situation, charges incurred by operators have only a limited impact on the project's financial appraisal results. The income, according to recommendations of Jaspers experts [13], are not included in the economic analysis.

The amount of investment expenditure for the modernization and the structure of railway lines depends on the designed maximum train speed. In the feasibility study, an investment expenditure that referred to track work, engineering work, traffic control and signaling, overhead lines, design and supervision, the acquisition of land, earth works, ERTMS, and other costs were estimated. It is planned that the total value of the expenditure on modernization and structure of the "Y line" Warszawa – Łódź – Poznań/Wrocław will amount to around 5.4 billion euros [20]. This amount can be increased by expenses for railway junction modernization (these expenses have already been considerably incurred) and construction of the tunnel in Łódź (around 230 million euros).

Railway line maintenance and operating costs are a reflection of the negative cash flow, generated in the period of operating the project, that burden the infrastructure manager. Determining the cost level correctly affects not only results of the planned investment's financial and economical effectiveness evaluation, but also the financial sustainability of the project. In case of projects generating income, there is also an impact on the level of the European Union's funding.

In the absence of generally accepted, standard methods of forecasting the infrastructure's operating and maintenance costs, in the relatively long project life (25–30 years), heuristic methods are applicable. They are based on the qualitative assessment of facts, experts' intuition, and an individual associational scheme that is a specific kind of understanding and predicting. The cost level is also determined by:

- level of complexity and technical diversity of the infrastructure;
- geographic conditions;
- operational burden [18], [11].

Forecasting maintenance and operations costs of railway infrastructure, they took into account recommended by Jaspers experts and resulting from reference system of infrastructure manager the division that include:

- fixed maintenance costs;
- variable maintenance costs;
- traffic operation (traffic management) costs;
- administrative costs connected with the project.

Unit values of mentioned cost categories were included according to the experience of infrastructure managers who operate the high-speed lines. In case of fixed maintenance costs, traffic operation, and administrative costs, total cost is a product of unit costs and track length. Total variable maintenance costs were calculated as a product of unit costs and the line burden resulting from the number of trains and their weight. Value of forecasting maintenance and operations costs are presented in Table 20.1.

Projecting the income for the infrastructure manager a maximum level of track access charges, which the operator could cover at the estimated level of the sales revenue from the cartage service and covering operating costs, that include financial and operating cost and

Table 20.1 Forecast of maintenance and operations costs of high-speed line Warszawa – Łódź – Poznań/Wrocław (thousand euro)

Costs	2030–2039	2040–2048
Fixed maintenance costs	29,770.0	29,770.0
Variable maintenance costs	16,020.2	17,298.8
Traffic operation	4761.2	4761.2
Administrative costs	3322.6	3322.6
Insurance costs	5442.4	5442.4
Total	**59,316.4**	**60,595.0**

Source: Own study based on [20]

generating the determined net margin (at 10% level) which would lead to conducting a profitable business, was determined. Analysis showed that the maximum, acceptable by operators, level of charge for the access to the infrastructure amounts around 12.85 euro/train-km. This value was compared to charges on chosen high-speed rail lines in Europe (Figure 20.5). In the analysis, it was emphasized, summoning the work [2], that in different European countries different systems and mechanisms of access to infrastructure charges are applied. It causes the incomplete comparability of the charges' levels [20], [15]. As for level of charges accomplished on other high-speed lines, it may be accepted, that this level is rational and acceptable. The projected access charge is about three times higher than at present on lines with the best parameters in Poland, i.e. on modernized conventional lines.

Total revenues was calculated as a multiplication of track access charges (euro/train-km) and railway traffic forecast (train-km/year). They amount from 160 million euro in the first year of the operating period up to 187 million euro in the final year.

Financial benefits included in the effectiveness analysis of the high-speed rail structure includes the financial residual value. This value was calculated on the basis of the income-based method. This value was estimated in amount of 2.3 billion euros.

Considering presented assumptions, financial effectiveness indicators of Warszawa – Łódź – Poznań/Wrocław high-speed lines structure were calculated. Calculation results are presented in Table 20.2. Results of the effectiveness evaluation from the operator's point of view and consolidated analysis results, calculated according to total cash flows of the infrastructure manager and the operator, were also included in it. For comparative purposes, evaluation results conducted in the study were also presented [15]. In this study, only consolidated analysis was conducted. However, two investment variants were taken into account, i.e. the structure of the new railway line adapted to speed of 350 km/h and the modernization of existing lines and their adaptation to speed of 200 km/h.

Values of indicators presented in Table 20.2 confirm the financial inefficiency of the high-speed rail structure. It is confirmed by FNPV/C < 0 and FRR/C < FDR. This development of indicators' value is typical for infrastructure investments. It means that the project needs financial support from public funds. However, covering the investment expenditure from public funds is sufficient for ensuring the project's financial sustainability, as in each year of analysis the income from track access charges are higher than maintenance and operations costs.

The conducted calculations confirm the effectiveness of conducting transport activity by the operator. Financial rate of return is relatively high (12.8%). Assessment of the financial

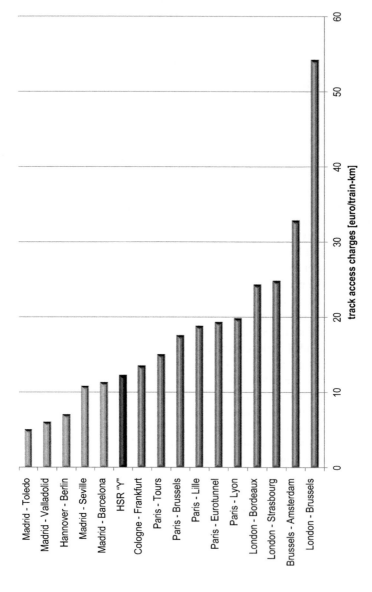

Figure 20.5 Track access charges in chosen high-speed lines in Europe

Source: Own study based on [20]

Table 20.2 Financial effectiveness indicators of Warszawa – Łódź – Poznań/Wrocław high-speed line structure [15]

Indicator	Effectiveness according to the point of view of [20]			Effectiveness according to [15] for maximum speed	
	infrastructure managers	operator	consolidated	200 km/h	350 km/h
FNPV [billion euros]	−4.3	0.5	−4.2	−2.57	−3.76
FRR [%]	−2.4	12.8	−1.6	lack	−8%
Financing gap [%]	77.1	0	72.0	94	88

Source: Own study based on [20], and [15], [10]

sustainability show that cumulated net cash flows in each year of analysis, even at the necessity of the capital repayment from credits for purchase of railway rolling stock, are higher than zero. By forecasted cash flows, the operator doesn't need support for the purchase of railway rolling stock.

20.4. Economic effectiveness of the high-speed rail structure in Poland

The main economic benefits from the structure of the high-speed rails are the following:

* savings of travel time;
* savings in vehicle operating cost;
* savings in costs of accidents;
* savings in costs of impact on environment, including costs of climate changes;
* savings in costs of noise.

Time as the abstract category, substantially elusive, has no value in economic terms. However, it has a determined value for everyone, whether the person is aware of it or not. Time passes by independently of the person's will who cannot speed it up or slow it down. In the life of every person time passes irretrievably; it belongs to limited resources and cannot be replaced by anything else.

It is possible, however, to substitute its use. So time is the highest and most irreplaceable good that should be constantly examined, controlled, and protected from mindless waste [3]. In this situation, the basic choice of every person concerns sharing time between rest and work. Giving up one hour of rest and starting work, it is possible to receive the additional quantity of consumer goods.

The planned modernization or structure of railway adapted to high-speed will enable shortening the travel time both for people travelling earlier by conventional trains, as well as for people taken over by the railways from other means of transport (passenger cars, buses, planes). The attractive travel time will also cause newly generated traffic.

In the European Union, a significant number of road accidents and fatalities from these accidents is a relevant social and economic problem. Actions undertaken through the European transport policy [22] simply seek to improve the existing situation, in particular reduction in the number of fatalities from road accidents. An increase in the speed of trains, enabling taking over passengers from passenger cars, is an essential instrument of fulfilling this purpose. As the level of security in train is much higher than in case of the journey by passenger cars, an increase of rail in transport market share causes a reduction in the number of accidents and their effects and costs.

A crucial factor of indirect benefits obtained as a result of infrastructure investments is the reduction of operating costs of passenger cars. Increasing train speed contributes, as already mentioned, to taking over a portion of passengers from the road transport. In consequence, exploitation work carried out by road vehicles will decrease. That leads towards savings in operating costs of these vehicles. However, instead of using one's own car, using the train causes the need to incur additional costs for the purchase of tickets. That can also be connected with other expenses like purchase of the public transport ticket for the journey to the railway station. When forecasting savings of vehicle operating costs, cost reduction by expenditure resulting from a different way of travelling should be done [4].

The improvement in the state of the natural environment and the counteraction against climate change are one of the most important areas in the European Union's policies. Standards applicable in the European Union in the scope of environmental protection are classified as the most restrictive in the world. At present, the most important problems include a battle against climate change, biodiversity conservation, reduction of health problems resulting from the environmental pollution, and more responsible use of natural resources. A crucial factor leading to an increase of emission of harmful substances into the atmosphere is an increasing level of consumption.

The present consumption model leads to the intensity of negative environmental impact. That is caused by the growth of expenses on consumption categories related to the strong environmental impact, in particular for transport and householders' energy consumption [21]. That leads to the consistent growth in passenger transports. The greatest participation in this growth regards motor transport. In 1995–2015, this kind of transportation grew by 831.6 billion pas/km, while the ecological rail transport only grew by 77.9 billion pas/km. The increase of harmful substances emission, including greenhouse gases and particulates, are the consequence of the growth in motor transport. Greenhouse gas emission in the motor transport (including cargo transportation) grew from 765 million tons of the CO_2 equivalent in 1995 to 845.3 million tons in 2014, i.e. by 10.5% [19].

Detailed theoretical bases of forecasting savings in the travel time, reductions of costs of road accidents, and savings in operating costs of passenger cars and the impact on the environment were described in [4]. In feasibility studies of railway lines carried out in recent years in Poland, including the modernization and the structure of high-speed rail, unit costs estimated as part of the HEATCO project [9] adapted to Polish conditions Jaspers experts are used [13]. The total economic benefits of the modernization and construction of Warszawa-Łódź-Poznań/Wrocław high-speed line project are presented in Table 20.3.

Economic benefits presented in Table 20.3 show the significant advantage of the variant of new high-speed line construction adapted to speed of 350 km/h. Total economic benefits are over twice as high as the case of the modernization enabling train movement with the speed of 200 km/h.

The described project's economic benefits, corrected investments expenditures and operating costs, and the residual value the project's economic effectiveness indicators were calculated. Their values are presented in Table 20.4.

The indicators presented in Table 20.4 confirm the economic effectiveness of the structure of the high-speed rail system of railways in Poland. It is indicated by ENPV > 0, ERR >

Table 20.3 Economic benefits of the modernization and construction of Warszawa – Łódź – Poznań/Wrocław high-speed line [15]

	Value [million euro]		
	Option 350 km/h		Option 200 km/h
	by [20]	by [15]	by [15]
savings of travel time	5134.8	8114.4	2748.3
savings in costs of accidents	207.3	299.0	108.9
savings in vehicle operating cost	8055.9	10,573.2	6943.9
savings in costs of impact on environment	1885.3	n/a	n/a
savings in costs of noise	9122.4	n/a	n/a
Total	**24,405.7**	**18, 986.6**	**9801.1**

Note: Results of forecasting economic benefits in studies [20] and [15] aren't fully comparable. In the first study the time horizon amounts 30 years, in the second one 40 years.

Source: Own study based on [20] and [15]

Table 20.4 Economic effectiveness indicators of the construction and modernization Warszawa – Łódź – Poznań/Wrocław line [15]

	Option 350 km/h		Option 200 km/h
	by [20]	by [15]	by [15]
ENPV [mln euros]	1088.5	2560.1	1340.6
ERR [%]	6.3	9.3	9.6
B/C Ratio	1.2	1.7	1.71

Source: Own study based on [20] and [15]

SDR, and B/C Ratio > 1. It means that this project deserves support from public funds. The size of involved resources will be appropriate to predicted benefits. However, special attention should be given to the results' incomparability in studies [20] and [15]. Different time horizon causes that the modernization of the line could be more profitable than the structure of the new one. However, such a conclusion is incorrect. With the same assumptions made in study [15], the effectiveness of the structure of the new line is much higher.

In the forecast of economic benefits, they concentrated on high-speed rail transportation. However, benefits that would occur in other segments of rail market, in particular in agglomerative and regional transports, were omitted. Their inclusion would cause the considerable increase of the economic effectiveness of the project.

20.5. Financing sources for building high-speed rail lines in Poland

The characteristics of the train infrastructure – in particular the high capital intensity of infrastructure investments, a long period of its construction and use, and meaning of the infrastructure for country's social and economic development – make it necessary to apply the effective funding scheme of infrastructure investments. Characteristic for this system, irrespective of detailed solutions accepted in various countries, is dominating participation of public means.

In Poland, financing source of train infrastructure investments are resources from:

- the state budget;
- self-government units budget;
- Railway Funds;
- rail infrastructure manager, including received from the capital market;
- European Union funds.

In Poland, the high economic effectiveness of the structure of high-speed rails and the simultaneous financial inefficiency says that public means should be the main financing source. In recent years, the main financing source of infrastructure investments in the rail transport were resources from the Cohesion Funds, complemented by a grant from the state budget, Railway Funds, and investment credits granted by the European Bank.

The level of works progress of the high-speed rail structure in Poland shows that the largest part of investment expenditures will be borne after 2020. In this situation, the current essential financing source will be Connecting Europe Facility (CEF) established in 2013 [17]. The CEF shall enable projects of common interest to be prepared and implemented within the framework of the trans-European networks policy in the sectors of transport, telecommunications, and energy. In particular, the CEF shall support the implementation of those projects of common interest that aim at the development and construction of new infrastructures and services, or at the upgrading of existing infrastructures and services, in the transport, telecommunications, and energy sectors. It will give priority to missing links in the transport sector. The CEF will also contribute to supporting projects with a European added value and significant societal benefits that do not receive adequate financing from the market. In the transport sector, the CEF will support projects of common interest, as identified in Article 7(2) of Regulation (EU) No 1315/2013 [16]. In the list of TEN-T core networks was included the high-speed Warszawa – Łódź – Wrocław/Poznań line.

Financial means from the Cohesion Fund and Connecting Europe Facility constitute non-refundable investment's financing source. Their availability after 2020 will depend on the financial situation of the European Union, which is relevant for obtaining budget incomes, and in consequence of allocation of resources between all sorts of funds. The amount of these allocated funds for individual projects will be established according to the principles of subsidiary and proportionality. So providing the national contribution will be significant. In Poland, the state budget and Railway Fund will be its source.

The Railway Fund established in Bank Gospodarstwa Krajowego pursuant to the Act on the Railway Fund, dated 16 December 2005, has been operated within the structures of the bank since 2006. The fund is financed from the inflows at 20% of the income from a fuel fee imposed on motor fuels and gas introduced to the market, used in combustion engines, paid by the producers and importers of motor fuels (increased additionally by the amount of PLN 100 m a year, which will be allocated to the fund until 2015).

Rail infrastructure investments can be financed from the infrastructure manager's own funds. These include equity (surplus from operating activities and the net profit, as well as means that increase the share capital) and outside capital (means received from the capital market).

The existing system of setting charges for the access to infrastructure in Poland and its financing from public means reduces the possibility of the investment project implementation from the equity. The level of the income obtained by the rail infrastructure manager (PKP Polskie Linie Kolejowe SA – PKP PLK S.A.) from the operational activity (including subsidy) does not cover their costs. In this situation the financial surplus can be generated from the cost of depreciation of fixed assets in the part that constitutes the basis of setting charges of the access to rail infrastructure.

The long period of the train infrastructure operating causes depreciation costs in following years to represent only a minor part of incurred investment expenditure. Financial surplus from the depreciation may represent only a small part of means for covering the investment expenditure constitute. However, since PKP PLK S.A. has generated losses from the operational activity the part of the depreciation, the costs as not constituting cash outflows must be allocated to cover operational expenditures. Possible growth in charges of the access to the train infrastructure would have a negative impact on the competitiveness of the rail transport. In consequence, train carriers could limit their offer, which would lead to a reduction of the infrastructure manager revenues.

A special financing source train of the rail infrastructure investments are long-term credits granted by the European Investment Bank (EIB). This bank is an entity aimed to ensure financial sources for projects which contribute to achieve European Union goals. The EIB grants loans on favorable terms for projects supporting EU goals. These projects include, inter alia, investments in the rail transport. Credit costs depend on Euro Interbank Offered Rate (EURIBOR). At present the rate displays negative values. Including EIB's very low profit margins (often equaling 0%), a loan from this bank is a very attractive investment financing source. However, it is necessary to provide resources for capital payments. Also, negative differences in exchange rates of currency can be a relevant problem (EIB grants a loan in euro, and the infrastructure manager obtains it in PLN).

Analyses carried out in recent years concerning investment financing source related to the structure of the high-speed rail system have also included a possibility of use the Public-Private Partnership concept (PPP) [10]. This concept is a form of long-term cooperation of the public and private sector, whose purpose is to achieve mutual market and social benefits from the undertaking. The PPP is based on the fact that private investors participate in execution of public investments and in providing the public services. Parties of an agreement share the responsibility, the risk, costs, and future profits.

The Public-Private Partnership idea uses the most important advantages of the public sector units and private entities. The administrative authority with its control competence offers them stability and comfort in conducting the business and risk reduction in conducted activity including investment. Advantages of the private sector are mainly a possibility of delivering additional capital, a greater management efficiency, better identification of needs, and optimal allocation of resources and reduction of political interference.

In Public-Private Partnership projects, cash flows generated by the project are a primary source of a loan repayment and remuneration for the private investor. It is important that conditions of loan repayment are adapted to value of cash flows. However, invoked results of financial analyses for the Warszawa-Łódź-Wrocław/Poznań line show that obtaining sufficient financial surplus may be very difficult. Therefore, for the success of the line structure, state involvement in financing the operating and maintenance costs is a must. The size of this commitment must provide not only the repayment of contracted loans and the remuneration for the private investor, but also the profitability for the completion of transportations by operators. Rescheduling of expenditure essential for the structure of infrastructure would be a benefit of the public partner of applying the Public-Private Partnership concept. It would lead to faster creation of the high-speed rail system and a reduction in current budget expenses. In consequence, in a short period of time it would lead to achieving the improvement of the level of sustainable development [1], [5].

20.6. Summary

Research concerning possibilities and usefulness of construction the high-speed rail adopted so far have confirmed its economic attraction. The implementation of these investments is an essential instrument in creating a competitive and resource-efficient transport system. Generated by the project of the new "Y line" Warszawa – Łódź – Wrocław/Poznań structure, social benefits are relevant to invested means from public funds.

Comparative analyses concerning the modernization of existing railway line and their adaptations to the speed of 200 km/h confirm economic effectiveness of such a project. However, economic benefits are almost three times lower than in case of the structure of a new line adapted to the speed of 350 km/h.

In the evaluation of the feasibility study [20], conducted by external experts of the Transport Research Institute, the appropriateness of adopted methodological assumptions and reliability of analyses' results were confirmed. The experts considered that the methodology used to estimate the passenger traffic is appropriate to such a feasibility study. They accepted these results as credible because they are based on cautious assumptions and use pragmatic proven methods. All the economic development has been cautiously estimated, and at each step the most prudent assumptions have been made. The level of induction of traffic seems very low in comparison with results obtained in others countries [14].

Adopting careful macroeconomic assumptions is justified by the specificity of infrastructure investments in the rail transport, in particular by the long operating period of infrastructure objects. Preparing forecasts for a period of 30–40 years is burdened with the high level of risk. In case of a growth in the economy, an income level of households increases, and in consequence so does mobility. Passengers are willing to pay more for services with a greater level that meet their transit needs. Since income level is positively correlated with the value of time, a demand for transports carried out by high-speed rails grows`.

Over a long period, there exists a risk of economic crises. Appearance of business cycles is an immanent feature of a market economy. In case of long recessionary periods, demand for transport services may decrease. That may constitute a threat of achieving the desired effectiveness of investment projects associated with the structure of high-speed rails.

In spite of adopting prudent macroeconomic assumptions and a low level of newly generated traffic, results of the economic effectiveness confirm the usefulness of construction of the new high-speed Warszawa – Łódź – Wrocław/Poznań line and support the investment from public funds. Additional consideration of benefits from the regional and agglomerative transportation, omitted in feasibility studies, could improve the project's economic effectiveness. Research conducted by the Railway Institute in Warsaw show that on this line a sufficient traffic capacity for regional trains transportation will exist, as will agglomeration trains in the railway junctions.

Infrastructure investments, including the structure of high-speed rail lines, effects positively on the value of the real estate. In particular, it regards areas in the railway stations' neighborhood, in which high-speed trains stop. It generates additional advantages both for these real estate owners and for local self-government bodies, which obtain additional budget revenue from adjacent fees. It also supports a growth in the economy, because the real estate is increasingly being used for business activity. It can lead, however, as experiences in various countries confirm, to social dissatisfaction. Public means of taxpayers are contributing to the increase in the affluence of a small group of property owners and business entities located near rail stations. While on the other hand investments carried out by these entities are the source of budget additional revenues and create new jobs, supporting the growth in the economy and improving the quality of life in household. They can also be the source of additional demand for high-speed rail services.

A positive effect on the development of an economic activity around high-speed rail stations requires pursuing the rational spatial planning policies. Planning the development of complementary transport systems is particularly important. High-speed rails can also generate the additional demand both for urban and extra-urban transport services, providing transport for stations of high-speed rails. That may lead for obtaining additional advantages in public transport enterprises, as well as at carriers providing intercity bus transport and by conventional trains. The possible occurrence of such benefits requires drawing up the methodology of the benefits' distribution among the complementary projects. According to the principle of comparability in the cost-effectiveness calculation, only those effects (intentional, pointless, measurable, and non-measurable) that are an effect of incurred expenditure and only those expenditures (financial and nonfinancial) that are necessary for achieving established effects should be taken into consideration. In practice, the precise division of these benefits can be impossible. Meanwhile, such a division can be essential to justify implementation of complementary projects, in particular carried out with the support of European Union funds, or within the concept of public-private partnership.

Main investment expenditures for building a new high-speed rail line will be incurred after 2020. As the Warszawa – Łódź – Wrocław/Poznań line constitutes the fragment of TEN-T core network, it may get funding from the Connecting Europe Facility.

The subsidy from CEF can be granted not only for construction works, but also on studio works of priority projects. List of pre-identified projects on the core network in the transport sector includes Annex I of Regulation (EU) No 1316/2013. Among them there are studies for high-speed rail Belarus border – Warszawa – Poznań – German border in Core network North Sea – Baltic Sea corridors. So a completion of the further preparatory works would be intentional, according to recommendations formulated by experts from the Transport Research Institute.

Bibliography

[1] A sustainable future for transport. European Communities, 2009.
[2] Barrón de Angoiti, I.: *Study on Infrastructure Charges for High-Speed Services in Europe*. Paris: UIC, 2008.
[3] Bieniok, H.: *Time management*. Wydawnictwo Akademii Ekonomicznej, Katowice, 1999.
[4] Dyr, T., and Kozubek, P.R.: *Ocena transportowych inwestycji infrastrukturalnych współfinansowanych z funduszy Unii Europejskiej [Assessment of Infastructural Transport Investments Co-financed by European Union's Funds]* . Spatium, Radom, 2013.
[5] Dyr, T.: European Transport Policy and Strategy for Sustainable Development. *Central European Review of Economics & Finance*, 2012, 2, 1.
[6] Dyr, T., Pomykała, A., and Raczyński J.: Finansowanie rozwoju sieci TEN-T z instrumentu "Łącząc Europę" [Financing the Development of TEN-T Network from Connecting Europe Facility Funds]. *"Technika Transportu Szynowego,"* 2015, no.4.
[7] Guidance on the Methodology for Carrying Out Cost-Benefit Analysis. Working Document No. 4, European Commission, Directorate-General Regional Policy, 2008.

[8]   Guide to Cost-Benefit Analysis of Investment Projects Structural Funds, Cohesion Fund and Instrument for Pre-Accession. European Commission, Directorate General Regional Policy, 2008.

[9]   Heatco: Developing Harmonised European Approaches for Transport Costing and Project Assessment. Project funded under the 6th Framework Programme, coordx`inated by the University of Stuttgart, European Commission, 2006.

[10] Koncepcja organizacji budowy i eksploatacji linii dużych prędkości w Polsce [The Concept of the Organisation and Operation of High-Speed Lines in Poland] . A study drawn up by CNTK, Warszawa, 2007.

[11] Merkert, R.: A Transaction Cost Perspective on the Organization of European Railways. 11th World Conference on Transport Research, Berkeley, June 2007.

[12] Narodowe Strategiczne Ramy Odniesienia 2007–2013. Wytyczne w zakresie wybranych zagadnień związanych z przygotowaniem projektów inwestycyjnych, w tym projektów generujących dochód. Ministerstwo Rozwoju Regionalnego [National Strategic Framework of Reference 2007–2013. Guidelines regarding selected issues connected with the preparation of investment projects, including projects generating profit. Ministry of Regional Development] , Warszawa 27 September, 2011.

[13] Niebieska, księga: Sektor kolejowy: Infrastruktura i tabor [Blue Paper: Railway Sector: Infrastructure and Rolling Stock]. Warszawa: Jaspers, December, 2008.

[14] *Peer Review Warszawa-Łódź-Poznan/Wroclaw High-Speed Rail Feasibility Study Preferred Route.* Paris: Transport Research Institute, 14.12.2011.

[15] Przygotowanie budowy linii dużych prędkości. [Preparing the Construction of High Speer Lines]. A study drawn up by Consortium Egis Poland sp. z o.o., Ernst & Young Corporate Finance sp. z o.o. and DHV POLSKA sp. z o.o., Warszawa, 2012.

[16] Regulation (EU) No 1315/2013 of the European Parliament and of the Council of 11 December 2013 on Union guidelines for the development of the Trans-European Transport Network and repealing Decision No 661/2010/EU. OJ L 348, 20.12.2013, pp. 1–128.

[17] Regulation (EU) No 1316/2013 of the European Parliament and of the Council of 11 December 2013 establishing the Connecting Europe Facility, amending Regulation (EU) No 913/2010 and repealing Regulations (EC) No 680/2007 and (EC) No 67/2010. OJ L 348, 20.12.2013, pp. 129–171.

[18] Stalder, O.: *International Benchmarking of Track Cost.* Paris: UIC, 2001.

[19] *Statistical Pocketbook 2016: EU Transport in Figures.* Luxembourg: Publications Office of the European Union, 2016.

[20] Studium Wykonalności dla budowy linii kolejowej dużych prędkości "Warszawa – Łódź – Poznań/Wrocław." Analizy finansowe i ekonomiczne. [Feasibility Study for Building High-Speed Line Warszawa – Łódź – Poznań/Wrocław. Financial and economic analyses]. A study drawn up by Consortium: Ingenieria IDOM Internacional S.A. and Biuro Projektów Komunikacyjnych w Poznaniu Sp. z o.o., Warszawa, 2013.

[21] Sustainable Consumption and Production. European Environment Agency, 2007.

[22] WHITE PAPER Roadmap to a Single European Transport Area – Towards a Competitive and Resource Efficient Transport System. COM(2011), 144.

[23] Wstępne studium wykonalności budowy linii dużych prędkości Wrocław/Poznań – Łódź – Warszawa. [Preliminary Feasibility Study for Building High-Speed Line Wrocław/Poznań – Łódź – Warszawa.]. Centrum Naukowo-Techniczne Kolejnictwa, Warszawa, 2006.

[24] Wstępne studium wykonalności dla przedłużenia linii dużych prędkości Warszawa – Łódź – Poznań/Wrocław do granicy z Niemcami w kierunku Berlina oraz do granicy z Republiką Czeską w kierunku Pragi. [Preliminary Feasibility Study for Extending High-Speed Line Warszawa – Łódź – Poznań/Wrocław to the Border with Germany towards Berlin and to the Border with the Czech Republic towards Prague]. A study drawn up by Consortium: IDOM Ingenieria y Consultoria SA and IDOM Inżynieria Architektura i Doradztwo sp. z o. o., Warszawa, 2015.

Index

Note: Page numbers in italic indicate a figure and page numbers in bold indicate a table on the corresponding page.

Printed and bound by CPI Group (UK) Ltd, Croydon, CR0 4YY

24/10/2024

01778286-0008